Atomic Force
Microscopy for Biologists

Second Edition

Atomic Force Microscopy for Biologists

Second Edition

GEORGE GREEN LIBRARY OF
SCIENCE AND ENGINEERING

Victor J. Morris
Andrew R. Kirby
A. Patrick Gunning

Institute of Food Research, Norwich Research Park, UK

Imperial College Press

Published by

Imperial College Press
57 Shelton Street
Covent Garden
London WC2H 9HE

Distributed by

World Scientific Publishing Co. Pte. Ltd.
5 Toh Tuck Link, Singapore 596224
USA office: 27 Warren Street, Suite 401-402, Hackensack, NJ 07601
UK office: 57 Shelton Street, Covent Garden, London WC2H 9HE

British Library Cataloguing-in-Publication Data
A catalogue record for this book is available from the British Library.

ATOMIC FORCE MICROSCOPY FOR BIOLOGISTS
Second Edition

ISBN-13   978-1-84816-467-3
ISBN-10   1-84816-467-X

Printed by FuIsland Offset Printing (S) Pte Ltd, Singapore

This book is dedicated to Martin Murrell and Mark Welland who introduced us to probe microscopy, with whom we have subsequently spent many happy hours delving into the workings of our AFM, and to Christina, Gloria and Yvonne, Ellie and Alex for their patience and understanding during the writing of this book.

In memory of "Ginny" Annison and Christopher Gunning.

CONTENTS

ACKNOWLEDGEMENTS

The authors wish to acknowledge the kind co-operation of those who provided figures for inclusion in this book; individual accreditation is given in the relevant figure captions. The beauty of the images stands not only as testament to the dedication of the researchers, but also to the power and rapid progress of the technique. Alan Mackie and Pete Wilde gave helpful advice on the topics discussed in chapter 5 and provided some of the schematic figures used. Nicola Woodward and Rob Penfold contributed sections on the discussion on colloid probe microscopy in Chapter 9. We wish to thank World Scientific Press / Imperial College Press for giving us the opportunity of writing this book, and in particular Joy Quek and Sook Cheng Lim on the editorial staff for their almost limitless patience! Finally, we acknowledge the Biotechnology and Biological Sciences Research Council for providing funding for our research through its core grant to the Institute of Food Research.

CHAPTER 1

AN INTRODUCTION

The atomic force microscope (AFM) is perhaps the most versatile member of a family of microscopes known as scanning probe microscopes (SPMs). These instruments generate images by 'feeling' rather than 'looking' at specimens. This novel mode of imaging results in a magnification range spanning that associated with both the light and the electron microscope, but under the 'natural' imaging conditions normally associated with just the light microscope. The potential to image biological systems in real time, under natural conditions, with molecular, or even submolecular resolution is clearly of interest to biologists. Thus, since their inception, SPMs have had an obvious appeal to biologists and biophysicists. Since these first studies in the early 1980s publications describing biological applications of SPM have grown extremely rapidly. One of the aims of this book is to look at what impact these techniques have made in the biological sciences, and to try to assess their future potential.

Scanning probe microscopy began in the early 1980s when Binnig and Rohrer revolutionised microscopy through the invention of the scanning tunnelling microscope (STM). The importance of this discovery was recognised through the award of the Nobel prize in Physics in 1986. The STM is the first of this large and growing family of probe microscopes, which sense the structure of a surface by scanning it with a sharp probe and measuring some form of interaction between the surface and the probe. The development of the STM arose from an interest in the study of the electrical properties of thin insulating layers. This led to an apparatus in which the probe-surface separation was monitored by measuring electron tunnelling between a conducting surface and a conducting probe. A few years later Binnig and colleagues (1986) announced the birth of the second member of the SPM family - the atomic force microscope, also known at the scanning force microscope (SFM). In the late 1980s commercial STMs became available. Commercial AFMs began to appear in the early 1990s and have evolved through several generations. Refinements and new types of SPMs have appeared and will undoubtedly continue to be developed in the future. Particular developments of importance in biological research are combinations of probe microscopes with light or electron microscopes, cryo AFMs, scanning ion conductance microscopes (SICM) and scanning near field optical microscopes (SNOM).

SPMs are not strictly microscopes: they visualise surfaces by 'feeling' or sensing them with a sharp probe. Conventional (far field) microscopes image by collecting radiation transmitted through, or reflected from the sample. The ultimate resolution is diffraction limited and depends on the wavelength of the radiation.

Thus light microscopes are limited to a resolution of ≈ 200 nm. Higher resolution images of biological materials can be obtained using high energy electrons in the electron microscope (EM). Despite recent advances in the development of environmental EMs molecular resolution still requires that specimens are examined under vacuum or partial vacuum. Electron microscopists have developed many elegant preparative methods to preserve the 'native' structure of biological materials. SPMs image by a different mechanism (Fig. 1.1) and different criteria determine their resolving power.

Figure 1.1 Scanning electron microscope image of an AFM tip used to probe the structure of the sample surface. Magnification approximately x 10,000. The probe 'feels' the sample surface. Image courtesy of Paul Gunning.

In SPM images are obtained by measuring changes in the magnitude of the interaction between the probe and the specimen surface as the surface is scanned beneath the probe. Hence the resolution will depend on the sharpness, or apparent sharpness of the probe tip, and the accuracy with which the sample can be positioned relative to the probe. SPMs are capable of 'atomic' resolution on flat surfaces and such resolution can be achieved in gaseous or liquid environments. For macromolecules atomic resolution is only possible for simple molecules in which each atom is in intimate contact with the surface of a flat substrate. However SPMs do allow sub-molecular resolution on most biopolymers under 'natural' conditions. Thus the attractive potential of SPMs for biologists is the ability to visualise molecular processes under natural or physiological conditions. They offer the resolution of most commercial electron microscopes but under the experimental conditions familiar to the light microscopist.

The first biological studies were made using STM. The tunnelling current decays exponentially with increasing separation between the surface and the probe. A change in probe-sample separation of atomic dimensions will lead to an order of magnitude change in tunnelling current. This means that tunnelling effectively occurs from the atom on the tip nearest the surface, and the probes behave as if they are atomically sharp. Thus STMs offer the highest resolution obtainable by SPM. However, the rapid decay in the tunnelling current basically restricts investigations to the study of thin interfaces, or individual biopolymers. For larger biological systems the probe-sample separation would become too large and any tunnelling current would be expected to be too small to detect. Furthermore, the sample surface needs to be conducting and this usually means coating the biological sample, offsetting the main advantages of the SPM method. With the AFM there are no such restrictions on the size of the specimen that can be examined, and biological samples ranging in size from individual molecules to cells or tissues can be, and have been, imaged in their native state. AFMs, and refinements such as cryo-AFM, have become the preferred SPM methods in biology. It was originally believed that SPMs were non-invasive techniques. In practice sample damage and displacement plagued the early uses of STM and AFM. Understanding and overcoming these problems has led to reliable and reproducible methods of imaging. Emphasis has passed from validation of the microscopes to their use to study biological problems.

Other types of SPM have and will continue to become important in biology. A likely candidate, now available commercially, is the scanning ion conductance microscope (SICM). The basic principles of SICM and appropriate biological applications will be discussed. There is a growing appreciation of the value of combining AFM with conventional optical or confocal microscopes, or of using them in conjunction with surface techniques such as surface plasmon resonance, and the basis of some of these combined microscopes will also be described. The early generation of combined microscopes suffered in resolution when compared to stand alone AFMs. However, the latest generation of combined instrument offer the resolving power of a stand alone AFM with the versatility of combined optical microscope for locating and characterizing biological samples. It is still difficult to obtain high-resolution images on soft biological samples such as cells, and the newer technique of scanning ion conductance microscopy and advances in both far-field and near-field optical microscopy are worthy competitors. As probe microscopy has matured the emphasis has moved from just obtaining images to a more general use of the technique to solve problems in biology. Increasingly probe microscopes are seen as tool kits for studying biological systems: applications range from generalized mapping of parameters such as elasticity, friction or charge, specialized affinity mapping of surfaces and the use of probe microscopes to modify and manipulate biological systems. Increasingly the ability to measures forces between molecules at the single molecule level under natural conditions is rivalling imaging applications of AFM: applications range from the characterization of individual molecules, studies of molecular interactions and their roles in friction and adhesion at both molecular and cellular levels. This book will concentrate on biological applications of AFM. The advantages and limitations of the technique will be assessed. The literature in this area is vast and increasing rapidly. Thus it is not possible to reference all of the published papers on a particular topic. Rather we have tried to cite books, recent reviews and selected research papers. The papers have been chosen to emphasise a point, or to provide a route to the literature in this area. The choice is not meant to imply priority and omission of papers is purely the result of the limitations of space and time and is not necessarily a reflection on the quality of the publication.

What do we wish to achieve in writing this book? One aim is to introduce the AFM, to describe the type of apparatus available and how it is used. A second aim is to look at the types of biological samples which have been studied, to look at how successful these studies have been, and to assess whether the use of AFM has generated new knowledge or understanding in these areas. In general terms we wish to look at what can and what cannot be done, if it is possible to do it then how is it done, and at what has been done and where things may go in the future. Who do we hope will benefit from reading this book? We hope that the book will provide a good resource base for literature on biological applications of AFM. The information presented should allow the reader to critically evaluate published present and future AFM data on biological systems, to decide whether AFM is likely to be useful in their areas of interest and, for new recruits to AFM, provide a basis for deciding what sort of technique to invest in, its inherent limitations, how to avoid and recognize artifacts and thus how to optimise its use to solve problems.

CHAPTER 2

APPARATUS

2.1. The atomic force microscope

Despite its rather grandiose title, the atomic force microscope is probably one of the easiest forms of microscopy to understand at a rudimentary level. However, the first thing to consider when thinking about how an AFM works is that all notions of conventional microscope design need to be disregarded, since it has no lenses of any kind. In fact the AFM images samples by 'feeling' rather than by 'looking'. A good analogy is a blind person feeling objects with their fingers and then building up a mental image of what they touch. Like the blind persons' fingers this method of imaging can produce an exquisitely detailed picture, not just of the topography of the surface being touched but also of its texture or material characteristics; soft or hard, springy or compliant, sticky or slippery. This latter aspect of AFM will be discussed in more detail elsewhere (sections 3.3 and 7.1.2). For now let us concern ourselves only with the topographical information that the AFM provides and how this is achieved optimally in modern instruments. The schematic in Fig. 2.1 illustrates the main features of an atomic force microscope.

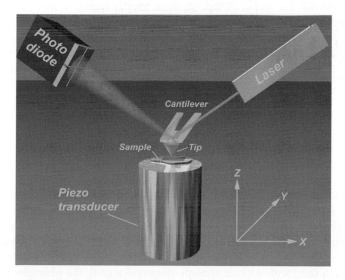

Figure 2.1. Schematic representation of the atomic force microscope.

The first and arguably most important part of the AFM is the stylus, or tip, which does the 'feeling', and an actual AFM tip is shown in the electron micrograph in Fig. 2.2.

5

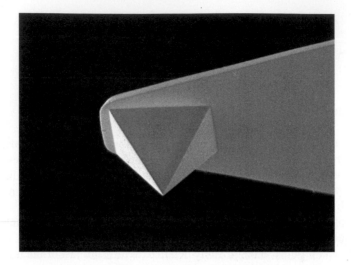

Figure 2.2. Scanning electron micrograph (SEM) of an AFM tip and cantilever. To give some sense of scale the base length of the pyramid is 4 μm. Image taken by Paul Gunning.

The tip consists of a micro-fabricated, extremely sharp spike mounted on the end of a cantilever. This tiny assembly is bonded to a glass chip to allow easy handling. The sharpness of the spike, or tip, as it is more usually called, determines the resolving power of the instrument. The cantilever on which it is mounted allows the tip to move up and down as it tracks the sample, in the same way that a record player stylus tracks a record (or rather used to in the days before CDs!). Furthermore, the cantilever typically has a very low spring constant (also referred to in physics text books as force constant) enabling the AFM to control the force between the tip and the sample with great precision. The cantilever-tip assembly is generally made of silicon or silicon nitride, these materials being both hard, and so wear resistant, and ideally suited to micro-fabrication.

The second crucial feature of the AFM is the scanning mechanism. Simply having a very sharp tip is of no use whatsoever without a means of accurately positioning it relative to the sample surface. This is done by means of a piezoelectric transducer (Fig.2.3). The principle is the same as in a piezoelectric gas-lighter - namely that when a crystal of piezoelectric ceramic is squeezed it produces a potential difference (i.e. a bias voltage) large enough to generate a spark. If this process is reversed and a potential difference is applied to the piezoelectric ceramic, it will expand. This motion is incredibly reproducible and sensitive such that, with a clean enough electrical signal, the piezoelectric ceramic can be made to move with an accuracy of atomic dimensions. This provides the AFM, and indeed all probe microscopes, with the accuracy of positioning sample or tip that they require. There are many different geometries which will be discussed later (section 2.2), but a generic layout is shown in Fig. 2.1 in which the sample is mounted on top of the piezoelectric transducer. The motion of the sample can be controlled in the three orthogonal directions x, y and z and these are assigned to three channels in the instrument control electronics. The sample can now be positioned in extremely close proximity to the AFM tip (often actually in contact) using the z channel, and then raster-scanned (i.e. line by line) using the x and y channels, in order to build up an image of a selected area of the sample surface.

The final feature of the instrument is the detection mechanism. The motion of the tip as it traverses the sample must be monitored. Again there are several different ways of doing this which are described in detail later (section 2.5) but, for the sake of simplicity, the most common method, the optical lever system, is illustrated in the schematic in Fig. 2.1. A laser beam is focused onto the end of the cantilever, preferably directly over the tip, and then reflected off onto a photodiode detector. In modern instruments the photodiode is split into four segments. As the tip moves in response to changes in the sample topography during scanning, the angle of the reflected laser beam changes, and so the laser spot falling onto the photodiode moves, producing changes in intensity in each of its quadrants. This surprisingly simple system, which is in effect a mechanical amplifier, is sensitive enough to detect atomic scale movement of the tip as it traverses the sample. The difference in laser intensity between the top two segments and the bottom two segments produces an electrical signal which quantifies the normal (up and down) motion of the tip, and the difference between the laser intensity in the left and right pairs of segments quantifies any lateral or twisting motion of the tip. Thus frictional information can be distinguished from topographical information.

When the sample is scanned the topography of the sample causes the cantilever to deflect as the force between the tip and sample changes. In the simplest operating mode (for more details on modes of operation see section 3.2) the cantilever deflection is maintained at a constant predefined level by a control loop which moves the sample or tip in the appropriate direction at each imaging point. In this mode of operation the feedback mechanism is crucial to generating an image. The x, y and z displacements of the piezoelectric scanner are recorded and displayed to produce an image of the sample surface.

2.2 Piezoelectric scanners

Modern AFMs use one of two basic types of scanning mechanism: there are some that scan the sample, and others that scan the tip. Both however, rely upon piezoelectric transducers. The *piezoelectric effect* is the generation of a potential difference across the opposite faces of certain crystals (piezoelectric crystals) as a result of the application of stress. The electrical polarization produced is proportional to the stress and the direction of the polarization changes if the stress changes from compression to tension. The *reverse piezoelectric effect* is the opposite phenomenom and the one that AFM scanners employ. If the opposite faces of a piezoelectric crystal are subjected to a potential difference, the crystal changes shape. The piezoelectric materials used in AFMs are ceramics, generally of the so-called PZT type (lead zirconate titanates). In the early days of scanning probe microscopy, the geometry of the scanner had three blocks of piezoelectric ceramic arranged in a tripod, moving the tip or sample in the three orthogonal directions x, y and z. These days piezoelectric tripods have been superseded by tubes of piezoelectric ceramic materials (or in some cases a combination of tripods and tubes) which were first introduced by Binnig and Smith in 1986 (Binnig and Smith, 1986). This geometry has many advantages over the tripod arrangement, but the principal one is that larger scan ranges are possible with what is a far more compact and symmetrical geometry of the scanner.

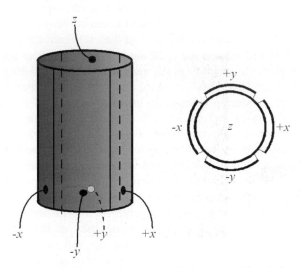

Figure 2.3. Schematic representation of a piezoelectric tube scanner. Although the tube moves in only the three orthogonal directions x, y, z there are actually five electrodes $+x$, $-x$, $+y$, $-y$ and z required to achieve this motion.

The scanner tube consists of a thin-walled hard piezoelectric ceramic which is polarised radially. Electrodes are attached to the internal and external faces of the tube, and the external face of the tube is split into quarters parallel to the axis, as shown in the schematic in Fig. 2.3. By applying a bias voltage between the inner and all of the outer electrodes the tube will expand or contract (i.e. move in the z direction). If a bias voltage is applied to just one of the outer electrodes the tube will bend, i.e. move in the x and y directions. To make this bending even more pronounced, in order to improve the scan range, the outer electrodes are arranged opposite each other in terms of direction ie. $+x$ opposite $-x$, meaning that if the tube is biased with $+n$ volts on one electrode and $-n$ volts on the opposite electrode, the resultant bending of the tube will be twice as much as that produced by simply biasing one electrode. A detailed mathematical study of the dynamics of tube scanners has been given elsewhere (Taylor 1993). Tube scanners have two main disadvantages. The first is that the motion of the end of the tube, which drives the tip or sample in the x and y directions, traces out an arc rather than a straight line, leading to an effect known as 'eyeballing' when large scans are carried out. A flat surface appears as if it is part of the surface of a sphere - hence the term. This effect can be corrected with image processing software. The second problem is that of scan speed. Tubes cannot move quickly and, if very fast processes are to be studied, an alternative scanner arrangement consisting of small stacks of piezoelectric ceramics is used. Other problems relating to the non-linearity of piezoelectric scanners are discussed in the calibration section (section 2.8) in this chapter.

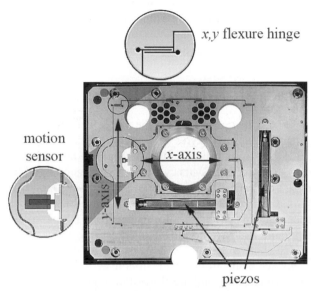

Figure 2.4. *X, Y* Flexure scanner mechanism from an MFP-3D AFM, seen from below. Two independent piezo elements push against metal arms which are connected to the nested scanner plates (*x* in the centre and *y* outside it). The nested plates are hinged in a manner which confines their motion to purely orthogonal directions for each impulse. Note the incorporation of motion sensors to validate scanner accuracy. Image courtesy of Asylum Research, (Santa Barbara, CA USA).

A new geometry of piezoelectric scanner is emerging in modern high-performance AFMs; the so-called 'flexure-stage'. This rather ingenious design places the piezoelectric elements within a nested cradle-like mechanism, usually machined out of a single piece of thin alloy sheet. Because it is machined out of a single piece of material the flexure stage has significantly improved mechanical and thermal stability when compared with traditional multi-part scanner designs. As can be seen in figure 2.4 the piezoelectric elements push against separate arms cut into the metal sheet causing flexure at the fulcrum region at the base of each arm. The layout of the hinges on the corners of each of the nested plates confines the subsequent motion of the respective plate to purely orthogonal directions. This arrangement isolates the motion of each piezoelectric element and so eliminates axis cross-talk, allowing the scanner to trace out the selected area in a much more faithful manner than a tube scanner ever could. The flexure design also makes it easier to incorporate independent motion sensors which are used in closed-loop feedback correction circuits to ensure scan accuracy (see section 2.8.1). Another significant advantage of flexure stages is that the delicate piezoelectric elements are protected by virtue of their position within the mechanism. In this arrangement the piezos are not directly load-bearing and so are less prone to the mechanical damage inflicted by over-enthusiastic practitioners. Their cosy home also protects them to some extent from the dreaded liquid-spilling hordes of first year undergraduates (you know who you are!). Another significant advantage of this arrangement is that, by virtue of their flat shape *x/y* flexure stages are relatively easy to integrate with conventional light microscopes, leaving the sample area almost as before (most will happily accommodate glass slides and Petri dishes).

2.3. Probes and cantilevers

The heart of an AFM is the tip since this is the part which interacts with the sample. Rather like fitting bald worn-out tyres to a Ferrari, fitting a blunt tip to even the most sophisticated AFM produces results which are at best disappointing and, at worst, worthless. Quality and consistency are the key to good AFM tips. This means that even since the very early days of AFM the need to have specially manufactured tips was recognised (Binnig *et al* 1987) and the home-built variety were quickly abandoned! Modern AFM tips and cantilevers are made by micro-fabrication, using many of the techniques that have been developed for integrated circuit manufacture, such as lithographic photo-masking, etching and vapour deposition. Cantilevers and tips are nearly always made from either silicon or silicon nitride. They can be conducting or non-conducting, and are thus often coated with another material. If optical sensing methods are to be used to monitor the cantilever deflection then they are coated with a thin gold layer to improve their reflectivity or, if magnetic sensitivity is required a ferromagnetic coating may be applied. Commercial AFM tips are usually supplied in 'wafers' containing a few hundred tips. Each tip needs to be separated according to the manufacturer's instructions. A detailed description of the manufacture of tips and cantilevers is given by Boisen and co-workers (Boisen *et al* 1996).

2.3.1. Cantilever geometry

There are two basic geometries for the cantilever on which the tip is mounted and these are shown in the scanning electron micrograph in Fig. 2.5. The triangular geometry, usually referred to as a 'V' shaped lever, was originally designed to minimise torsional motions, or twisting of the cantilever whilst scanning a sample, and is generally the lever of choice for purely topographical imaging. More recently the brothers John and Raymond Sader have used a scale model to confirm a theoretical prediction (Sader 2003) that rectangular levers are actually torsionally stiffer than V-shaped levers (Sader and Sader, 2003).

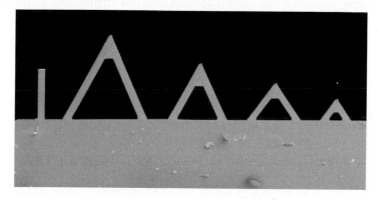

Figure 2.5. SEM image of AFM cantilevers. Magnification approximately 100 x. Image courtesy of Paul Gunning.

Irrespective of geometry, the force contribution (F) from the bending of the cantilever which acts on the sample is determined by the following simple equation, known as Hooke's Law:

$$F = -ks \qquad (2.1)$$

where k is the spring constant of the cantilever and s is its displacement. The minus sign indicates that the force acts in the opposite direction to the displacement of the cantilever. The spring constant (k) increases with lever thickness but decreases with cantilever length.

Q: " Since longer cantilevers have lower spring constants, are they the best ones to use for imaging delicate biological samples with low forces, in order to get better resolution?"
A: "Whilst they do allow lower forces to be achieved, in practice this factor alone often doesn't lead to better resolution due to their poorer sensitivity"

A general rule for both cantilever geometries is that longer cantilevers are less sensitive if the optical beam detection method is employed (which is the case for nearly all present day AFMs). The relationship between the angular displacement of the reflected laser beam (θ) and cantilever length (*l*) is:

$$\theta \propto \frac{1}{l} \qquad (2.2)$$

i.e. for a longer cantilever a given displacement deflects the laser beam through a smaller angle than it would if the cantilever were shorter. Consequently the laser spot moves a smaller distance over the face of the photodetector, producing a smaller output signal for the control loop. A practical application of this lack of sensitivity is that longer cantilevers are better for imaging rougher samples, because larger displacements of the tip produce less angular deflection of the laser beam, reducing the possibility that it will miss the photodiode detector, thus inactivating the feedback mechanism with the inevitable loss of control this would cause. However, very low spring constant cantilevers have the disadvantage of low resonant frequencies (typically 7-9 kHz) which makes them less stable and more difficult to use if the scan conditions are not chosen carefully. The principle is that a cantilever should have a resonant frequency at least ten times higher than the fastest scan speed it will encounter during imaging, otherwise it may be excited into resonant oscillation causing loss of image quality (see section 3.6.2). Additionally, these long very soft levers are also susceptible to ambient noise.

A solution to the lack of sensitivity and instability of long soft cantilevers is to make them shorter. Normally this comes at the price of higher stiffness, which is not such good news for force spectroscopy applications. However, significant progress has been made in the fabrication of very short (thus very sensitive) cantilevers with low spring constants (Olympus, Japan). These changes have had a dramatic impact on the force sensitivity routinely available to modern AFM instrumentation. In theory the shorter one can make a cantilever the better will be the performance. However, there is a trade-off; the shorter the cantilever the smaller will be its linear range when a traditional split-photodiode detector is used to collect the reflected beam. Experimental data obtained using a 12 μm cantilever demonstrated that the photodiode signal deviated from a purely linear response at lever displacements of only 115 nm, and that the upper segment of the detector was saturated at 182 nm (Schäffer 2002).

Since the cantilever is the spring in the system it determines the categories that AFM probes are divided into, contact, non-contact, Tapping and force modulation. These are imaging modes in their own right and they are described in detail later (Table 2.1 and section 3.2). However, a brief description is given in table 2.1 at the end of this chapter, which gives typical values of spring constant (k) and resonant frequency (f_r) and explains why particular levers are used in particular applications.

2.3.2. Tip shape

The actual shape of the tip used in AFM is an important consideration, and choice of tip shape is closely linked to the properties of the sample under study. There are all manner of different tips available commercially, each with a specific function. First of all they can simply be divided into two important categories; high and low aspect ratio tips. In deciding which type is needed for any given sample a concept familiar in other forms of microscopy is relevant, namely depth of field. If a large depth of field is required, such as when a sample is rough, then high aspect ratio probes are needed, as illustrated in Fig. 2.6.

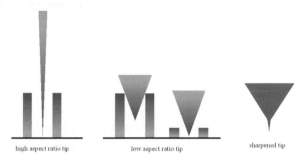

high aspect ratio tip   low aspect ratio tip   sharpened tip

Figure 2.6. Schematic representation of different aspect ratio AFM tips. On the left is a high aspect ratio tip suitable for rough samples, in the middle low aspect ratio tips which are limited to studies on relatively flat specimens, and on the right is a sort of hybrid tip known as a 'sharpened' tip.

Having said this it should be born in mind that AFM is actually quite limited in its depth of field, and is not a suitable form of microscopy for very rough samples, irrespective of the tip aspect ratio. It is obvious from Fig. 2.6 that when sample roughness approaches, or exceeds the height of the tip, proper imaging is impossible. If, however, the sample is flat then a low aspect ratio tip is suitable (Fig. 2.6). Some low aspect ratio tips, referred to as 'sharpened' tips, have high aspect ratio sections at the apex, in order to improve resolution on relatively flat samples with only modest increases in cost (Fig. 2.6). The three most common tip shapes are shown in Fig. 2.7, and these are; pyramidal (left), isotropic (middle) and 'rocket-tip' (right). From a practical point of view the importance of different tip shapes can be defined by the 'opening angle' of tips: which is a way of defining the aspect ratio at the business end of the tip. Put simply, this factor determines to what extent the AFM images of a sample are 'softened' or blurred by tip shape, low opening angles giving sharper images (this effect known as probe-broadening is discussed in more detail in section 2.8.2).

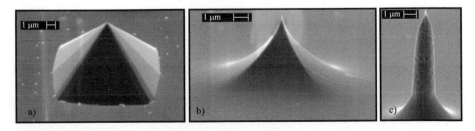

Figure 2.7. SEM image of common tip shapes. Reproduced with permission of the authors and publisher, Institute of Physics publishing, from Boisen *et al* 1996.

The rougher a specimen the lower the opening angle of the tip should be in order to avoid excessive tip convolution in the images. Note that the opening angle should not be confused with tip sharpness, which is usually defined as the radius of curvature at the apex of an AFM tip - for example a pyramidal tip can be just as sharp at its actual apex as a rocket tip.

2.3.3. Tip functionality

One of the great strengths of the AFM comes from its ability to interact with a sample rather than simply produce an image. This potential can be further enhanced by functionalisation of AFM tips. By coating the tips with certain materials the force between tip and the sample can be varied in a myriad of possible ways. This can be used to investigate particular types of tip-sample interaction or to map selected sites on surfaces (see section 7.1.2.). Coating can be used to enable a particular chemical sensitivity to be obtained (Frisbie *et al* 1994). In a significant refinement of this approach, chiral sensitivity was demonstrated using a functionalised AFM tip (Mckendry *et al* 1998). Functionalised tips have obvious implications for the application of AFM to biology, and it is now possible to purchase tips with defined functionality from several specialist labs (Bioforce, NT-MDT). Many of the functional attributes are based on attaching biological molecules to the tips. They range from fairly simple hydrophilic or hydrophobic coatings to more sophisticated functionality such as antibody or antigen coated tips, and ligand or receptor coated tips. Another development which shows promise is the attachment of carbon nanotubes to AFM levers (Dai *et al* 1996). This gives a flexible, high aspect ratio tip with an extremely well characterised geometry at its apex, i.e. a C_{60} molecule, which is amenable to further chemical functionalisation (Wong *et al* 1998). This promise has been demonstrated in dramatic fashion recently with the application of carbon nanotube functionalised tips to resolve the structuring of water layers above a biological membrane surface (Higgins *et al* 2006a). In this study discrete jumps in force, at spacing consistent with the dimensions of single water molecules, were observed as a nanotube tip approached a phospholipd bilayer.

2.4. Sample holders

For imaging in air the sample on its substrate is simply mounted on a small metal disc. It is best to stick the sample to the disc using double sided conducting adhesive tape or silver Dag^{TM}, in order to prevent movement during imaging and the build up of static charge on the sample. However, imaging in air is not generally the best option for biological samples and most of the time a liquid cell is used. Having said that, with modern AFMs which generally feature automatic cantilever tuning routines, often the easiest way to get started on an unknown sample is to use an ac mode in air such as Tapping.

2.4.1. *Liquid cells*

The feature which really sets the AFM apart from the electron microscope is its ability to operate under liquids at unparalleled resolution, although pioneering new techniques in light microscopy are beginning to close this gap (Hell 2007). In order to do this a liquid cell is needed. All liquid cells, irrespective of their design, basically perform three functions; contain the sample, contain the liquid, and provide a stable optical path for the laser beam which is reflected off the back of the cantilever. Obviously if the optical beam detection method is being used the beam cannot simply pass through a liquid-air interface since it will be refracted 'all over the place' due to the movement of the liquid surface. The solution to this problem is to use a glass sighting plate which is submerged in the liquid. A general layout of a liquid cell is shown in Fig. 2.8. Some liquid cells are sealed by means of compressing an 'O-ring' between the sample holder and the tip holder, others are sealed with a latex membrane, and some are left open. Sealed cells are more 'fiddly' to set up but they prevent evaporation of the liquid and allow flow of liquids through the cell during an experiment. Open cells on the other hand make flow of liquid through the cell difficult, but partial exchange of liquid or addition of liquid is still possible. The reason for adding, exchanging or flowing liquids through the cell whilst imaging is to allow the study of dynamic events, for example enzymatic catalysis. It is best to begin with a stable static system before the reagent is added so that a clear starting point is defined.

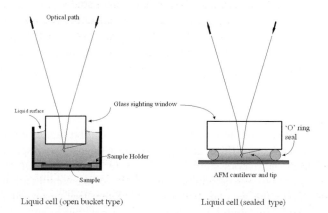

Figure 2.8 Bucket type (left) and 'O-ring' type (right) liquid cells using optical detection.

The liquid cell should be made of an inert material and glass, PTFE or stainless steel are the materials usually used; the choice depends upon the nature of the sample and liquid being used. Another requirement is that the sample is fixed to the bottom of the cell. This is normally done using a push-fit washer for plastic and glass cells, but magnetic washers can also be used. Glue may be used to fix the sample to the bottom of the cell but it is the least satisfactory option, carrying the risk of sample contamination. Recently significant improvements in liquid cell design have taken place and most of the major AFM manufacturers offer excellent temperature control and, in many cases, specialised liquid chambers for live cell imaging.

2.5. Detection methods

During the embryonic stages of AFM development several different methods were used for detecting the motion of the AFM tip as it traversed the sample. As a form of historical record a brief description of these methods is given here. Broadly speaking the methods fall into two categories. These are optical and electrical methods.

2.5.1. Optical detectors: laser beam deflection

Laser beam deflection has now become the most common detection method used in modern commercial AFMs, and it is no coincidence that it is also the simplest, cheapest and most versatile. A generic layout is shown in the schematic diagram in Fig. 2.9. The principle is familiar to every unruly schoolchild who, sitting at the back of a classroom on a sunny afternoon, when they'd rather be somewhere else, has used their watch-face to redirect the sun into the teacher's eyes. Although distance equals safety for the transgressor it also increases the difficulty of hitting the target, and herein lies the crux of this method; namely mechanical amplification. Very small movements of the reflecting surface of the watch-face are translated into quite large displacements of the reflected sunlight that dazzles the teacher. Replace the watch-face with a micro-fabricated tip, the sunlight with a laser beam and the teacher's eyes with a photo-detector and you have an AFM sensor! In AFM, however, the distance between the reflector (cantilever) and the target (photo-detector) is not in fact used to determine the amplification factor, because a greater distance results in a more diffuse, and therefore less intense, laser spot on the photo-detector, lowering the overall signal to noise ratio. The amplification in an actual AFM optical beam set-up is determined by the size of the cantilever; the shorter the cantilever the larger the angular displacement of the laser beam. This mechanism was pioneered by Meyer and Amer (Meyer and Amer, 1988) and, like every good idea, seems obvious with the benefit of hindsight. What perhaps was less obvious was the sensitivity this method could offer. Meyer and Amer demonstrated that by constructing small cantilevers the smallest theoretical measurable displacement was approximately 4×10^{-4} Å. In practice the displacement sensitivity is limited by random thermal excitation of the cantilever on which the tip is mounted but, with proper design, this is easily good enough to detect atomic scale displacement. The photo-detector is usually a simple photodiode, a semiconductor device which turns light falling on it into an electrical signal such that, as the incident light becomes brighter, the electrical signal increases. The photodiode is split into four sections enabling both normal and lateral motions of the tip to be differentiated. By comparing the relative intensity of the reflected laser light in each quadrant an approximate quantification of tip displacement can be achieved. However, for more precise measurement of tip displacement, such as is necessary if one is principally interested in performing force-distance spectroscopy on single molecules, then linear position sensitive detectors are a better choice of detection method (Pierce *et al* 1994).

In addition to cantilever length, laser spot diameter has recently been shown to play a significant role in the sensitivity of the optical beam detection technique (Schäffer 2005). This is because in reality the laser spot is not a point-like entity of infinitesimally small dimensions, as assumed in the original theoretical treatments, but rather has a finite size and non-circular shape. This means that the lever actually behaves more like a curved mirror than a planar surface, which changes the mathematics of light reflection considerably. Somewhat contrary to expectation, larger spot diameters will produce a lower thermal noise level in the deflection signal than a small, more tightly focussed spot on the same cantilever (Schäffer 2005). This effect can be as large as a 4-5 fold reduction in the thermal noise for the case of a pinned-cantilever (i.e. when it is actually in contact with a rigid sample surface). Note, however, that on a typical soft biological sample this ideal can only be met with a very soft cantilever, since the sample may posses a similar effective spring constant to the cantilever (≈ 0.1 Nm^{-1}). If, by an inappropriate choice of the lever and sample they do have similar spring constants then the cantilever will behave as if it were free, even at moderately high force set-points of ≈ 1nN, and so the thermal noise (that caused by Brownian bombardment of the lever by surrounding gas or liquid molecules) will be much higher than it would be if optimal levers are used. Of course there is a limit; the spot diameter shouldn't become larger than the cantilever itself otherwise light will spill over the edges of the lever, enhancing the effect of other noise sources. In practice the optimal situation for minimising thermal noise appears to be a soft cantilever combined with a laser spot size approximately equal to its length (Schäffer 2005). As luck would have it (or rather careful design in many cases) most commercial AFMs have spot diameters which approximately match the length of the shortest commercially available cantilevers (38 μm).

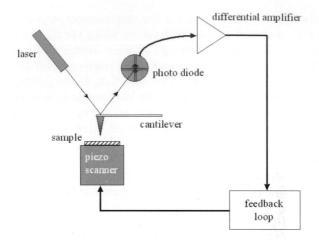

Figure 2.9. Laser beam deflection detection method.

In the last few years there has been a rapid development in light emitting diode technology (these-days everything is fitted with a sexy blue LED) and a new class of device, the super-luminescent diode (SLD), has emerged. These provide similar light intensity to laser diodes but, unlike a laser, their light output has a low coherence, which offers significant advantages for AFM detection. Low coherent light sources virtually eliminate the periodic noise commonly seen in force spectra obtained with laser diodes. As AFM instrumentation and cantilever technology has improved over the years so the lower limits of force detection have improved. Hence the elimination of noise from whatever source has become a critical issue, and SLDs are set to become the industry standard for modern high performance AFMs.

2.5.2. Optical detectors: Interferometry

This was initially the obvious choice for detecting the very small displacements of the AFM tip, since it is a well-established technique for high precision measurement of small displacements invented as long ago as 1880 by Michelson (Hecht 1987). Its implementation in AFM is shown schematically in Fig. 2.10. The phase change of the laser beam reflected from the back of the AFM cantilever is measured relative to a standard using an optical interferometer. This phase shift represents the relative displacement of the cantilever. It has advantages over optical beam deflection. This approach has the ability to cope with large deflections and has a superior signal to noise ratio. However, in practice it leads to an instrument which is harder to setup, requires an optical table to achieve sufficient vibration and acoustic isolation, and is susceptible to thermal drift and variation in the laser frequency. Despite these problems it was successfully implemented in AFM by Erlandsson and coworkers (Erlandsson *et al* 1988) and several essentially interferometric detection methods have been developed by others (Martin *et al* 1987; Schönenberger *et al* 1989; Rugar *et al* 1988; 1989). The technique continues to be used in a few specialised research AFMs.

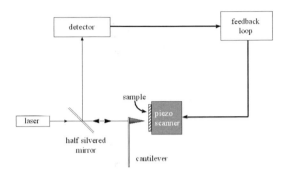

Figure 2.10. The Interferometer detection method.

2.5.3. Electrical detectors: electron tunnelling

In this approach an STM sensor is used to measure the displacement of the AFM tip. This is shown schematically in Fig. 2.11. The cantilever must be conducting, or coated with a conducting material (usually gold) and it then forms the 'sample' for the STM. The tunnelling current varies exponentially with STM tip-cantilever separation and is used to monitor the motion of the AFM tip. Because of this high sensitivity to tip-'sample' separation it is necessary, in order to have any sort of useful working range, for the tunnelling current to be used in a feedback loop and to maintain constant tip-'sample' separation, i.e. between the STM tip and the AFM cantilever.

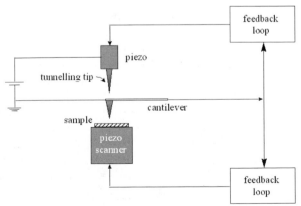

Figure 2.11. Electron tunnelling detector.

The technique has two drawbacks, which ironically are both related to its extreme sensitivity. The first is that STM feedback control and vibration isolation are problematic. The second is that the STM sensor is sensitive to the roughness of the cantilever, so that if the tunnelling site on the cantilever changes due to lever flexure during imaging, then the image produced is a composite of the actual sample surface, plus the topography of the back of the cantilever. This was the method used in the first ever AFM which was built by Binnig, Quate and Gerber in 1986 (Binnig *et al* 1986) but it is not used in any modern instrument.

2.5.4. Electrical detectors: capacitance

A small metal plate, typically 300 x 300 μm, is attached to the back of an AFM cantilever, with a second plate attached to the end of a piezoelectric transducer (this is to give it a better dynamic *z* range) with a plate separation of about 1mm (Göddenhenrich *et al* 1990). A schematic layout of a capacitance force sensor is shown in Fig. 2.12. The capacitance depends on the separation between the plates and so provides a measure of cantilever displacement. It is very compact, UHV compatible and easy to configure, but it is susceptible to temperature-induced drift of the reference capacitor used in the measurement circuitry. Another disadvantage is that the spring constant of the cantilever varies with capacitance. As in the case of interferometric displacement sensing, capacitance sensing has found greater application for monitoring the displacement of piezoelectric scanners in closed loop feedback systems rather than as the primary means of sensing AFM tip motion.

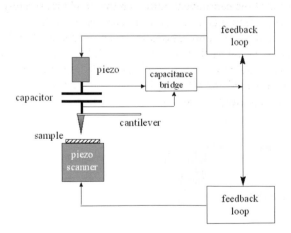

Figure 2.12. Capacitance detection method.

2.5.5. Electrical detectors: piezoelectric cantilevers

In this method the AFM cantilever is made from, or coated with, a piezoelectric material (Tansock and Williams, 1992). The principle is illustrated in the schematic diagram shown in Fig. 2.13.

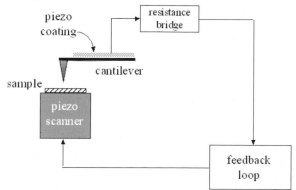

Figure 2.13. Piezoelectric cantilever detection method.

As the cantilever flexes in response to changes in sample topography, the bending of the piezoelectric material generates a potential difference which corresponds to the displacement of the lever. Flexure of the piezoelectric material also causes its resistance to change, which provides an alternative means of detection. This method has obvious benefits for examining light sensitive samples and this, combined with its reasonable sensitivity, makes it the most popular alternative to optical detection methods in modern commercially available AFMs. One disadvantage is the relatively high cost of the cantilevers, although this might reasonably be expected to come down as production increases. Piezoelectric resistive levers only really occur in a very limited number of applications these days. Undeniably the most exciting example of these is the AFM that is onboard NASA's Phoenix Lander, which was until very recently being used to image mineral grains scooped up from the Martian surface - until the dreaded extra-terrestrial technical gremlins struck and off-world tech. support is not currently offered by any of the major AFM manufacturers…

2.6 Control systems

2.6.1. AFM electronics

Every modern AFM has a digital control system (Wong and Welland, 1993; Baselt *et al* 1993). This consists of four elements which are illustrated in the schematic diagram shown in Fig. 2.14. The first element is the digital control electronics, in the early days this took the form of a digital signal processor (DSP) card, which resided in the image acquisition computer.

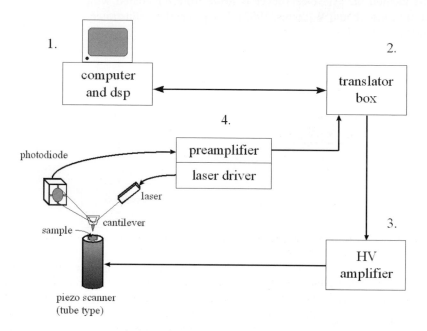

Figure 2.14. Typical AFM control system showing the main components.

The DSP performed all of the signal processing and calculations involved in the operation of the AFM in real-time. These days many of the functions performed by the DSP cards are now handled by the computer's CPU directly since they are easily fast-enough to keep up with the demands of the rest of the instrument. The digital processor, whatever form it takes, is linked to the second element of the electronics which, for want of a better description, we will call a translator box. This translator box performs the conversion of the digital signals sent from the DSP card to an analogue form in order to run the microscope's scanner mechanism and, unsurprisingly, incoming analogue data signals from the microscope head are converted to digital form for the DSP card. Basically the translator box is akin to the CD player in your living room, it is full of digital to analogue converters (DAC) and analogue to digital converters (ADC) and usually has an independent power supply for reasons of noise reduction.

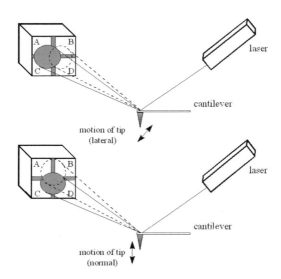

Figure 2.15. By splitting the photodetector into four segments lateral motion or twisting of the cantilever (top) can be distinguished from normal or vertical motion (bottom).

The third part of the control electronics is the high voltage (HV) amplifier. The input of this amplifier is fed from the analogue signal which comes from the DACs in the translator box. It amplifies this low voltage signal to produce a high voltage signal, typically ± 150V, that drives the piezoelectric scanners described earlier. The final pieces of the jigsaw are the laser/SLD driver electronics (assuming the AFM uses an optical detection method) and the raw data signal preamplifier. These functions are usually combined in one box, or area of the control electronics, which is well shielded from the digital translator and high voltage amplifier electronics, both of which generate electrical noise. This noise (high speed switching noise from the digital translator box and mains frequency noise from the high voltage amplifier) can be a real problem if it finds its way onto the incoming low voltage raw data signal. The laser/SLD driver circuitry provides power for the AFM's light source and, because light output stability is an important requirement for the laser/SLD and detection circuitry, it has its own feedback loop which varies the power supply to maintain a constant light intensity. The details of the raw data signal preamplifier obviously vary according to the type of detection system used but, whichever system, it amplifies the small output signal which comes direct from the AFM cantilever displacement sensor. For the optical beam method used by most AFMs, in addition to simple amplification, it also has a series of summing and difference amplifiers to produce an output signal for the translator box, which defines either normal or lateral motion of the AFM tip (Fig 2.15).

    A new development in electronics which has found application in many modern instruments, including the latest generation of AFMs, is the field programmable gate array chip (FPGA). In simple terms this device can be thought of as an integrated circuit (IC) whose function can be programmed by software. The advantage of this over traditional fixed-function ICs is that the instrument is no longer hardware limited, since the FPGAs can be continuously upgraded as new techniques are developed. This has obvious advantages for an instrument such as the AFM where new modes of operation continue to appear at a dizzying pace.

2.6.2. Operation of the electronics

The function of each part of the AFM's control electronics has been described in the last section. This section describes how they work together whilst the instrument is actually scanning a sample. The whole system actually works in a feedback or control loop so that the tip can accurately track the surface of the sample in a controlled manner at all times. The most obvious demonstration of this is when the AFM is running in the so-called 'constant force mode' (section 3.2), and it is easiest to describe this in terms of contact mode imaging. In contact mode imaging (also referred to as dc mode) the AFM tip actually touches the surface of the sample just like a phonograph stylus playing a record. The tip is brought into contact with the sample surface and then driven in the z direction using the piezoelectric scanner mechanism by a predefined amount, causing the cantilever on which the tip is mounted to be deflected. This predefined level of cantilever deflection is determined in the instruments software and is known as the operating set-point of the instrument. The set-point is maintained at a constant level throughout the scan by a feedback loop which is controlled by the DSP card. The scanning is achieved by sending drive signals to the $\pm$ x, y electrodes of the piezoelectric scanner such that the area to be examined is scanned in a raster fashion i.e. the sample or tip moves along one line in x and then moves up a line in y and so on. When the tip encounters an object the cantilever begins to bend and the feedback loop adjusts the z channel of the piezoelectric tube to move the sample or tip in the appropriate direction to return the cantilever to its deflection set-point. This process means for a homogeneous sample that, since the cantilever deflection is maintained at a fixed value, so also is the force between the tip and the sample, and this mode of operation has the name constant force imaging. The image is generated by plotting the z correction signal from the feedback loop against x and y. Height information in this mode is, therefore, derived from the feedback correction signal and not directly from the AFM tip. One can also operate in a mode where the cantilever is allowed to bend freely in response to sample topography, and this mode is known as the variable deflection mode. In this case the 'height' data in the image comes directly from the AFM cantilever and tip displacement. In practice even this mode uses some feedback control but the gain of the loop is set to a very low value. Whilst there are several other imaging modes, which are described in more detail in the next chapter, there is a third, in between state of affairs, known as 'error-signal' mode imaging. Error-signal mode imaging is described here because it is a special case in that it provides a way of overcoming a fundamental limitation of the feedback loop used in the AFM. In the error-signal mode the feedback loop is operational but at a reduced level, meaning that the deviation of the cantilever from its set-point (i.e. the 'error-signal') is relatively large on steep gradients, but negligible on small gradients, and so by recording the cantilever deflection directly it produces an image which enhances fine detail at the expense of coarse detail. This mode is particularly useful for rough samples such as whole cells, where the overall shape is of little interest but information on fine surface detail is required. The principle is that the feedback loop does just enough to remove the low frequency background (coarse detail) information from the image, leaving the high frequency (fine detail) information to be displayed.

2.6.3. Feedback control loops

The feedback control loop used in the AFM can be varied by the user through the gain setting of the instrument. The basic function of a controller is to maintain a predefined set-point; an everyday example is the thermostat unit in the central heating system in your house. The most basic form of control is so-called on-off control. For the heating system in your house say this would mean simply switching the heating off when the desired temperature is reached and then switching it on again when the temperature drops. This is shown graphically in Fig. 2.16. It is obvious from the graph that by using this method it is never actually possible to maintain the set-point. Rather the signal (temperature in the case of a heating system) continually hunts around it, overshooting then undershooting the required value. Clearly this would be a highly unsatisfactory way to try to control the motion of the AFM tip or scanner.

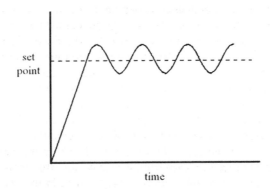

Figure 2.16. On-Off or 'bang-bang' control; the set-point is never achieved.

The AFM control system should be as accurate as possible, but also as fast as possible, in order to give it a sufficiently high bandwidth. This is achieved by introducing additional terms into the control signal of which three types are commonly used; proportional, integral and derivative (PID) controls.

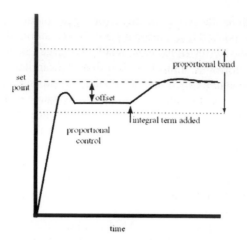

Figure 2.17. Response of a controller with proportional and integral terms added.

Proportional control simply amplifies the error between the set-point and the measured value in order to establish the size of the correction signal required. It does this by defining a proportional band which is some percentage of the total span of the controller, the actual percentage being the gain of the system; for example if the proportional band is 25% then the proportional gain is 4, a band of 5% would translate to a gain of 20 and so on. High gain therefore translates to a narrow width for the proportional band. The width of the proportional band determines the magnitude of the response to an error signal. Proportional control leads to a response as shown in Fig. 2.17, where the signal settles to a point which is slightly lower (or higher) than the set-point, i.e. there is an offset. In theory the width of the proportional band could then be reduced by increasing the proportional gain to reduce the offset. In practice, however, this leads to a highly unstable system because, if the signal moves outside the new narrower band there will be an inevitable reversion to on-off control, i.e. oscillation.

A better way of removing the offset is by adding an extra term which integrates the deviation from the set point over a small time period (Fig. 2.17). It is important to remember though, that if this time period is made too small, then the control signal could shift faster than the piezoelectric scanner can actually respond, and so oscillation will occur. Finally a derivative term proportional to the rate of change of the control signal is sometimes applied to reduce the tendency of the control loop to erratic behaviour, since it will respond quickly to large deviations. Incidentally, derivative control is also the most beneficial for recovering from small perturbations from the set-point. For an AFM control loop the proportional gain responds quickly to small features on the sample surface, and the integral gain helps maintain an accurate set-point (it cannot respond to small features). Derivative gain reduces unwanted oscillations, but can exaggerate high frequency noise if set to excessive values. Some AFM systems may have only a single gain parameter that the user can adjust but the instrument's software will then implement this using an algorithm which varies all three gain controls. In many systems the user has access to only the first two forms of control (or gain) and a good rule of thumb for setting these is that proportional gain should be set 10-100 times higher than integral gain. Some systems allow access to all three gain controls and so the following points should be kept in mind:

• Too high a proportional gain gives control loop oscillation.
• Too low a proportional gain gives sluggish control.
• Too high an integral gain gives over eager response and instability.
• Too low an integral gain slows down both the approach to set-point on start-up and return to set-point after a disturbance.

Although derivative action helps to achieve stability an excessive derivative gain may bring instability by over eager correction. It is probably the most difficult of the gain terms to set correctly. It is therefore usually set to zero except when scanning in non-contact (ac) modes of operation. Finally, the practical effects of instrument gain settings on actual AFM images is demonstrated in the next chapter (section 3.6.2).

2.6.4. Design limitations

Operation with a feedback control loop implies certain limitations on performance. These are both electronic and mechanical. In terms of electronic effects factors such as the bandwidth (frequency response) and stability of amplifiers, speed of analogue to digital conversion, or the sampling rate as it is normally called, all place limits on the operation of the AFM. In practice the bandwidth of a properly designed AFM system as a whole should be limited by the mechanics of the piezoelectric scanner rather than the electronics, because a tip traversing a very rough surface will require extremely fast correction signals from the feedback electronics to maintain a constant deflection, and there is a limit to how fast the piezoelectric scanner will respond to such high frequency signals. Bandwidth is an important consideration in any design of AFM since it places limits on the level of detail in an image. This is because fine detail in an image represents high frequency information.

Piezoelectric tubes and flexure stages have a relatively low resonant frequency, particularly in the *x, y* directions, typically around 5-10 kHz, so that they cannot be driven too quickly or resonant oscillation will occur with an inevitable loss of stability. Therefore they have limited bandwidth. The scan range of a piezoelectric tube increases with length, but its resonant frequency decreases with length, so we have a "Catch-22" situation. Longer tubes enable larger scans but have to be operated at slower speeds, which is not a problem in itself since when performing large scans it is wise to scan slowly to keep the tip velocity within reasonable bounds. However, a low resonant frequency makes them prone to unwanted resonant oscillation if the scan speed is increased. This means that even when scanning a smaller area, in order to produce a higher magnification image, the scan must be done slowly. Another factor which follows the same downward trend of piezoelectric tube stability versus length is their sensitivity; since long piezoelectric tubes deflect through a larger distance per volt, any noise on the driving signals causes a larger unwanted movement of the scanner for a longer piezoelectric tube. This causes a problem if scanning a very small sample such as a single molecule; the deviation of the AFM tip away from the set-point will be small, and so the *z* feedback voltage correction signal will be correspondingly small, meaning that the signal to noise ratio deteriorates. This implies that no one piezoelectric tube scanner is ideal for every sample, but rather the scanner should be chosen carefully to match the particular sample. So, for example, if one wishes to image large rough samples such as whole cells then a large scan range piezoelectric tube must be used but the ultimate resolution will not be as high, but, if one wishes to study single molecules on a flat surface such as mica then a smaller scan range tube is better, and will provide higher resolution and permit higher scan speeds. For this very reason the commercial AFMs that still use them have interchangeable piezoelectric-tube scanners.

2.6.5. Enhancing the performance of large scanners

Although one is forced to use a large scan range piezoelectric tube for large samples there are ways of improving their performance so that high magnification is still possible. After all what use is an AFM that can image a whole cell but then resolves little more detail than that obtainable from an optical microscope? Fortunately the signal to noise problem associated with long, large scan range piezoelectric tubes can be improved by scaling down unwanted headroom on the high voltage amplifier that drives the scanner when scanning small areas. This is done by scaling down the *input* signal to the high voltage amplifier, using either potential dividers to simply divide the output signal from the DACs in the translator box, or by reducing the voltage swing of the DACs with software control from the DSP card. This is easier and safer than trying to divide the output of the HV amplifier directly.

2.7. Vibration isolation: thermal and mechanical

Any high resolution measuring device requires a stable environment to operate at its optimum and the AFM is no exception to this rule. The two most important factors which must be controlled are temperature and vibration. AFM design tries to reduce the effect of thermal gradients through the careful selection of materials in its construction, i.e. ones with similar coefficients of thermal expansion, but nevertheless don't necessarily believe any salesperson who may claim that air-conditioning isn't needed! Indeed some AFM practitioners advocate allowing periods of 1-2 hours after inserting a sample into an AFM before commencing imaging so that instrumental drift is negligible. Fortunately control over temperature is relatively straight-forward to achieve with air-conditioning. A unit capable of controlling the room temperature to within 1°C is sufficient.

The AFM by its very nature is sensitive to very small displacement of the sample or tip. This of course means that vibration isolation is crucial to obtaining high resolution images. Of course AFM design strives to reduce the effect of mechanical vibration by making the microscope construction as compact and rigid as possible, in a similar way to that used for high performance optical microscopes. Just like the piezoelectric tube scanner and the AFM cantilevers, the desired effect is to move the resonant frequency of the microscope head itself as far as possible from the frequency of typical mechanical vibrations which may affect it, such as building vibrations (15-20 Hz). What is this 'resonance' stuff all about? A useful comparison here is how a musical instrument such as a guitar works. The sound produced by plucking a string is actually quite small, but the acoustic cabinet or body of the instrument works by having its resonant frequency very close to the frequency of the strings so that it oscillates sympathetically, increasing the amplitude of the oscillation, and hence amplifying the sound; an effect known as *resonance*. Anyone who has strummed an unplugged electric guitar and then compared this to strumming an acoustic guitar will know that the difference in sound levels is like 'night and day'. The frequency of oscillation of the strings is identical in both cases, assuming that they are both in tune, but the resonant frequency of the solid electric guitar body is not matched to the strings and so no (mechanical) amplification takes place. So you can imagine that AFM design has the completely opposite objective to musical instrument design; external vibration should never be amplified. In addition to this, even without resonant amplification by the instrument, mechanical vibrations can still cause problems for the AFM. The solution to this is to mount the instrument on some form of mechanically-isolated platform. These range from optical tables isolated from the ground with air-bearings to heavy stone plinths hanging from bungee cords. Actually the second solution is simpler, cheaper and provides better isolation from sideways (shear) movement since it can swing freely, a factor which is particularly useful if the AFM is not sited in a ground floor laboratory. Having said all of the above most AFMs will work surprisingly well simply sitting on a sorbothane™ mat on a sturdy table, but ultimately very high resolution may elude you.

Active vibration isolation platforms have become commonplace since the first edition was written. These work in the same way as noise-cancelling headphones. They incorporate a motion sensor which 'listens' to the background vibration coming from the building. This is used to generate a signal which is exactly out of phase with the incoming vibration to drive a series of actuators underneath the instrument bearing platform of the unit. This has the effect of cancelling out the incoming vibrations and the microscope is isolated. Their principal advantage is that they have a much smaller footprint than their snooker table-sized predecessors!

2.8. Calibration

There always comes a time when one is required to make some quantitative measurement of the AFM images. After all an AFM image provides us with a three dimensional map of the sample surface, chock-full of useful information waiting to be unlocked. The use of AFM has now moved firmly from the stage of simply asking what can be imaged by AFM to using it as a tool for solving problems, so more is being asked from the data it gathers. This is generally a good thing, but it does mean that now, more than ever, calibration of the instrument is very important. There are a growing number of papers concerned with using AFM in metrology (measurement science), and this area of research is helping to produce standards which can assess the reliability of such measurement (see the UK's National Physical Laboratory web-page for example). There are potential problems with using AFM to make measurements of samples: some of the problems are scanner related and some are probe-related.

2.8.1. Piezoelectric scanner non-linearity

Although piezoelectric scanners can move with incredible precision to allow high-resolution imaging they are inherently non-linear. This non-linearity is negligible for small displacements but, for large scans, it can become a problem. Here large scans mean more than 70% of the full scale displacement of the scanner. The most common configuration of scanner, the piezoelectric tube, suffers from hysteresis in its forward and backward traces as shown in Fig. 2.18.

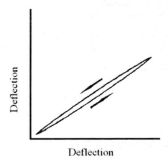

Figure 2.18. Hysteresis in the response of a piezoelectric tube scanner between forward and reverse directions.

Furthermore, the scanner can be non-linear with respect to scan speed, and the scanner always moves faster in one direction than the other as it rasters over the sample, meaning that the calibration factors can be slightly different for the 'fast' and 'slow' direction. The scanners are also susceptible to creep, which is a sort of relaxation which occurs under constant stress. Most modern commercial AFMs feature 'closed-loop' feedback systems, which independently monitor the motion of the scanner using interferometry, capacitance, strain-gauge or linear variable differential transformer (LVDT) sensors, and then correct its motion for non-linearity. Thermal drift can also affect the scanner, but in this respect piezoelectric tubes are far superior to piezoelectric tripod scanners. Finally the electrical and physical characteristics of piezoelectric scanners change with age, particularly if the scanner is not used for long periods of time.

2.8.2. Tip related factors: convolution

AFM tips are not infinitely sharp and so every image will be a convolution of the actual surface topography and the shape of the tip. The degree to which tip shape imposes itself on an image is governed by the nature of the surface being examined, i.e. it is sample dependant.

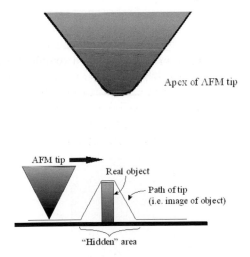

Figure 2.19. Tip convolution: because the AFM tip is not infinitely sharp (top) the image of an object is smeared out by the profile of the tip. The degree of this convolution is dependant upon both the tip shape and the height of the object being imaged (bottom). Note also that there are regions of the surface near a tall feature which are hidden from the tip.

When a rough surface is being scanned the edges of the tall features will produce a mirror image of the side-walls of the tip rather than of the object itself as illustrated in Fig. 2.19. For flat surfaces the degree of convolution is greatly reduced since the tip should approximate to a hemisphere at its apex. Nevertheless, even small objects such as single molecules of DNA for example, which are about 2 nm in diameter, are subject to an effect known as 'probe-broadening'. This effect is just the same as that illustrated in Fig. 2.19, but, in this case, only the hemispherical region at the tip apex touches the molecule, so that the image of the molecule does not look angular in cross section as is seen for a tall feature, but the width of the molecule is greatly overestimated. Even with accurate calibration of the instrument probe-broadening will *always* occur, meaning that, unless one has access to good tip-deconvolution software, it is better to use the heights of symmetrical features as a measure of their sizes, since the displacement of the tip in the z direction is unaffected by probe-broadening. A word of warning though, soft biological samples can be compressed by the tip (Weisenhorn *et al* 1993) and, if imaging is done in aqueous liquids, then electrostatic effects can alter apparent heights (Müller and Engel, 1997). A useful method for obtaining accurate height and volume measurements on soft samples, based upon force mapping to avoid set-point induced variability, has been suggested (Jiao and Schäffer 2004).

The geometry of the tip shape can affect the level of broadening or tip convolution, particularly for taller objects. This means that the scan angle of the tip with respect to the sample will have an effect on the level of probe-broadening seen in the image. Taking, for example, pyramidally-shaped tips, this is because opposite corners of the pyramidal tip are spaced further apart than opposite faces as shown in Fig. 2.20. Things can be complicated even further by the fact that the AFM tip can change shape during imaging by becoming contaminated by the sample, or by being blunted due to wear. In both cases this will affect measurements to some degree because the degree of probe broadening or tip convolution will alter, and either effect can be asymmetric.

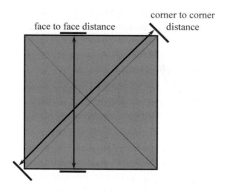

Figure 2.20. Pyramidal tip seen from above. The distances face-to-face and corner-to-corner form the bases of two triangles, the apexes of which define the sharpness of the tip in the direction of scanning. Obviously the larger base length corner-to-corner means a larger angle at the apex and so the tip is effectively 'blunter' in this direction.

2.8.3. Calibration standards

Standards for calibration of the AFM in x, y and z are commercially available and, by using a combination of these standards multiple point calibration can be carried out in each dimension, which is important because of piezoelectric non-linearity. Calibration samples are generally designed to reveal tip shape (Fig. 2.21c), and to minimise probe broadening effects, so that calibration in the x and y dimensions are as accurate as possible (Fig. 2.21a,b).

Standards are generally made from silicon and so are incompressible and stable. However they can cause tip wear. Other useful materials which can be used for calibration include polystyrene latex, colloidal gold spheres (both available from any microscope accessory supplier) and tobacco mosaic virus (TMV). TMV is a cylindrical shaped virus particle with such a well defined diameter (18 nm) that it is used as a calibration specimen for the transmission electron microscope. It is also relatively hard, in biological terms, and so compression is not a problem as long as the imaging force is not too high. TMV will not damage the AFM tip like a silicon standard may and this makes it ideal for calibration of the z axis of the AFM scanner, particularly for biological applications.

Figure 2.21 AFM deflection mode images of calibration gratings from NT-MDT (Russia) designed to determine (a) height and (b) lateral calibration constants. (c) Arrays of silicon spikes are used to determine tip shape.

2.8.4. Tips for scanning a calibration specimen

A few general points describing how to go about the calibration of an AFM are listed below.

• Choose the appropriate calibration standards i.e. within the size range of your samples. It is pointless to calibrate the z dimension with a 1 μm step if you are measuring samples of only 2 nm in height.

• Choose the right gain setting for the feedback loop, particularly when calibrating z. Too much gain and feedback overshoot will distort the data. Too little and underestimation of heights will occur.

• Scan at different speeds in order to determine over which range of scan speeds your calibration factor is reliable.

• Scan several areas of different sizes (or heights) to give a multi-point calibration factor.

•   Perform forward (trace) and backward (retrace) scans to check for frictional effects, piezoelectric creep and scanner consistency.

Finally, although many modern commercial instruments have sophisticated (and in some cases automated) software algorithms for performing calibration, an appreciation of the fundamental problems associated with measurement by AFM is useful if things go wrong. After all the odd manual check of your instrument's calibration following the above procedure doesn't take very long and it is better than relying upon blind faith! Taking all of the factors mentioned above into account it is obvious that measurement from AFM images is not a straightforward process and, if qualitative data is required, all of the potential systematic errors need to be considered and, if possible, quantified.

2.9. Integrated AFMs

AFMs are often combined with other instruments and some examples are described below. This list is not intended to be exhaustive, since there are many highly specialised custom built AFMs, but there are a few instruments of particular interest to biologists which justify a brief description here.

2.9.1. Combined AFM-Light microscope (AFM-LM)

By far the most common integrated instrument is the combined AFM-light microscope. One of the main drawbacks of AFM is its slow image acquisition rate; a typical scan may take 1-2 minutes. This means that finding an area of interest on a specimen surface is not as easy as in other forms of microscopy, where one can track around quickly until the eye spots a recognisable feature. By combining the AFM with a good optical microscope this problem can be overcome. Most of the major manufacturers offer combined instruments although they are of course rather more expensive than stand-alone AFMs. They nearly all take the form of an inverted optical microscope with the AFM mounted on top of the sample stage, giving it unimpeded access to the sample. This configuration also allows the sample and AFM tip to be seen by the optical microscope at all times. The samples can be presented on a glass slide or Petri-dish for imaging in liquids and require no special treatment. When the AFM uses optical beam detection methods, filters can be placed in the optical path to prevent laser light entering the user's eyes through the eyepieces.

The past decade has seen significant improvements in the integration of AFMs with inverted optical microscopes, with thoughtful software and hardware modifications being developed. Examples include allowing AFM scan areas to be defined using the optical images, overlaying of AFM and optical data, the use of phase contrast illumination and the use of high numerical aperture (NA) condensers. Furthermore, combinations of AFM with RAMAN, confocal laser scanning microscopes (CLSM) and total internal reflectance (TIRFM) are now commercially available. Combined AFM-LMs are excellent for large biological samples, such as whole cells or biological tissue, where there is an overlap in the magnification ranges required, and areas of interest can be located quickly and accurately with the light microscope before doing an AFM scan. Many of the techniques used in optical microscopy, such as staining and fluorescence, should be available on most combined commercial instruments to aid location and identification of structural features of interest, but some of the contrast enhancing modes which rely upon overhead illumination may not be possible because of the location of the AFM head. This is a point worth checking before purchasing a combined instrument, although not really an issue with the latest generation of combined instruments. In terms of absolute resolution the combined instruments used to be noticeably inferior to stand-alone AFMs due to the compromises needed in the design, but in the last few years this problem has been virtually overcome. However, stand-alone AFMs will probably always represent the ultimate in resolving power. Finally at least one company has built a very compact AFM that can be fitted to any light microscope in place of one of the objective lenses via a standard screw mount (DME A/S, Denmark).

2.9.2. 'Submarine AFM' — the combined AFM-Langmuir Trough

Interfacial phenomena are an area of major interest in biology and one of the principle methods for their investigation is the Langmuir trough (see chapter 5). Perhaps the most important example of an interfacial system in biology is the cell membrane. In animal cells this is composed of a phospholipid bilayer and it controls many aspects of cellular function. The AFM is of course an interfacial technique itself, usually operating at a solid-liquid interface and so a combination of the two methods is a very attractive proposition. Just such an instrument has been built (Eng *et al* 1998). In a Langmuir trough the surface active molecules of interest are assembled at a planar air-water interface of reasonable area, in order to produce an interfacial film, and allow various physical parameters to be measured. In order to acquire an *in-situ* AFM image of the interfacial film the tip must approach from below, otherwise capillary forces would simply drag it through the air-water interface and into the bulk liquid. Hence the instrument was labelled a 'submarine AFM'. There is at present, however, a limit to what can be imaged in this manner; the interfacial film has to be relatively rigid, the required rigidity being governed by the softness (spring constant) of the AFM cantilever. If the cantilever has too high a spring constant the AFM tip will simply push through the interfacial film without deflecting. This necessitates high packing densities of the surface active molecules to produce an interfacial film in the 'solid-phase', a situation which would not be found in healthy natural cell membranes.

2.9.3. Combined AFM-surface plasmon resonance (AFM-SPR)

Although AFM is capable of imaging dynamic events on a molecular scale, it is limited in its ability to provide quantitative data of such events due to factors such as scan size, and number of scans per unit time. This means that whilst the AFM is excellent for providing a description on a highly detailed scale it is less useful for measuring average properties of a system as a whole. Surface plasmon resonance (SPR), on the other hand, is a technique which is routinely used for quantifying the dynamics of biomolecular interactions, by monitoring changes in surface refractive index over relatively large areas of a sample. An excellent review of SPR can be found in the article by Silin and Plant cited at the end of this chapter (Silin and Plant, 1997).

Plasmons, or plasma waves, are charge density oscillations which propagate in a plasma; in the case of metals this is the 'gas' of free electrons. Plasmons which can occur at the interface of a metal and a dielectric material, usually gold and water in commercial instruments, are called surface plasmons. The general principle of operation is that these plasmons can be excited into resonance by an incoming light beam. As this occurs the some of the light's energy is adsorbed providing a means of quantifying the extent of plasmon resonance. It turns out that the angle at which resonance occurs is sensitive to the refractive index of the interfacial region above the metal surface, and this changes as material is added - the thicker an adsorbed layer of molecules on the gold surface the greater will be the angle of the incoming light beam required to excite plasmon resonance. In fact the technique is extremely sensitive to changes in optical thickness, and a resolution of better than 0.1 nm can be achieved. This translates to a detection limit of 0.5 ng. cm^{-2} surface concentration of adsorbate. Although it has exceptional height resolution, its lateral resolution is only about 5 μm, and so AFM adds the element of exquisite lateral resolution to the combined instrument.

Therefore the combination of AFM with SPR has provided a powerful and unique method for the quantitative study of dynamic surface events at the molecular level (Chen *et al* 1996). The capabilities of the AFM-SPR were demonstrated with studies of polymer degradation and protein adsorption. In both experiments the SPR provided qualitative kinetic data on surface dynamics and the AFM provided detailed spatial information on the nature of the events which were invisible to the SPR.

2.9.4. Cryo-AFM

Although the AFM derives much of its advantage over conventional electron microscopy by its ability to operate in ambient conditions, a low temperature AFM has been developed to enable higher resolution to be obtained on biological systems

(Han *et al* 1995). The benefits of working at low temperature are that molecular motion is frozen and the molecules have a much higher mechanical strength, some 1000-10,000 times that of a hydrated protein at room temperature for example, and so tip-induced distortion of the sample should be negligible. Basically the instrument consists of an AFM suspended in a cryostat containing liquid nitrogen. Operation of the AFM in the liquid nitrogen vapour allows temperatures in the range 77 to 220 K to be achieved. It was noted that below 100 K high resolution images of immunoglobulins, DNA and red-blood cell ghosts were possible.

Table 2.1. Summary of cantilever properties and uses.

Type of cantilever	k (N.m^{-1})	f_r (kHz)	Comments
contact (dc mode) usually V shaped	0.01-1.0	7-50	soft levers required for minimisation of force. Small levers required to prevent unwanted resonant oscillation and give maximum sensitivity.
non-contact (ac mode) V shaped and rectangular	0.5-5	50-120	ac mode means that the lever is oscillated near resonance so a high Q factor is important. Best achieved with a stiffer lever than for dc mode.
tapping (ac mode) rectangular	30-60	250-350	Very stiff lever required to give high Q factor and to overcome capillary adhesion between tip and surface if working in air. Note: for Tapping in liquid a softer lever is used.
force modulation (ac mode) rectangular	3-6	60-80	Used for mapping the compliance of a surface by applying a variable force during scanning and measuring the response using ac techniques. Therefore a high Q lever is important for sensitivity. Choice of spring constant is dictated by sample moduli; some degree of sample deformation must be possible or the measurement will be meaningless (you can't measure the deformability of a football by pressing it with a feather!)

What is Q-factor? See figure 3.5 in the next chapter.

References

Albrecht, T.R, Akamine, S, Carver, T.E and Quate, C.F. (1990). Microfabrication of cantilever styli for the atomic force microscope. *J. Vac. Sci. Technol. A* **8**, 3386-3396.

Baselt, D.R, Clark, S.M, Youngquist, M.G, Spence, C.F. and Baldeschwieler, J.D. (1993). Digital signal control of scanned probe microscopes. *Rev. Sci. Instrum.* **64**, 1874-1882.

Binnig, G. and Smith, D. (1986). Single-tube three-dimensional scanner for scanning tunnelling microscopy. *Rev. Sci. Instrum.* **57**, 1688-1689.

Binnig, G, Quate, C.F. and Gerber, Ch. (1986). Atomic force microscope. *Phys. Rev. Letts.* **56**, 930-933.

Binnig, G, Gerber,C, Stoll E, Albrecht, T.R. and Quate, C.F. (1987). Atomic resolution with the atomic force microscope. *Europhys. Letts.* **3**, 1281-1286.

Boisen, A, Hansen, O. and Bouwstra, S. (1996). AFM probes with directly fabricated probes. *J. Micromech. Microeng.* **6**, 58-62.

Chen, X, Davies, M.C, Roberts, C.J, Shakesheff, K.M, Tendler, S.J.B. and Williams, P.M. (1996). Dynamic surface events measured by simultaneous probe microscopy and surface plasmon detection. *Anal. Chem.* **68**, 1451-1455.

Dai, H, Hafner, J.H, Rinzler, A.G, Colbert, D.T. and Smalley, R.E. (1996). Nanotubes as nanoprobes in scanning probe microscopy. *Science* **384**, 147-150.

Eng, L.M. Seuret, Ch, Looser, H. and Günter, P. (1996). Approaching the liquid/air interface with scanning force microscopy. *J. Vac. Sci. & Technol. B* **14**, 1386-1389.

Erlandsson, R, McClelland, G.M, Mate, C.M. and Chiang, S. (1988). Atomic force microscopy using optical interferometry. *J. Vac. Sci. & Technol. A* **6** 266-270.

Frisbie, C.D, Royzsnyai, L.W, Noy, A, Wrighton, M.S. and Lieber, C.M. (1994). Functional group imaging by chemical force microscopy. *Science* **265**, 2071-2074.

Göddenhenrich, T, Lemke, H, Hartmann, U. and Heiden, C. (1990). Force microscope with capacitive displacement detection. *J. Vac. Sci. & Technol. A* **8** 383-387.

Han, W, Mou, J, Sheng, J, Yang, J, and Shao, Z. (1995). Cryo-Atomic Force Microscopy: a new approach for biological imaging at high resolution. *Biochemistry* **34**, 8215-8220.

Hecht, E. (1987). Interference. In *Optics*, pp. 333-388. Addison-Wesley Publishing Company, Reading, Massachusetts.

Hell, S. (2007) Breaking the barrier: fluorescence microscopy with diffraction-unlimited resolution. *Chromosome Research* **15**, 55-55.

Higgins, M.J, Polcik, M, Fukuma, T, Sader, J.E, Nakayama, Y, and Jarvis, S. (2006a). Structured Water Layers Adjacent to Biological Membranes. *Biophys. J.* **91**, 2532-2542.

Higgins, M.J, Proksch, R., Sader, J.E, Polcik, M., McEndoo, S., Cleveland J.P and Jarvis S.P. (2006b) Noninvasive determination of optical lever sensitivity in atomic force microscopy. *Rev. Sci. Instrum.* **77**, 013701.

Jiao and Schäffer (2004). Accurate Height and Volume Measurements on Soft Samples with the Atomic Force Microscope. *Langmuir* **20**, 10038-10045.

Martin, Y. and Wickramasinghe, H.K. (1987). Magnetic imaging by force microscopy. *Appl. Phys. Letts.* **50**, 1455-1457.

Martin, Y, Williams, C.C. and Wrickamasinghe, H.K. (1987). Atomic force microscope force mapping and profiling on a sub 100Å scale. *J. Appl. Phys.* **61**, 4723-4729.

Mckendry, R.A, Theoclitou, M, Rayment, T. and Abell, C. (1998). Chiral discrimination by chemical force microscopy. *Nature* **14**, 2846-2849.

Meyer, G. and Amer, N.M. (1988). Novel approach to atomic force microscopy. *Appl. Phys. Letts.* **53** 1045-1047.

Müller, D.J. and Engel, A. (1997). The height of biomolecules measured with the atomic force microscope depends on electrostatic interactions. *Biophys. J.* **73**, 1633-1644.

Pierce, M.L, Stuart, J.K, Pungor, A, Dryden, P. and Hlady, V. (1994). Specific and non-specific adhesion force measurements using AFM with a linear position sensitive detector. *Langmuir* **10** 3217.

Rugar, D, Mamin, H.J, Erlansson, R, Stern, J.E. and Terris, B.D. (1988). Force microscopy using a fiber-optic displacement sensor. *Rev. Sci. Instrum,.* **59** 2337-2340.

Rugar, D, Mamim, H.J. and Guethner, P. (1989). Improved fiber-optic interferometer for atomic force microscopy. *Appl. Phys. Lett*s. **55** 2588-2590.

Schäffer, T.E. (2002) Force spectroscopy with a large dynamic range using small cantilevers and an array detector. *J. Appl. Phys.* **91**, 4739-4746.

Schäffer, T.E. (2005) Calculation of thermal noise in an atomic force microscope with a finite optical spot size. *Nanotechnology* **16**, 664-670.

Schönenberger, C. and Alvarado, S.F. (1989). A differential interferometer for atomic force microscopy. *Rev. Sci. Instrum.* **60** 3131-3134.

Silin, V. and Plant, A. (1997). Biotechnological applications of surface plasmon resonance. *Trends Biotechnol.* **15**, 353-359.

Tansock, J. and Williams, C.C. (1992). Force measurement with a piezo-electric cantilever in a scanning force microscope. *Ultramicroscopy* **42-44** 1464-1469.

Taylor, M.E. (1993). Dynamics of piezoelectric tube scanners for scanning probe microscopy. *Rev. Sci. Instrum.* **64**, 154-158.

Weisenhorn, A.L, Khorsandi, M, Kasas, S, Gotzos, V. and Butt, H-J. (1993). Deformation and height anomaly of soft surfaces with an AFM. *Nanotechnology* **4**, 106-113.

Wong, S.S, Joselevich, E, Woolley, A. T, Cheung, C.L. and Lieber, C.M. (1998). Covalently functionalised nanotubes as nanometre-sized probes in chemistry and biology. *Nature* **394**, 52-55.

Wong, T.M.H. and Welland, M.E. (1993). A digital control system for scanning tunnelling microscopy and atomic force microscopy. *Measurement Sci. Technol.* **4**, 270-280.

Useful information sources

John Sader's online spring constant calculator - http://www.ampc.ms.unimelb.edu.au/afm/

NT-MDT Co., Zelenograd Research Institute of Physical Problems, 103460 Moscow, Russia. http://www.ntmdt.ru/

DME A/S, Herlev, Denmark, http://www.dme-spm.dk

Dimensional and spring constant and tip shape calibration advice/service from the National Physical Laboratory (NPL), Teddington UK: http://www.npl.co.uk/server.php?show=ConWebDoc.583

CHAPTER 3

BASIC PRINCIPLES

3.1. Forces

How does the AFM actually record images and how can we ensure that they are of good quality? As has been mentioned briefly in chapters 1 & 2, the AFM generates images by 'feeling' the sample surface. The force between the tip and sample will vary if the sample is scanned beneath the tip or the tip scanned over the sample. Changes in force are sensed by the tip which is attached to a flexible cantilever. Depending on whether the cantilever is sensing repulsive or attractive forces, different imaging modes can be applied (section 3.2). The operation of the AFM depends on monitoring forces between the tip and sample and, in this section, the different types of forces likely to be encountered in biological systems will be introduced. Finally, once an image can be obtained, there are sections on recognising artifacts, common problems, and image processing techniques.

As the name 'Atomic Force Microscope' suggests the important interactions between the tip and sample are due to one or more forces. What are these forces, and what are their origins?

3.1.1. The van der Waals force and force-distance curves

Using a classical model, the electrons within a substance are in continual motion, and travel extremely rapidly. Quantum physics treats them as waves rather than particles. Whilst a given substance may appear electrically neutral over conventional periods of time, over extremely short periods of time, say a snapshot, the distribution of electric charge due to the electrons is not perfectly symmetrical. This gives rise to subtle charge imbalances known as 'dipoles' or 'multipoles'. Each molecule therefore exhibits a slightly different distribution of charge within a given snapshot, which is also dependent on the number of electrons that the molecule possesses. The charge imbalance in one molecule can electrically induce a similar imbalance in a neighbour. The net result being that the slightly positive end of one molecule will be attracted to the negative end of another neighbouring molecule; this is the origin of the van der Waals force. Normally these effects are completely masked by the very much stronger electrostatic force. However, it is important to stress that van der Waals forces are present in all materials, even those that are electrically neutral.

It is possible to characterise both attractive and repulsive parts of the force-distance relationship between the tip and sample by modelling their interaction. This involves looking at the variation of the potential energy of a particle, say at the apex of the AFM tip, due to its interaction with a discrete particle on the surface of the sample. As their separation (r) changes, so does the value of the potential energy, which can be described mathematically by the pair-potential energy function $E^{pair}(r)$. A special case of the well known 'Mie' pair-potential energy function is used to model this behaviour, called the 'Lennard-Jones' or '12-6' function:

$$E^{pair}(r) = 4\varepsilon\left[\left(\frac{\sigma}{r}\right)^{12} - \left(\frac{\sigma}{r}\right)^{6}\right] \qquad (3.1)$$

Where ε and σ are constants that depend on the material.

Incidentally, σ is approximately equal to the diameter of the atoms involved, and is sometimes called the 'hard sphere diameter'. Fig. 3.1 illustrates the variation of the pair-potential energy between two atoms. The $1/r^{12}$ term accounts for the steep increase in $E^{pair}(r)$ at small separations i.e. when $r < \sigma$ where the atoms strongly repel each other due to the Pauli Exclusion Principle. The $1/r^{6}$ term is responsible for the slower change in the attractive behaviour at relatively large separations, where the van der Waals force dominates.

Imagine what happens when you gradually approach this page of text with your hand and then push down onto its surface. You will not experience the initial attractive forces because they are only significant at very small separations – too small to be discernible in this case. But when your hand actually rests upon the page, and you try and push it down further, the repulsive forces abruptly stop you from making any further progress.

With some understanding of this force, let's replace the notion of using your hand as a primitive AFM tip with the real thing. Remember that the tip is suspended on the end of an extremely flexible cantilever. At relatively large separations, say a few hundred nm, any attractive forces between the atoms in the sample and those found in the very end of the pyramidal AFM tip are too small to exhibit a significant effect. In addition, the springy nature of the cantilever ensures that no deflection is apparent (refer to Fig. 3.2, position 1). Yet as the separation is reduced the force between tip and sample rises rapidly. This happens even if the tip and sample are both electrically neutral. So although inter-ionic interactions need not occur the cantilever can still start to bend under the influence of attractive van der Waals forces; position 2. In fact, this happens over an extremely short distance from actual contact and the AFM tip appears to land on the surface of the substrate almost at once. This phenomenon is commonly known as 'jump to contact' or 'snap in'. At this point the tip lands on the surface of the sample, and the cantilever exhibits a significant bend in an attempt to pull the tip back off the sample; position 3. As the tip is pushed even closer towards the sample the bend on the cantilever straightens out again, as shown by position 4.

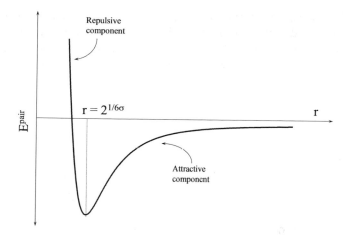

Figure 3.1. Schematic diagram showing the variation of the pair-potential energy (E^{pair}) with separation (r) between two atoms as described by the Lennard-Jones function.

This will always happen unless the sample is not very rigid and is easily deformed. When the cantilever is approximately straight the force exerted by it on the sample is nearly zero (position 5). This is usually the ideal condition for general imaging.

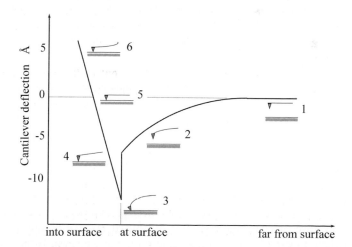

Figure 3.2. An idealised force-distance curve, illustrating how the bend of the cantilever changes with distance from a rigid sample in a vacuum. Note that this diagram shows a magnified region, close to contact. Some features are exaggerated in order to clearly differentiate between the different stages of cantilever bending. A more typical experimental force-distance curve is illustrated later in Fig. 3.4.

However if the tip is damaged this will not be true due to the presence of an 'adhesion force' as discussed later in section 3.1.3. It also follows that if the tip is forced further towards the sample the cantilever starts to bend in the opposite direction, as shown by position 6.

3.1.2. The electrostatic force

The Electrostatic, or Coulombic, force present in ionic bonds has by far the largest physical influence of any of the intermolecular forces that we shall consider here. Visualise two oppositely charged ions, q_1 and q_2, in a vacuum. At a short separation (r) they are attracted to each other. The force between them follows the well known Coulomb force law, and is proportional to $1/r^2$.

$$F = \frac{1}{4\pi\varepsilon_o} \cdot \frac{q_1 q_2}{r^2}$$

(3.2)

The constant ε_o is referred to as the 'permittivity of free space'.

As the ions are brought closer together the attractive force between them rises sharply. Eventually the outer shell electrons around each ion interact, and the force between the two ions becomes repulsive. This is due to the Pauli Exclusion Principle and the fact that the surrounding electrons now do a relatively poor job of screening the interaction between the nuclei at very small separations. You may be familiar with this notion as 'core repulsion', and at this point the ions cannot be pushed any further together without a relatively large input of energy.

3.1.3. Capillary and adhesive forces

A point contact with a small radius of curvature, and resting on a surface, acts as an ideal nucleation site for the condensation of water vapour present in the air. Unfortunately a typical AFM tip has a radius of curvature of around 30 nm, and so fits this criterion perfectly when used to image in contact mode. In addition a layer of water will condense on the sample surface at normal relative humidity (RH). This means that when imaging in air the tip will be pulled down towards the sample by a strong liquid meniscus, giving rise to the so-called 'capillary force', which 'glues' the tip to the sample, see Fig. 3.3. This is a major problem because the capillary force is independent of the instrument settings, and therefore cannot be easily compensated for by the operator. The result is that the overall imaging force can now be large enough to destroy or move delicate samples over the substrate. Although it is possible to circumvent this effect by enclosing the AFM head in a sealed box fed with dry air to reduce the RH, this is hardly convenient as it hinders access to the instrument.

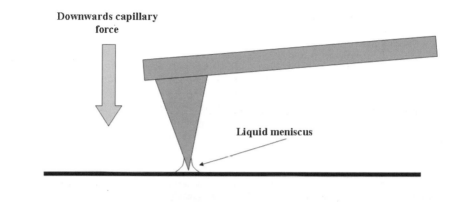

Figure 3.3. When imaging in air a large capillary force pulls the tip towards the sample (Top). ESEM images of the liquid meniscus taken a different levels of humidity (Bottom) courtesy of Brandon L. Weeks. Dept. of Chemical Engineering, Texas Tech University.

The presence of capillary force in contact mode sets a practical scan size limit, beyond which scanning smaller areas is unlikely to produce more information, because the sample will be damaged by the increased density of imaging points. As a rule of thumb, this scan size limit occurs at about 1 μm square when imaging in air with the tip in contact with the sample. Eliminating capillary forces has perhaps proved to be the most important step in successfully imaging many biological specimens. This can be achieved by imaging under liquid or using Tapping mode (section 3.2.2.)

Obviously AFM tips do not last forever. With time they become blunt, and they can also become contaminated with small amounts of the sample. Both these effects lead to a greater contact area between the tip and sample and, ultimately, to the presence of an adhesive force. This is a major problem when studying small objects such as discrete molecules since they are easily damaged by high forces. However, it is not so critical when studying large objects, because they can invariably withstand higher imaging forces. Fortunately, it is possible to check for the presence of adhesion by generating a force-distance curve (see chapter 9 for more information on force curves) and examining it for any asymmetry in its approach and retract portions, see Fig. 3.4 below.

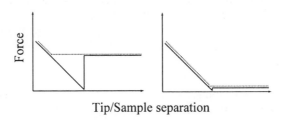

Tip/Sample separation

Figure 3.4. Two typical schematic force-distance curves. The curve on the left shows the presence of a large adhesion force whereas the curve on the right is reasonably ideal for most imaging situations. Dashed line: tip approach, solid line: tip retreat.

When adhesion forces become a significant problem the only real solution is to start afresh with a new tip. Unfortunately, the quality of new tips cannot always be guaranteed, and certain tips produce better images than others, even from the same wafer!

3.1.4. Double layer forces

This type of force arises when imaging under aqueous media, where mica (a common AFM substrate) is negatively charged and can attract oppositely charged ions from solution. This causes them to cluster at the solid-liquid interface leading to the formation of a positively charged layer, often known as an 'ionic atmosphere', since it can be visualised as being something like the earth's atmosphere. The electrostatic potential decays exponentially away from the surface, and the "thickness" of this ionic atmosphere is known as the 'Debye length' ($1/\kappa$). For low surface potentials (ψ_0), the potential at a distance (ψ_x) is related to the Debye length through the Debye-Hückel approximation:

$$\psi_x \approx \psi_0 e^{-\kappa x} \tag{3.3}$$

For a complete explanation of the Debye length, double layers and many other surface and interfacial phenomena, see Israelachvili, 1985.

Generally, the greater the ionic strength of the imaging medium, the lower the electrostatic repulsion that an approaching AFM tip experiences (see section 6.4.2).

Therefore, it is possible to electrically shield the AFM tip from the influence of the relatively large forces present when imaging in some liquids, such as buffers. This is usually achieved by adding a small quantity (a few mM) of a salt, containing divalent metal ions, to the imaging liquid. However, you should be aware that the addition of an excessive quantity of metal ions (~2 M) is undesirable as it can exaggerate the effects of the so-called 'hydration force' (Butt, 1991).

3.2. Imaging modes

There are many imaging modes available when using AFM. These are often differentiated by the nature of the force interaction involved in each case. Each mode has its own specific advantages and disadvantages. Below is an outline of some of the most important ones.

3.2.1. Contact dc mode

In this case the AFM tip is brought into direct contact with the surface of the sample, and in many AFMs the sample scanned beneath it (the tip-sample interaction is repulsive in nature). This process can be performed either in air or under a liquid, such as a biological buffer. The value of the pre-set imaging force is adjusted in the instrument software, which is equivalent to performing the entire scan with the cantilever bent by a small but fixed amount - hence this is known as 'constant force mode'. The greater the extent to which the cantilever is made to bend, the higher the imaging force that the sample experiences. Of course, a similar level of bending on a cantilever with a large force constant will produce a higher resultant force than that from a cantilever with a low force constant. By adjusting the force it is possible to vary the image contrast and/or reduce damage to the sample. Aspects of constant force mode are also discussed in section 2.6.2.

   One rather nice feature of this mode is that special tips are not required; in fact almost any reasonably flexible tip will do to get started. The most common tips used are perhaps those with a force constant of around 0.4 Nm^{-1}, and thankfully these are reasonably cheap. By employing contact dc mode under liquid the capillary force can be eliminated, allowing much greater precision in the control of the applied forces. Solvents such as propanol and butanol are popular, although greater understanding of ionic interactions has permitted the use of a wide range of biological buffers.

3.2.2. Ac modes: Tapping and non-contact

Another way of avoiding the problems caused by the capillary layer is to use the longer range attractive forces to monitor the tip-sample interaction. These attractive forces are weaker than the repulsive force detected in contact dc mode, and consequently different techniques are required to utilise them. There are two main types of ac mode imaging: the first is often known as 'Tapping', whilst the second is usually referred to as true non-contact mode.

An important point is that in the early days these ac imaging modes used to be significantly more complex for the novice user because of the need to set up the instrument. Therefore, it was generally harder to get good images using ac modes than with the traditional contact mode. However, in recent years the improvements in software – hardware communications, particularly with the advent of all-digital electronics, have made ac mode imaging very much easier.

Tapping in air

In this mode a relatively stiff rectangular or 'beam type' cantilever is used when the instrument is operated in air. The purpose of using this mode is to prevent the AFM tip from being trapped by the 'capillary force' caused by the extremely thin film of water surrounding samples in air. Here the cantilever is deliberately excited by an electrical oscillator to amplitudes of up to approximately 100 nm, so that it effectively bounces up and down (or taps onto the surface) as it travels over the sample. The glass chip on which the AFM tip is mounted is simply secured with a blob of glue, or a spring clip, to a small block of piezoelectric material which is excited directly with the signal from a precision signal generator. In addition to eliminating the effects of capillary adhesion, Tapping mode also reduces the lateral (side to side) force on the sample because the tip spends less time on the surface of the sample. This is useful because it means that delicate samples such as networks of molecules can be imaged without severe distortion or damage from these shear forces.

Tapping under liquid

In this case there is no 'capillary force' to cause imaging difficulties because the sample is immersed under a liquid, so a super stiff cantilever is not required. The cantilever can be driven into oscillation indirectly, for example it can be excited by applying a small sinusoidal electrical signal onto the z-channel input of the high voltage amplifier. This causes the main piezoelectric scanner to vibrate up and down in the vertical (z) direction, whilst still performing its normal task of responding to signals from the control loop. Therefore the sample and the liquid surrounding it begin to vibrate. This vibration is communicated to the cantilever, which is immersed in the liquid, by viscous coupling. Alternatively a small piezoelectric oscillator can be attached to the outside of the liquid cell or integrated into the tip holder and used to excite the cantilever more directly. This method is often favoured because the vibrational coupling is more efficient.

Q: *"If there is no capillary force present when imaging under liquids, why would we want to use Tapping mode when traditional contact mode is simpler to set up?"*

A: *Even under low force instrument conditions, contact mode exerts a significant shear or lateral force on the sample. Although for many samples this is not a problem, if we want to look at a sample that is weakly attached to the substrate, dynamic, or even living, high lateral forces can scrape it off the substrate - remember that the sample may have a poor affinity for the substrate.*

Cantilevers coated with a magnetic material can be excited using a nearby electromagnet, this is known as 'magnetic ac mode' or MACmode™. The electromagnet can, for example, be a few loops of copper wire wrapped around the liquid cell. It is then driven by a signal generator. The principal advantage of using this method is that it is easier to find the true resonant peak of the cantilever, because it is not convoluted with signals being reflected off the sample holder or mixed up with acoustic modes of the tip-cell assembly.

True, non-contact ac mode

In this imaging mode the tip never actually touches the surface of the sample. The cantilever oscillates with an amplitude of a few nm whilst hovering above the sample The relatively long range van der Waals attractive force between the sample and AFM tip produces a damping effect on the oscillating cantilever, and therefore its amplitude of oscillation is reduced as it approaches the surface.

Q: *"So I now know what the difference is between the various ac modes, but what are the advantages and disadvantages of using true non-contact ac mode?"*

A: *"Since in true non-contact ac mode there is obviously no contact with the sample, the forces exerted on it are extremely low. This results in very little deformation and shear. More importantly, the image contrast can be remarkably high producing striking images. In fact non-contact mode is used to achieve true atomic resolution with the AFM.*

It is usually necessary to choose a fairly flexible cantilever with a low resonant frequency, say around 30 kHz. In fact, a contact dc mode (liquid) cantilever will often do. This is because the relatively weak van der Waals force would have little dampening effect on a stiff cantilever. If the AFM is not configured optimally, right from the outset there is a good chance that the tip will hit the sample at some point during the scan. When imaging in air this is disastrous because the capillary force will easily trap such a flexible cantilever, leaving it 'glued' to the sample, and abruptly halting its oscillation. If this happens repeatedly the end of the pyramidal tip can be damaged or contaminated by fragments of the sample. In other words, the tip can easily be blunted by inadvertent high speed contact with the sample. A blunt tip is useless in any AFM mode because it invariably produces poor resolution images.

True non-contact imaging is relatively unpopular because super soft cantilevers are now available so that regular Tapping in liquid will in any case yield minimal sample damage. Additionally, slow scan speeds must be used to prevent accidental tip damage and it will not track rough samples well."

Tuning the cantilever

In order to oscillate the cantilever it is necessary to find its natural resonant frequency (f_r) i.e. the frequency where it is most sensitive to excitation. Modern instruments have a frequency sweep function built into the software. Fig. 3.5 shows such a typical sweep with a resonant peak at 321.6 kHz. Notice that there is only a single peak which is almost symmetrical and quite 'sharp'. This is because the cantilever is stiff and the sweep is performed in air.

A measure of the 'sharpness' of the resonant peak is known as its quality factor, or Q-factor.

$$Q = \frac{f_r}{\Delta f} \tag{3.4}$$

Where Δf is the 'width at half height'

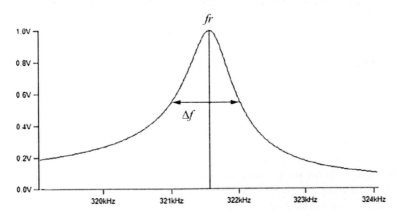

Figure 3.5. A frequency sweep of an Olympus AC160 cantilever in air showing the resonant frequency (f_r) and the width at half height (Δf). The Q-factor in this case is about 320.

A cantilever with a reasonably high Q factor will have a sharply defined peak with its shoulders displaying a steep gradient. The greater the gradient, the more sensitive the cantilever is to changes in drive frequency and interactions with the sample surface. These are exactly the characteristics that make a good AFM sensor.

Whilst this is a simple and straightforward procedure in air, driving the cantilever into oscillation will not work at all well if you want to measure its resonant properties under liquid. This is because the motion of the vibrating cantilever is damped by the viscosity of the liquid. Additionally, there are signal reflections and resonance contributions from the sample holder itself. Fortunately it is possible to 'listen' to the natural thermal vibrations of the undriven cantilever. These occur because the air/liquid molecules are in continuous collision with the cantilever through Brownian motion.

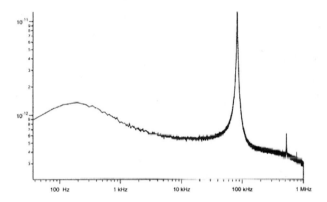

Figure 3.6. A thermal spectrum of an Olympus bio-lever 'mini' recorded in air. The fundamental is at about 80 kHz, with accompanying Eigen frequencies occurring at about 140 kHz and above.

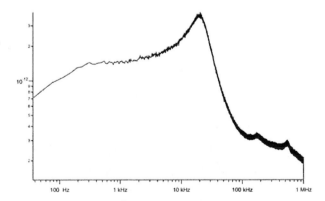

Figure 3.7. A thermal spectrum of the same lever recorded in water. The corresponding peaks shift to lower values and they become broadened, indicating lower Q-factors. Note the reduced peak amplitudes.

As you can see from comparing the graphs in Fig. 3.6 and Fig. 3.7 the various resonant peaks loose their sharpness when the cantilever is immersed under liquid. They become broadened and their Q-factors are greatly reduced, commonly down to single figures. In order to perform Tapping in liquid it may be necessary to drive at one of the higher frequency modes (the second mode often works well but it is worth experimenting with the others).

Influence of drive frequency

Because the terminology used to describe the various ac imaging modes is not particularly formalised there is a degree of confusion among casual users about the important distinctions between them. For example the term 'non-contact' is often used to described virtually everything that is not dc mode, which is unhelpful for the beginner. In order to try to clarify this situation the following section provides detailed descriptions of the distinction between Tapping and true non-contact mode in terms of the underlying physics, and in terms of actual operation of the two modes in AFM. Intuitively you might think that it is best to drive the cantilever at precisely its resonant frequency. However this is not the case as it can cause image artifacts. For Tapping in air or under liquid the drive frequency should be a little *less* than the resonant frequency. For true non-contact operation the drive frequency should be a little *more* than the resonant frequency.

Tapping in air or in liquid - Far from the sample the cantilever oscillates in free space, the drive frequency is deliberately set to be slightly lower than the natural resonant frequency of the cantilever. As the cantilever is brought closer to the sample it experiences an attractive force. The cantilever 'feels' heavier because of this and its resonant frequency drops. The drive frequency and the resonant frequency are now closer together than they were before. This increases the efficiency of the energy transfer so that the cantilever oscillates with a greater amplitude than it did in free space Fig 3.8.

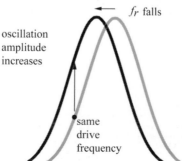

Figure 3.8. In tapping mode the resultant amplitude of oscillation *increases* as the cantilever approaches the surface.

The oscillation amplitude continues to increase as the cantilever approaches the sample surface until such a point as the tip 'taps' on the surface, at which point it reduces back to a lower value. The harder the tap the bigger the reduction in cantilever amplitude. Thus the set-point should be lower than the free amplitude; how much lower will dictate how hard the tip taps on the sample.

True, non-contact - Far from the sample the cantilever oscillates in free space, the drive frequency is deliberately set to be slightly higher than the natural resonant frequency of the cantilever. As the cantilever is brought closer to the sample it experiences an attractive force. The cantilever 'feels' heavier because of this and its resonant frequency drops. The drive frequency and the resonant frequency are now further apart than they were before. This decreases the efficiency of the energy transfer so that the cantilever oscillates with a smaller amplitude than it did in free space Fig 3.9. The oscillation amplitude continues to decrease as the cantilever approaches the sample surface. By not allowing the oscillation to drop right down to zero we can maintain the tip a small distance above the sample without actually touching onto the surface. Thus the set-point should be chosen to be only slightly below the free amplitude value.

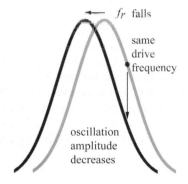

f_r falls

same drive frequency

oscillation amplitude decreases

Figure 3.9. In true non-contact mode the resultant amplitude of oscillation *decreases* as the cantilever approaches the surface.

In summary:

For Repulsive mode use:	For Attractive mode use:
Standard Q-factor tapping cantilever	Very high Q-factor cantilever
Drive frequency set below f_r	Drive frequency set above f_r
Large oscillation amplitudes	Small oscillation amplitudes
Hard tapping force between tip and sample (i.e. set-point a fair bit lower than the free amplitude)	Tiny force between tip and sample (i.e. set-point only slightly below the free amplitude)

Advantages of attractive mode imaging: *very high resolution, high contrast images, with minimal force exerted on the sample.*

Disadvantages of attractive mode imaging: *if there is any thermal drift then it is almost impossible to stay in the attractive regime. The result is often observed as image distortion or the tip disengaging with the sample. Also it does not work well with rough samples. It is easier to set up in air than in liquid because of the necessity for high Q-factor cantilevers.*

3.2.3. Deflection mode

In order for the tip to track accurately across the sample, whilst still exerting a low force, it is necessary for the gain of the system control loop (i.e. the control or feedback loop) to be set as close as possible to its optimum value. However, deflection mode the gain is deliberately set to a relatively low value in order to make the control loop response extremely sluggish. In this case the image is certainly not recorded at a constant force. In fact the AFM tip literally crashes into the features on the sample, rather than lifting gently over them. So what is recorded by the AFM in this case is a force map (or 'force image') of the sample, with large features represented by regions of high force, since they cause a greater degree of cantilever bend. The origin of the name 'deflection' mode comes from the fact that the deflection of the cantilever is used to generate image contrast. A more subtle version of deflection mode is the so-called 'error-signal' mode which is defined in section 2.6.2. Basically the feedback loop is allowed just enough gain to iron out the low frequency background undulation in sample topography, but not enough to cope with the high frequency changes caused by fine structure. For a tall or rough sample this produces an image in which the surface detail is enhanced at the expense of the overall shape of the object being imaged (Fig. 3.10).

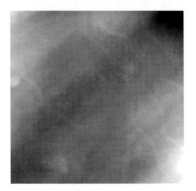 

Figure 3.10. A comparison of topography (left) and error signal images (right) taken over the same area of nettle cell wall material. The higher frequency information available in the error signal image yields considerably more detail, although height information is lost. Scan sizes: 2.5 x 2.5 μm.

Q: *"I understand how the error signal mode is set up, but it seems to represent the exact opposite of what we are trying to do in order to get a decent image. Does it really have any practical value?"*

A: *"Yes. The error signal mode is an extremely useful technique, particularly when used to image rough samples, such as in Fig 3.10. There are many samples that are simply too rough for the AFM to ever get a good quality image without continuously adjusting the gain throughout the scan. In this case, the force exerted on the sample is going to be fairly high anyway because the gain is never at its optimum value. In standard imaging modes the sample (or tip) is moved vertically by the piezoelectric scanner under instruction from the control loop. Although this is the traditional way of dynamically controlling the imaging force, it has its flaws. In particular, this electromechanical technique has to move a relatively heavy mass. If the sample is rough, yet fairly rigid, the actual imaging force is not so critical. By setting the gain of the control loop to a lower value we are effectively restricting the response of the piezoelectric scanner and sample in the z direction. This means that the cantilever will deflect away from its setpoint to a larger extent than before and there will be more fine detail in the error signal image than in the equivalent topography image. Remember the cantilever has a much lower mass than the scanner/sample combination, ergo it's frequency response is much higher. Its main application is for imaging rough yet fairly rigid samples such as cells and bacteria."*

3.3. Image types

This section discusses different ways of displaying the data and is somewhat dependent on what sample properties you are interested in measuring. Image types, as opposed to imaging modes, can often be combined during a single scan. For example, it is relatively easy to obtain frictional data on a sample, whilst at the same time recording its topography.

3.3.1. Topography

This is by far the most common way of recording images using the AFM. The information used by the instrument software to create an image is simply the vertical and horizontal movements of the piezoelectric scanner. Ideally the whole scan should be performed with a constant applied force i.e. a constant bend on the cantilever. The way of compensating for any excessive cantilever bend is to move the sample or tip together or apart via the piezoelectric scanner. From this type of image it is possible to measure the heights of objects in the image, by using 'line profile' software. Whilst this may sound obvious, it is actually worth stressing because other image types use differing contrast mechanisms, so that the height information is not recorded.

3.3.2. Frictional force

This type of operation is also known as lateral force imaging. Remember from chapter 2 that the face of a typical photodiode detector is split into four areas, or quadrants. In most imaging situations the difference signal between the top two and the bottom two quadrants is all that is needed to determine the extent of vertical cantilever deflection. In frictional force imaging, however, the side to side difference signal is used to determine the twisting behaviour of the cantilever under lateral forces. That is, the difference between the laser intensity received by the two left hand side quadrants is compared to that received by the two on the right hand side. When scanning an area of a sample with a significant frictional component, the AFM tip is restricted in its motion by the lateral force, so twisting of the cantilever occurs and the friction is detected. This can be very powerful, revealing information not available in the topography image, as shown in figure 3.11.

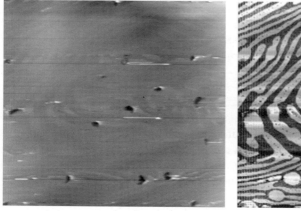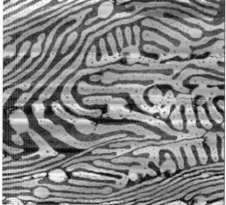

Figure 3.11. Topography (left) and friction (right) image of a phase-separated phospholipid-protein mixed monolayer on mica. Scan size: 10 x 10 μm.

3.3.3. Phase

Phase, or more specifically phase lag, is a quantity that can be recorded in the ac Tapping mode. In this mode the control loop uses the drop in amplitude of the oscillating cantilever to determine the vertical movement of the piezoelectric scanner.

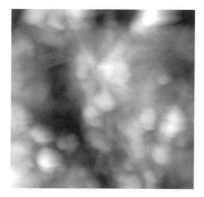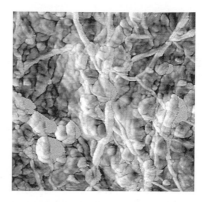

Figure 3.12. An example of the power of phase imaging illustrated using a two-part epoxy resin sample that was deliberately poorly mixed. The conventional topography image (left) shows little information because the sample is essentially flat. However the simultaneously recorded phase image (right), is considerably more detailed and highlights areas with different elastic properties. Scan sizes: 4 x 4 μm.

In addition, when the tip actually strikes the sample, its phase of oscillation is disturbed and it is no longer precisely in step with the phase of the electrical oscillator that is driving it. This is principally because each time the tip strikes the sample it transfers a small amount of energy to it. Just how much energy depends on the viscoelasticity of the sample. Fig. 3.12 shows a synthetic sample prepared from epoxy adhesive, note how much more detail there is in the phase image when compared to the somewhat unimpressive topography image. When you first start using the phase technique you might like to try this sample for yourself, since it is easy to prepare, and has the bonus of being virtually indestructible! Mix together a small amount of two-part epoxy adhesive - but ensure that you do it poorly! By doing this the hardener and the adhesive do not form a homogeneous mix, so that when the mixture sets some areas are more elastic than others. However, before this occurs, sandwich the mixture between two sheets of PTFE to produce a surface that is flat enough to image. After its cure time, peel back the PTFE to reveal the rigid epoxy slab.

A reasonable level of contrast in a phase image is primarily dependent on there being at least two components in the sample with sufficiently different viscoelastic properties. The ideal material would perhaps be something like small rigid metal particles embedded into a flexible polymer matrix. A word of warning though, interpretation of contrast in phase images is not a trivial matter, since as mentioned earlier the interaction between the oscillating AFM tip and sample can hop between different modes abruptly during the scan. This can cause large changes in the phase signal which are not necessarily related to the sample, but rather to the ratio of amplitude, damping and resonant properties of the combined system.

3.4. Substrates

In order to image a material by AFM it needs to be deposited and secured onto a rigid substrate. The most common types of substrates used in biological AFM experiments are discussed below.

3.4.1. Mica

This is the most popular AFM substrate, particularly for studying individual molecules, partly due to its wide availability and low cost. It consists of a series of thin, flat crystalline plates that can easily be split apart ('cleaved') by inserting a pin at its edge, or even using adhesive tape. The result is a truly fresh surface that has not been exposed to the atmosphere since it was originally created millions of years ago. In addition, mica sheets are atomically flat over large areas (typically a few microns), which is essential if molecular resolution is to be achieved. There are many different types of mica, which can be distinguished from each other by noting which metal ions they contain. Perhaps the most common type used as an AFM substrate is known as 'muscovite' mica [(K $Al_2(OH)_2$ $AlSi_3O_{10}$)]. One point that you should remember is that mica is negatively charged in aqueous liquids at neutral pH.

3.4.2. Glass

Glass, usually in the form of polished coverslips, is an ideal substrate for imaging larger samples, such as cells, where molecular resolution is often unnecessary. Glass coverslips can be surprisingly flat considering their low cost, and their roughness can be as low as a few nanometres over areas of several microns. It is nevertheless wise to rinse each coverslip in isopropanol, or acid, before use in order to remove any contaminants. Cells and bacteria can be cultured directly onto the coverslips, although any remaining media may need to washed away carefully before imaging. Cleaning generally promotes better attachment to the glass and therefore helps to ensure that the sample is not desorbed when imaging under aqueous media (for detailed surface cleaning protocols see chapter 5). Glass is also negatively charged under aqueous liquids at neutral pH.

3.4.3. Graphite

This material has been used since the early days of STM, where the substrate had to be conducting. This is clearly unnecessary for AFM studies unless you need to simultaneously acquire STM data. Graphite can be cleaved using adhesive tape, although some practice is necessary if you are to avoid removing thick layers. One point to consider is that graphite is extremely hydrophobic, so that samples deposited from aqueous solution both 'wet and spread' very poorly. However, graphite can still be useful in AFM, particularly for performing control experiments where it is suspected that the sample conformation is affected by interaction with the mica surface.

3.5. Common Problems

3.5.1. Thermal drift

This is generally only apparent when imaging under a liquid. If there is a significant difference in temperature between the liquid and the liquid cell, then thermal currents can evolve. A typical cantilever is composed of a layer of reflective gold applied onto silicon nitride. If the gold layer happens to be unusually thick, this composite can behave like a bimetallic strip and bend with temperature. This happens because each layer has a different coefficient of expansion, and therefore expands by different amounts at any given temperature. In addition, internal stresses created during manufacture can produce substantially larger thermal bending of cantilevers from different batches (Radmacher *et al* 1995). Thermal drift is best avoided by leaving both the instrument and the sample to equilibrate for a short period of time after filling the liquid cell. Since the presence of thermal drift depends partly on the thickness of the gold layer, you may find that it affects a whole wafer of tips, but then does not occur at all with tips from a different wafer.

Thermal drift is apparent in two forms: Firstly, a variation in the bend of the cantilever over time so that the sample either gets scraped off the substrate or the cantilever continually pops off the surface while imaging. Secondly, x-y drift of the sample so that the image appears slewed. Very slow scans highlight the presence of thermal drift.

3.5.2. Multiple tip effects

Even more frustrating is a very common effect known as a 'double tip'. As the name implies, the sharp point at the very apex of the tip can sometimes be accompanied by others, usually as the result of wear, damage or contamination (Fig. 3.13). In theory there can be any number of extra tips but most commonly only two or three are actually observed. This produces copies of the image, offset by a distance equal to the gap between the tips, typically a few tens of nm. Whilst this sounds like a serious source of artifacts, it can be reassuringly easy to determine when a double tip is present because the image adopts an obvious symmetry; Fig. 3.14 illustrates two examples. For discrete features the double-tip artifact is easy to spot (Fig. 3.14 left). However for a sample with more complex topography, such as the one seen in the right hand image in figure 3.14, the effect can be harder to recognize. Nevertheless, careful examination of the image reveals an unnatural repetition of features which are the tell-tale sign of a double-tip. If the double tip is caused by contamination it can sometimes be rectified by scanning over a large area at high speed. This can remove the contaminants, improving subsequent images.

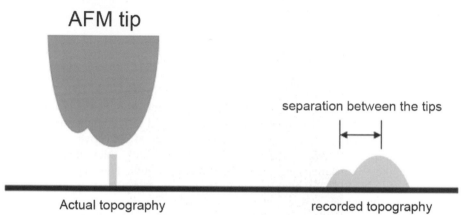

Figure 3.13. Ideally the apex of the AFM tip should terminate with just a few atoms. However tip damage and contamination means that this is rarely the case and multiple tips can be present. Because AFM images will always be convolved to some extent by tip shape (see section 2.8.2) the recorded image for a double tip will show two 'apparent' features for every actual feature the tip tracks over.

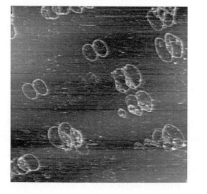 

Figure 3.14. Examples of multiple tip images; one is obvious the other more subtle. Left: circular configuration of a polysaccharide arabinoxylan extracted from wheat, scan size: 2 x 2 μm. Right: Valonia cellulose, scan size: 2 x 2 μm.

3.5.3. The 'pool' artifact

This type of artifact can occur when using Tapping mode. It is so-named because it gives the appearance that the tall features in the image are partly immersed under liquid, as in Fig 3.15. It occurs when the drive frequency is set too close to the resonant frequency of the cantilever. At resonance (f_r) the rate of change of the amplitude is zero i.e. the sensitivity of the cantilever at that precise point is zero – not a good place to be! Essentially the 'pool' artifact is caused by the instrument's control loop hopping between attractive and repulsive modes.

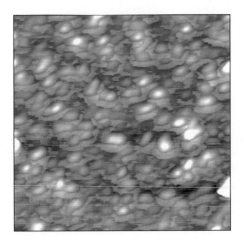

Fig 3.15. An example of the 'pool' artifact taken on a polyelectrolyte multilayer. Scan size 800 x 800 nm.

3.5.4. Optical interference on highly reflective samples

If the laser spot is poorly focussed on the cantilever then a significant amount of laser light can spill over onto the surface of the sample. On very reflective samples e.g. gold or silicon, some of this light can be reflected back towards the photodiode detector. Since the laser is a coherent light source, constructive and destructive interference can occur between the beam reflected off the cantilever and the overspill reflected off the sample. The result is that the AFM image exhibits a fringe pattern instead of a flat background (Fig. 3.16). Although this is not a huge problem when imaging, it can present a serious issue when trying to acquire force-distance data because the large apparent amplitude of the fringes ruins the signal to noise ratio so that small forces cannot be resolved. It is for this reason that many newer instruments use a low coherence Super Luminescent photoDiode (SLD) source instead of a conventional laser.

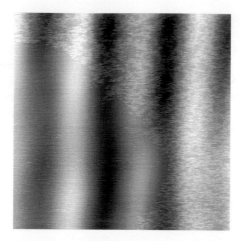

Figure 3.16. Topography image of a protein film. Scan size 5μm x 5μm. The vertical lines are interference fringes caused by overspill of the laser light from the cantilever.

3.5.5. Sample roughness

A typical pyramidal tip is only around 3 μm tall, which places a strict limitation on the roughness of any sample that the AFM can successfully image (Fig. 3.17). If the sample is very rough and possesses features taller than 3 μm, then the flat underside of the cantilever will touch the sample instead of the actual tip. This makes an unwelcome contribution to the imaging process, and is effectively like using the instrument with the ultimate blunt tip. This is unfortunate because many interesting samples are simply too rough to image by AFM.

Figure 3.17. This SEM image of a 200 μm long cantilever highlights the importance of ensuring that the sample is not excessively rough. It also illustrates that the tip is incredibly small. Image courtesy of Paul Gunning.

One solution to this problem is to partially embed the sample into a matrix in order to effectively reduce its height (Thompson *et al* 1994). Examples of this approach are discussed in sections 4.4.4 & 7.11.1. Alternatively, it is possible to purchase taller high aspect ratio tips (such as those outlined in section 2.3.2) although they are somewhat more expensive than ordinary tips. In addition, their use assumes that the tip height really is the limiting factor, but this is not necessarily true. Most piezoelectric scanners have a relatively modest vertical range of only a few μms, so this factor can also determine the maximum allowable sample roughness of a sample that can successfully be imaged.

It is possible, in some circumstances, to use 'sectioning' techniques developed for histology. By encasing the sample in a resin, then slicing it into fine layers with a microtome, it may be possible to image it using AFM, see section 7.11.1. This of course also gives the AFM access to internal structure in large samples.

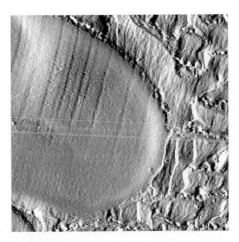

Figure 3.18. A starch granule embedded in an resin matrix and then thinly sliced using a microtome. The sample surface is now considerably smoother and can be imaged successfully, although there are obviously some disadvantages in choosing this method. The vertical striations are artifacts caused by small flaws in the edge of the microtome blade and 'knife judder'. Scan size: 17μm x 17 μm.

The principal disadvantage with this method is that the AFM is easily powerful enough to resolve any flaws in the microtome blade which may cause image artifacts (Fig. 3.18). However, when sectioned in this way the sample can acquire a degree of anisotropy which can usefully provide extra contrast (see Fig. 7.23).

3.5.6. Sample mobility

Samples are generally immobile when imaged in air, since they are dried onto the substrate. However this situation changes when imaging under liquids, especially aqueous buffers. Most biological materials have a high affinity for water and swell or become mobile in its presence. This is obviously a major problem when imaging in aqueous media because the sample can desorb off the substrate, and therefore become impossible to image. The main weapon to combat this problem is to use charge screening techniques as discussed under *Double layer forces* (section 3.1.4). Successful imaging in natural media is a challenging problem and a variety of solutions have been devised depending on the sample characteristics.

3.5.7. Imaging under liquid

Working in liquids is generally more difficult than working in air, but the points listed below should help you avoid some of the common pitfalls that await the inexperienced user!

Good thermal equilibrium is crucial when working in liquids as convection currents can severely upset the stability of the AFM. If possible, store the liquids in the air-conditioned, temperature controlled, laboratory where the AFM is sited so that their temperature is not greatly different from the instrument itself. If using a buffer which needs to be stored in a refrigerator, take it out and place it in the laboratory for a couple of hours before you begin the experiment. Be careful when filling the cell, as leakage can be very expensive, as well as dangerous. Many instruments have the high voltage electrodes (typically up to 200 V) that are connected to the piezoelectric device, sited underneath the liquid cell. Open 'bucket' type cells need particular care, although even so-called 'sealed' cells can leak if they are filled too hastily. The technique of sandwiching a drop, without the O-ring in place, can also be susceptible to leakage if the surface tension of the liquid is reduced by the addition of surface-active compounds, such as detergents, lipids or proteins. Many AFMs are now available sited on an inverted optical microscope, thus using a conventional glass slide as a sample platform. It is possible to reduce the likelihood of spillage by using Teflon coated glass slides such as that shown in Fig. 3.19. Alternatively a circle can be drawn around the sample area using a device known as a 'PAP pen'. This creates a barrier of hydrophobic 'ink' which is somewhat like paraffin wax when dry.

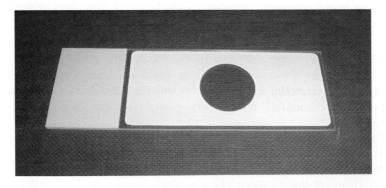

Figure 3.19. Teflon coated slides (Electron Microscopy Sciences, Hatfield, PA, USA) are useful for preventing leaks into expensive AFM scanners.

Liquid cells can be difficult to 'wet' due to the high surface tension of aqueous liquids. This is a particular problem for bucket type cells where the liquid sits as a discrete hemispherical drop. However, this can usually be circumvented by filling the cell, and then gently agitating it before securing it into the instrument. Obviously it is absolutely essential that all the components are scrupulously clean. With sealed cells air bubbles can sometimes become trapped between the tip and glass sight-plate during filling. This can manifest itself in one of two ways. Firstly, it becomes impossible to find the reflected laser spot when the liquid has been added, even after relentless adjustment of the photodiode. Secondly, the reflected laser spot can be found but appears to 'flicker', producing a rapidly varying output signal from the photodiode, making the instrument impossible to operate.

3.6. Getting started

If you have just acquired an AFM and are unsure where to begin, then this section is principally aimed at you. It is not intended to be a *'tour de force'* of operating AFM. Rather, its main purpose is to quickly get you up to speed by giving you a few tips and covering the most common pitfalls, without you having to read through a vast amount of literature. Obviously all instruments are different in their exact layout, and just as importantly, their software is likely to vary considerably. Therefore, it will be impossible to discuss specific features that are unique to individual instruments. However, assuming you have got to grips with the basics i.e. putting in the tip, aligning the laser, and getting onto the surface you are well placed to start on a real biological sample.

3.6.1. DNA

DNA is an excellent sample to begin learning about the AFM for a variety of reasons:

1. It is extremely well characterised, so you should have some idea about what you expect it to look like, and if its observed size and shape actually seems reasonable.

2. It is widely available. Therefore you should have no difficulty in obtaining a relatively pure sample.

3. It is a fairly delicate sample, so you can quickly see the results of correctly setting the instrument parameters, such as the force and gain, in order to obtain a

good quality image. Then you can learn about what happens when you vary one of the imaging parameters to obscure values.

4.   Although delicate, DNA is surprisingly robust when compared to many other discrete molecules, which is due to the fact that it forms a double helix. This means that it should not prove impossibly difficult for the beginner to image, which could easily be the case if you were to choose a sample at random.

You will need:

- A purified ~ 1 μg mL^{-1} suspension of DNA, and definitely not more concentrated than 10 μg mL^{-1} in pure water or low salt (< 2mM) buffer.

- A short cantilever of about 100 μm in length, with a force constant of approximately 0.4 Nm^{-1}

- Some propanol or butanol, preferably redistilled. This is to be used as the imaging medium, and is considerably easier than starting with an aqueous buffer since the DNA does not re-dissolve from the mica surface when imaged under alcohol. In fact alcohols are widely used as precipitants in DNA preparations.

- A small piece of mica, say 10 mm square, which has been freshly cleaved by inserting a sharp pin or something similar, at one edge.

- It is also worthwhile getting copies of the following papers, as both give useful information for imaging DNA: Hansma *et al* 1992, and Li *et al* 1992.

If you are fortunate enough to have an AFM system with a range of scan heads that cover different scan sizes, then you will generally be better off choosing one of the smaller ones. This is because there is no advantage in performing scans over about 5 μm since any individual molecules will then be too small to be seen. Additionally a smaller scan head will typically have a better signal to noise ratio than a larger one. Using one of the freshly exposed mica surfaces, deposit a small drop of the DNA suspension onto its centre. You should use a clean micropipette and apply no more than 5 μL. Make sure that the mica really is freshly cleaved, so do not use it if it is more than a few minutes old, otherwise organic molecules from the atmosphere could interfere with the adsorption of the DNA. Next, using tweezers, pick up the mica and gently swirl the drop around by hand for about 20 seconds to produce a reasonably even coverage. This is to prevent the drop from drying down into one tiny, but extremely concentrated spot, leaving no molecules anywhere else. Finally leave to air dry for about 10-15 minutes at room temperature.

After securing the sample in the liquid cell, you will probably need to refer to the instrument documentation in order to determine how much propanol, or butanol, to use as the imaging medium. About 200 µL is probably about average, but check anyway as you could damage the piezoelectric scanner from leakage caused by overfilling as discussed earlier. After filling the liquid cell do not be surprised if you loose track of the laser spot. As air and alcohol have different refractive indices, the laser path can change quite dramatically after introducing the liquid. You will therefore need to readjust the position of the laser and/or the photodiode, so that it once again captures the reflected laser spot.

You are now ready to approach the surface of the sample. Most systems tend to have some combination of automatic and manual sample approach. So this step should be fairly straightforward, assuming the default instrument parameters are reasonable. If they are not you may observe one of the following:

- Severe oscillation of the tip as it sits on the surface. This is generally caused by having the gain parameter(s) set too high, so that the control loop is excessively sensitive and constantly overshoots in its attempts to stabilise the tip position.

- The indicator showing the laser intensity received at the photodiode becomes very dim or changes to a low value. Congratulations! You have probably just "crashed the tip" (self explanatory). No matter, we have all done it. The number one cause of this is approaching the sample far too rapidly - in which case even a properly set up control loop could not react fast enough to prevent the tip from colliding with the sample. Alternatively, the control loop gain may have been set to a ridiculously low value by a previous user, causing the vertical movement of the piezoelectric scanner to be extremely sluggish.

- The instrument hunts around but never quite seems to settle the tip onto the sample. This is probably due to the value of the force set point in the software being too low (i.e. too close to the null-point) resulting in a 'false' engagement of the tip. Try increasing it a little.

- You successfully get onto the surface but, sometime later, perhaps part way through a scan, the tip pops off the surface again. Again try increasing the force a little. If this happens repeatedly, then you are probably experiencing thermal drift. In this case you will have to leave the system to equilibrate for at least 20 min.

After successfully getting the AFM tip onto the surface of the sample, modest scan sizes should be selected, say 1 or 2 µm. This will enable you to ascertain the degree of molecular coverage over the mica substrate. You should aim to adjust both the scan line density and speed so that a whole image composed of 256 lines is acquired in about 1.5 to 2 minutes, thus ensuring that tip is not moving so fast that it simply rips through the sample. Slower scan speeds are obviously needed when acquiring images with a line density of more than 256.

If you are fortunate you will find that the DNA molecules are reasonably evenly spaced across the mica substrate. However, you may find that there is a degree of aggregation leading to some areas of bare mica, with other areas containing relatively large clumps of molecules. If the second description fits your initial observation, you are probably better off preparing a new deposit, taking care to gently swirl the mica in the air when held with tweezers as described earlier. However, if you repeatedly observe aggregates after preparing new deposits, it is likely that the stock solution has aggregated. This can happen when the sample is frozen or stored at low temperatures.

Gentle heating in a water bath at around 40-50 °C can often provide sufficient energy to break up aggregation. Alternatively, excitation of the sample in an ultrasonic bath for a few minutes can work well or UV nicking to relax the chains if plasmids are used.

3.6.2. Troublesome large samples

Normally with a conventional microscope, the larger the sample, the easier it is to image. Surprisingly, AFM turns this rule on its head, principally because of the problems associated with tip convolution and surface roughness - such as ensuring that there is adequate clearance between the underside of the cantilever and the sample. For extremely large samples, there is no justification in using AFM when traditional forms of microscopy would be easier to apply, and almost certainly yield better results. However, it is sometimes desirable to obtain high resolution data over a small region of a relatively large sample, for example, receptors on a cell surface - yes, in AFM terms a cell really is a fairly large sample! In this case AFM can be useful provided that the tip is not continuously forced to climb over outrageously large features. The packing together of particulate samples into a monolayer ensures that the tip simply skips across the very top of the sample without having to travel back down to the bare mica substrate between features. Hence the apparent roughness is reduced. Materials that fall into this group of larger samples include cells, bacteria, starch granules and latex spheres, amongst others. As they are all discrete, separate particles, up to a few μm in size, the same general rules apply to imaging them. The various schemes used to minimise roughness for cellular materials are discussed in detail in section 7.1.1.

Firstly, and most importantly, concentrate on preparing your sample so that it forms a confluent monolayer in order to reduce the sample roughness. If the sample forms stacks of particles it will be impossible to get good quality, meaningful images. In the case of bacteria, as with cells, the form of the monolayer deposit can usually be controlled by manipulating the culture conditions (Gunning *et al* 1996). For particulates such as starch granules, apply a thin dusting of the sample onto double sided adhesive tape attached to the sample holder, followed by a gentle blast with a jet of air to remove any excess. However, do not attempt to image the larger varieties with diameters > 5 μm, as AFM is simply not the appropriate technique unless you resort to embedding. In the case of latex spheres, these can usually be encouraged to form monolayers by the addition of a small amount of detergent to the stock suspension. Of course it is still possible to image isolated samples of the type discussed here but this depends on them being securely attached to the substrate.

The importance of ensuring that samples that are composed of large particulates are as close to a continuous monolayer as possible cannot be stressed too heavily. Image distortion problems that occur by not taking this advice seriously can be summarised as follows:

- Mistracking and sporadic jumping of the AFM tip - indicated by streaky images.

- Artifacts resulting from severe image convolution. For example curved surfaces can appear severely flattened.

- Subsequent images taken over the same area yield different results - Isolated particles, not locked into a continuous layer, can be displaced by the tip - microscopic football!

Here are some useful guidelines to consider when imaging this class of material:

- Begin imaging with conservative scan sizes, say around 2 μm. Then if you encounter an unexpectedly large surface feature it won't bring about the end of the experiment through tip damage.

- Most large samples are surprisingly resilient, so achieving ultra low forces is no longer such a priority.

- If you can afford them, use high aspect ratio tips.

- Scan more slowly than usual. The control loop has more work to do in negotiating large features, and reducing the scan speed acts rather like increasing the gain of the control loop, but without causing 'gain ripple' shown in Fig. 3.20. Note that when sizeable objects are present in the image gain ripple tends to occur at their leading edge.

- Try using error signal mode.

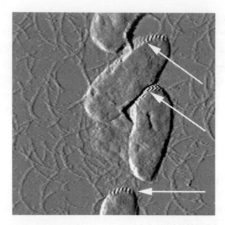

Figure 3.20. An error signal image taken on a portion of a bacterial biofilm displaying rippling artifacts (arrowed) at the leading edge of the bacteria due to excessive control loop gain. Note that the scan direction (left to right) is indicated by the presence of shadows on the far edge of the bacteria. Scan size: 6 x 6 μm.

3.7. Image optimisation

3.7.1. Grey levels and colour tables

This is simply a way of depicting image information (be it in terms of height, force, or phase etc) as regions of different brightness. The usual convention employed for a topographical (height) image is to set the lowest point as black, the highest as white, with everything else in between as a linearly changing shade of grey. It is fairly standard practice to utilise 256 different grey levels (0-255) in an image, although the definition of whether level 0 represents black or white does vary between manufacturers. This is not a problem as it is a trivial matter to invert the image to obtain correct contrast when swapping between different software packages.

It is also possible to artificially colour the image, either in the instrument software itself, or through an image enhancement package such as PhotoShop® (Adobe Systems Inc). This can be achieved by applying a monochrome colour table onto the image, for example green linearly graduated through to black. However, colour tables containing more than one colour can be difficult to implement unless you are fairly experienced in using this type of package. The advantage of applying colour to an image is that the eye is far more sensitive to changes in wavelength (colour), than it is to the intensity variation in greyscale images.

3.7.2. Brightness and contrast

This is probably the easiest way to optimise an image, either in the instrument software or by using a separate image enhancement package. The effects are relatively subtle but worthwhile, particularly prior to printing. Adjustment of either parameter beyond +/- 15% is to be avoided.

3.7.3. High and low pass filtering

This is fairly self-explanatory. The application of a high pass filter removes the lower frequency information from the image. This is particularly useful if you want to flatten the image background, because it appears like a rolling landscape, without loosing any of the fine detail. Low pass filtering, on the other hand, removes everything but the rolling landscape.

3.7.4. Normalisation and plane fitting

When an image is recorded by AFM it is acquired gradually, line by line. Each individual line can contain up to 256 grey levels, although subsequent lines may not necessarily attribute the same grey level to the same height, force or phase value. Therefore, it is necessary to ensure that a given grey level represents the same height throughout that particular image - this process is called 'normalisation'.

Plane fitting or 'Plane subtraction' involves using a least squares technique over the whole image to calculate an imaginary plane. This is performed because the sample surface is almost certainly not perfectly parallel to the X-Y plane of the AFM scanner. By subtracting this imaginary plane from the original image we are effectively removing any underlying tilt. This technique is usually followed by a normalisation procedure as described above.

3.7.5. Despike

Small bright white spots in the image, which can be single pixels, are very often produced by tip jumps. A tip jump occurs when the AFM tip is forced violently upwards and away from the sample, due to momentarily getting stuck on a surface feature. This adversely affects the resultant contrast because the tip jump is easily the tallest part of the image, so that all the interesting parts of the image are compressed into just a few of the darker grey levels. The despike routine searches for small areas that have wildly varying gradients - a spike, which can then be removed by replacing it with the median value of the pixels that make up its immediate neighbours. The image is then subjected to a normalisation procedure.

3.7.6. Fourier filtering

This technique has unfortunately earned itself a bad reputation in many branches of image processing. This is unjustified since it is an extremely powerful tool when used correctly. Problems usually only begin to appear when the technique has been misapplied, or used in an extreme manner. When it is used incorrectly it is quite literally possible to generate something out of nothing. Fourier filtering is generally only ever useful for optimising images containing periodic information, such as atomic lattices or ordered strucures.

Introductions to Fourier filtering tend to be highly mathematical since the technique itself is necessarily mathematical and, unfortunately, complex. This approach is rather indigestible, particularly if you have not come across it before, and is certainly outside the aim of this book. Fortunately, it is possible to get a good grasp of the technique by considering it purely from the optical perspective, which is the approach adopted here. To get a sense of how this technique operates, consider an experiment where a narrow beam of monochromatic light from a laser falls onto a narrow slit. As the beam exits on the other side of the slit, it spreads out, or 'diffracts'. This effect can be captured by projecting it onto a screen where it is displayed as a number of bright fringes, known as a 'diffraction pattern'. By varying the width of the slit, the number of slits present, or their orientation, it is possible to significantly alter the diffraction pattern. The interesting point is that there is an inverse spatial relationship between the separation of the fringes in the diffraction pattern and the width of the slit that caused them. In more general terms this can be expressed as: "Fine detail leads to widely spaced fringes - or fringes at the outer edge of the diffraction pattern".

The Fast Fourier Transform (FFT) routine, generally present in AFM software, provides an automatic mathematical method of producing an equivalent diffraction pattern for any given digital image, although it is now usually referred to as a 'power spectrum' or 'frequency domain'. Two examples are shown above their respective masks in Fig. 3.21. Highly periodic information in the image causes bright spots to appear in the power spectrum. Once again, any spots at the edge of the power spectrum are caused by fine detail (high frequency) present in the original image, whereas spots closer to the centre of the power spectrum are from coarser detail (low frequency) in the image. It is now possible, by using the FFT software, to edit the power spectrum, which provides an opportunity to remove any non-periodic random noise before reapplying the Fourier transform. The reapplication of the Fourier transform is known as a 'reverse' or 'back' transform. Hopefully, the result of the reverse Fourier transform should be a somewhat clearer image. This is where the danger lies; if the power spectrum is very noisy it could simply be that the original image is noisy, <u>or</u> that the image contains no significant periodic information. If the power spectrum is now edited injudiciously to enhance any areas that appear even remotely periodic, then the reverse transform will always produce an image with strong periodic structure, whether it was actually there or not!

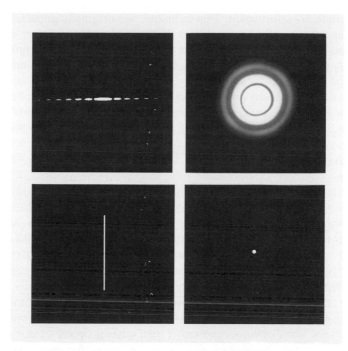

Figure 3.21. Two examples of power spectra (top), and the masks from which they were produced (bottom).

3.7.7. Correlation averaging

This image processing technique, which has its roots in electron microscopy, is frequently applied to periodic structures such as 2D-protein crystals or even isolated macromolecules. Initially a unit feature is identified which is repeated throughout the image. A number of these unit features are then chosen and aligned, for example by rotation. By summing over several unit features an averaged representation can be produced which has an improved signal to noise ratio, and therefore an improved contrast. This technique has been used in studies on membrane proteins as discussed in section 6.3.5.

3.7.8. Stereographs and anaglyphs

A stereograph, or a 'stereo-pair', is simply two images of a subject taken and viewed from slightly different viewpoints. This is in order to simulate the subtle variation in perspective observed by the left and the right eye. When the images are placed side by side and then viewed through a stereoscope, a simple frame containing a collector lens for each eye, the brain superimposes them and a single image is produced containing crude 3D information. This technique was originally popular in Victorian times but has subsequently been used for SEM, TEM and, to a lesser extent, AFM images (Shao and Somlyo, 1995). In the case of AFM data, the original image is electronically processed to produce two subtly different images which appear to have been acquired at different viewpoints.

An anaglyph is very similar except the offset images are tinted with red and cyan and the image is viewed with glasses containing the respective colour filters. The 3D effect is much stronger in an anaglyph but any colour purity is lost.

3.7.9. Do your homework!

The simple fact that AFM literature contains images means that you should take special care of how they are finally reproduced. It is absolutely vital that you take the trouble to maximise the impact of your images, as it can dictate whether your research gets read or passed over. Unfortunately this does not end with using image optimisation techniques such as those described above. Even if you start out with excellent, high contrast, images there is no guarantee that they will appear in print exactly as they were submitted. Surprisingly most publishers actually concentrate more on the text (spelling, grammar, style and so on), rather than the images. This is partly because it is the printers who have the final say in what appears on paper. Many readers see library photocopies rather than original copies or reprints, so if the originals are poor - the photocopies will be terrible, sometimes just a black box. This problem is becoming less frequent with the availability of electronic reprints. Dedicated microscopy journals nearly always do the work justice, but to reach the widest audience you may do better by targeting a journal according to the sample, or problem under investigation.

Colour images are generally reproduced well, although not all journals offer this option, and those that do usually charge. In order to ensure good quality image reproduction; **demand to see image proofs**, don't accept fax or photocopies. Additionally it is always worthwhile studying back issues of journals, paying attention not only to the quality of the images but also to the actual paper used to print the image.

References

Butt, H-J. (1991). Measuring electrostatic, van der Waals, and hydration forces in electrolyte solutions with an atomic force microscope. *Biophys. J.* **60**, 1438-1444.

Gunning, P.A, Kirby A.R, Parker. M.L, Gunning, A.P. and Morris, V.J. (1996). Comparative imaging of *Pseudomonas putida* bacterial biofilms by scanning electron microscopy and both DC contact and AC non-contact atomic force microscopy. *J. App. Bact.* **81**, 276-282.

Hansma, H.G, Vesenka, J, Siegerist, C, Kelderman, G, Morrett, H, Sinsheimer, R.L, Elings, V, Bustamante, C. and Hansma, P.K. (1992). Reproducible imaging and dissection of plasmid DNA under liquid with the atomic force microscope. *Science* **256**, 1180-1184.

Israelachvili, J.N. (1985). *Intermolecular and Surface Forces*. Academic Press.

Li, M-Q, Hansma, H.G, Vesenka, J, Kelderman, G. and Hansma, P.K. (1992). Atomic force microscopy of uncoated plasmid DNA: nanometer resolution with only nanogram amounts of sample. *J. Biomo. Struct. & Dynamics* **10**, 607-617.

Radmacher, M, Cleveland, J.P. and Hansma, P.K. (1995). Improvement of thermally induced bending of cantilevers used for atomic force microscopy. *Scanning* **17**, 117-121.

Shao, Z. and Somlyo, A.P. (1995). Stereo representation of atomic force micrographs: optimizing the view. *J. Microscopy-Oxford* **108**, 186-188.

Thompson, N.H, Miles M.J, Ring S.G, Shewry P.R. and Tatham A.S. (1994). Real-time imaging of enzymatic degradation of starch granules by atomic force microscopy. *J. Vac. Sci. & Technol. B* **12**, 1565-1568.

Weeks, B.L., Vaughn, M.W. and DeYoreo, J.J. (2005). Direct Imaging of Meniscus Formation in Atomic Force Microscopy Using Environmental Scanning Electron Microscopy. *Langmuir* **21**, 8096-8098.

CHAPTER 4

MACROMOLECULES

4.1. Imaging methods

To study isolated macromolecules by AFM it is necessary to confine them to a suitable surface or interface. Ideally, in order to gain the full advantages of the AFM, one would wish to investigate these molecules under natural conditions: usually in aqueous or buffered environments. Finally, the imaging process itself should not damage or displace the molecules. Achieving all these conditions simultaneously is very difficult and certain compromises need to be made. The methodology used will depend on the type of molecule under study and the sort of information required on molecular structure. It is convenient to discuss first methods of imaging in general terms.

4.1.1. Tip adhesion, molecular damage and displacement

In the earliest studies on macromolecules, originally through the use of scanning tunnelling microscopy, the major problems were sample damage, artifacts and irreproducibility. The origins of these problems are common to both STM and AFM. Most of these problems arose because samples were imaged in air, after adsorption to a suitable substrate. This is because, except at low relative humidity, a thin layer of water will be present on the sample surface and, depending on its radius of curvature, on the probe tip. When these two surfaces are brought sufficiently close together then these liquid layers will coalesce. The result is an adhesive force which effectively glues the tip to the surface (section 3.1.3). When the tip and surface are scanned relative to each other it is difficult to lift the probe over the deposited molecules. The tip either tears through the molecules, or pushes them across the surface. At high adhesive forces the molecules can be pushed across the surface and it was this which led to irreproducibility in imaging: molecules tended to be seen only when they became trapped at defects on the sample surface. Groups of molecules tended to be seen aligned at steps or other features. Not only did sample displacement account for irreproducibility, but also probably the high level of artifacts reported in early studies. If molecules are only seen when they are trapped at defects then the chance of observing a defect will be similar to that of seeing a molecule. Unfortunately some defects can give images which can be mistaken for macromolecules: the most publicised examples are the STM images of grain boundaries, which can be mistaken for helical molecules such as DNA (Clemmer and Beebe, 1991). The earliest methods of imaging, many of which were developed empirically, were designed to reduce molecular mobility and prevent sample damage and displacement.

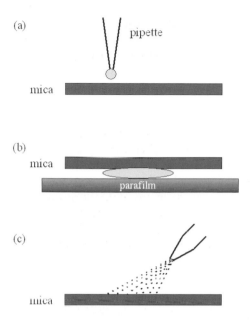

Figure 4.1 Schematic diagram showing different methods for depositing macromolecules onto substrates. (a) Drop deposition, (b) sandwich method and (c) spray deposition.

4.1.2. Depositing macromolecules onto substrates

Biopolymers are normally prepared as aqueous solutions and then deposited or spread onto a suitable substrate. The substrate needs to be flat with respect to molecular dimensions, clean, cheap and easy to prepare. The most commonly used substrate is mica. Mica is a non-conducting layered material. It is cheap and can easily be cleaved, usually with a pin or sometimes with cellotape, to produce clean, atomically flat surfaces up to even millimetres in size. The commonest form of mica is Muscovite $[KAl_2(OH)_2AlSi_3O_{10}]$. The minimum step size which can be observed on the surface is the thickness of an individual layer (1 nm) and the hexagonal lattice constant within the layers, which can be used for calibration, is 0.52 nm. The commonest methods for introducing the molecules onto the substrate are illustrated in Fig. 4.1. A pipette can be used to place a drop of solution onto the surface (Fig. 4.1a), a drop of solution can be formed on a hydrophobic surface, such as Parafilm, and the mica touched onto the surface of the drop (Fig. 4.1b), or the solution can be sprayed as an aerosol onto the substrate (Fig. 4.1c). With drop deposition the spreading on the substrate can be enhanced by addition of small quantities of surfactant to the aqueous solution. Samples can be sprayed directly from aqueous solution, or in the presence of glycerol. The glycerol is not essential, and is probably only present as a preservative or cryoprotectant for the sample, rather than to increase sample viscosity. In any case any glycerol present needs to be removed (e.g. by drying in vacuum) in order to prevent it coating the molecules, or contaminating the tip. The samples are then normally air dried for about 10 minutes before further preparation or imaging. Mica is a hydrophilic surface. An alternative hydrophobic surface is highly oriented pyrolytic graphite (HOPG). HOPG is also a simple layered structure and flat, clean surfaces can be produced by cleavage using cellotape. Graphite surfaces tend to contain more steps than that found on mica, and common defects on HOPG are grain boundaries. The layer spacing for graphite is 0.355 nm and the minimum step size is 0.669 nm. The true hexagonal lattice spacing for carbon atoms in the layers is 0.142 nm. However, alternate layers are staggered and the spacing of equivalent carbon atoms in each layer is 0.246 nm. Larger macromolecular complexes can be imaged after deposition onto rougher substrates: for example chromosomes are usually spread on glass and collagen fibres can be imaged on substrates as rough as filter papers. The roughness of the substrate must be small compared to the 'height' of the sample.

4.1.3. Metal coated samples

This approach borrows methodology used in transmission electron microscopy (TEM) and adapts it for use in probe microscopy. Rather than generating and imaging metal replicas, metal coated samples can be examined directly by AFM. The molecules are deposited onto a substrate, usually mica, air dried, inserted into a vacuum coating unit, evacuated and metal coated. The metal coated samples can be imaged directly, or stored for imaging at a later time. Provided the initial deposition conditions are correct then the surface concentration will be sufficiently low to permit visualisation of individual macromolecules (Fig. 4.2). The metal coating freezes molecular motion and protects the underlying molecules from damage or displacement. Thus images can be obtained under constant force (dc) conditions in air. Information can be obtained on molecular size and shape for whole populations of molecules. Even if the metal coating is fairly uniform it will still be difficult to obtain information on the heights, or diameters, of the molecules. The major disadvantages of this approach are possible denaturation of the molecules during sample preparation, and the loss of resolution due to the finite size of the metal grains coating the sample. Molecular distortion during preparation is a problem well studied by electron microscopists, and the elegant methodology developed in this field can be adapted for AFM studies. Thus samples can be freeze dried, or thin specimens cooled very rapidly and the water subsequently removed by sublimation below the glass transition temperature. These methods can be used to preserve the native structure of the molecules. By optimising coating conditions, and suitable choice of the coating material, it is possible to minimise the size of the metallic grains. However, this factor will always limit the potential resolution achievable by AFM.

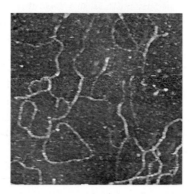

Figure 4.2. AFM image of a metal-coated fibrous polysaccharide (xanthan) imaged under butanol. Scan size 1.6 x 1.6 μm.

4.1.4. Imaging in air

To obtain images of molecules in air it is necessary to eliminate the adhesive forces. One approach is to remove the surface layers of moisture. This can be done by subjecting the sample to a low vacuum or by storing the sample in a desiccator under controlled low relative humidity (typically RH < 40%). The samples should then be imaged in the AFM at low RH: the AFM head is enclosed and flushed with dry nitrogen. Under these conditions it is possible to achieve imaging forces in the dc constant force mode which are high enough to generate contrast in the image, but not too high to cause sample damage or displacement (typically between 1-10 nN). An alternative approach is to image in the Tapping mode. Imaging can be improved by working under an atmosphere of dry air, helium or nitrogen. The cantilevers are normally operated at a frequency just below (say 0.1-2 kHz below) the resonant frequency of the tip-cantilever assembly. Figure 4.3 shows a Tapping mode image in air for a polysaccharide sample. It has been noticed in our laboratory that, under similar deposition conditions, the concentration of polymers observed on the surface by Tapping is often higher than that seen when imaging in the dc constant force mode. It is possible that deposition leads to several layers: a most tightly bound primary layer covered with less tightly bound secondary layers, which are then displaced by the less controlled dc contact mode imaging conditions. Reports in the literature suggest that lower concentrations are often used for studies made by Tapping mode. Imaging under Tapping mode also tends to reveal more 'dirty' substrates. Again debris on the surface may more easily be displaced or picked up by the probe in the dc mode and thus is not seen in the image.

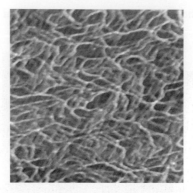

Figure 4.3. Tapping mode image of an entangled array of the polysaccharide xanthan imaged in air. Scan size 1.2 x 1.2 μm.

4.1.5. Imaging under non-aqueous liquids

Another approach for eliminating adhesive forces is to image under liquids. The chosen liquid displaces the water layer allowing the imaging force to be controlled during imaging. It is common to image under non-solvents such as alcohols. A variety of alcohols can be used including butanol, propanol and propan-2-ol. These appear to function by displacing water, thus inhibiting molecular motion on the substrate surface, and by preventing desorption from the surface. Imaging can be carried out using either dc constant force conditions or Tapping mode. Under dc conditions it is still necessary to precisely control the imaging force. Figure 4.4 shows the effect of varying the imaging force on the quality of the image. In Fig. 4.4a when the imaging force becomes too low (< 1 nN) the molecules merge into the background. Fig. 4.4b shows the effect of allowing the imaging force to increase during scanning. At sufficiently high imaging forces (> 10 nN) evidence appears for the damage and displacement of the molecules. Under optimum conditions reproducible images with good contrast are obtained (Fig. 4.4c). Figure 4.5a shows the force-distance curve corresponding to the image shown in Fig 4.4c. The force distance curve is perfectly reversible with no evidence of adhesion. Even under these conditions the quality of the image will begin to decay with time; typically over periods of hours, although the timescale involved is very sample dependent and can be as short as minutes. Figure 4.4d shows such an image in which the quality has begun to decay, and Fig. 4.5b shows the corresponding force-distance curve. The decay in image quality is related to the appearance of an increasing adhesive force, presumably arising as the tip accumulates molecules or other debris from the sample surface. Images can also be obtained by Tapping under liquid conditions. Setting the imaging conditions is slightly more difficult because the single resonance frequency of the tip-cantilever assembly is usually replaced by a range of resonance frequencies the nature of which will depend on the geometry of the liquid cell. The chosen conditions are normally about 0.3-2 kHz below the second harmonic.

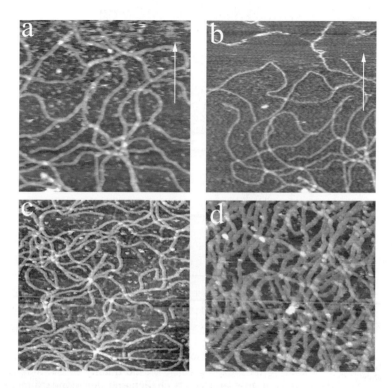

Figure 4.4. Contact mode images of xanthan polysaccharide imaged under alcohol. (a) Force decreases from the bottom (3 nN) to the top of the image. The force is about 1 nN at the arrow and image contrast fades. Scan size 600 x 600 nm. (b) The force increases from the bottom (3 nN) to the top of image. The force at the arrow is about 10 nN and damage and displacement occurs. Scan size 700 x 700 nm. (c) Optimised contrast. Scan size 1.2 x 1.2 μm. (d) Image degradation due to adhesive force. Scan size 700 x 700 nm.

4.1.6. Binding molecules to the substrate

Instead of simple physisorption of molecules onto substrates like mica it is possible to specifically modify the substrate in order to improve molecular attachment. This is particularly important if subsequent imaging is to be in aqueous or buffered media, and as a method of avoiding molecular distortion or denaturation on drying. The main approach involves the use of self-assembled monolayers on the substrate. The background literature in this field is vast and only the general features appropriate to scanning probe microscopy will be considered here. The earliest attempts to modify substrates were made for STM studies which require a conducting substrate, and concerned modification of epitaxially-grown gold layers on mica.

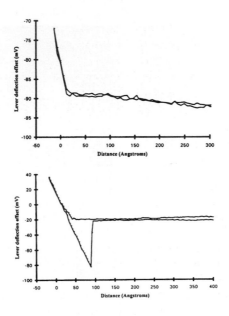

Figure 4.5. Force-distance curves for the images shown in Fig. 4.4. Top corresponds to Fig. 4.4c and bottom corresponds to Fig. 4.4d.

The growth of flat islands of gold on mica is challenging, but standard methods are published in the literature (Chidsey *et al* 1988; Clemmer and Beebe, 1992). An approximately 200 nm thick gold layer is thermally evaporated onto freshly cleaved preheated (300 ^{0}C, overnight) mica. Small plates or discs of coated mica (about 1 mm in size) are cut or punched. The gold coated mica fragments are derivatised by incubating immediately with a solution of an alkane-thiol derivative. The derivatised mica is removed, washed, dried under nitrogen and then used immediately for immobilising the biopolymer. The biopolymer solution, in an appropriate buffer, is placed as a drop on a piece of Parafilm and the activated coated mica placed on top of the drop. The assembly is covered to prevent evaporation and allowed to incubate for a suitable period of time. The biopolymer coated sample is washed with buffer and placed directly in the AFM liquid cell without drying, and then imaged using dc contact, ac or Tapping modes. Different types of thiol derivatives can be used to bind different types of biopolymers. Thus, for example, proteins, phospholipids or molecules containing amino sugars can be covalently attached (Fig 4.6) using dithiobis (succinimidylundecanoate) (DSU) (Wagner *et al* 1994; 1996). Anionic biopolymers can be bound by use of 'positively' charged 2 dimethylaminethanethiol (Allison *et al* 1992a; 1992b). The limitation of this method for AFM is the difficulty in producing large flat gold surfaces.

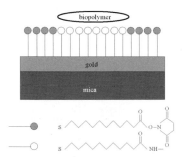

Figure 4.6. Schematic diagram showing the use of derivatised thiols for binding biopolymers to gold coated mica. The process is illustrated for DSU.

An alternative is to derivatise mica surfaces directly. The main constituent of mica is SiO_4 tetrahedra and thus this can be achieved by covalent attachment of silanes onto the mica surface (Bhatia *et al* 1989). Covalent attachment of 3-aminopropyltriethoxysilane (APTES) can be used to generate a positively charged surface which will bind anionic biopolymers (Lyubchenko *et al* 1992a; 1992b). The mica is freshly cleaved and placed at the top of a desicator containing a solution of the silane. The desiccator is briefly evacuated, purged with dry argon, and the reaction allowed to proceed for about 2 hours. The silane solution is removed, the desiccator again briefly evacuated and then purged with dry argon to allow storage of the activated mica until use. The biopolymers are attached (Fig. 4.7) by either immersing the substrate in the biopolymer solution or by placing a drop of the solution on top of the substrate. The surface is then rinsed and dried either under vacuum or dry gas.

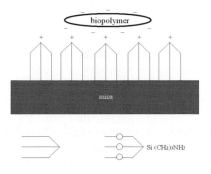

Figure 4.7. Schematic diagram illustrating the use of silane derivatives to bind biopolymers to mica. The process is illustrated with APTES.

By omitting the drying step it is possible to prepare samples for imaging which have been maintained in a liquid environment. By using other silane derivatives it is possible to selectively bind particular biopolymers.

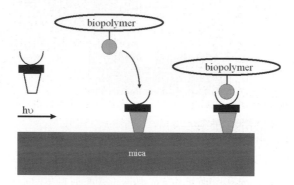

Figure 4.8. Schematic diagram illustrating the use of photoactive bifunctional reagents for binding biopolymers to mica. The reagent is first bound to the mica and then captures the biopolymer by binding a target site on the biopolymer.

Photochemical reactions can be employed to immobilise biopolymers (Luginbuehl and Sigrist, 1998) onto bare substrates such as mica, or to bind macromolecules to the thermochemically-derivatised mica described above. Direct attachment to bare substrates can be carried out using heterobifunctional photoreagents such as trifluoromethyl aryldiazirines (Fig. 4.8). The target groups for attachment of biopolymers are primary amines or thiols. Freshly cleaved mica is rinsed with ethanol and dried under nitrogen. A solution of the photoactive agent is dropped onto the substrate surface and the solvent evaporated under vacuum at room temperature. The substrate is agitated and irradiated with a suitable light source: at 350 nm carbines are generated from the trifluoromethyl diazirenes which covalently link to the surface. Excess crosslinker is removed by rinsing and sonication and the substrates dried under mild vacuum. Biopolymers are attached by incubating a solution of the biopolymer with the substrate, washing and sonicating, and then imaging under buffer. An alternative approach is to use these photocrosslinkers to attach biopolymers to derivatised substrates containing covalently attached alkylthiols or alkylamines. The experimental conditions need to be optimised, particularly in order to avoid aggregation of the crosslinkers due to photopolymerisation reactions. Assessment of non-specific binding to these substrates can be made by using separate controls, or photo masking techniques.

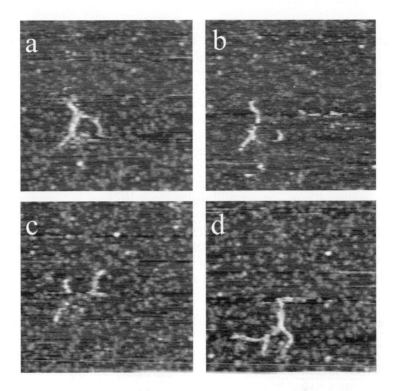

Figure 4.9. Tapping mode AFM image of a water soluble wheat pentosan polysaccharide imaged under 10 mM HEPES buffer plus 2 mM $ZnCl_2$ (pH 7). Scan size 600 x 600 nm. Sequence of images showing 'motion' of the molecules.

4.1.7. Imaging under water or buffers

Imaging molecules and molecular processes under water or buffers is perhaps one of the most exciting aspects of the use of AFM. Although it was at first felt that the molecules would need to be covalently attached to the substrates to prevent their desorption, damage or displacement there is now growing evidence that these restrictions need not be as severe as first thought. For fibrous molecules such as nucleic acids or polysaccharides it is difficult to achieve low enough applied imaging forces in order to image in the dc mode. Presumably the weakly attached physically adsorbed molecules are displaced from the surface by the lateral or frictional forces imposed by the probe during scanning. The use of the Tapping mode certainly makes imaging much easier. Imaging can also be improved by enhancing adsorption or inhibiting desorption of the molecules from the surface. For example the samples can be air dried onto mica, vacuum dried or even washed with alcohols prior to imaging under buffer. Fig. 4.9 shows a sequence of images of a water soluble polysaccharide which has been deposited onto mica and then imaged by Tapping under buffer. Not only can the molecule be clearly resolved but successive images reveal an apparent motion of the molecule on the mica surface. As will be explained in sections 4.2.1 & 4.4.1 this type of motion is the result of the molecules trying to desorb from the surface. However, the ability to obtain such images of molecules in motion has allowed the production of 'molecular movies' of biological processes. In the case of certain physically adsorbed proteins the binding to substrates such as mica is sufficiently strong, and the proteins structures are sufficiently resilient to distortion, to allow imaging of individual proteins under water or buffer even in the dc contact mode, provided minimal normal forces are applied (section 4.5.1). The use of Tapping mode has now become the standard method for imaging macromolecules under liquid. As will be shown in sections 4.5.1 and 6.2.2 methods are slowly being developed to allow complex membrane proteins to be examined under physiological conditions in environments close to their natural state. These advances are slowly realising one of the original aims of applying probe microscopy to biological systems.

4.2. Nucleic acids: DNA

DNA is the biopolymer most studied by scanning probe microscopy methods. In fact, the first published STM image of a biopolymer was of a DNA molecule deposited onto a coated Si wafer (Binnig and Rohrer, 1984). DNA was one of the first, if not the first biopolymer to be imaged by AFM. It is not difficult to understand why DNA has received such attention. DNA is a well characterised biopolymer of exceptional biological importance. Its characteristic size and shape make it a good model system for evaluating the quality and reliability of SPM images. The availability of linear DNA, circular plasmids, well defined restriction fragments, together with the possibility for synthesising specific sequences of defined structure and length, make these molecules ideal candidates for study. The early studies of DNA, primarily by STM, laid the foundations for routine imaging by AFM; now the method of choice for studies on DNA. Indeed the studies on DNA have probably played a pivotal role in the development of general methodology for imaging biopolymers. The driving force for studies on DNA lies in the potential high resolution of the AFM under natural or physiological conditions. In the early days the possibility for using AFM methods for automatic sequencing or manipulation of DNA provided an impetus for studies on nucleic acids. Other goals include the prospects of studying how small changes in sequence, or the binding of small molecules (drugs, carcinogens), alter the structure of the molecule, and the ability for characterising, under physiological conditions, the many biologically important protein-DNA interactions.

4.2.1. Imaging DNA

It is now possible to obtain reliable and reproducible AFM images of DNA: the method used depending on the type of information required. Basically there are two types of experiment which are of importance. Firstly, static studies where the molecules are immobilised, and the interest is in determining size, shape or details of molecular structure, or molecular interactions. Secondly, dynamic studies in buffered or physiological media where the molecules are partially immobilised and there is an interest in imaging the molecules in motion.

The most common substrate for AFM imaging of DNA is mica or derivatised mica, although a few studies have been made on graphite (Brett and Chiorcea, 2003; Klinov *et al* 2003; 2006). An advantage with a conducting substrate is the ability to apply potential differences to aid adsorption. However, the historical difficulties associated with STM and artifacts on graphite may explain why few studies have been made using this material as a substrate. In the case of mica the molecules are generally drop deposited onto a freshly-cleaved surface from solution and in the early studies were then air dried prior to imaging. The major problem in obtaining reliable images of such immobilised DNA molecules appears to be tip damage or displacement due to 'adhesive forces' (sections 3.1.3 & 4.1.5). Although 8 - 10 nm resolution images have been obtained in air (Bustamante *et al* 1992; Vesenka *et al* 1992a) the most reliable and reproducible images were originally acquired by imaging under alcohols (Hansma *et al* 1992a; 1992b; Lyubchenko *et al* 1993), where the adhesive forces are eliminated, and the applied force can be controlled during imaging.

Figure 4.10. Phase images showing motion of a ϕX-174 DNA molecule. Scan size 2 x 2 μm. Reproduced with permission of the authors and Oxford University Press from Argamen *et al* 1997.

The development of the Tapping mode provides an alternative method of eliminating adhesive forces and has allowed imaging of DNA in air, gaseous media such as dry helium, water and buffered media. This is now the method of choice for imaging DNA. Innovations to enhance imaging include feedback algorithms that recognise individual molecules and restrict the scan area to the local environment of the molecule (Andersson, 2007), shear force microscopy (Antognozzi *et al* 2002), and the use of active quality-factor control (Humphris *et al* 2000a; 2001). Since DNA is soluble in aqueous media the trick in imaging involves enhancing the binding of the DNA to the mica. This can be achieved in several ways. Air drying the sample, vacuum drying or washing the deposit with alcohols before imaging in the aqueous medium appears to improve the surface coverage. Addition of certain divalent cations to the imaging medium enhances binding without the need for a drying stage. The roles of monovalent ions (Ellis *et* al 2004) and divalent ions (Zheng *et al* 2003) in attaching DNA loosely to mica to allow molecular motion is discussed in more detail by Pastre and co-workers (2003). The use of Tapping mode in water and buffered medium has allowed real time imaging of DNA molecules in motion (Gallyamov *et al* 2004a; 2004b, Greenlife *et al* 2007) and their interactions with various enzymes (section 4.2.3).

Phase imaging has also proved useful for studying DNA under static and dynamic (Fig. 4.10) conditions. For impure or 'dirty' samples, where the DNA is present within a thick contamination layer on the mica, the phase images highlight the stiffer, less compressible nucleic acids, which are not resolved in the topographic images (Hansma *et al* 1997). Phase imaging allows molecular resolution at lower forces and also at higher scan rates; permitting faster processes to be studied (Argaman *et al* 1997).

4.2.2. DNA conformation, size and shape

Are the images of DNA adsorbed to mica representative of the 'natural' state of the molecule, or does the adsorption perturb or denature the molecules?

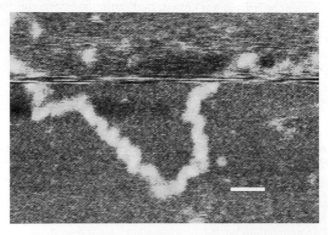

Figure 4.11. Right handed DNA helix imaged using Tapping mode AFM. The double-stranded DNA has been adsorbed to mica under aqueous buffer and then imaged under propanol. The spacing of the helix turns is comparable to the spacing of the major groove of B-DNA. Scale bar 10 nm. Reproduced with permission of H.G. Hansma and based on Hansma, 1996.

In general the resolution achieved is not sufficient to visualise the helical structure directly, although, quite often, exceptional images show structural features compatible with the turns of the helical structure (Fig. 4.11). Thus the images obtained under butanol (Hansma and Hansma, 1993), or propanol (Hansma *et al* 1995), are consistent with the adoption of the helical structure. Measurements of the length of cyclic plasmids deposited from aqueous solution and then imaged under propanol are consistent with the adoption of the B-DNA helical structure (Hansma *et al* 1993a). The highest spatial resolution images of DNA (Fig. 4.12) have been obtained for samples deposited onto a cationic bilayer as close packed arrays and then imaged in dc contact mode under aqueous buffer (Mou *et al* 1995a). The images revealed a right handed helix of pitch 3.4± 0.4 nm, in good agreement with the known pitch of the double helix for B-DNA. The use of cryo-AFM (Zhan, Y *et al* 1996; Han *et al* 1995) offers the prospect of improved resolution but, as yet, it has not proved possible to consistently resolve the helical structure (Han *et al* 1995). The diameter of the DNA helix is known and measurements of molecular widths or heights should help to confirm the presence of the helical structure. Measurements of widths are complicated by probe broadening effects. When probe broadening effects are estimated then the sizes observed are generally consistent with the expected helical structure, and the measured values in close packed arrays, where probe broadening effects should be minimised, approach the values expected for the helical conformation.

In general measurements of heights are smaller than expected. The values vary with sample preparation conditions and with the applied force. Measured heights tend to be smaller than expected for the dimensions of the helix. The explanation for this discrepancy is not fully understood at present. Given the evidence above for retention of a 'native' helical conformation the most likely explanation is that the helices are compressed during scanning and there is some experimental evidence in support of this assertion (Li *et al* 2003). Certainly changes in height consistent with changes in structure or conformation have been reported: AFM studies (Thundat *et al* 1994) of relaxed and stretched DNA molecules revealed lower heights for the stretched molecules and comparative studies on single, double and triple helical DNA (Hansma *et al* 1996) showed smallest heights for the single stranded form and largest heights for the triple helical form. Changes in sequence also lead to changes in measured height: the height of G-DNA, containing guanine tetrad motifs (G quartets), was close to the expected value obtained from X-ray and NMR data, suggesting that these molecules are less compressible than B-DNA (Marsh *et al* 1995). Indeed height measurements can be used to probe unusual features of DNA molecules. For example, measurements of length and height have been used in the characterisation of enzyme-resistant DNA fragments, and have been taken to suggest that such structures are unusual multi-stranded forms of DNA (Li, J *et al* 1997). Studies using cryo-AFM will eliminate thermal motion and may yield more reliable molecular dimensions. Force measurements on individual DNA macromolecules at cryogenic temperatures suggest that they are substantially stiffer than at room temperature (Han *et al* 1995;

Zhang, Y *et al* 1996). A method has been proposed for estimating a radial Young's modulus for DNA from Tapping mode data (Lin *et* al 2007). Height measurements should be more realistic in cryo-AFM. At present the measured heights appear to be almost 1 nm smaller than room temperature values, although there is good evidence that this is, at least in part, because the molecules are buried in a thin layer of frozen solution on the mica, and better measurements of heights will await development of freeze-etching methods for exposing the DNA molecules. Shear force SNOM measurements of the height of double stranded dsDNA yield higher values ($\approx$ 1.4 nm) than AFM measurements and are closer to the expected diameter ($\approx$ 2 nm) of the DNA helix (van Hulst *et al* 1997). In this case the reduced heights are attributed solely to dehydration effects. The measurement of sample heights in aqueous media will be dependent on sample-tip and substrate-tip interactions (Muller and Engel, 1997).

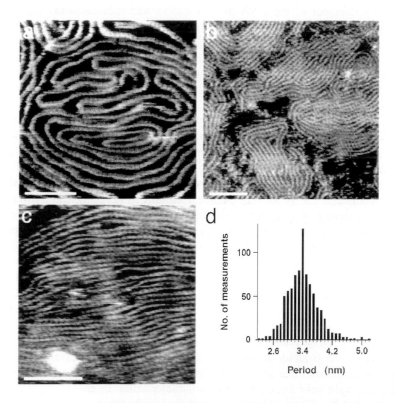

Figure 4.12. High resolution images of three DNA samples (a) pBR 322 (4.36 kb), (b) pBR 325 (6 kb) and (c) HaeIII restriction fragments of φX174. Data reproduced from Mou *et al* 1995 with permission. This reference contains full details of the preparation and imaging methodology. All three specimens showed periodic modulations (measurements recorded in the histogram (d)) of mean value 3.4 ± 0.4 nm which is consistent with the pitch of the DNA double helix.

For example, if the local charge distribution on the molecule is different to that of the substrate then the force-distance curves will be different for the sample and surface. Thus the estimated 'height', measured under 'constant force' conditions, will be not be simply determined by the diameter of the molecule. Electrostatic interactions can be screened (Butt, 1992a; 1992b) and under these conditions better estimates of height should be obtained. However such effects do not appear to have been systematically studied for DNA.

Images of DNA can provide information on molecular size and stiffness of the molecules. Sizing of DNA is an important and widely used tool in molecular biology; lengths of DNA molecules are used in restriction mapping, +/- screening, fingerprinting and genotyping. It has been shown that contour lengths determined by AFM correspond to lengths expected from the sequence and number of base pairs (Hansma and Hoh, 1994; Bustamante and Rivetti, 1996; Shao *et al* 1996; Thundat *et al* 1993). Measurements of contour length have been used to infer a novel A-helix for certain types of DNA (Borovok *et al* 2007). However, the lengths of the molecules bound to mica are found to be sensitive to the type of cation species used for binding the DNA to the mica and optimum and standardised methods need to be used for sizing. Methods have been suggested for automating the recognition, sizing and analysis of DNA (Fang *et al* 1998; Ficarra *et al* 2005a; 2005b) and these methods have been compared with the conventional use of electrophoresis methods (Fang *et al* 1998). Although this approach appears to be technically feasible it is not widely used at the present time. A problem in developing fully automated systems for biological systems is the contamination of the tip, which introduces adhesive forces and limits the working lifetime of the tip, and methods for exchanging or cleaning tips would need to be devised. Automated imaging would have potential for mapping sequence specific decoration of DNA molecules: examples may include chemical modification, hybridisation with short oligomers, or binding of sequence specific proteins, such as restriction enzymes, for health care or forensic screening applications. Improvements in the accuracy of the measurement of DNA contour lengths has allowed discussion of the effects of deposition conditions on the size and conformation of DNA (Rivetti and Codeluppi, 2001; Sanchez-Sevilla *et al* 2002).

The shape adopted by the molecules on the substrate is determined by sequence specific factors, which dictate the stereochemistry of the molecule. For the simplest type of polymer the monomeric units can rotate freely about the inter-monomer linkages and the molecules adopt a random coil configuration. The overall size of the polymer depends on the number and size of the monomer units. Stereochemical restrictions on free rotation about the linkages, such as the adoption of a helical conformation, increase the overall size of the molecule. The simplest treatment of such effects for extended polymer chains defines molecular stiffness (or extension) in terms of a parameter called the persistence length L_P. The persistence length is a measure of the separation of points within a chain which are randomly distributed in space and is usually measured by light scattering (Burchard, 1994), or by electro-optic methods such as transient electric birefringence or dichroism (Hagerman, 1988). For polyelectrolytes there are essentially two contributions: charge-charge repulsions, which enhance stiffness at low ionic strength, and can be eliminated by screening to reveal a second 'intrinsic' contribution. Persistence lengths can be determined for DNA by AFM (Bustamante and Rivetti, 1996; Rivetti *et al* 1996; Hansma *et al* 1996, 1997; Moukhtar *et al* 2007). The procedure involves measuring the angle (θ) between tangents to the molecule at points separated by a contour length (L). The reciprocal of the slope of a plot of the arithmetic mean of the square of the angle $<\theta^2(L)>$ versus L gives the persistence length L_P. These experimental plots are not linear and the value of L_P depends on the size range studied, possibility accounting for the variability in measured L_P from sample to sample. The L_P values for a given sample have been found to be substrate dependent, suggesting that surface interactions affect the shape and extension of the molecules. Such surface interactions mean that the L_P values found by AFM may not be equivalent to those determined for DNA in solution. An alternative approach is to estimate the contributions of individual combinations of base-pairs to the local flexibility of the DNA: a combination of x-ray and NMR data has been used to calculate local flexibility and the results validated for DNA by AFM (Marilly *et al* 1995). Provided the surface interaction with the DNA make only a small contribution to the molecular extension then measurements of shape by AFM can still be used to investigate structural changes of the DNA molecules. In fact differences in L_P have been observed for different types of DNA: single stranded DNA (ssDNA) has been seen to be more flexible than normal double stranded DNA (dsDNA) itself more flexible than double stranded poly (A)$^+$ (Hansma *et al* 1997). Measurement of overall size provides an alternative assessment of molecular stiffness. Changes in shape can be used to investigate the effects of molecular interactions such as ligand binding to DNA. Some early examples of this approach are the use of AFM to observe the straightening of abnormally bent kinetoplast DNA (the mitochondrial DNA from trypanosomes and related parasitic protozoa) due to the addition of the ligands distamycin and the microgonotropen (MGT-6b), the bending of normal DNA on addition of MGT-6b (Hansma *et al* 1994) and site specific oligonucleotide binding to DNA (Potaman *et al* 2002). Since these early studies AFM is now widely used to probe DNA interactions with ligands (Pope *et al* 1999, 2000; Wang *et al* 2000; Kaji *et al* 2001; Pastre *et al* 2005). The advantage of using microscopy to study molecular shape is that it allows localised changes within individual molecules, or variations of structure or conformation for molecules within a population, to be visualised and quantified. The likely advantage of the use of AFM over the use of EM is that the sample preparation is simpler and the samples can be examined under more natural conditions.

   An alternative approach to the measurement of the stiffness of DNA molecules, and fibrous molecules in general, is to use the AFM to measure force-distance curves, and to analyse such data to determine the stiffness of the molecules. This type of study provides interesting information on conformational changes and ligand-DNA or protein-DNA interactions. The reviews by Giannotti and Vansco (2007) and by Strick and co-workers (2003) are a useful summary of force spectroscopy of single molecules. The methods and analysis of such data is discussed in more detail in chapter 9. In this type of experiment the molecules are usually deposited onto a substrate, picked up by the tip and the force-extension curve measured. In the simplest type of experiment the molecules can be physically adsorbed to the substrate. For DNA it is possible to derivatise the 3' and 5' ends of the molecule to allow selective attachment to the substrate and tip assembly. DNA molecules can be attached using a variety of covalent linkages directly to the substrate or tip, or through the use of flexible spacer molecules such as polyethylene glycols. Having a linker moves the molecules further from the surface and can help to eliminate substrate-molecule interactions but it may complicate data analysis through the need to quantify stretching effects of the linker. Attachment can also involve specific interactions such as the attachment of biotinylated DNA to streptavidin-coated surfaces (Zhang *et al* 2005). For small extensions of DNA the force-extension curve can be described by a random coil model (Smith *et al* 1992) or a worm-like chain model (Bustamante *et al* 1994b; Marko and Siggia, 1995; Bouchiat *et al* 1999; Strick *et* al 2003; Wiggins and Nelson, 2006; Wiggins *et al* 2006) which then includes entropic and also enthalpic contributions through the use of a Kuhn length. This approach offers the prospect of analysing molecules over a wide range of ionic strengths not easily accessible in a purely microscopic study. Certainly studies of the stretching of DNA molecules using optical tweezers at different ionic strengths have led to changes in persistence length (Wenner *et al* 2002). At larger extensions DNA shows first a cooperative transition followed by a further irreversible transition at higher extensions. The position of the first plateau in the force-extension curve depends on the nature of the base-pairs and is attributed to a transition from the normal double helix to a stretched form of DNA (S-DNA). The transition at higher extensions is attributed the dissociation or melting of the double helix into single chains. The extension of DNA is sequence dependent for both sDNA and dsDNA (Lankas *et al* 2000; Ke *et al* 2007; Kuhner *et al* 2007; Chen *et al* 2008) and studies and analysis have been made on the behaviour of long (Strunz *et al* 1999; Clausen-Schaumann *et al* 2000; Netz, 2001; Williams *et al* 2001; Cocco *et al* 2002; 2004; Danilowicz *et al* 2003; Krautbauer *et al* 2003; Storm and Nelson, 2003; Harris, 2004; Voulgarakis *et al* 2006) and short (Pope *et al* 2001; Seol *et al* 2007) strands of DNA. Despite the importance of the stretching transition in relation to natural processes such as replication and transcription the detailed nature of the B- to S- transition, and the analysis and modelling of the stretching of DNA remain an area of active debate (Charvin *et al* 2004; Conroy and Danilowicz, 2004; Harris, 2004; Morfill *et al* 2007; Bornschlog and Rief, 2008; Luan and Aksimentiev, 2008; Wei and van de Ven, 2008; Whitelam *et al* 2008; Alexandrov *et al* 2009; Calderon *et al* 2009; Santosh and Maiti, 2009). Studies and analysis have also been made on the stretching of sDNA (Zhang *et al* 2001; Dessinges *et al* 2002; Ou-Yang *et al* 2003; Tkachenko 2004; Li *et al* 2005; Cui *et al* 2006) and the second transition for dsDNA at high extensions is attributed to the melting out and extension of single strands of DNA. The mechanical spectrum of DNA molecules is affected by the binding of ligands and this process can be studied by AFM or other single molecule techniques such as optical tweezers (Krautbauer *et al* 2000; 2002a; Schotanus *et al* 2002; Husale *et al* 2002; 2008; Eckle *et al* 2003; Ros *et al* 2004; Sischka *et al* 2005; Zhang *et al* 2005; Wang *et al* 2007b).

Processes which involve changes in the shape of the DNA molecules can be studied. Examples of this type of study are supercoiling (Tanigawa and Okada, 1998; Lyubchenko and Shlyakhtenko, 1997; Samori *et al* 1993; Henderson, 1992) of DNA, DNA damage (Murakami *et al* 1999; Lysetska *et al* 2002; Yan *et al* 2002; Li *et al* 2003; Sui *et al* 2004; 2005; Yavin *et al* 2004; Psonka *et al* 2005; Yang *et al* 2005; Uji-I *et al* 2006; Jiang *et al* 2007), DNA-drug interactions (Rampino, 1992; Berge *et al* 2002; 2003; Krautbauer *et al* 2002b; Onoa and Moreno 2002; Pastushenko *et al* 2002; Viglasky *et al* 2003; Zhu *et al* 2004; Tseng *et al* 2005; Mukhopadhyay *et al* 2005; Cheng *et al* 2005; Adamcik *et al* 2008; Banerjee and Mukhopadhyay, 2008; de Mier-Vinue *et al* 2008), and studies of DNA condensation and transfection (Fang and Hoh, 1998; Hansma *et al* 1998; Golan *et al* 1999; Dame *et al* 2000; Andrushchenko *et al* 2001; Liu *et al* 2001; 2005; Iwataki *et al* 2004; Saito *et al* 2004; Chim *et al* 2005; Danielson *et al* 2005; Wittmar *et al* 2005; Limanskaya and Limanskii, 2006; Patnaik *et al* 2006; Sansone *et al* 2006; Volcke *et al* 2006; von Groll *et al* 2006; Guillot-Nieckowski *et al* 2007; Lee *et al* 2007; Limanskii 2007; Mann *et al* 2007; 2008; Ruozi *et al* 2007; Ahn *et al* 2008; Hou *et al* 2008; Stanic *et al* 2008). A more general review of uses of AFM in drug delivery studies is given by Tumer and co-workers (2007). The observed shape of DNA molecules provides information on structural features of DNA molecules such as telomeric loops (Chen *et al* 2001), knotted DNA structures (Lyubchenko *et al* 2002), intermediate structures which can be formed during replication and recombination (Shlyakhtenko *et al* 2000; Yamaguchi *et al* 2000a; 2000b) and can be used to visualise the structure of unusual complexes such as polysaccharide-polynucleotide complexes (Bae *et al* 2004; Sletmoen and Stokke, 2005) or structural changes associated with processes such as Alzheimer's disease (Moreno-Herrero *et al* 2004a; 2004b).

Because of the widespread use of AFM it is becoming difficult to find reviews dedicated to the use of AFM to probe DNA. Some useful reviews include Poggi and co-workers (2002; 2004), Lyubchenko (2004), Hansma and co-workers (2004) and Greenleaf and co-workers (2007). These reviews also provide an indication of the breadth of usage of AFM particularly in the use of DNA as a tool for assembling nanostructures.

4.2.3. DNA-protein interactions

A major area of interest is the study of DNA-protein binding (Lyubchenka *et al* 1995; Hansma, 1996; Kasas *et al* 1997 a, 1997b; Hansma *et al* 1995) and the consequent effects on DNA condensation (packaging) and processing. Because of the extensive literature on AFM studies on DNA, and DNA-protein interactions, there are a number of good review articles covering this research area. The articles by Lyubchenka and coworkers (Lyubchenka *et al* 1995) and Bustamante and coworkers (Bustamante *et al* 1993) are particularly useful, and provide detailed descriptions of methods for imaging DNA, and DNA-protein complexes. Some examples of the types of DNA protein complexes studied by AFM are given below.

A growing area of interest is the use of force spectroscopy methods to probe interactions between DNA and ligands and, in particular, to probe interactions between DNA and peptides (Okada *et al* 2008) or proteins (Bartels *et al* 2003; Jiang *et al* 2004; Kuhner *et al* 2004; Qin *et al* 2004; Raible *et al* 2004; Yu *et al* 2006; Basnar *et al* 2006; Montana *et al* 2008; Okada *et al* 2008; Zhang *et al* 2008). Changes in contour length can be used to infer conformational changes such as looping, the effects of protein binding on DNA melting or duplex formation can be probed, and information can be obtained on the binding between complementary structures immobilised on tip and substrate. For example information can be obtained on protein binding through effects on the stretching of DNA (Williams MC *et al* 2008), by studying specific binding between peptides (Sewald *et al* 2006) or proteins (Bartels *et al* 2003; Jiang *et al* 2004; Kuhner *et al* 2004; Qin *et al* 2004; Yu *et al* 2006) and DNA, or by analysing the cross-linking of individual DNA strands by proteins (Krasnoslobodtsev *et al* 2007).

RNA polymerase complexes

This is a major area of interest and the purpose of these studies is to investigate the transcription of DNA. RNA polymerase is a high molecular weight protein (465 kD) which should be relatively easy to observe attached to DNA, and studies (Bustamante *et al* 1993; Zenhausern *et al* 1992b; Hansma *et al* 1999; Thompson *et al* 1999; Rivetti *et al* 1999; Crampton *et al* 1999; 2006; Maurer *et al* 2006) have shown that the enzyme can be visualised bound to DNA and that AFM can be used to study the structures of complexes involved in transcription of DNA. In addition to studying the binding of individual RNA polymerases to DNA it has been possible to investigate the formation of polymerase-activator complexes (Maurer *et al* 1999) and collision events in convergent transcription (Crampton *et al* 2006). Comparative EM and AFM images (Zenhausern *et al* 1992) revealed two RNA polymerase enzymes specifically bound to sites on circular plasmid DNA. Early studies of the binding of RNA polymerase to DNA as a transcription complex were made in air, after drying the sample onto mica (Rees *et al* 1993; Bustamante *et al* 1993). Changes in shape or the contour length of the DNA can be used to infer conformations of the complexes formed with DNA. The activity of polymerases can be assessed by monitoring the extent of replication of DNA: replicated and unreplicated DNA can be distinguished through the differences in their persistence lengths. Information can also be obtained on the conformation of RNA generated by the polymerase. The ability to image DNA and DNA-protein complexes under water or buffers has led to spectacular advances in probing transcription and transcription complexes. Key advances were the use of selected cations for binding DNA onto mica, and the use of Tapping mode for imaging the molecules (Thomson *et al* 1996a; Hansma and Laney, 1996). Under buffered conditions it is possible to observe DNA molecules in motion (Fig. 4.10) (Hansma *et al* 1995; Shao *et al* 1995). Time lapse films show changes in shape of linear or circular DNA. In some images parts of the molecules are missing, suggesting that these regions have detached from the surface, and that their motion in solution makes them impossible to image. This would seem to suggest that the preparation conditions force the DNA onto the surface and that the motion is a reflection of the desorption-adsorption of parts of the molecule, as the entire molecules are trying to escape from the surface into the buffered medium. This belief is reinforced by the observation that some DNA molecules eventually completely disappear and new molecules appear on the surface (Hansma *et al* 1995). The ability to image under natural conditions has allowed imaging of the assembly of RNA polymerase-DNA complexes (Guthold *et al* 1994; Hansma *et al* 1999; Thompson *et al* 1999) and real time visualisation of RNA polymerases transcribing linear dsDNA (Fig. 4.13) and single stranded ssDNA circular templates (Hansma, 1996; Kasas *et al* 1997a; 1997b). The rate of the transcription process was slowed by using a low nucleotide triphosphate (NTP) precursor concentration. It has been possible to visualise the translocation of DNA templates by the RNA polymerase, to image stalled complexes by controlling the supply of specific NTPs and, after initiating its binding to the mica, to visualise the RNA produced in the process. The demonstration that changes in protein shape during enzyme activity can be detected by AFM as fluctuations in height (Radmacher *et al* 1994), coupled with the development of 'tracking' methods for observing such changes on individual proteins (Thomson *et al* 1996b), mean that it is possible to detect changes in the shape of the polymerase during the transcription process.

Chromatin

An understanding of the molecular structure of chromatin is of importance in explaining mechanisms of gene replication and expression in the cell. AFM is a tool well suited to the investigation of DNA condensation induced by complex formation with nucleoproteins. In fact AFM studies of reconstituted (Allen *et al* 1993a; Vesenka *et al* 1992b) and native (Zlatanova *et al* 1994; Martin *et al* 1995; Leuba *et al* 1994; Fritzsche *et al* 1994; 1997; Allen *et al* 1993b) chromatin have played an important role in resolving controversies on chromatin structure. Chromatin is an important structural unit of chromosomes. The basic units of chromatin are nucleosomes; 10 nm particles composed of 146 bp of DNA molecule wrapped around the histone octamer. Recent studies (Woodock *et al* 1993) had raised doubts about the proposed regular 'solenoid-like' (Finch and Clug, 1976) or 'twisted-ribbon like' (Woodock *et al* 1984; Bordas *et al* 1986) geometry for the chromatin fibres. The AFM data suggests that the fibres consist of an irregular 3D array of nucleosomes. In studies of marsupial spermatozoa by AFM differences in the organisation of chromatin packaged by histones or protamines has been detected (Soon *et al* 1997): the nucleoprotamine particles appearing as tighter bundles than the nucleohistones particles. Isotonic cell lysis has been used to prepare and image largely intact chromatin from chicken erythrocytes (Fritzsche *et al* 1994; Fritzsche and Henderson, 1996a; 1996b) and human B lymphocytes (Fritzsche *et al* 1995a; 1995b; 1997). Such studies partially preserve the spatial relationships between neighbouring chromatin sites permitting investigation of the proposed compartmentalisation of chromatin in interphase nuclei.

Other DNA-protein complexes

AFM studies of protein complexes involved in DNA recombination and repair include investigations of the binding of DNA to single stranded DNA binding (SSB) protein (Hansma HG *et al* 1993b; Hamon *et al* 2007) and RecA protein (Lyubchenko *et al* 1995; Hansma *et al* 1995; Seong *et al* 2000; Umemura *et al* 2001; 2005; Sattin and Goh, 2004; 2006; Shi and Larson, 2005; 2007; Li *et al* 2006; Guo *et al* 2008; Sanchez *et al* 2008) to DNA. SSB binds to single-stranded DNA and is important in DNA replication and recombination. The AFM images show a uniform coating of the protein along the DNA chain. The Rec proteins bind to both ssDNA and dsDNA. Studies have been made of the assembly and disassembly of Rec-DNA complexes and the intermediate states involved in assembly including real time images of the processes for Rec-DNA complexes formed in the presence of ATP. Higher resolution images were achieved through the use of carbon nanotubes as tips. Studies have also been made of competitive binding and further ordering of RecA and RecN-ssDNA filaments in the presence of ssDNA-binding proteins (Shi and Larson, 2007; Sanchez *et al* 2008).

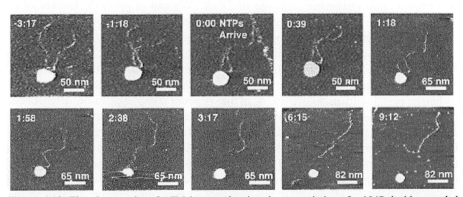

Figure 4.13. Time lapse series of AFM images showing the transcription of a 1047 double stranded DNA template by an RNA polymerase molecule. The first two images show that before addition of ribonucleoside 5'-triphosphates (NTPs) the DNA is mobile on the mica substrate. The following 6 images after NTP addition, from time 0.00 onward are sequential and show 'reading' of the DNA molecule until release (02.38). Data reproduced with permission from Kasas *et al* 1997b. Copyright (1997) American Chemical Society.

DNA mismatch repair is important for the maintenance of genomic stability and AFM has been used to probe the rcognition and binding specificity of MutS proteins for DNA (Zhang *et al* 2002; Yang *et al* 2005; Jia *et al* 2008). There are a number of specific protein-DNA interactions in which conformational constraints on binding can lead to bending, unwinding or overwinding of the DNA. Such interactions are of considerable interest, and there is a reservoir of background crystallographic, spectroscopic and biochemical data on these complexes. Not surprisingly a number of these complexes have been probed by AFM. Examples include studies on the kinking induced by the binding of DNA glycosylase to undamaged regions of DNA (Chen *et al* 2002), the specific and non-specific binding of Cro protein, a key regulatory protein in the growth of λ-phage, and the resultant bending of the DNA (Bustamante *et al* 1994a; Eire *et al* 1994). AFM revealed the bending of DNA on interaction with the bacteriophage φ 29 connector protein (Valle *et al* 1996) or transcriptional repressor protein TtgV (Guazzaroni *et al* 2007) and U-shaped complexes induced by binding of ParR to *E. coli* plasmid R1 centromere *parC* (Hoischen *et al* 2008). The binding of antibodies to incorporated Z DNA sequences (Pietrasanta *et al* 1996) has been observed, as has the stiffening or straightening of DNA due to the binding of Fur repressor protein (Le Cam *et al* 1994).

DNA looping is believed to play an important part in the molecular interactions involved in the regulation of the expression of prokaryotic and eukaryotic genes, of site specific recombination, and of DNA replication. The phenomenon arises when individual proteins or protein complexes bind to two separated sites on the DNA molecule, thus linking these sites together and forming loops in the DNA molecule. Here AFM complements or extends EM work. There are a growing number of AFM studies investigating looping in relation to transcription and enzymatic cleavage of DNA (Wyman *et al* 1995; Becker *et al* 1995; Yoshimura *et al* 2000; Vimik *et al* 2003; Heddle *et al* 2004; Shin *et al* 2005; van Noort *et al* 2004; Lushnikov *et al* 2006a; Kaur and Wilks, 2007; Pavlicek *et al* 2008). For example in the case of heat-shock transcription factor 2 (HSF) looping was observed due to cross-linking of HSF binding sites on the DNA. The use of AFM provided more information than previous EM studies because measurements of height permitted the stoichiometry of the binding complexes to be determined: linkage is considered to occur due to the interaction of HSF trimers bound to specific sites on the DNA (Wyman *et al* 1995). Multi-enzyme complexes have also been observed in AFM studies (Becker *et al* 1995) of DNA looping using the transcription complex (PIC). PIC binds to the promoter region and looping only occurs in the presence of Jun protein, which binds to a distant AP-1 site. It was shown that deletion of the AP-1 site prevents looping and protein binding to the DNA, suggesting that binding of Jun protein stabilises the PIC complex.

AFM has been used to visualise enzymatic interactions involved in replication and degradation of DNA. Images have been obtained showing the binding of a DNA polymerase to ssDNA (Yang *et al* 1992) and DNA replication has been observed (Hansma, 1996; Argaman *et al* 1997). Phase imaging has allowed the time dependent replication of DNA to be visualised directly in a buffered environment (Argaman *et al* 1997). The ssDNA appears as globular structures attributed to base pairing within parts of the molecule. With time elongated dsDNA molecules appear in the images. Another exciting example of real time imaging is the visualisation of enzymatic degradation of DNA and condensed DNA complexes (Bezanilla *et al* 1994; Hansma, 1996; Ikai, 1996; Abdelhady *et al* 2003; Yokokawa *et al* 2006a).

The primary, secondary and tertiary structures of DNA may be of importance in protein binding. AFM can be used to identify specific sites of binding to DNA, the types of complexes formed, or the role of DNA structure on binding (Ohta *et al* 1996; Jiao *et al* 2001; Moreno-Herrero *et al* 2001; Medalia *et al* 2002; Seon *et al* 2002; Argaman *et al* 2003; Gaczynska *et al* 2004; Kim *et al* 2004; Qu *et al* 2004a; 2004b; Chang *et al* 2005; Lysetska *et al* 2005; Poma *et al* 2005; Murakami *et* al 2006; Maskin *et al* 2007). The use of AFM has provided evidence for the importance of such effects in the regulation of heat-shock protein production (Ohta *et al* 1996). The *Staphyloccus aureus* heat shock operon (HSP70) contains an inverted-repeat sequence, between the promoter and the first structural protein (ORF37) gene, designated CIRCE (controlling inverted repeat of chaperone expression) which is believed to form a stem-loop structure. AFM images of the promoter region have revealed a protrusion, positioned correctly for the CIRCE region, and interpreted as a novel pairing of stem-loops. The ORF37 protein has been shown to bind to this structure, providing a molecular basis for a 'feed-back' process of regulation of heat shock gene expression (Ohta *et al* 1996). AFM has revealed the novel 'beads-on-a-string' complexes formed by tau proteins with DNA (Qu *et al* 2004a; 2004b). Phase mapping has been used to map the location of immobilised oligonucleotides using protein functionalised tips (Kim *et al* 2004) and time-lapse imaging was used to reveal intermediate stages in the association/disassociation of p53-DNA complexes (Jiao *et al* 2001)

4.2.4. Location and mapping of specific sites

The tagging of specific sites on DNA offers a route to locating and mapping these regions. This is just the type of microscopic study which should be well suited to the use of AFM. Combination of such labelling with a rapid and automated procedure for collecting and analysing images, of the type described earlier for sizing DNA (section 4.2.2), would provide a powerful physical tool for mapping DNA. A variety of tags can be envisaged, some of which have been studied experimentally. Labelling may involve chemical modification, hybridisation of short oligomers to create triple helical regions (Fang *et al* 1998), cause kinking of the DNA structure (Potaman *et al* 2002) or invove the binding of site specific proteins. Certainly the binding of sequence specific proteins (Schleif and Hirsh,

1980) and the identification of probes hybridised to ssDNA (Wu and Davidson, 1975) have been used as aids in the mapping of DNA by TEM.

As discussed earlier, proteins can be imaged bound to DNA and thus can be used as markers. In the case of hydrolytic enzymes it may be possible to modify them at their active sites to allow binding but not cleavage. An example of such a study is the mapping of individual cosmid DNAs (Fig. 4.14) by direct AFM imaging (Allison *et al* 1997).

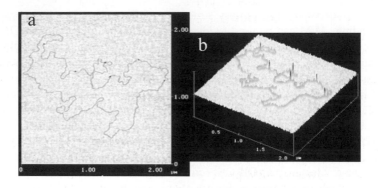

Figure 4.14. Mapping of a 35 kb cosmid clone with a mutated *Eco*RI endonuclease. (a) Image showing the 6 *Eco*RI binding sites. (b) Computer-generated image emphasising the binding sites. Data provided by the authors and publisher and based on Allison *et al* 1997.

It has been shown that wild type EcoRI endonuclease can be visualised bound to DNA and the precision of the mapping using mutant enzymes, able to bind specifically but not cleave, has been assessed by mapping well characterised λ DNA, and then applied for mapping a cosmid clone. Similar mapping studies have ben made using a variety of other enzymes (Lushnikov *et al* 2004; Lonskaya *et al* 2005; Lushnikov *et al* 2006b).

Biotin-labelled nucleotides have been used as a basis for binding sequence specific tags to DNA (Murray *et al* 1993; Shaiu *et al* 1993a; 1993b). Streptavadin gold conjugates have been used as labels to target biotinylated nucleotides incorporated into linear or cyclic DNA (Shaiu *et al* 1993a; 1993b). A chimeric protein fusion between streptavadin and two IgG binding domains of *Staphylococcal* protein A was used as a label to identify biotin labels at the ends of DNA fragments (Murray *et al* 1993). Force spectroscopy can be used to monitor hybridisation of complementary DNA strands immobilised on the tip and surface (Mazzola *et al* 1999) and this type of interaction can be used to characterise different stages of DNA chip preparation (Rouillat *et al* 2005).

Many proteins involved in transcriptional regulation of gene expression have been cloned and their functionalities are well characterised. Such proteins can be used as probes for specific sequences of nucleotides. An example is the investigation by AFM of the binding of protein Ap2 to DNA (Nettikadan *et al* 1996). The distance of these binding sites from the ends of the molecule, as measured from the AFM images, were found to be consistent with the known nucleotide sequence data. A low level of non-specific binding was also noted in these studies.

Antibodies also provide high specificity. An anti-Z DNA IgG antibody has been used to label the left handed Z DNA conformation of a $d(CG)_{11}$ insert in a negatively supercoiled DNA plasmid (pAN022) for imaging by AFM. The antibody binding caused bending of the DNA, and the position of the bound antibodies was consistent with known nucleotide sequence, and the adoption of the B DNA helix (Pietrasanta *et al* 1996). Location and mapping on chromosomes is discussed later in section 4.2.5.

The pros and cons of using SPM methods for sequencing DNA are discussed elsewhere (Hansma and Hansma, 1993; Morris, 1994). As discussed earlier the prospects for automating the AFM procedures seems good but, at present, the resolution achievable with AFM would need to be improved by at least an order of magnitude in order to achieve the level required for sequencing. Although AFM has good prospects as a tool for high resolution mapping it seems unlikely that it will challenge standard biochemical methods of sequencing in the near future. A major problem with biological systems will always be tip contamination. However, AFM probes have been used to dissect DNA molecules into smaller fragments (Henderson, 1992; Hamsma *et al* 1992b; An *et al* 2005; 2007) and there are suggestions that, following such dissection, DNA fragments can become attached to the probe, or deliberately picked up by the probe and can be isolated, transferred or amplified by PCR (Lu *et al* 2006a; 2006b; Zhang *et al* 2007). Related studies of this kind on chromosomes are discussed in the next section (section 4.2.5). An ingenious method for sequencing is based on monitoring translocation of DNA through nanopores. This approach was first reported by Kasianowicz and co-workers (1996). Attempts were made to sequence DNA by noting changes in ion currents as DNA was passed through nanopores (Akeson *et* al 1999; Fologea, 2005) but the signals were weak and collective motion of bases precluded detection of single bases. It has been suggested that nanopores threaded around DNA molecules could be selectively attached to AFM tips and pulled along the molecule to sequence the structure (Ashcroft *et al* 2008; Qamar *et al* 2008). The practical feasibility of the approach has been demonstrated by using such a system to measure the force required to unzip hairpins in DNA (Ashcroft *et al* 2008) but computer simulations suggest that, although individual base pairs should be detectable, the pulling rates achievable at the present time are too slow to prevent thermal fluctuations, precluding individual base-pair discrimination (Qamar *et al* 2008).

Even if AFM is not used for DNA sequencing then it may have applications related to gene therapy. As discussed earlier (section 4.2.2) it has been shown that AFM has potential for assaying DNA condensation, which is important for different modes of gene therapy. A variety of shapes are seen to be adopted by the DNA, including short rods and toroids. These studies show that AFM can be used to assess optimum condensation, potentially related to the efficiency of uptake by the receptor. In adition reports are beginning to emerge on the successful use of plasmid coated tips and AFM to transfect cells (Cuerrier *et al* 2007; Han *et al* 2008).

4.2.5. Chromosomes

Cytogenetics, the study of chromosomes, is basically a visual science. Established microscopic methods, such as light and electron microscopy, have been widely used to study chromosomes. The use of hypotonic methods results in good spreading of chromosome preparations allowing the determination of the number and morphology. Introduction of banding and hybridisation techniques has permitted the identification and mapping of chromosomes. Goals in cytogenetics are improved specification and localisation of probes, and more rapid analysis. AFM has potential in both these areas and has joined the range of microscopic techniques used to study chromosomes. The chromosomes are normally spread on glass and can be imaged in air or under liquid.

AFM is ideally suited for investigating the protein-DNA interactions involved in the formation of nucleosomes (Zhao *et al* 1999; Zhang S *et al* 2003; Doyen *et* al 2006; Takashima *et al* 2008), the assembly and folding of nucleosomes to form chromatin fibres (Leube and Zlatanov, 2002; Zlatanov and Leube, 2003; Woodcock 2006; Lahr *et al* 2007) and the subsequent condensation and assembly of chromatin fibres (Wang *et al* 2005; Cano *et al* 2006; Das *et al* 2006; Hirano *et al* 2008). In addition to visualising the 'beads on string' model for the nucleosomes it has been possible to image the dynamics of nucleosome reconstitution, investigate the specific roles of particular proteins in nuclesome remodelling, study the roles of histone mutations and variants, and such AFM studies have led to improved models for hydrated chromatin structures.

Polytene chromosomes are an amplified form of interphase chromosomes present in the nuclei of specific dipteran cells. They are substantially larger than normal mammalian metaphase chromosomes and contain densely condensed banded regions, and loosely condensed interbanded and puff regions. These features can be imaged in the AFM, the bands appearing as high regions, and the interbands and puffs as low or flat regions (Li *et al* 1996; Jondle *et al* 1995; Mosher *et al* 1995; Vesenka *et al* 1995; Puppels *et al* 1992). The resolution obtained depends on the type of tip used, but features as small as 1 nm have been reported in some images (Jondle *et al* 1995). Colloidal gold as a calibration standard has been used for reconstructing images of chromatin (Fritzsche *et al* 1996). By increasing the imaging force it has been possible to dissect polytene chromosomes. The image quality was found to decrease after dissection, and this was attributed to the adsorption of chromatin fragments onto the tip. It is reported that DNA can be recovered from the tip and amplified by PCR methods (Jondle *et al* 1995). More detailed and controlled studies of this type have been made on metaphase chromosomes.

Metaphase chromosomes imaged by AFM have revealed structures similar to those reported by light and electron microscopy. Cytogenetic abnormalities, expressed as changes in the length of the chromosome (McMaster *et al* 1994), or bi-armed chromosomes (Oberleithner *et al* 1996), which are believed to play crucial roles in tumour development, have been identified by AFM. AFM has been used to image human (Heckl, 1992; Murakami *et al* 2001; Ushiki *et al* 2002; Hoshi *et al* 2004; 2006; Kimura *et al* 2004; Yangzhe *et al* 2006) and Chinese hamster (DeGrooth and Putman, 1992; Wu *et al* 2006) chromosomes. Wu and co-workers have reviewed attempts to visualise chromosomes during various mitotic phases within Chinese hampster ovary cells (Wu *et al* 2006). Metaphase chromosomes consist of a 30 nm fibre folded to form a tandem array of radial loops, which are packaged into a fibre of overall diameter between 200-250 nm. High resolution AFM images of the surface of the chromosomes revealed structural features in the size range of 30-100 nm which may correspond to the loops of the 30 nm fibre (DeGrooth and Putman, 1992). Other authors (Winfield *et al* 1995; McMaster *et al* 1996a; 1996b; Hoshi *et al* 2006) have reported features as small as 10-20 nm which could correspond to individual nucleosomes. AFM has been used to probe the mechanisms of condensation.

Banding is used to classify, or karyotype, chromosomes by microscopy. AFM images of plant chromosomes revealed that the C and N bands are observed as regions of high relief on a slightly collapsed chromosome structure (McMaster *et al* 1996a; 1996b; Winfield *et al* 1995). It has been possible to identify features equivalent to G banding patterns in untreated human metaphase chromosomes (Musio *et al* 1994) and to use this for classifying the chromosomes. This suggests that, at least in this instance, the high resolution of the AFM has allowed an intrinsic banding pattern to be visualised which, for viewing by other microscopic methods needs to be enhanced by accumulation of stain. Combined AFM and SNOM studies have been made on human (Wiegrabe *et al* 1997; Kimura *et al* 2002) and barley chromosomes (Ohatani *et al* 2002) demonstrating the higher resolution achievable using near field, rather than far field optical microscopy for chromosome classification. Although chemical banding is the basis of karyotyping it has been shown by EM that the relative volume of metaphase chromosomes is chromosome specific, and thus can be used for classification (Heslop-Harrison *et al* 1989). The volumes of plant (McMaster *et al* 1996a) and human chromosomes (Fritzsche and Henderson, 1996c) have been measured using AFM. The height of air dried chromosomes increases with rehydration in aqueous buffer (DeGroth and Putman, 1992; Fritzsche *et al* 1994; 1996c; Rasch *et al* 1993), and this swelling is accompanied by changes in their viscoelastic properties, as measured by AFM (Fritzsche and Henderson, 1996b). For plant chromosomes it has been shown that the volumes determined on dried specimens can be used for classification (McMaster *et al* 1996a). Hybridisation techniques have been developed for mapping chromosomes. AFM has been used for localisation of an *in situ* hybridisation probe on a human chromosome (Putman *et al* 1993b; Rasch *et al* 1993). Biotinylated DNA probes were used for mapping, and the specific sites visualised by detecting the changes in topography induced by a peroxidase-diaminebenzidine reaction. In studies on cereal chromosomes the D genome specific probe pAs 1 was detected using AFM to image changes in height due to biotin-avidin-fluorescein isothiocyanate complexes formed as a consequence of using a standard fluorescent *in situ* hydridisation (FISH) procedure (McMaster *et al* 1996). SNOM images of FISH studies of human chromosomes have demonstrated detection of the maxima in fluorescence at nanometre resolution raising the possibility of detecting individual probe molecules (van Hulst *et al* 1997).

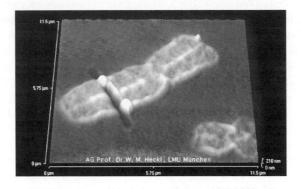

Figure 4.15. Microdissection of a human chromosome 2 using the AFM. Image provided by Prof. W.M. Heckl and reprinted with permission. Methodology is as described by Thalhammer *et al* 1997.

The ability to directly dissect and collect DNA from cytogenetically recognisable regions on chromosomes offers the potential for producing *in situ* hybridisation probes, the generation of band specific libraries for chromosomes, and mapping of chromosomes using such probes. It has been demonstrated that AFM can be used as a microdissection tool for extracting DNA from selected sites on human chromosomes (Thalhammer *et al* 1997; Iwabuchi *et al* 2002; Oberringer *et al* 2003; Tsukamoto *et al* 2006a; 2006b; Yamanaka *et al* 2008). Microdissection of chromosome 2 was achieved by raising the imaging force for a single line scan (Fig 4.15). The resultant DNA attached to the probe was isolated, amplified by a modified PCR procedure and used, via a FISH study, to demonstrate specificity for the cut region of chromosome 2. The use of AFM offers the prospect of isolating smaller probes than those obtainable with current methods such as dissection with glass needles or laser cutting, allowing a more detailed mapping of chromosomes. AFM tips have been refined to improve the micro-machining of chromosomes or DNA (Tsukamoto *et al* 2006a; 2006b; Yamanaka *et al* 2008) and the possibility of using AFM to extract DNA-histone complexes from chromosomes has been demonstrated (Sun *et al* 2002).

4.3. Nucleic acids: RNA

RNA is the Cinderella of the nucleic acids: far less data is available, possibly due to the much greater difficulties involved in the handling of RNA, and in adsorbing RNA to mica. RNA is an attractive candidate for study by AFM because of the complex three-dimensional structures adopted by RNA and RNA-protein complexes. Images of RNA released by osmotically-shocked viruses, such as influenza virus (Shao and Zhang, 1996), have been obtained by cryo-AFM. Early studies describe methodology for binding RNA to chemically-modified mica (Lyubchenko *et al* 1993a) and report studies on reovirus dsRNA (Lyubchenko *et al* 1993a; 1993b). The molecules were found to show convoluted shapes and, in some instances, compact structures. Data has been reported which is interpreted as showing ternary structure for ribosomal RNA (Wu and Liu, 1997). AFM images of ssRNA deposited from water onto mica showed a globular structure. Treatment with formamide and deposition onto silanised mica captured the linear denatured structure (Fritz *et al* 1997). Data has also been reported for polyA and polyG ssRNA (Smith *et al* 1997; Hansma *et al* 1996) using the Tapping mode for air dried samples on mica. The polyA samples show a heterogeneous population of sizes with 'blob-like' structures located at the ends of the RNA strands. The heights of the 'blobs' are larger than the remaining ssRNA and may represent regions of base-pairing or helical RNA. PolyG samples were very different in appearance, showing amorphous 'gel-like' aggregates. Similar amorphous aggregates are reported for mRNA and this has prevented studies of the binding of polyA-binding protein (PABP) to the polyA tails of mRNA (Smith *et al* 1997). PABP-RNA binding has been observed for polyA; PABP was observed to bind randomly, and the 'blobby' features seen on the isolated RNA disappeared, suggesting an alteration of the RNA structure on protein binding (Smith *et al* 1997). Formation of protein dsRNA complexes are important molecular events in host defence against viral infections. The vaccina virus inhibitor (protein p25) of protein kinase (PKR) is characterised by unusual binding specificity to dsRNA. AFM studies of p25-dsRNA complexes have revealed binding of the protein to the RNA and evidence for p25 induced aggregation of the RNA. The authors (Lyubchenko *et al* 1995) speculate that such condensation may prevent PKR accessibility to dsRNA thus inhibiting the dsRNA dependent induction of interferon synthesis. Ternary complexes of both linear and cyclic DNA with RNA polymerase and the nascent RNA have been imaged (Kasas *et al* 1997a; 1997b; Hansma *et al* 1993a; Rees *et al* 1993; Zenhaussern *et al* 1992b) and the transcription process observed in real time (Fig. 4.13) (Kasas *et al* 1997b). Protein-RNA complexes have also been visualised by AFM (Fritzsche *et al* 1997; Kiselyova *et al* 2001). The continued improvements in imaging methods are making it easier to observe the complex structures formed by RNA molecules and the dynamic interchanges between these structures (Hansma *et al* 2003; Janus and Yarus, 2003; Giro *et al* 2004; Kuznetsov *et al* 2005; Kuznetsov and McPherson, 2006). Studies have been made on isolated RNA and RNAs in membrane supported on hydrophobic mica substrates (Janus and Yarus, 2003). Degradations by RNase A has been used to discriminate between DNA and RNA released from retroviruses and to generate fragments of RNA (Kuznetsov and McPherson, 2006) reflecting the patterns of secondary structure. Antibodies have been visualised bound to dsRNA showing that they can be used to identify the location of specific sites (Bonin *et al* 2000).

AFM has been used to image intercellular RNA after cytochemical detection (Kalle *et al* 1996). Rat 9G and HeLa cells were hybridised with haptenised probes for ribosomal and messenger RNA. These were detected with peroxidise-labelled antibody and visualised with diaminebenzidine (DAB). The DAB precipitate can be detected by AFM, although the resolution is just marginally better than that obtained by optical microscopy, even on dried cells. Crystallisation of tRNA has been followed by AFM (Ng *et al* 1997) and this is discussed further in section 6.1.3.

The use of force spectroscopy is emerging as a tool for probing the folding and unfolding transitions of RNA (Zhang 2005; Vilfan *et al* 2007; Woodside *et al* 2008). Force spectroscopy can also be used to probe important RNA-protein interactions, such as mechanisms by which helicases can scan RNA molecules (Marsden *et al* 2006). The article by Vilfan and co-workers discusses a 'toolbox' for single molecules studies of RNA by various methods: it discusses procedures for producing DNA constructs for surface immobilization and ways of inhibiting RNA digestion.

4.4. Polysaccharides

Polysaccharides are ideal candidates for study by AFM. There are good reasons for studying polysaccharides. They are vital structural components of plant cell walls, starch and other cell wall polysaccharides are important food components, and polysaccharide extracts from bacterial, land plant and algal sources are widely used industrially, with cellulose being perhaps the most important example.

Many polysaccharides are of importance in understanding the development and treatment of disease: particular examples are the role of hyaluronic acid in the stiffening of joints with age and as a consequence of diseases such as rheumatoid arthritis, the role of bacterial alginate in cystic fibrosis, and the continuing use and development of bacterial polysaccharides as vaccines. Polysaccharides secreted by bacteria are of wide interest because they are important virulence factors for plant and human pathogens, they promote bacterial adhesion to surfaces and, in addition to medical uses as vaccines, these polysaccharides are used commercially as industrial polymers. Like nucleic acids polysaccharides are fibrous polymers but, unlike nucleic acids, they are often complex irregular structures. The structures may be branched or multiply branched. There are examples of natural block copolymers, and minor substituents in polysaccharide chains can radically alter their shape, conformation, or mode of interaction with other biopolymers. Most of these heterogeneous features are difficult to study by current biophysical methods and probe microscopy should play an important role in this area.

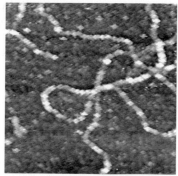

Figure 4.16. AFM image of a metal coated xanthan polysaccharide. Scan size 700 x 700 nm. The image shows the restricted resolution caused by the finite size of the metal grains.

4.4.1. Imaging polysaccharides

The earliest probe microscopy studies on polysaccharides used STM. These data were generally fairly poor, and failed to produce new information on polysaccharide structure (Morris, 1994). The challenges to imaging polysaccharides by AFM are similar to those faced in imaging nucleic acids. The methodology developed to image polysaccharides parallels the methods developed for nucleic acids. The major problem which had to be solved in order to obtain reproducible, reliable images was the need to eliminate adhesive forces, and the resultant damage and displacement of the molecules. The difficulties arose because the molecules were either sprayed or drop deposited onto mica, air dried and then imaged in air. The simplest solution, first developed for STM studies (Wilkins *et al* 1993) and then applied to AFM studies (Gunning *et al* 1995; Kirby *et al* 1996a), was to deposit the molecules onto mica either by spraying (Wilkins *et al* 1993) or drop deposition (Gunning *et al* 1995; Kirby *et al* 1996a) and then to metal coat the sample. The metal coating prevents damage or displacement of the molecules. The samples can be stored and imaged at leisure, and provide information on molecular size and shape (Fig. 4.2). The major disadvantages of this approach are the need to evacuate the samples and the limitation on resolution imposed by the grain size of the metallic coating (Fig. 4.16). The next simplest solution is to image under a liquid thus eliminating the adhesive forces. As with nucleic acids it is convenient to image under liquids such as alcohols which inhibit desorption of the molecules from the substrate (Kirby *et al* 1996a). Provided the applied normal force is controlled, in order to optimise image contrast and to prevent molecular damage or displacement (section 4.1.5), then high quality images (figure 4.4c) can be obtained reliably and reproducibly. Image quality slowly decays with time due the gradual appearance of an adhesive force as the probe tip accumulates debris from the sample surface (section 4.1.5, figure 4.5b). An alternative method for imaging polysaccharides is to use non-contact ac methods. Samples sprayed or drop deposited onto mica, and then air dried, can be imaged in air by Tapping or non-contact mode atomic force microscopy (NCAFM) (Gunning *et al* 1996a; Cowman *et al* 1998a; McIntire and Brant, 1997a). By using Tapping it is also possible to image polysaccharides in aqueous or buffered environments. Fig. 4.9 shows an image of a fibrous water soluble pentosan deposited onto mica and imaged by Tapping under buffer. Close inspection of the image shows that parts of the polysaccharide chain appear to be missing (Figs. 4.9b, c). This is because the polysaccharide is actually trying to desorb from the surface and those parts (loops), which are sampling the bulk medium, are moving too fast to be imaged. The visible parts (trains) of the molecule are those sections which are still in contact with the mica. The result of this process is an apparent motion of the molecule on the mica surface. This is shown in Fig. 4.9 which shows several 'stills' from a sequence of images taken at different times. With time different sections of the molecule desorb, and then re-absorb in a different position, leading to changes in shape of the molecule. Eventually the molecule finally wins the struggle and is able to completely desorb from the surface. This sequence of images can be linked together in order to produce a molecular movie of this molecular motion. The ability to image polysaccharide molecules under natural conditions offers the prospect for studying molecular interactions, or processes such as the enzymatic breakdown of complex carbohydrate structures.

4.4.2. Size, shape, structure and conformation

Two main methods are used for imaging polysaccharides. In the first case the molecules are drop-deposited onto mica, air dried and then imaged in the dc contact mode under alcohols (Kirby *et al* 1996a). In the second case the molecules are sprayed onto mica, air dried and then imaged in the ac non contact mode (McIntire and Brant, 1997a).

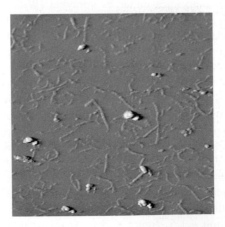

Figure 4.17. Error signal mode AFM image of a high methoxyl tomato pectin molecule showing evidence of branching. Scan size 1 x 1 μm. The molecule was imaged on mica under butanol.

In the instances where both techniques have been used to study the same systems there has been good agreement (Kirby *et al* 1996a; Gunning *et al* 1996b; 1997; McIntire and Brant, 1997a; 1997b). The AFM provides information on the shape and size of the molecules. Random coil polysaccharides such as dextrans are highly mobile and the AFM images (Tasker *et al* 1996; Frazier *et al* 1997a; 1997b) show globular structures. It is possible to image dextrans passively adsorbed to mica although the published images are of thiolated derivatives bound as monolayers. The size of the 'blobs' varies as expected with molecular weight (Tasker *et al* 1996) and swelling/deswelling was observed with hydration/dehydration of the monolayers (Frazier *et al* 1997a). AFM studies have been used to complement 'in situ' monitoring of dextran degradation with dextranase (Frazier *et al* 1997b).

As the persistence length of the molecules increases the molecules become more extended, less mobile and easier to image. The images reveal more detail about the molecular structure. Xyloglucam molecules appear as semi-flexible rods showing a low degree of branching (Ikeda *et al* 2004a). Pectic polysaccharides, generally considered as linear polymers, show a percentage of branched structures (Round *et al* 1997; 2001; Kirby *et al* 2007). The level of branching detected is too low to be detectable by conventional chemical or enzymatic methods of structural analysis, so the AFM images are providing new information on molecular structure. Figure 4.17 shows an AFM image of a population of 'green tomato' pectin showing the presence of both linear and branched structures. Infrequent branching of polysaccharide backbones has also been reported for arabinoxylans (Adams *et al* 2003) and amylose (Gunning *et al* 2003) and quite complex multi-branched structures have been observed for a surface-active soyabean polysaccharide (Ikeda *et al* 2005). The images obtained can be analysed in detail to give information such as distributions of contour length, branch length, number of branches per molecule, or branch separation, and these data can be used to investigate the effects of chemical or enzymatic modifications of the pectins.

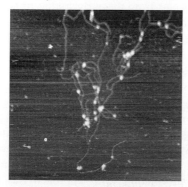

Figure 4.18. AFM image of entangled hyaluronan molecules. (Left) Deposited onto mica and imaged under butanol. Scan size 2 x 2 μm. Image obtained in authors laboratory. (Right) Tapping mode in air image of a hyaluronan chain showing simple crossing points. Scan size 1.2 x 1.2 μm. Data reproduced from Cowman *et al* 1998a with permission of the authors and the Biophysical Society.

Pectins are important constituents of the plant cell wall and play important roles in elongation and growth. A detailed understanding of pectin structure and interaction is of importance in the construction of molecular models of cell walls and for explaining their biological function. There are a few studies on carrageenans, an analogous structural polysaccharide found in algal cell walls (McIntire and Brant, 1997b; Kirby *et al* 1995; Morris *et al* 1997; Gunning *et al* 1998). Some of these studies show individual molecules (McIntire and Brant, 1997b) but the others (Kirby *et al* 1996a; Morris *et al* 1997; Gunning *et al* 1998) reveal aggregated structures which are dealt with in section 4.4.3. The published images of individual kappa carrageenan molecules (McIntire and Brant, 1997b) are highly extended, even though in the sol state prior to deposition they should be in the denatured 'coil' state suggesting, either that spray deposition extends the molecules, or that they reform the helical state on deposition. Images of the extracellular matrix polysaccharide hyaluronan can be obtained after deposition onto mica either by dc contact mode imaging under butanol (Fig. 4.18) or by Tapping mode in air (Fig. 4.18) (Cowman *et al* 1998a; 1998b). At moderately high stock concentrations the molecules form network structures indicative of the tendency of the molecules to associate in solution (Gunning *et al* 1996c). At sufficiently high dilutions individual stiff elongated structures can be seen (Cowman *et al* 1996a; 1996b). It is also possible to image a variety of more complex structures corresponding to various levels of self-association (Cowman *et al* 1996a).

The polysaccharides which have been studied most are the very stiff helical bacterial polysaccharides: xanthan (Meyer *et al* 1992; Gunning *et al* 1995; 1996a; Kirby *et al* 1995a; 1996a; McIntire and Brant, 1997b; Capron *et al* 1998), acetan (Gunning *et al* 1995; Kirby *et al* 1995b; 1996a), the acetan variant CR1/4 (Ridout *et al* 1998), gellan (Gunning *et al* 1996b; 1997; McIntire and Brant, 1997a; 1997b; Ikeda *et al* 2004b), curdlan (Jin *et al* 2006: Ikeda and Shishido, 2005) and scleroglucan (schizophyllan) (McIntire *et al* 1995; McIntire and Brant, 1997a; 1997b; 1998; Brant and McIntire, 1996). Apart from the early studies of Meyer and coworkers (Meyer *et al* 1992), which showed periodic structures with no clearly identifiable individual molecules, the recent studies on xanthan, acetan, CR1/4 and scleroglucan show distributions of highly extended stiff fibrous biopolymers. As well as just imaging the molecules it is possible to quantify the data by generating contour length distributions for the polysaccharides (McIntire and Brant, 1997b; 1998; Ridout *et al* 1998) providing information on molecular size and polydispersity. If the conformation of the molecule is known, then its mass per unit length is known, and the contour length distributions can be converted into molecular weight distributions. Conversely, if the conformation is not known, then the AFM data can be combined with light scattering data to determine the mass per unit length, and hence the molecular conformation. This approach was used to demonstrate a double helical conformation for the polysaccharide CR1/4 (Ridout *et al* 1998). Fairly accurate data can be obtained by imaging several hundred molecules and, although the measurement of contour lengths is tedious, the whole process is probably as quick as the use of gel permeation chromatography (GPC).

The use of GPC is often unreliable because many of these polysaccharides are prone to aggregation. As has been mentioned for hyaluronan (Cowman *et al* 1996a) it is possible to monitor and analyse unusual structures formed by intra-molecular association (McIntire *et al* 1995; Brant and McIntire, 1996; McIntire and Brant, 1997a; 1997b; 1998).

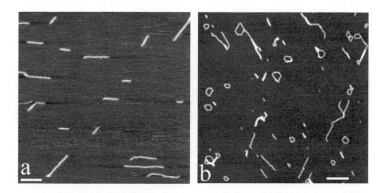

Figure 4.19. AFM images showing the transition between (a) extended helical and (b) cyclic forms of the polysaccharide scleroglucan. (a) Scale bar 200 nm. Data shown was provided by McIntire and Brant and published with permission.

In particular AFM has been used to monitor, record and analyse the linear triple helix to circular triple helix transition for scleroglucan (McIntire *et al* 1995, Brant and McIntire, 1996; McIntire and Brant, 1997a; 1997b; 1998). Fig. 4.19 shows a mixture of both linear and cyclic scleroglucan together with some hairpin structures. These types of cyclic structures are seen with polysaccharides which can adopt multiple helical structures. If the helical structures are first denatured, and then allowed to reform, then either intra- or inter-molecular association may occur. The cyclic structures result from intra-molecular association (Brant and McIntire, 1996; McIntire and Brant, 1998). Inter-molecular association will be discussed later in section 4.4.3.

The elongated structures of these bacterial polysaccharides suggest that they retain their helical structure when deposited onto mica. At the present time no detailed analysis has been made to determine whether meaningful values for persistence length can be extracted from such images. As for DNA measurements of molecular width are generally too large, due probe broadening effects, and can not be used to determine the conformation of the polysaccharide. If reasonable estimates are made for the size and shape of the probe tip, then the measured widths for polysaccharides such as xanthan are consistent with the dimensions expected for the helical form. If arrays of polysaccharides are imaged then the periodicity of the 'lattice' is measured. In this case the probe broadening effect disappears and the true width of the molecules is measured (Kirby *et al* 1995b). The widths measured for the acetan polysaccharide were consistent with the expected diameter of the helix. In addition it was also possible to observe (Fig. 4.20a) a periodicity along the polysaccharide molecules consistent with the pitch of the known helical structure (Kirby *et al* 1995b). Similar results can be obtained for aggregates of other bacterial polysaccharides such as xanthan (Fig. 4.20b)

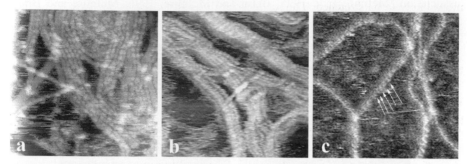

Figure 4.20. AFM images of polysaccharides showing evidence of the helical structure. (a) Acetan microgel scan size 140 x 140 nm, air dried and imaged under butanol. (b) Xanthan aggregate scan size 200 x 200 nm, air dried and imaged under 1,2 propanediol. (c) Individual xanthan molecules scan size 135 x 135 nm, air-dried, evacuated and then imaged under butanol. Turns of the helix are indicated by the white arrows.

It is occasionally possible to observe the helical structure on individual polysaccharides (Fig. 4.20c). This tends to happen when 'double tipping' occurs: presumably the sharper points on the end of the tip responsible for 'double tipping' allow higher resolution images of the polysaccharide chain. In the absence of direct visual evidence for a helical structure then measurements of heights are generally more realistic tests of the presence of the polysaccharide helix than measured widths. Typically height values of the order of about 66% of the expected helical diameter are usually obtained (Kirby *et al* 1996a; McIntire and Brant, 1997b). The reason for this effect is still not known, but it is undoubtedly the same as that seen with nucleic acids (section 4.2.2). The most likely origins of this effect are the compression of the molecule during scanning, or the breakdown of the assumption that imaging at constant force directly yields heights of molecules. If the charge density on the molecules is different to that of the substrate then the displacement of the probe to maintain constant cantilever deflection (constant force) may not be directly relatable to molecular thickness. These issues are addressed in more detail in the publications by Muller and coworkers (Muller and Engel, 1997; Muller *et al* 1999).

Measurement of force-distance curves for individual molecules provides a measure of molecular elasticity. This approach has been used to investigate the elasticity and conformational changes of individual polysaccharide chains (Abu-Lail and Camesano, 2003; Brant, 1999; Fisher *et al* 2000; Giannotti and Vancso, 2007; Rief *et al* 1997; Sietmoen *et al* 2003; Yuasa 2006; Zhang QM and Marszalek, 2006; Zhang W and Zhang X, 2003). In the simplest experiments the molecules are drop deposited onto a substrate and the tip is used, rather like a fishing rod, to pick up the molecules and then stretch them. This is easy to do but it is not certain whether single molecules or groups of molecules are attached to the tip, or at what point on the molecule any attachment occurs. The best approach is to derivatise the ends of the molecules so that they can be specifically attached to a substrate and then collected by the tip. In these experiments the whole molecule is more likely to be stretched. The simplest analysis is the use of theories for worm-like coils or freely-jointed chain models (Bouchiat *et al* 1999: Abu-Lail and Caesano, 2003) which allows a calculation of the persistence length. When the contour length is large compared to the persistence length then the stretching will be largely entropic in nature with enthalpic effects appearing at high extensions. The transition region will depend on the molecular weight and the detailed secondary structure of the polysaccharide. Data has been reported for polysaccharides such as native and derivatised dextrans (Rief *et al* 1997; Li, H *et al* 1998; Frank and Belfort, 1997), cellulose derivatives (Li, H *et al* 1998; Kuhner *et al* 2006), extracellular polysaccharide from *Pseudomonas atlantica* (Frank and Belfort, 1997), heparin (Marszalek *et al* 2003), amylose and modified amyloses (Lu *et al* 2004; Zhang QM and Marszalek, 2006; Zhang QM *et al* 2006), alginate (Williams *et al* 2008), curdlan (Zhang L *et al* 2003) and xanthan (Li H *et al* 1998). Analysis of the data on dextrans at low extensions suggested Kuhn lengths (L_K = $2L_P$) of 0.6 nm (Rief *et al* 1997). Force spectroscopy can follow unwinding of helical structures and distinguish between different helical structures (Zhang L *et al* 2003). In the case of xanthan, measurements were made on native and denatured xanthan. Denatured xanthan behaved similarly to the structurally related carboxymethylcellulose, whereas native xanthan showed quite different behaviour, which was attributed to helix formation (Li H *et al* 1998). Investigations of the force-distance curves of a number of polysaccharides (amylose, pullulan, dextran, pectin and methylcellulose) have considered the role of the pyranose ring in determining the elasticity of the polymer chain (Marszalek *et al* 1998; 2002). For dextran, pullulan and amylose it was shown that the enthalpic component of the elasticity was eliminated by periodate cleavage of the glucose rings, suggesting that force-induced distortion of the sugar ring and a chair-boat conformational transition determine this contribution. In the case of polysaccharides such as pectin or amylose which adopt ordered helical secondary structures, periodate oxidation will inhibit helix formation and this will contribute to the loss of the enthalpic contribution. Whereas polysaccharides such as dextran, pullulan, amylose, heparin and pectin all show enthalpic contributions to their elasticity, at least at large extensions, carboxymethylcellulose and methylcellulose behave entropically (Li, H *et al* 1998; Marszalek *et al* 1998): in the cellulose derivatives the pyranose rings are already fully extended and cannot contribute further to extension of the chain. A comparative study of λ, ι and κ-carragennans showed how the presence of an anhydride bridge can alter the conformation and conformational transitions of the polysaccharide (Xu *et al* 2002) on extension. It has been suggested that the characteristic transitions associated with sugar rings can be used as a fingerprint for the polysaccharide (Lee *et al* 2004; Li *et al* 1999; Marszalek *et al* 2001: Zhang W and Zhang X, 2003). Alginate is an example of a natural block copolymer and force spectroscopy has been explored as a tool for characterising heterogeneity by monitoring the signatures of the different conformational changes from the different 'blocks' within the polysaccharide (Williams *et al* 2008). Rather than probing forced conformational changes of the molecules by direct extension it is also possible to induce oscillations of molecules through the use of magnetically oscillated cantilevers (Humphris *et al* 2000a; Kawakami *et al* 2005) or through analysis of thermally-driven oscillations (Kawakami *et al* 2004) of the AFM cantilevers with attached polysaccharide molecules.

In addition to studying the force-distance curves for isolated polysaccharides it is possible to probe the behaviour of polysaccharides on cell or particulate surfaces and to investigate their roles in particulate adhesion to surfaces. Studies include the investigation of the role of extracellular polysaccharides in the attachment of *Pseudomonas putida* KT 2442 cells to surfaces (Abu-Lail and Caesano, 2002; Camesano and Abu-Lail, 2002), the material properties of surface polysaccharides on diatoms (Higgins *et al* 2003) and germinating fungal spores (van der Aa *et al* 2001) and the adhesion of cholesterol-pullulan nanogels to surfaces (Lee and Akiyoshi, 2004).

Force-distance curves can also be used to probe specific interactions between molecules by measuring the detachment energy or force required to pull the bonded structures apart. For these types of studies it is important that the molecules are appropriately covalently attached to a suitable substrate and to the tip-cantilever assembly. Force spectroscopy has been used to probe the specific interaction of xyloglucans with cellulose (Nordgren *et al* 2008) and the detachment of the C-5 epimerase AlgE4 from mannuronan chains: the data provided information on the specificity of different domains within the emzyme for the carbohydrate substrate and on the mode of action of the enzyme (Sletmoen *et al* 2004). This type of study can be employed to probe the bioactivity of carbohydrates. In particular, the role of dietary carbohydrates in cancer progression and metastasis is an emerging field of clinical importance. Generally health claims for dietary components are usually based on epidemiological studies and there is often only limited information available to allow understanding of their mode of action. However, for modified pectin it has been proposed (Nangia-Makker *et al* 2002) that the anti-cancer role arises due to fragments of the polysaccharide binding to and inhibiting the various roles of the mammalian lectin galectin 3 (Gal3). Dettmann and co-workers (2000) used force spectroscopy and surface plasmon resonace to measure ligand (lactose) binding to galactose-binding proteins. Force spectroscopy studies have been used to show specific binding of pectin-derived linear galactans to Gal3, thus identifying the nature of the bioactive fragment and confirming the molecular mode of action (Gunning *et al* 2009). In addition to studying discrete molecule interactions it is also possible to map carbohydrate-lectin interactions on cellular systems under physiological conditions (Touhami *et al* 2003: Gunning *et al* 2008). Xyloglucan-cellulose interactions important in plant cell wall assembly and elongation have been probed by force spectroscopy through studies on xyloglucan molecules tethered between cellulose substrates and an AFM tip (Morris S *et al* 2004). Studies on synthetic polymer gels have shown how AFM can be used to probe molecular motion through gel networks. Polyethylene glycol (PEG) molecules terminated with thiol groups were introduced into polyacrylamide gels. Gold coated AFM tips were used to pick up thiols at the surface of the gel and to extract the PEG polymers (Fig. 4.21). This allowed investigation of the extraction of the PEG molecules as a function of the cross-link density of the gels (Okajima *et al* 2007).

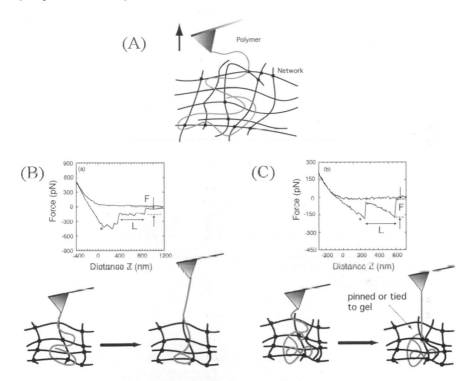

Figure 4.21. Extraction of single polymer molecules from gels. (A) Thiolated PEG molecules (grey) incorporated into polyacrylamide gels networks (black) are picked up by a gold coated AFM tip and pulled. (B) Sometimes when the PEG is pulled out of the network plateaus due to reptation are seen in the force curves. (C) Sometimes the PEG is pinned or tied into the network, and elastic stretching of the PEG is seen in the force curves. Image kindly supplied by Takaharu Okajima (Hokkaido University).

The direct imaging of protein-polysaccharide complexes can provide new insights into the functionality of polysaccharide extracts and new information on the heterogeneity of polysaccharide structures. Most polysaccharides are not surface-active. There are, however, a few polysaccharide extracts that do show useful surface activity. These include gum Arabic, sugar beet pectin and water-soluble wheat pentosans (arabinoxylan) extracts. All of these extracts are considered to contain protein-polysaccharide complexes which are believed to contribute, at least in part, to the surface activity of the extracted polysaccharide. AFM has the potential to visualise and characterise protein-polysaccharide complexes but, in practice, the extracts are often difficult to image. During drop deposition and air drying the protein components will collect at the air-water interface, partially denature and associate resulting in aggregated structures. Gum Arabic extracts contain arabinogalactans, an arabinogalactan–protein complex and a glycoprotein. Drop deposition of gum Arabic onto freshly-cleaved mica yields aggregated structures which can be disrupted in the presence of the surfactant Tween 20, allowing small individual molecules to be seen. Since Tween 20 would not only disrupt protein aggregates but also displace complexes from the mica, it is likely that the molecules seen are the arabinogalactan fraction (Ikeda *et al* 2005).

Water-soluble arabinoxylans and sugar beet pectin extracts contain protein which is difficult to remove by conventional separation methods. Where images have been obtained of the entire extract there is evidence for the existence of protein-polysaccharide complexes. For arabinoxylans an enzymatic treatment to remove protein eliminated the complexes and yielded images of the polysaccharide component alone (Adams *et al* 2003). In the case of sugar beet pectin ~ 67% of the pectin molecules in the extract were found to contain what appears to be (Kirby *et al* 2006a; 2007) a protein molecule attached to the end of the polysaccharide chains.

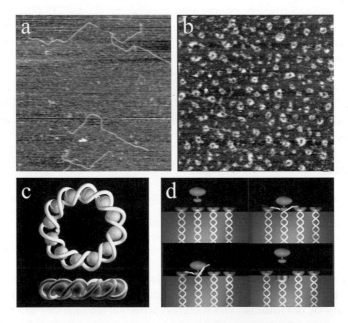

Figure 4.22. Topography AFM images of (a) amylose molecules scan size 600 x 600 nm, (b) circular amylose-starch binding domain complexes, scan size 1200 x 1200 nm, and (c) a model of the complex. The schematic models in (d) suggest a mode of action for the starch binding domain in docking the enzyme glucoamylose onto a starch crystal. For further details see Gunning *et al* (2003) and Morris *et al* (2005).

The direct observation of complexes provides a basis for understanding the types of structures they could form at air-water or oil-water interfaces and thus their role in stabilising foams and emulsions. Many polysaccharide structures show intra- and inter-molecular heterogeneity. Where proteins bind specifically to certain sequences on the polysaccharide chain they can be used to assess heterogeneity. Visualisation of the binding of the protein to the polysaccharide chain will confirm the existence of particular substituents or perhaps indicate evidence for block structures. Examples of proteins that could be used for this type of study are antibodies, non-hydrolytic enzymes, inactivated hydrolytic enzymes or carbohydrate binding domains. The specific binding of a mannuronan C-5 epimerase to mannuronan substrates has been visualised (Sletmoen *et al* 2004) as part of a study using force spectroscopy to probe mechanisms of epimerisation. Inactivated xylanases have been used to map structure on arabinoxylans (Adams *et al* 2004). The binding can be defined by plotting the distribution function of nearest neighbour separations. The theoretical distribution function is difficult to calculate even for random structures (Adams *et al* 2004) and the use of this type of approach to test models of biosynthesis or mechanisms of enzyme action would probably require simulation studies to predict the form of the distribution function. Imaging can occasionally provide unique information on the mode of action of an enzyme. The starch degrading enzyme glucoamylase 1 is a multi-domain enzyme and the starch binding domain (SBD) is essential to enable breakdown of crystalline starch. Mixtures of SBD and the linear starch polysaccharide amylose form unique circular complexes (Fig. 4.22). Modelling of the complex suggests that the SBD should be capable of locating and binding to the ends of double helical amylosic chains presented on certain faces of starch crystals, facilitating cleavage of the amylosic chains, thus providing new insights into the mechanistic role of the SBD (Giardina *et al* 2001; Morris *et al* 2005).

4.4.3. Aggregates, networks and gels

Interactions between polysaccharides play an important role in determining the functional behaviour of the polysaccharides in their native biological state or when used as industrial additives. AFM provides a useful means of studying such interactions. Thus the complex structures formed by intra-molecular association of hyaluronan (Cowman *et al* 1996a) are probably alternatives to the more complex inter-molecular association formed at higher concentrations (Gunning *et al* 1996c). Hyluronan is responsible for the useful viscoelasticity and lubrication of joints. In diseases such as arthritis the molecular structure is broken down. One form of proposed treatment is the injection of hylan, a cross-linked hyaluronan 'microgel'. This increases the viscoelasticity of the joint and, by their very nature, the cross-linked structures are more resistant to breakdown. AFM images believed to be of a hylan microgel (Gunning *et al* 1996c) showed a complex point cross-linked network structure with parts of individual hyaluronan molecules extending from the aggregated microgel. Here the AFM provided new information on a complex heterogeneous aggregate which could not easily be obtained by other means. Furthermore, AFM could be used to monitor the breakdown of such structures with time during trials of their use as a treatment for arthritis. Similar types of structures are formed by bacterial polysaccharides such as xanthan (Morris 1998a; Morris *et al* 1999) and are responsible for determining the solubility and thixotropy important for their use as thickening and suspending agents. A number of AFM studies on xanthan have reported evidence for molecular aggregation (McIntire and Brant, 1997b; Capron *et al* 1998) and it is generally quite difficult to prepare true solutions of polysaccharides such as xanthan: usually aggregates have to be removed by centrifugation or filtration.

Figures 4.5c and 4.23 show AFM images of xanthan prepared under different conditions. In Fig. 4.5c the xanthan has been dried down onto mica as an entangled solution of individual molecules. The ends of the molecules can be seen and the bright spots in the image show the doubling of height when one molecule lies on top of another.

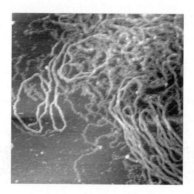

Figure 4.23. AFM image of a xanthan microgel particle, air dried onto mica and imaged in the dc contact mode under butanol. Scan size 1.4 x 1.4 μm. Figure 4.21b shows that the aggregate is based on interacting helical molecules.

Figure 4.23 shows an image of a xanthan microgel. These structures are formed when the polysaccharide is concentrated, precipitated or dried under conditions where the helical structure is partially denatured. The process of concentration promotes further helix formation and inter-molecular rather than intra-molecular linkages are formed and consolidated on drying. The result is the microgel particles shown in Fig. 4.23. The solubility of xanthan is determined by the ease of swelling of these particles. Aqueous xanthan preparations are actually dispersions of these swollen particles and it is this which dictates the rheology of the sample. Despite elaborate attempts to clarify acetan samples it was the presence of small amounts of microgel which gave rise to the molecular arrays which allowed imaging of the acetan helix (Kirby *et al* 1995b). The microgels are continuous branched aggregates within which it is difficult to locate ends of individual molecules (Fig. 4.23). The observation of the helical structure in the acetan microgels suggests that helix formation is the molecular basis of the inter-molecular association.

Similar forms of aggregation are believed to be responsible for the gelation of polysaccharides. Gels are often considered as useful models for the natural biological roles of these polysaccharides. The slimy extracellular polysaccharides secreted by bacteria play roles in preventing dehydration and in assisting adhesion to surfaces. In plants polysaccharides play an important role in determining the structural integrity of the cell wall. It is possible to follow this type of polysaccharide-polysaccharide association, to test mechanisms of gelation and to investigate the long range structures of gels. This type of study is best illustrated through studies on the gelation of gellan gum (Gunning *et al* 1996b; 1997). The strong tendency for gellan molecules to associate means that it is in fact quite difficult to image individual molecules: AFM studies of gellan sols generally show aggregated structures (Gunning *et al* 1996b; 1997; McIntire and Brant, 1997a; 1997b; Morris 1998b; Ikeda *et al* 2004b; Noda *et al* 2008).

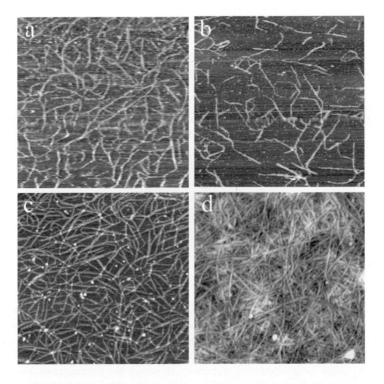

Figure 4.24. AFM images of gellan gel precursors, a gellan network and gel. (a) Aggregates of TMA gellan molecules. Scan size 1 x 1 μm. (b) Gel precursors formed from potassium gellan. Scan size 800 x 800 nm. (c) Aqueous gellan network. Scan size 800 x 800 nm. (d) Aqueous acid-set gellan gel. Scan size 2 x 2 μm. The samples a-c were deposited from aqueous sols onto mica, air dried and then imaged under butanol. For sample d the gel was set on mica and imaged wet under butanol. Data based upon Gunning *et al* 1996b; 1997.

This can be turned to advantage and used to investigate gelation mechanisms. By studying such aggregates (Fig. 4.24) it is possible to investigate the effects of the 'coil-helix' transition and selective cation binding on molecular association (Gunning *et al* 1996b; 1997; Morris, 1998b; Ikeda *et al* 2004b; Noda *et al* 2008). In the absence of gel-promoting counterions helix formation alone results in intermolecular association and aggregates (filaments) of constant height and width (Fig. 4.24a). When gel-promoting cations are present side-by-side aggregation of these filaments occurs resulting in branched fibrous structures of variable width and height (Fig. 4.24b). Deposition onto mica from more concentrated sols results in the formation of aqueous films. The network structures formed in these films can be visualised (Gunning *et al* 1995; 1996b; 1997; Morris, 1998b; 1998c) (Fig. 4.24c) and can be regarded as models for structures formed in 3D gels.

The networks are continuous branched fibrous structures consistent with the modes of association deduced from the studies on the gel precursors. The only ends of molecules which can be seen are short embryonic stubs which have not grown into larger branches. Imaging 3D gels is more difficult because the gel surfaces appear to deform during scanning blurring the images. Gellan gels can be prepared directly on the mica and the gel surface imaged under butanol. Scanning the gels rapidly makes it is possible to minimise distortion and reveal some details of the molecular structure of the gel (Morris, 1998b), although the images are poor showing periodic striations; a typical artifact of scanning too quickly. By setting gellan gels at acid pH it is possible to produce very stiff gels (1.2% gel, shear modulus $\approx 10^4$ Pa). For these stiff gels the distortion is minimal and the molecular network within the hydrated gel can be seen (Gunning *et al* 1996b; 1997; Morris, 1998b) (Fig. 4.24d). The fibrous network observed in the hydrated gel is similar to that seen in the hydrated film. Similar fibrous structures have been reported for potassium and caesium set bulk gellan gels (Ikeda *et al* 2004). The use of AFM has confirmed mechanisms of gellan gelation deduced from physical chemical studies, and provided new insights in the long range structure within the gels and the molecular origins of the elasticity of the gels (Gunning *et al* 1996b; 1997; Morris, 1998b). Gellan can be considered as a model system for studying thermoreversible polysaccharide gels and the behaviour of the gellan system appears to be typical of other polysaccharide systems. Thus reported studies on iota and kappa carrageenans also reveal aggregated gel-precursors and aqueous films containing fibrous networks (Kirby *et al* 1996a; Morris *et al* 1997; Morris *et al* 1998b; 1998c; Funami *et al* 2007). Semi-refined carrageenans are extracted from algal cell walls using a milder extraction process, which doesn't completely remove all the cellulose component of the cell wall. The AFM images of semi-refined carrageenan (Gunning *et al* 1998) show interpenetrating networks in which the carrageenan and cellulose components can be easily distinguished (Fig. 4.25). The stiffer, thicker cellulosic fragments of the disrupted cell wall appear brighter than the thinner carrageenan molecules.

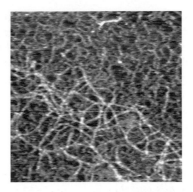

Figure 4.25. AFM image showing a mixed network of carrageenan and cellulose molecules. The cellulose molecules are stiffer and thicker than the carrageenan molecules and appear brighter in the image. Scan size 700 x 700 nm. Reproduced with permission from Gunning *et al* 1998.

In AFM images of mixtures of xyloglucans and gellan it was possible to distinguish between the molecules on the basis of height and flexibility and it was possible to distinguish both polysaccharides in microgels formed by breaking weak bulk gels (Ikeda *et al* 2004a). AFM has been used to image unperturbed gels of the related algal polysaccharide agarose under TBE (tris-borate-EDTA). The studies (Pernodet *et al* 1997; Maaloum *et al* 1998) revealed fibrous bundles of agarose but were mainly focussed on a detailed study of the pore distribution of the gels. The deduced variation of pore size with agarose concentration was found to be consistent with that deduced from electrophoretic mobility measurements using reptation models (Maaloum *et al* 1998). Curdlan forms heat-set gels. Preliminary studies have been made on the microfibrillar structure of curdlan sols and the heat-induced association formed on heating (Ikeda and Shishido, 2005).

Pectin and alginate are examples of polysaccharide block copolymers. In the case of alginate gelation can be induced by association of blocks of guluronic acid (G) blocks due to cooperative binding of calcium ions, suggesting that the gels may be rubber-like, consisting of extended junction zones (associated G blocks) linked by essentially random chains. The AFM images of alginate films and microgels suggest a fibrous structure the nature of which depends on the guluronic acid calcium ratio and can be influenced by the addition of free G blocks, shown to modify the rheology of the gels (Decho 1999: Jørgensen *et al* 2007). Similar fibrous structures have also been observed by AFM in other gelling polysaccharides such as pectin (Fishman *et al* 2007).

A detailed study of the surface roughness of synthetic polymer gels has been reported by Suzuki and coworkers (Suzuki *et al* 1996; 1997; 1998). These studies include observations of changes in surface structure attributable to temperature-induced phase transitions of the gels. An alternative method for gelling polysaccharides is chemical cross linking. Power and coworkers have used AFM to study gels formed by cross linking hydroxypropylguar with borate (Power *et al* 1998a; 1998b). Gels were air dried before imaging and then imaged using dc contact or Tapping mode under butanol or in air, and Tapping mode images in air were found to give the best images (Power *et al* 1998b). The authors report changes in gross morphology for gels prepared under quiescent conditions or under different shearing regimes. Given that the gels are supposed to be formed by point cross linking of a flexible polysaccharide, then the fibrous structures observed are surprising, even for the quiescent gels. The fibres between branch points appear to be too extended and too broad to be sections of individual polysaccharides. However, it is possible that in these xerogels the polysaccharide chains do become highly extended and rigid and the widths are artificially enhanced due to probe broadening effects.

4.4.4. Cellulose, plant cell walls and starch

Cellulose

Cellulose is the main structural component of plant cell walls. It is also one of the few polysaccharides to be biosynthesised in a crystalline form. Within the cell wall it occurs as microfibrillar structures of varying degrees of crystallinity. There are a number of AFM studies on isolated cellulose fibres (Hanley *et al* 1992; 1997; Baker *et al* 1997; 1998; van der Wel *et al* 1996; Kuutti *et al* 1994) some of which report high resolution images and analysis of the surface structure (Hanley *et al* 1992; 1997; Baker *et al* 1997; 1998; Kuutti *et al* 1994). The sizes of the fibres are consistent with those observed by electron microscopy, provided probe broadening effects are taken into consideration. Comparative EM and AFM images of microfibrils from *Micrasterias denticulata* provided evidence for a right handed helical twist of the microfibrils (Hanley *et al* 1997). The highest resolution images have been obtained for *Valonia ventricosa* cellulose crystals where it has been possible to resolve individual glucose molecules and features attributable to the hydroxmethyl groups (Baker *et al* 1997). Use of AFM to investigate the crystal structure of cellulose is discussed in more detail in section 6.1.1.

Certain bacterial species also produce cellulose and studies of cellulose production by *Acetobacter xylinum* have played an important role in understanding the genetics and biosynthesis of cellulose. The bacterial cellulose filaments appear to be more ribbon-like (Gunning *et al* 1998) than those extracted from plants (Fig. 4.26). Both the plant and the bacterial cellulose exist in the cellulose I form but so far no high resolution AFM images have been obtained for bacterial cellulose.

Plant cell walls

Cellulose is the main structural component of plant cell walls and it is also possible to image the cellulose microfibril network in cell wall fragments (Kirby *et al* 1996b: Round *et al* 1996; Morris *et al* 1997; van der Wel *et al* 1996; Thimm *et al* 2000; Davies and Harris, 2003). Extracts of root hair cell walls from *Zea mays* and *Rhaphanus sativus* have been visualised after extraction of the cell wall matrix material with H_2O_2/HAc (van der Wel *et al* 1996). The samples were imaged in air after air drying onto glass slides although poly-L-lysine coated glass slides were also used as substrates for imaging these cell walls under water (van der Wel *et al* 1996). The size and orientation of the microfibrils was consistent with that observed for platinum/carbon coated specimens imaged by EM. Similar information has been obtained on cell wall microfibrils from the sporangia of *Linderina pennispora* (McKeown *et al* 1996). For plant tissues such as carrot, potato, apple or Chinese water chestnut it is possible to isolate cell wall fragments by ball milling, wash them free of cytoplasmic components and then deposit them onto mica (Kirby *et al* 1996b; Round *et al* 1996; Morris *et al* 1997).

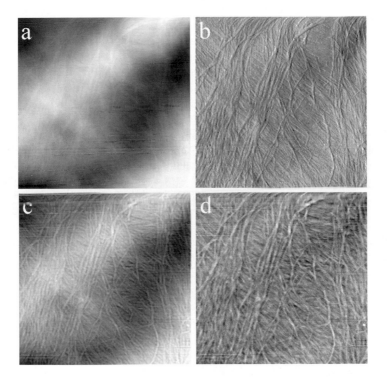

Figure 4.26. Composite set of images showing the effect of different processing steps on AFM images of hydrated Chinese water chestnut plant cell walls. Scan size 2 x 2 μm. (a) Topographic image: the bright and dark bands reflect the roughness of the cell wall surface. (b) Error signal mode image of the region shown in 'a'. This mode of imaging emphasises the 'high frequency' molecular structure within the image. (c) High pass filtered image corresponding to the region shown in 'a'. (d) Background subtracted image of the region shown in 'a'. This form of processing selects the molecular structure and projects it onto a flat plane. The image resembles that shown in 'b' but in this case the heights of the molecules can be measured. The figure is based on data presented in Kirby *et al* 1996b and Round *et al* 1996 and is reproduced with permission.

A problem using the AFM alone is locating fragments which are not buckled, but lay flat on the mica surface and use of a combination AFM/optical microscope are an advantage in this type of study. In the early studies it was also difficult to image these fragments under water because they tend to float off the mica (Kirby *et al* 1996b). For flat fragments it was possible to blot off excess liquid and then image by dc contact mode in air before the fragments dehydrated. The method of preparation presents the face of the cell wall previously adjacent to the plasma membrane (Fig. 4.26). The images reveal layered arrays of cellulose microfibrils whose orientation varies as one looks down through the cell wall towards the middle lamella. The surface of the fragment is rough and the images show light (high) and dark (low) regions apparently devoid of molecular structure (Fig. 4.26a). The microfibrillar structure is more clearly visible in the brighter (higher) regions: in these regions the eye can more easily perceive differences between grey levels whereas the whole image can contain more shades of grey than the eye can distinguish. The error signal mode image of the same area of surface reveals more detail (Fig. 4.26b). This form of imaging effectively represses the low frequency background curvature of the surface emphasising the high frequency molecular information. Although the error signal mode image displays molecular detail it is not strictly a 'real' image. The heights represent momentary changes in force and are effectively a differential function of the topographic image. To generate an image showing fine detail in a measurable form it is necessary to process the topographic image (Round *et al* 1996; Morris *et al* 1997). High pass filtering improves the image (Fig. 4.26c) but a better approach is to subtract the low frequency background curvature from the topographic image effectively projecting the molecular detail onto a flat plane (Round *et al* 1996; Morris *et al* 1997) (Fig. 4.26d). The background function can be generated by locally smoothing the topographic image to remove the fine (high frequency) detail. The level of detail seen by AFM in the cell wall preparations is comparable to that seen by EM studies.

Woody tissues are lignified and phase images of wood pulp fibres reveal bright patches, attributed to residual lignin, which are invisible in the normal topographical images (Hansma *et al* 1997). The lignin is considered to be more hydrophobic than the cellulose thus giving rise to the difference in contrast. For non-graminaceous plants the AFM images reveal the cellulose fibres within the cell wall, but it is not possible to image the other components of the cell wall, such as the xyloglucan tethers between the fibres, and the interpenetrating pectin network. This is presumably because these structures undergo rapid thermal motion and are deformed on scanning by the tip. However, it is possible to selectively and sequentially remove other cell wall components and observe the effects on the cellulose networks (Davies and Harris, 2003; Kirby *et al* 2006b). Because the cellulose fibres are fairly close-packed the AFM can be used to monitor the inter-fibre spacing in the hydrated cell wall fragment. Removal of the pectin network leads to shrinkage, consistent with a role for the polyelectrolyte pectin network in swelling or de-swelling the cell wall (Kirby *et al* 2006b). Cellulose fibrils have been observed for a range of plant species including bamboo (Yu, Y *et al* 2008), maize (Ding and Himmel, 2006), straw (Yan and Zhu, 2003; Yan *et al* 2004; Yu, H *et al*

2008), and Norway Spruce wood cells (Fahlen and Salmen, 2002; Zimmermann *et al* 2006). An interesting and novel study involved the direct demonstration of binding between xyloglucan and cellulose using single molecule force spectroscopy studies (Nordgren *et al* 2008).

The advantage of imaging under aqueous conditions is that it offers the prospect of probing enzymatic degradation of cell walls, an area of importance for understanding utilisation of plant waste material, gut fermentation of cell wall material and natural degradative processes such ripening. At present the only reported AFM studies of cellulase-cellulose interactions are low resolution imaging of the effect of adding either cellobiohydrolase I (CBH I), or CBH I inactivated at the catalytic site, to cotton fibres (Lee *et al* 1996). The surface morphology was reported to remain unchanged after addition of the inactivated enzyme whereas some disruption of the microfibrillar structure was seen as a result of adding the active enzyme. Another goal of this type of study would be to be able to image cell wall structure in intact, isolated plant cells under physiological conditions.

AFM is starting to be used for investigation of structural changes in plant, algal, fungal and bryophyte cell walls during biological processes such as growth (Ma *et al* 2005; 2006; Zhao *et al* 2005; Mine and Okuda, 2007; Kaminsky *et al* 2008; Wyatt *et al* 2008) or elongation (Marga *et al* 2005), and for studies on the processing of cell wall material (Fava *et al* 2006; Wang *et al* 2007a).

Starch

Starch is the major storage polysaccharide in plants. It is biosynthesised as large spheroidal semi-crystalline granules. Starch can be dissolved in solvents such as DMSO and fractionated into essentially two extreme types of polysaccharides: an essentially linear high molecular weight (about 10^6 D) polymer called amylose and a very high molecular weight, highly branched polysaccharide called amylopectin. At present there are a few reported high resolution AFM studies on the starch polysaccharides. However, by forming novel helical inclusion complexes it has been possible to image amylose molecules, revealing the nature of the small percentage of branched molecules, and thus new information on the nature of the branching (Gunning *et al* 2003). At present, to our knowledge there are no AFM images of individual amylopectin molecules. There is also interest in imaging the intact starch granule. Starch granules range in size from µms to 100's µms depending on their botanical source. This raises problems in the use of the AFM. When such objects are deposited onto a flat surface they cause the effective surface roughness to become comparable to, or larger than the height of the probe (Fig. 4.27a). Currently AFMs are designed to image very flat surfaces at high resolution and the axial ratio of the normal pre-fabricated tips is quite small. One approach to imaging rough surfaces is to use higher aspect ratio probe tips. However, such tips are often brittle and can easily snap if they 'hit' objects on the surface during scanning.

The development of new 'flexible' probe tips, based on materials such as carbon nanotubes, may provide a solution to this problem. Even then the scan range available in the 'z' direction will be limited by the lateral scan size of the image, making high resolution imaging of rough surfaces very difficult. A practical solution to this problem is to embed the starch granules (Thomson *et al* 1994: Baldwin *et al* 1996; 1997) thus decreasing the effective roughness of the surface (Fig. 4.27b). This approach has been used in a real time investigation of the enzymatic degradation of wheat starch granules (Thomson *et al* 1994). The starch granules were dusted onto mica, metal coated and then further coated with carbon. After soaking in water it was possible to strip of the mica with cellotape exposing the surfaces of the granules. The preparation could then be imaged in an aqueous environment. Examination of the surface of the granule revealed various surface features ranging in size between 50-450 nm. With granules with exposed surface pores, or those whose surfaces were cracked, possibly as a result of milling, it was possible to follow, in real time, breakdown of the starch granules due to the action of the enzyme α-amylase.

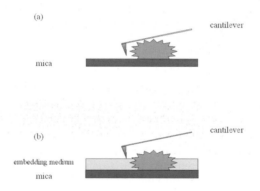

Figure 4.27. Schematic diagram illustrating the difficulties encountered in imaging large objects. (a) Large starch granules will contact the cantilever forcing the tip off the surface. (b) By embedding the starch granules the effective roughness of the surface is reduced and the exposed surface of the granule can be imaged.

The types of pits formed are consistent with those previously observed by electron microscopy (Frannon *et al* 1992). Differences in surface morphology have been reported between wheat and potato starch granules and surface protrusions have been attributed to the ends of ordered clusters of amylopectin sidechains (Baldwin *et al* 1997). There are a growing number of studies on the nature of the surface of starch granules, but fewer studies on the internal structure of the granule.

In the early studies on starch granule structure the granules were embedded in a penetrating resin (Baker *et al* 2001), set into blocks and then cleaved to cut open the granule and reveal the interior. To obtain reliable information on starch granule structure it is necessary to employ a non-penetrating resin, because use of a penetrating resin reduces contrast and can generate artifacts that make it difficult to unambiguously visualise internal structure (Ridout *et al* 2002).

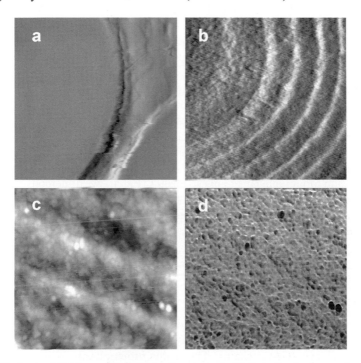

Figure 4.28. *In situ* AFM images of starch in dry pea seeds. (a) Part of a starch granule exposed on cutting open the seed, error signal mode, scan size 10 x 10 μm. (b) Controlled hydration of the exposed face of the granule reveals growth rings, error signal mode, scan size 20 x 20 μm. (c) High-resolution topography image of the blocklet structure within the granule, scan size 1.7 x 1.7 μm. (d) High-resolution phase image of the region shown in part c showing that the blocklets are distributed continuously throughout the granule, scan size 1.7 x 1.7 μm. For further details see Parker *et al* (2008).

Although the cut face of the block is flat, contrast in the images is induced by controlled wetting of the exposed cut face of the granule (Ridout *et al* 2004): selective absorption of water in the amorphous regions of the granule leads to swelling and softening of these areas, emphasising the presence of harder crystalline structures within the granule (Fig. 4.28). The methodology can be used to examine changes in granule structure due to selective mutations in biosynthesis (Ridout *et al* 2002; 2003; 2004; 2006; Parker *et al* 2008) and such studies have identified novel crystalline structures in high-amylose pea starches that drastically modify the functional properties of the starches (Bogracheva *et al* 1999).

In dry seeds the starch is naturally embedded within the seed. The seeds can be cut open and this allows *in situ* high-resolution imaging of starch within seeds (Parker *et al* 2008). The methodology can be used, together with other microscopy techniques, as a functional tool for screening libraries of natural or induced starch mutants. Imaging of starch in seeds allows the changes in starch structure to be monitored during development and growth, and permits characterisation of the heterogeneity in starch structure within granules, within cells, and throughout cells within the seed, which is seen to arise in natural or induced biosynthetic mutants. The earlier studies on the internal structure of starch granule were made using first generation hybrid AFM/optical microscopes (Ridout *et al* 2002; 2003; 2004; 2006) and detail in the images was limited. The use of second generation hybrid AFM/optical microscopes has substantially enhanced image quality (Parker *et al* 2008): these studies have confirmed a blocklet model of starch structure (Gallant *et al* 1997) originally proposed from TEM studies. The AFM studies have dispelled a widely-held belief, deduced from early microscopy and recent small angle x-ray and neutron scattering studies, that starch granules contain alternating amorphous and semi-crystalline growth rings (Donald *et al* 2001). In particular, a comparison of high-resolution topographical and phase images (Fig 4.28c, d) demonstrates that the entire granule is semi-crystalline, and that the banding arises due to differences in the localised heterogeneous distribution of amorphous (amylose) background material in the granule within which the crystalline blocklets reside (Parker *et al* 2008).

4.4.5. Proteoglycans and mucins

Proteoglycans are members of a superfamily containing more than 30 molecules that perform a variety of biological functions. The molecules are a particular type of fibrous glycoprotein composed of a protein core to which is attached glycosaminoglycan (GAG) chains. The major proteoglycan found in cartilage is aggrecan and studies on this polymer illustrate the type of information that can be obtained on these complex biopolymers. Bovine fetal and mature nasal cartilage aggrecan molecules have been imaged on mica (Ng *et al* 2003). The mica surfaces were functionalised with 3-amino propyltriethoxysilane (APTES) and the electrostatic interaction with the negatively charged GAG chains was used to retain the aggrecan molecules on the substrate. The resultant images (Fig. 4.29) are superb showing the non-glycosylated N-terminal region and resolving individual GAG chains, allowing the influence of GAG chain length and spacing on the stiffness of the 'bottle-brush-like' structures to be investigated. Reconstruction of AFM images is generally difficult because of probe broadening effects and because the AFM only generates a surface profile. For the cylindrically-symmetric aggrecan molecules it has been possible to correct for probe broadening and use other biophysical information to allow 3D reconstruction of the AFM data on the molecules: width measurements identified different protein glycosylation sites along the molecules and contour length measurements indicated specific protein sites of enzymatic cleavage of the molecules (Todd *et al* 2003).

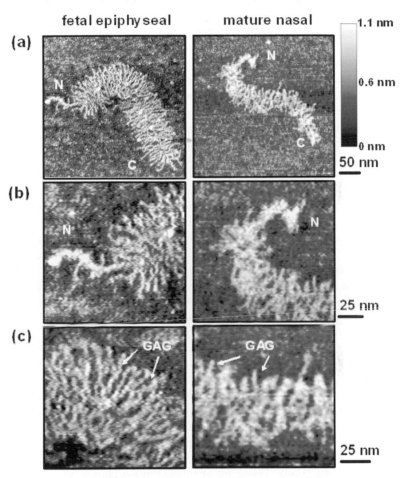

Fig 4.29 High resolution AFM topography images of an individual (a) fetal epiphyseal and mature nasal aggrecan monomer. (b) Images of the core protein visible in the N-terminal region on both of the monomers. (c) GAG chains which are clearly visible in the CS brush region on both the mature and fetal monomers and appear to be shorter for the mature nasal aggrecan. Images reproduced from Ng *et al* 2003 with permission from Elsevier.

Lateral force spectroscopy has been used to investigate the shearing of aggrecan molecules end-attached to surfaces (Han *et al* 2007a) and with attachment of the aggrecan molecules to tips it has been possible to probe nanoscale compressive interactions between aggrecan molecules under physiological conditions (Dean *et al* 2006; Han *et al* 2007b). These types of tribiology experiments seem ideally suited to the use of AFM. Other studies include imaging of the star-like branched structures of *Microciona prolifera* cell aggregation factor (MAF) responsible for inter-cell adhesion (Fritz *et al* 1997) and studies involving proteoglycan-collagen interactions (Dammer *et al* 1995; Raspanti *et al* 1996; 1997; 2002). Mucins are similar glycoproteins. Proteoglycans appear to be largely involved in structural aspects of tissue whereas the mucins seem to act as barriers and lubricants. Round and coworkers have used AFM to study the equilibrium conformations of mucins adsorbed to mica substrates (Round *et al* 2002; 2004; 2007). Inter and intra-mucin heterogeneity was attributed to post-translational modification of the glycosylation of the mucin and this was investigated through height measurements along mucins and the distribution of bound monoclonal antibodies. Force spectroscopy has been used to probe the interactions of mucin-derivatised tips with mica, and with mucin adsorbed onto mica (Berry *et al* 2001). Colloidal force spectroscopy has been employed to study interactions between muco-adhesive drug delivery vehicles and artifical mucin biomimics (Iijima *et al* 2008). Gastric mucins form gel-like structure at acidic pH and these structures are implicated in protecting the stomach from the acidic conditions. AFM studies have been used to probe the aggregation of gastric mucins at acidic pH in dilute solution and in thick gel-like films (Hong *et al* 2005). There are a small number of publications on salivary mucins including studies on the structure of the films, the role of different components in lubrication and the growth of cholesterol crystals on mucin substrates (Cardenas *et al* 2007; Berg *et al* 2004; Liao and Wiedmann, 2008). The use of AFM to study molecular aspects of lubrication and adhesion for complex biopolymers seems to be a promising area for future research (Misevic, 2001; Berg *et al* 2004).

4.5. Proteins

There are several reasons for imaging proteins at surfaces. Firstly there is interest in the intrinsic structure of the protein. Here the surface is considered to act as an inert carrier. Information is required on the overall size, shape and subunit structure of the protein. At a larger scale it is possible to follow molecular interactions involved in processes such as bioassembly or protein gelation. A second area of interest is protein surface interactions. These are of interest in a wide variety of applications from surface biocompatibility, biofouling, the cleaning of surfaces, through to development of biosensors and immunoassays. Here the interest is in how the protein binds to the surface and how such binding affects its biological functionality.

There is also a growing interest in directly observing the folding and unfolding of protein structures and force spectroscopy, alongside the technique of optical tweezers, is playing an important role in this form of study (Ng *et al* 2007). Information obtained on folding-unfolding transitions has been compared with that obtained from thermal (Cieplak and Sulkowska, 2005) and chemical (Carrion-Vazquez *et al* 1999) unfolding studies. At the time of writing there are relatively few proteins that have been investigated, although there are extensive studies on proteins such as titin (Rief *et al* 1997b; 1998; Wang *et al* 2001; Forman and Clarke, 2007; Sotomayor and Schulten, 2007; Linke and Grutzner, 2008; Oberhauser and Carrion-Vazquez, 2008). Titin is a 'mechanical protein' designed to generate, transmit or use mechanical forces. Other 'non-mechanical proteins' have been probed at the single molecule level including β-sandwich structures (Oberhauser *et al* 1998; Carrion-Vazquez *et al* 1999; Schwaiger *et al* 2004; Ng *et al* 2005), β-barrels (Brockwell *et al* 2003; Dietz and Rief, 2004; Perez-Jimenez *et al* 2006), β-spirals (Paananen *et al* 2006), α + β proteins (Yang *et al* 2000; Best *et al* 2001; Carrion-Vazquez *et al* 2003; Schlierf *et al* 2004; Ainavarapu *et al* 2005; Cecconi *et al* 2005; Lee *et al* 2006; Cao and Le, 2007; Sharma *et al* 2007), all α proteins (Rief *et al* 1999; Takeda *et* al 2001; Hertadi and Ikai, 2002; Gutsmann *et al* 2003; 2004; Janovjak *et* al 2003; Law *et al* 2003; Batey *et al* 2005; Bozec and Horton, 2005; Kreplak *et al* 2005; Guzman *et al* 2006; Paananen *et al* 2006; Brown *et al* 2007) and unstructured proteins and protein domains (Urry and Parker, 2002; Sarkar *et al* 2005; Leake *et* al 2006). Studies have also been made on some polypeptides and a small number of protein complexes (Cui and Bustamante, 2000; Schwaiger *et al* 2002; Ainavarapu *et al* 2005; Miller *et al* 2006). In addition to probing the folding and unfolding transitions of individual proteins it is also possible to study misfolding events (McAllister *et al* 2005; Oberhauser *et al* 1999; Yu JP *et al* 2008), protein-protein interactions (McAllister *et al* 2005; Paananen *et al* 2006) and protein-ligand interactions (Lee CK *et al* 2007; Borgia *et al* 2008; Ishii *et al* 2008). Novel studies suggest that differences in the mechanical properties of ordered and disordered regions of proteins can also be revealed through changes in shape of the disordered regions in successive high speed AFM images (Miyagi *et al* 2008). Compression of individual proteins can be used to measure mechanical properties (Afrin *et al* 2005; Ikai *et al* 2007; Parra *et al* 2007) and has been used to probe enzyme-ligand binding (Stali *et al* 2008).

4.5.1. Globular proteins

There is a large literature on the study of simple globular proteins by AFM (Silva 2005). A wide variety of methods have been used to immobilise proteins onto a wide variety of surfaces. However, most proteins adsorb fairly easily to surfaces such as mica and can be imaged under alcohols or aqueous conditions using dc or Tapping methods. The hydration-dehydration of adsorbed proteins can be monitored by AFM and other surface techniques such quartz crystal microbalance with dissipation (QCM-D) (Lubarsky *et al* 2007). Care needs to be taken in eliminating adhesive forces and in minimising damage and displacement of the molecules. It is also very important to prevent formation of multilayers on the substrate. The upper layers are easily displaced and adsorbed by the probe resulting in poor imaging conditions. There is a tendency to increase the concentration of the sample in order to improve images if few proteins are seen. However, if multilayers are present, and if this is causing difficulties with retaining the sample on the substrate then, paradoxically, the solution may be to use lower rather than higher protein concentrations in order to obtain better images. In general the images of proteins show globular structures with dimensions which, when corrected for probe broadening, are consistent with either individual proteins or protein aggregates. A correlation has been shown between the volume of native or denatured proteins as measured by AFM and molecular weight (Schneider *et al* 1998). Here probe broadening effects are compensated for by measuring diameters at half-height and calculating volume by treating the molecules as segments of spheres. It is not clear why the same calibration should apply for native and denatured proteins. However, the approach does allow discrimination between individual molecules and molecular aggregates.

Deliberate induced aggregation of proteins can be followed by AFM. Indeed studies of the dynamics of adsorption suggest that the process of adsorption triggers protein surface interactions that expose hydrophobic regions of the proteins favouring clustering of proteins (Kim *et al* 2002). A number of plant and dairy proteins can be induced to aggregate or gel (Tatham *et al* 1999; Humphris *et al* 2000; McMaster *et al* 2000; Mills *et al* 2001; Ikeda and Morris, 2002; Ikeda 2003; Gosal *et al* 2004; Tay *et al* 2005). Such processes can be followed by isolating multimers or gel precursors. Figure 4.30 shows a series of images following the association of the 7S soya protein β conglycinin. The sample has been heat treated at 100 °C for different lengths of time resulting in the formation of linear aggregates. The individual proteins are disc-like. Measurements of the height and length of the aggregates suggest they develop by the stacking of these discs into cylinders. By this sort of study of the behaviour of purified components it becomes possible to understand the functional behaviour of more complex protein isolates. A more dramatic example of protein aggregation is the classic and pioneering study of thrombin-induced fibrin polymerisation (Drake *et al* 1989). This was one of the first demonstrations of the use of AFM to monitor the dynamics of a biologically important process and provided new information on the mode of assembly in this complex system. There is still interest in the formation of fibrin networks and their mechanical properties (Blinc *et al* 2000; Brown *et al* 2007; Chtcheglova *et al* 2008; Lim *et al* 2008). A growing interest in the fabrication of nanostructures has led to an increasing number of AFM studies on the self assembly of model peptides (Dietz *et al* 2007; Wang *et al* 2007c; Ye *et al* 2008; Zhang, L *et al* 2008).

The AFM can be used in studies of proteins at surfaces or interfaces where the interest is in understanding how the proteins adsorb and also what effect the adsorption has on the subsequent functionality of the proteins. Typical applications of such studies can involve development of biosensors, biofouling and cleaning of surfaces or construction of biocompatible surfaces. AFM studies can be used to complement investigations by other methods, and to provide information on levels of protein adsorption, orientation of the proteins on the surface, aggregation and the formation of monolayers or multilayers. Information on protein aggregation and coverage does not require high resolution and can be inferred from measurements of surface roughness.

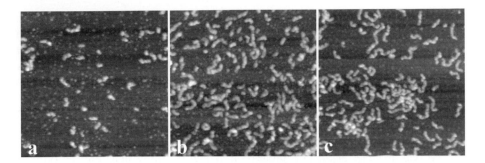

Figure 4.30. Sequence of AFM images illustrating aggregation of 7S soya protein. Scan size 1.2 x 1.2 μm. The protein solution was heated to 100°C and the aggregation process was followed with time. The solution was diluted, deposited onto mica, air dried and then imaged under butanol. (a) After heating for 2 minutes. Both individual proteins and linear aggregates can be seen. (b) After heating for 5 minutes and (c) after heating for 60 minutes. These proteins associate to form linear soluble aggregates.

Examples of studies on protein adsorption are investigations of ferritin adsorption (Caruso *et al* 1997; Davies *et al* 1994) and the binding of serum albumins to mica (Quist *et al* 1995; Mori and Imae, 1997) or proteins to more complex surfaces (Nakata *et al* 1996; Kowalczyk *et al* 1996). The nature and type of binding of enzymes to surfaces is of importance in the development and use of certain types of biosensors. In this case it should be possible to use AFM to study both the adsorption of the protein and the activity of the resulting surface. For example, AFM has been used to study the binding of glucose oxidase to gold surfaces (Quinto *et al* 1998) as a function of preparative conditions. It may also be possible to investigate the functional behaviour of such surfaces: conducting tips and the measurement of electron tunnelling could be used to study electron transfer processes, and the protein tracking methods, whereby enzymatic activity can be monitored by fluctuations in the height of enzymes (Radmacher *et al* 1994; Thomson *et al* 1996b), could be employed to map enzymatic activity on the surface in the presence of substrate. This type of study is similar to investigations of the efficiency of bound antibodies in immunoassays (section 4.5.2).

The adsorption and interaction of proteins at air-water and oil-water interfaces is of importance in understanding the stability of foams and emulsions. In this case AFM provide the only direct method for visualising the structures formed. Examples of such studies are given later in section 5.7.

An important goal in the use AFM to study globular proteins would be to provide detailed information on protein shape and internal structure. Clearly AFM is never going to rival the atomic resolution obtainable by X-ray diffraction or modern NMR methods. Electron diffraction patterns contain information on the atomic structure of proteins. In very high resolution electron microscopy it is possible to reconstruct images showing the secondary structure and its connectiveness within the protein. Such studies are few and, at present, require data on 2D crystals. However, there is the real prospect in the future of generating such images on individual proteins. The information required to construct such high resolution images is not present in the AFM data: the AFM generates only a surface profile modified by local factors such as charge or elasticity of the sample. However, the aim of the use of AFM would be to improve on the resolution obtained through the use of conventional electron microscopy, or to achieve comparable resolution but under natural imaging conditions. To obtain high resolution images the proteins need to be immobilised onto a flat substrate. Air drying proteins onto substrates such as mica and imaging in air is difficult for several reasons. The deposition process itself may partially denature the protein, motion within the protein structure will tend to blur the image and, unless the imaging force is carefully controlled, then the probe will distort the protein during scanning. For passively adsorbed proteins it may be possible to obtain information on overall shape and size and the images can often be interpreted in terms of known models of the protein structure. An example of this type of study is the Tapping mode images of ribosomes deposited onto mica (Wu *et al* 1997). For studies on isolated proteins there is the added complication of probe broadening effects which may further complicate interpretation of the images. Probe broadening effects can be reduced by organising the deposited proteins into ordered arrays. This also appears to reduce distortion or displacement of the proteins.

If the proteins bind strongly to mica, and resist disruption by the probe, then quite high resolution images can be obtained on individual proteins. Perhaps the best example of such studies is work on pertussis toxin (Yang *et al* 1994b). The intact pertussis toxin and the B-oligomers were deposited onto mica and imaged in the dc contact mode under water without passing through a drying stage. Even in the raw data it was possible to resolve 5 (2 large and 3 small) subunits in the B-oligomer and, by noting that the central pore of the B-oligomer is absent in the intact toxin, to fix the central location of the A-oligomer. The raw images were enhanced by correlation averaging and the result of such treatment is illustrated in Fig. 4.31. In these studies it was claimed that features as small as 0.5 nm were resolvable in the images. The pentameric structure observed by AFM differs from the heptameric structure deduced from X-ray diffraction studies (Stein *et ál* 1994). At present it is not clear why there is this discrepancy. However, this illustrates an important role which AFM can play in examining whether the protein structures found from X-ray data are appropriate for these molecules in solution. At present the only other method which could provide such information is NMR. It was also possible to use the AFM to investigate the stability of the toxin under different pH and temperature conditions (Yang *et al* 1994b).

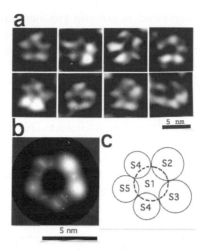

Figure 4.31. AFM images of pertussis toxin. (a) Eight representative images of the pertussis toxin D-oligomer. Two of the subunits appear brighter indicating that they are larger in size. (b) An averaged image formed from 300 individual images after alignment. (c) The proposed subunit structure of the molecule. Because it is not possible to distinguish between the subunits S2 and S3 it is not clear to which side the A-subunit is attached. Data reproduced from Yang *et al* 1994b with permission.

Another example of high resolution images of physically adsorbed proteins on mica is the study of *E. coli* chaperonin proteins (Mou *et al* 1996a; 1996b). High-resolution AFM images of both GroEL and GroES proteins were obtained in solution and the images were improved by chemical fixation using glutaraldehyde. The resolution achieved in the raw data was higher than that obtained by negatively stained EM, even after correlation averaging. The AFM was used to 'dissect' GroEL particles revealing information on the internal structure of this complex protein. In the case of these chaperonin proteins the structures obtained by AFM broadly agreed with those deduced from EM and X-ray diffraction studies, although combined EM and AFM data have been used to generate a new improved model for GroES. Chemical cross linking is a useful aid in imaging proteins. Not only does it improve the stability of the structure for imaging but, in addition, it may be used to trap a range of transient stages, allowing conformational changes to be examined at leisure. Fast scanning AFM has been used to image GroEL-GroES binding and also ATP/ADP induced conformational changes of individual GroEL proteins (Yokokawa *et al* 2006b). Despite the excellent studies described above, in general, in order to obtain images under aqueous or physiological conditions, it is often necessary to immobilise the proteins onto the substrate in some way.

A variety of approaches have been used to immobilise proteins for imaging by AFM. The most straightforward approach is to chemically attach the protein to the substrate. The general principles were discussed at the start of this chapter (section 4.1.6) and a few specific examples are given below. Many proteins contain thiols which can bind to gold-coated mica. Proteins such as BSA and gelatin have been thiol-derivatised in order to enhance binding to gold for AFM studies (Nakata *et al* 1996). Photocrosslinkers have been employed to attach HPI layers to glass substrates for AFM studies (Karrasch *et al* 1994). More specific tagging has also been employed. Bacteriorhodopsin molecules have been genetically modified, replacing a serine residue with a cysteine residue, in order to allow covalent attachment to gold for imaging by AFM (Brizzolarra *et al* 1997). A nickel chelating dipeptide was attached to the carboxyl terminus of the heavy chains of an IgG antibody, which was then bound to nickel chloride treated mica (Ill *et al* 1993). Insertion of targeted binding sites allows the orientation of the bound molecule to be defined but the modifications are complex, may alter the natural structure or function of the molecules, and thus will probably only be used as a last resort. As mentioned earlier the formation of ordered arrays of proteins often stabilises the molecules. An interesting variation of this approach described by Shao and coworkers (Shao *et al* 1996) is to stabilise sparsely populated deposits of large molecules from lateral motion by packing the intervening space with smaller molecules. This approach has been found to be successful in imaging low density lipoprotein using the smaller cholera toxin B-oligomer (Shao *et al* 1996). A particularly successful approach for imaging membrane proteins has been to reconstitute purified proteins into supported planar bilayers for study by AFM (Yang *et al* 1993a; 1993b). Using laterally polymerised phosphatidylcholine bilayers it was possible to image membrane bound cholera toxin in low salt buffer to a resolution of 1-2 nm (Yang *et al* 1993a; 1993b). For the B-oligomer (CTX-B) 5 subunits were resolved. In order to eliminate the possibility that the polymerisation may alter the function of the protein Mou and coworkers (Mou *et al* 1995b) have shown recently, through studies on cholera toxin, that AFM is fully capable of imaging membrane proteins on supported phospholipid bilayers of physiologically relevant species in solution. Surface features of the order of 1-2 nm can be resolved without the need for correlation averaging. In addition 2-D crystalline arrays can be grown directly on these model membranes and imaged by AFM (section 6.2.2). It is suggested that this methodology may be generally applicable for imaging membrane proteins, including integral membrane proteins, as long as supported bilayers can be made to incorporate the proteins (Mou *et al* 1995b). To this end these authors (Mou *et al* 1995b) suggest that direct fusion of protein-containing vesicles, and the spreading of these vesicles at air-water interfaces (Pearce *et al* 1992; Schindler 1980; Schurholz and Schindler, 1991), may provide a route to achieving this goal. It is possible to form 3D and 2D protein crystals and a number of membrane fragments are naturally occurring crystalline materials. The imaging of these types of material is described in chapter 6.

4.5.2. Antibodies

Antibodies are large, flexible multidomain proteins which have been well characterised by biophysical and biochemical methods. Because of their biological importance, and their use in immunolabelling and immunoassays, there is interest in characterising their structure and also their interactions with antigens and surfaces.

Most room temperature AFM studies have failed to match the resolution obtainable with electron microscopy (Parkhouse *et al* 1970). In general the 'molecules' appear to be globular, featureless and are variously ascribed to individual antibodies or aggregates (Lea *et al* 1992; Ill *et al* 1993, Yang *et al* 1994a; Fritz *et al* 1997; Harada *et al* 1997; Thimonier *et al* 1995). In the best images shapes can be identified which would be consistent with the expected structures (Fritz *et al* 1997; Harada *et al* 1997). The globular shapes are attributed to deformation of the molecules (Lea *et al* 1992) or molecular flexibility. High-resolution AFM images obtained by Tapping mode in the attractive regime reveal details of the 'Y' shape and the hinge region, whereas imaging in the repulsive regime appears to cause distortion and irreversible damage to the molecules (San Paulo and Garcia, 2000).

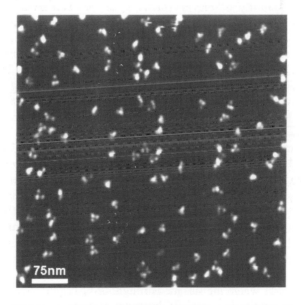

Figure 4.32. Cryo-AFM image of IgG (human IgG1), obtained at about 85°K. The characteristic Y shape of the antibody is clearly resolved. A range of conformations are visible in the images indicating the flexibility of the molecules. Data reproduced from Zhang, Y *et al* 1996 with permission of the authors and the Biophysical Society.

Drying the samples on mica has been found to freeze the molecular motion allowing the tri-nodal structure of antibodies to be revealed (Thomson, 2005). Lowering the temperature should also reduce molecular motion and higher quality images have been obtained by cryo-AFM (Han *et al* 1995; Zhang Y *et al* 1996; Shao and Zhang, 1996). Y shaped IgG (human IgG1) molecules are clearly visible with structural heterogeneity consistent with a flexible hinge region (Fig. 4.32).

Images of monoclonal IgA (mouse) revealed monomers and various multimers instead of the expected dimers. Both J chains and Fab domains can be seen in many of the molecules. Studies on monoclonal IgM (mouse) antibodies by AFM have revealed (Han *et al* 1995; Zhang Y *et al* 1996) a new conformation in addition to the accepted flat pentamer structure based on X-ray scattering data (Perkins *et al* 1991) and EM studies (Parkhouse *et al* 1970).

For immunoassays and biosensors there is considerable interest in the way in which antibodies are bound to surfaces, and how this affects antibody-antigen interactions. The adsorption of antibodies onto mica, modified mica, modified silicon oxide, gold and microtiter plates has been monitored by AFM (You and Lowe, 1996; Caruso *et al* 1996; Perrin *et al* 1997; Roberts *et al* 1995). After tip deconvolution it has been possible to discriminate between IgG and IgM antibodies on surfaces (Roberts *et al* 1995). Imaging can be used to assess the distribution of individual antibodies on the surface, assess problems due to antibody aggregation, or to study the stability of the interfacial structure. The nature of the binding mechanism (Kamruzzahan *et al* 2006; Ebner *et al* 2007), the length of spacers used to tether the antibodies (Cao *et al* 2007) and the orientation of the antibody, will affect the efficiency with which the antibodies bind antigens, and this can be studied by observing the antigen binding directly (Li *et al* 2002). These types of experiments have been used to assess methods of preparing biosensors or immunoassays, and for assessing their sensitivity (Perrin *et al* 1997; Caruso *et al* 1996; You and Lowe, 1996; Davies *et al* 1994).

Antibodies provide specific labels for identification and mapping. Immunolabelling techniques are well established in both light and electron microscopy. It is possible to use antibodies to probe and map specific antigens on individual molecules (e.g. DNA section 4.2.4), macromolecular complexes such as chromosomes (section 4.2.5), cells (Putman *et al* 1993a) or tissue (Saoudi *et al* 1994). In the case of individual molecules, or simple molecular complexes, the antibody is large, and easily recognisable. For flat layered structures, such as bacterial S layers (Ohnesorge *et al* 1992), it may still be possible to recognise the antibody-antigen complex directly, although care must be taken to discriminate between specifically bound and passively adsorbed material. In more complex systems, particularly where the surface is rough, or where there is a need to ensure specificity, some additional form of labelling is necessary. Procedures can be adapted from established immunolabelling methods. Thus gold labelled antibodies can be prepared and used directly, or gold labelled secondary antibodies can be used to locate antibody-antigen complexes (Mulhern *et al* 1992; Putman *et al* 1993a; Saoudi *et al* 1994). Gold labelling is not as straightforward in AFM as it is for EM studies. The gold labels may be confused with similar sized surface protrusions (Putman *et al* 1993a) or even compressed into the surface by the probe, making them difficult to spot (Mulhern *et al* 1992). The labelling procedure can be improved by generating larger particulate deposits: examples used with AFM include silver enhancement (Neagu *et al* 1994; Putman *et al* 1993a), peroxidise-labelled antibodies and their reaction with DAB (sections 4.2.5 and 4.3), or even fluorescently labelled complexes (McMaster *et al* 1996a; 1996b) (section 4.2.5). There may, however, be more sensitive types of labelling which could be used with the AFM: magnetic labels or tips coated with antibodies may lead to enhanced interactions and improved contrast for the labels. Antigen coated tips have been used to study antibody-antigen interactions for antibodies used in immunoassay systems (Allen S *et al* 1997). One area which does seem to require further research is the development of well established negative controls for immunolabelling studies by AFM. For relatively flat surfaces such as membranes it is possible to follow the dynamics of antibody binding in order to investigate the kinetics of the process as well as visualising the location of antigens (Keinberger *et al* 2004). Rather than visualising antibodies attached to surface antigens it is possible to couple antibodies directly to tips (Kamruzzahan *et al* 2006; Ebner *et al* 2007) and use affinity mapping to locate surface antigens. As well mapping cellular systems, or sites on heterogeneous biological systems (Avci *et al* 2004), the functionalised tips can be used as biosensors for probing the conformation of surface adsorbed molecules (Cheung and Walker, 2008), or for locating particular proteins in mixtures of surface adsorbed proteins (Soman *et al* 2008).

4.5.3. Fibrous proteins

The structure and organisation of fibrous proteins is a good topic for investigation by AFM: here AFM offers high resolution and the prospects of studying bioassembly under natural conditions. A number of such studies have been reported in the literature and some examples of this type of work are given below:

Muscle proteins

The proteins myosin and titin are important structural components of muscle. Studies on myosin have shown that the glycerol-mica method, widely used for preparing proteins for electron microscopy, can be used for AFM studies (Hallett *et al* 1995; 1996). The protein solution in 50% glycerol is sandwiched between mica sheets, the sheets pulled apart and then dried under vacuum to remove glycerol and water. The molecules can be imaged under propanol or water/propanol mixtures by dc contact or Tapping. The resolution is similar to that obtained by electron microscopy. For example, it is possible to observe the periodicity of the coiled-coiled α-helical myosin tail. Too high an imaging force was shown to damage or displace the molecules but, on one occasion, the probe appears to have separated the helical strands within the tail (Hallett *et al* 1995). Under propanol the heads appear to aggregate but, in water/propanol mixtures, they are separated and the images obtained are in agreement with those seen by electron microscopy (Hallett *et al* 1996). Cryo-AFM images (Zhang, Y *et al* 1997; Sheng *et al* 2003) of smooth muscle myosin clearly resolved the motor domains of the head and the pitch of the α-helical coiled-coiled tail. In addition it has been possible to obtain new information on the effects of thiophosphorylation on the tail structure and the flexibility of the head-tail junction of myosin maintained in the physiologically relevant 6S conformation. Detailed structural information on myosin can be obtained by cryo-AFM and this technique has been used to investigate the effects of thiophosphorylation on myosin structure (Zhang *et al* 1997).

For titin, which would normally adopt a coiled configuration, the molecules in 50% glycerol were dropped onto mica, subjected to centrifugation to extend the molecules, and then vacuum dried before imaging under propanol by dc contact or Tapping. Elongated structures, similar to that seen by electron microscopy, with a globular head and extended tail were observed by AFM (Fig. 4.33). Use of Tapping mode in liquid reveals some substructure suggesting that the 'tadpole-like' structures may in fact be assemblies of individual titin molecules (Hallett *et al* 1996).

Figure 4.33. AFM image of titin. Data reproduced from Hallett *et al* 1996 with permission.

It is the elastic properties of the molecules which are important for their biological roles in muscle and AFM studies have permitted the mechanical properties of individual molecules to be examined. The methodology for studying the mechanical properties of bioplymers is described in detail in chapter 9. Titin is the protein most extensively studied by force spectroscopy following the early pioneering work of Rief and co-workers (Rief *et al* 1997b), which was published alongside complementary experiments using optical tweezers (Kellermayer *et al* 1997; Tskhovrebova *et al* 1997). The titin molecules were allowed to adsorb onto gold and the tip dangled close to the surface to allow regions of the titin to physically adsorb to the tip, making it possible to obtain force-extension curves in phosphate buffered saline on individual molecules. There are now extensive studies on titin, and recombinant fragments, which has allowed a comprehensive determination and analysis of the forces required to unfold individual domains of the protein on extension, and also observation of protein refolding on relaxation (Rief *et al* 1998; Marszalek *et al* 1999; Wang *et al* 2001; Higgins *et al* 2006; Szymczak P and Cieplak, 2006; Forman and Clarke, 2007; Sotomayor and Schulten, 2007; Linke and Grutzner, 2008; Oberhauser and Carrion-Vazquez, 2008) and the I27 titin domain has become a model system for protein folding-unfolding studies. The molecule consists of a string of domains rather like beads on a string. Mechanical studies on titin and individual domains have been used to contruct a mechnical model for the protein and this has, in turn, been used to discuss the passive elasticity of intact myofibrils. Different domains unfold in different ways: at low extensions the unfolding arises from 'entropic elastic domains', whereas at high extensions, under non-physiological conditions, some partial and total unfolding of 'enthalpic domains' occurs. The latter are considered to play a role as shock absorbers preventing damge to the sarcomere. This type of study shows that the AFM has become so much more than just a microscope: it can also be used as a mini-laboratory for investigating mechanical properties at the molecular level.

Cytoskeleton proteins

There are reports of AFM studies on isolated actin filaments (Fritz *et al* 1995; Shi *et al* 2001; Zhang J *et al* 2004), spectrin molecules (Almqvist *et al* 1994; Zhang, P *et al* 1996), and microtubules (Fritz *et al* 1995; Vinckier *et al* 1995). For actin filaments cryo-AFM has revealed details of the helical structure of actin filaments and some aspects of their lateral association (Shao *et al* 2000). However, the main area of interest with these materials is in the use of AFM to probe the cytoskeleton structures of cells, and to study dynamic changes of these structures in living cells (chapter 7).

Collagen

Collagen is the most abundant structural protein found in connective tissue. It exists in a variety of morphological forms, and is a good example of a complex self-assembling biological structure. Individual collagen molecules are stiff coils 280 nm in length and 1 nm diameter. Studies on collagen molecules by cryo-AFM (Shattuck *et al* 1994) and normal AFM studies of segment-long-spacing (SLS) crystals of collagen (Fujita *et al* 1997) are starting to reveal variations in the structure (diameter) along the molecules, which may be of importance in aspects of their assembly in calcified tissue. Collagen molecules assemble into a range of fibrous structures and networks. The most well studied assembly product is the native fibril, characterised by a periodic banding pattern with a repeat of ≈ 68 nm. AFM has been used to visualise the D banding (Fig. 4.34a) and the detailed surface structure of native (Chernoff and Chernoff, 1992; Baselt *et al* 1993; Revenko *et al* 1994; Arogani *et al* 1995; Raspanti *et al* 1996; Yamamoto *et al* 1997; Paige *et al* 2001; Lin and Goh, 2002; Gutsmann *et al* 2003; Bozec and Horton, 2005; Heinemann *et al* 2007) and reconstituted collagen fibrils (Revenko *et al* 1994). Force spectroscopy has been used to characterise the mechanical properties of collagen fibrils (Gutsmann *et al* 2003; 2004; Bozec and Horton, 2005). The aggregation of monomers (Shattuck *et al* 1994;), and the various stable transient states in the *in vitro* assembly into fibres have been quantified using AFM, in terms of the amounts and structures of the various intermediates present at different stages of fibrillogenesis (Gale *et al* 1995; Paige and Goh, 2001).

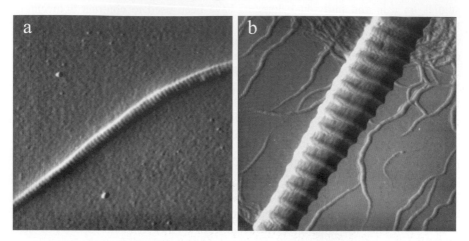

Figure 4.34. AFM images of (a) normal collagen fibril, scan size 3 x 3 μm showing the normal banding pattern, and (b) a fibrous long spacing collagen (FLS) fibril with a higher periodicity banding, scan size 3.5 x 3.5 μm. Data reproduced from Paige *et al* 1998 with permission of the authors and the Biophysical Society.

Real time AFM imaging has been used to follow the enzymatic degradation of type I collagen fibrils (Lin *et al* 1999; Paige *et al* 2002). Some studies have been made on the complexes generated in demineralised collagens (Habelitz *et al* 2002). Abnormal fibrillar structures such as fibrous long spacing collagen (FLS) are associated with various types of pathogenic conditions such as Hodgkin's disease, athlerosclerotic plaques, myeloproliferative disorder and silicosis. It has been suggested that interactions with other molecules, such as glycoproteins, may influence formation of FLS fibres. *In vitro* AFM studies of FLS fibre assembly (Fig. 4.34b), in the presence of glycoprotein, suggest an unique assembly process, rather than formation of normal fibres and their conversion into FLS (Paige *et al* 1998). In bones and other calcified tissue the final structure is based on the interaction between the collagen and deposited apatite. AFM Tapping mode studies of *in vitro* and physiologically calcified tendon collagen have suggested that surface structure of the fibril induces nucleation of apatite crystals, and that their subsequent growth does not markedly alter the fibril structure (Bigi *et al* 1997). Tapping mode AFM on dried collagen fibrils from rat tail tendon have been used to visualise proteoglycan bound to the collagen surfaces (Raspanti *et al* 1997). The distribution of the proteoglycans was determined by comparing images obtained for the native structure, samples treated with chondroitinase, and samples incubated with Cupromeronic blue, a copper phthalocyanin specifically designed to stabilise the anionic glycosaminoglycan chains. Such studies contribute to an understanding of this complex and vital bioassembly process.

Gelatin is basically denatured collagen. It has a wide variety of industrial uses which involve its ability to form gels and films.

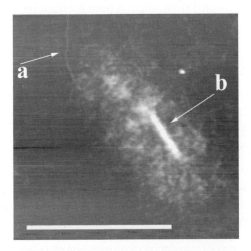

Figure 4.35. AFM image of a gelatin gel precursor. The junction zones labelled 'b' appear to be composed of filamentous structures labelled 'a', which are believed to be individual gelatin triple helices. Scale bar is 500 nm. Data is reproduced from Mackie *et al* 1998 with permission.

These network structures are usually formed by cooling hot solutions. The gelation mechanism is believed to involve a coil-helix transition: formation of the triple helical structure is believed to lead to intermolecular/aggregation and network formation. The initial gelation step is rapid, and is then followed by a slower stiffening of the network structure, usually attributed to a further level of aggregation, possibly involving reformation of a collagen fibre-like structure. AFM studies on gelatin films failed to reveal their molecular structure (Radmacher *et al* 1994; Haugstead *et al* 1993). This is generally thought to be because the films deform during scanning, blurring the image. By isolating gel precursors during the bulk gelation of gelatin it has been possible to use AFM to obtain clues about the gelation process (Mackie *et al* 1998). Fig. 4.35 shows a gelatin gel precursor which is believed to be the type of junction zone found in the gel. These appear to be aggregates of smaller fibres, believed to be reformed tropocollagen triple helical structures, and the aggregates do, on occasions, show periodicities reminiscent of those seen in collagen fibres. More information can be obtained on gelatin association and gelation by studying networks formed at air-water interfaces (Mackie *et al* 1998) and this is discussed in more detail in chapter 5.

Neurofilaments

Neurofilaments are important cytoskeletal components. They are unusual in that the filament is coated with sidechains. When imaged by AFM (Brown and Hoh, 1997) under aqueous conditions the sidechains are in motion and cannot be imaged directly. However, they reveal themselves by creating a zone around the molecules from which 'debris' deposited from solution onto the mica is excluded (Fig. 4.36).

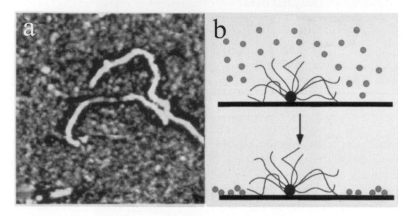

Figure 4.36. AFM Tapping mode image of (a) native neurofilaments adsorbed to mica and imaged in solution. The scan size is about 1.9 x 1.9 μm. The mobile sidearms create an exclusion region (black) around the molecule about 35-50 nm in width. (b) Schematic diagram illustrating how this exclusion zone is formed. Based on data reported by Brown and Hoh, 1997, with permission. Copyright (1997) American Chemical Society.

Such an exclusion zone is not seen with filaments lacking sidechains. Measurement of force-distance curves at each image point (force mapping - section 7.1.2) has revealed a repulsive interaction due to these sidechains. This 'entropic' repulsion has been proposed as a new mechanism for determining inter-filament spacing in nerve axons. AFM can be used to investigate the effects of phosphorylation on the unfolding of the sidechains (Aranda-Espinoza *et al* 2002). Clearly AFM can complement EM in the development of molecular models for the structure of nerves and the structural changes due to degenerative diseases. Studies of the interactions of neurofilaments with motor proteins such as dynein (Wagner *et al* 2004) have been made because of the importance in their transport within cells.

Amyloid fibres

The formation of fibrous protein deposits (amyloids) is associated with certain diseases. AFM is being used to characterise the fibrous structures (Gosal *et al* 2006) and to observe their growth and aggregation (Goldsbury *et al* 2005; Jansen *et al* 2005). Examples include studies on the early stages of fibre formation and aggregation that leads to the formation of neuritic plaques and vascular deposits associated with neurodegenerative disorders such as Alzheimer's disease (Shivji *et al* 1996; Stine *et al* 1996; Vieira *et al* 2003; Benseny-Cases *et al* 2007), Parkinson's disease (Hoyer *et al* 2004) and fibrous structures associated with diabetes (Kayed *et al* 1999; Goldsbury *et* al 1999; Green *et al* 2003; Marek *et al* 2007). Some proteins that are apparently unconnected with any particular disease also form fibrillar aggregates similar to amyloid fibrils under certain conditions. Such proteins and isolated peptide fragments are being investigated by AFM as part of studies to look at generic aspects of amyloid formation (Khurana *et al* 2003; Jenko *et al* 2004; Liu *et al* 2004; Ortiz *et al* 2007; Zerovnik *et al* 2007; Bochicchio *et al* 2008; del Mercato *et al* 2008; Hamada *et al* 2008; Kumar *et al* 2008; Sibley *et al* 2008). A similar type of peptide α helix β sheet transition of the cellular prion protein (PrPC) is associated the transmissible form of spongiform encephalopathies and AFM has been used in attempts to characterise individual PrP molecules (Kunze *et al* 2008). The high resolution attainable offers promise for understanding the structural origins of biological responses in this complex neurodegenerative disease.

The AFM can be used in studies of proteins at surfaces or interfaces where the interest is in understanding how the proteins adsorb and also what effect the adsorption has on the subsequent functionality of the proteins. Typical applications of such studies can involve development of biosensors, biofouling and cleaning of surfaces or construction of biocompatible surfaces. AFM studies can be used to complement investigations by other methods, and to provide information on levels of protein adsorption, orientation of the proteins on the surface, aggregation and the formation of monolayers or multilayers.

References: Selected Books and Reviews

Abu-Lail, N.I. and Camesano, T.A. (2003). Polysaccharide properties probed with atomic force microscopy. *J. Microscopy-Oxford* **212**, 217-238.

Alexandrov, B, Voulgarakis, N.K, Rasmussen, K.O, Usheva, A. and Bishop, A.R. (2009). Pre-melting dynamics of DNA and its relation to specific functions. *J. Phys.-Condensed Matter* **21**, article number 034107.

Binnig, G. and Rohrer, H. (1984). Scanning tunnelling microscopy. *Trends in Physics*, (eds. J. Janta and J. Pantollick) pp. 38-46.

Borgia, A, Williams, P.M. and Clarke, J. (2008). Single-molecule studies of protein folding. *Ann. Rev. Biochem.* **77**, 101-125.

Bornschlog, T. and Rief, M. (2008). Single-molecule dynamics of mechanical coiled-coil unzipping. *Langmuir* **24**, 1338-1342.

Brant, D.A. (1999). Novel approaches to the analysis of polysaccharide structures. *Current Opinion Struct. Biol.* **9**, 556-562.

Burchard, W. (1994). Light Scattering Techniques. In *Physical Techniques for the Study of Food Biopolymers*, (ed. S.B. Ross-Murphy), chap. 4. pp. 151-213. Blackie, London.

Bustamante, C, Keller, D. and Yang, G. (1993). Scanning force microscopy of nucleic acids and nucleoprotein assemblies. *Current Opinion Struct. Biol.* **3**, 363-372.

Bustamante, C, Eire, D.A. and Keller, D. (1994). Biochemical and structural applications of scanning force microscopy. *Current Opinion Struct. Biol.* **4**, 750-760.

Bustamante, C. and Rivetti, C. (1996). Visualising protein-nucleic acid interactions on a large scale with the scanning force microscope. *Ann. Rev. Biophys. Biomol. Struct.* **25**, 395-429.

Charvin, G, Allemand, J.F, Strick, T.R, Bensimon, D. and Croquette, V. (2004). Twisting DNA: single molecule studies. *Contemporary Phys.* **45**, 383-403.

Clemmer, C.R. and Beebe, T.P. (1992). A review of graphite and gold surface studies for use as substrates in biological scanning tunnelling microscopy studies. *Scanning Microscopy* **6**, 319-333.

Conroy, R.S. and Danilowicz, C. (2004). Unravelling DNA. *Contemporary Phys.* **45**, 277-302.

Engel, A. (1991). Biological applications of scanning probe microscopes. *Ann. Rev. Biophys. Biophys. Chem.* **20**, 79-108.

Firtel, M. and Beveridge, T.J. (1995). Scanning probe microscopy in microbiology. *Micron* **26**, 347-362.

Fisher, T.E, Marszalek, P.E. and Fernadez J.M. (2000). Stretching single molecules into novel conformations using the atomic force microscope. *Nature Struct. Biol.* **7**, 719-724.

Fritz, J, Anselmetti, D, Jarchow, J. and Fernandez-Busquets, X. (1997). Probing biomolecules with atomic force microscopy. *J. Struct. Biol.* **119**, 165-171.

Fritzsche, W, Takac, L. and Henderson, E. (1997). Application of atomic force microscopy to visualisation of DNA, chromatin, and chromosomes. *Critical Rev. Eukaryotic Gene Expression* **7**, 231-240.

Giannotti, M.I. and Vancso, G.J. (2007). Interrogation of single synthetic polymer chains and polysaccharides by AFM-based force spectroscopy. *CHEMPHYSCHEM* **8**, 2290-2307.

Gosal, W.S, Myers, S.L, Radford, S.E. and Thomson, N.H. (2006). Amyloid under the atomic force microscope. *Protein & Peptide Letts.* **13**, 261-270.

Greenlife, W.J, Woodside, M.T. and Block, S.M. (2007) High-resolution, single molecule measurements of biomolecular motion. *Ann. Rev. Biophys. & Biomol. Struct.* **36**, 171-190.

Hagerman, P.J. (1988). Flexibility of DNA. *Ann. Rev. Biophys. Chem.* **17**, 265-286.

Hansma, H.G. and Hansma, P.K. (1993). Potential applications of atomic force microscopy of DNA to the human genome project. *Proc. SPIE Int. Opt. Eng. (USA)* **1891**, 66-70.

Hansma, H.G. and Hoh, J.H. (1994). Biomolecular imaging with the atomic force microscope. *Ann. Rev. Biomed. Struct.* **23**, 115-139.

Hansma, H.G, Laney, D.E, Bezanilla, M, Sinsheimer, R. L. and Hansma, P.K. (1995). Applications for atomic force microscopy of DNA. *Biophys. J.* **68**, 1672-1677.

Hansma, H.G. (1996). Atomic force microscopy of biomolecules. *J. Vac. Sci. & Technol. B* **14**, 1390-1394.

Hansma, H.G, Kim, K.J, Laney, D.E, Garcia, R.A, Argaman, M, Allen, M.J. and Parsons, S.M. (1997). Properties of biomolecules measured from atomic force microscopic images: a review. *J. Struct. Biol.* **119**, 99-108.

Hansma, H.G, Kasuya, K. and Oroudjev, E. (2004). Atomic force microscopy imaging and pulling of nucleic acids. *Current Opinion Struct. Biol.* **14**, 380-385.

Harris, S.A. (2004). The physics of DNA stretching. *Contemporary Phys,* **45**, 11-30.

Hirano, Y, Takahashi, H, Kumeta, M, Hizume, K, Hirrai, Y, Otsuka, S, Yoshimura, S.H. and Takeyasu, K. (2008). Nuclear architecture and chromatin dynamics revealed by atomic force microscopy in combination with biochemistry and cell biology. *Pflugers Arch –Eur. J. Physiol.* **456**, 139-153.

Ikai, A. (1996). STM and AFM of bio/organic molecules and structures. *Surface Sci. Reports* **26**, 261-332.

Kasas, S, Thomson, N.H, Smith, B.L, Hansma, P.K, Miklossy, J. and Hansma, H.G. (1997). Biological applications of the AFM: from single molecules to organs. *Int. J. Imaging Systems Technol.* **8**, 151-161.

Lal, R. and John, S.A. (1994). Biological applications of atomic force microscopy. *Amer. J. Physiol.* **266** (*Cell Physiol.* **35**), C1-C21.

Lyubchenko, Y.L, Jacobs, B.L, Lindsay, S.M. and Stasiak, A (1995). Atomic force microscopy of nucleoprotein complexes. *Scanning Microscopy* **9**, 705-727.

Morris, V.J. (1994). Biological applications of scanning probe microscopies. *Prog. Biophys Mol. Biol.* **61**, 131-185.

Ng, S.P, Randles, L.G. and Clarke, J. (2007). Single studies of protein folding using atomic force microscopy. *Methods Mol. Biol.* **350**, 139-167.

Ou-Yang, Z.C, Zhou, H. and Zhang, Y. (2003). The elastic theory of a single DNA molecule. *Modern Phys. Letts.* **17**, 1-10.

Rief, M, Oesterhelt, F, Heymann, B. and Gaub, H.E. (1997). Single molecule force spectroscopy on polysaccharides by atomic force microscopy. *Science* **275**, 1295-1297.

Ros, R, Eckel, R, Bartels, F, Sischka, A, Baumgarth, B, Willking, S.D, Puhler, A, Sewald, N, Becker, A. and Anselmetti, D. (2004). Single molecule force spectroscopy on ligand-DNA complexes: from molecular binding mechanisms to biosensor applications. *J. Biotechnol.* **112**, 5-12.

Shao, Z, Yang, J. and Somlyo, A.P. (1995). Biological atomic force microscopy: from microns to nanometers and beyond. *Ann. Rev. Cell Dev. Biol.* **11**, 241-265.

Shao, Z. and Zhang, Y. (1996). Biological cryo atomic force microscopy: a brief review. *Ultramicroscopy* **66**, 141-152.

Shao, Z, Mou, J, Czajkowski, D.M, Yang, J. and Yuan, J-Y. (1996). Biological atomic force microscopy: what is achieved and what is needed. *Adv. Phys.* 45, 1-86.

Sietmoen, M, Maurstad, G, Sikorski, P, Paulson, B.S. and Stokke, B.T. (2003). Characterisation of bacterial polysaccharides: steps towards single-molecular studies. *Carbohydr. Res.* **336**, 2459-2475.

Silva, L.P. (2005). Imaging proteins with atomic force microscopy: an overview. *Current Protein & Peptide Sci.* **5**, 387-395.

Strick, T.R, Dessinges, M.N, Charvin, G, Dekker, N.H, Allemand, J.F, Bensimon, D. and Croquette, V. (2003). Stretching of macromolecules and proteins. *Reports Progr. Phys.* **66**, 1-45.

Tumer, Y.T.A, Roberts, C.J. and Davies, M.C. (2007). Scanning probe microscopy in the field of drug delivery. *Advanced Drug Delivery Rev.* **59**, 1453-1473.

Vilfan, I.D, Kamping, W, van den Hout, M, Candelli, A, Hage, S. and Dekker, N.H. (2007). An RNA toolbox for single-molecule force spectroscopy studies. *Nucleic Acids Res.* **35**, 6625-6639.

Wang, K, Forbes, J.G. and Jin, A.J. (2001). Single molecules measurements of titin elasticity. *Progr. Biophys. & Mol. Biol.* **77**, 1-44.

Woodcock, C.L. (2006). Chromatin architecture. *Current Opinion Struct. Biol.* **16**, 213-220.

Woodside, M.T, Garcia-Garcia, C. and Block, S.M. (2008). Folding and unfolding single RNA molecules under tension. *Current Opinion Chem. Biol.* **12**, 640-646.

Wei, H. and van de Ven, T.G.M. (2008). AFM-based single molecule force spectroscopy of polymer chains: Theoretical models and applications. *Appl. Spectrosc. Rev.* **43**, 111-133.
Yang, J. and Shao, Z. (1995). Recent Advances in Biological Atomic Force Microscopy. *Micron* **26**, 35-49.
Yuasa, H. (2006). Ring flip of carbohydrates: Function and applications. *Trends Glycosci. & Glycotechnol.* **18**, 353-370.
Zhang, Q.M. and Marszalek, P.E. (2006). Solvent effects on the elasticity of polysaccharide molecules in disordered and ordered states by single-molecule force spectroscopy. *Polymer* **47**, 2526-2532.
Zhang, W. and Zhang, X. (2003). Single molecule mechanochemistry of macromolecules. *Progr. Polym. Sci.* **28**, 1271-1295.
Zhang, X.W. (2005). Single-molecule RNA science. *Ann. Rev. Biophys. & Biomol. Struct.* **34**, 399-414.
Zlantanova, J. and Leuba, S.H. (20030. Chromatin fibers, one-at-a-time. *J. Mol. Biol.* **331**, 1-10.

Selected Research Papers

Abdelhady, H.G, Allen, S, Davies, M.C, Roberts, C.J, Tendler, S.J.B. and Williams, P.M. (2003). Direct real-time molecular scale visualization of the degradation of condended DNA complexes exposed to DNaseI. *Nucleic Acids Res.* **31**, 4001-4005.
Abu-Lail, N.I. and Camesano, T.A. (2002). Elasticity of *Pseudomonas putid*a KT 2442 surface polymers probed with single-molecule force microscopy. *Langmuir* **18**, 4071-4081.
Adams, E.L, Kroon, P, Williamson, G. and Morris, V.J. (2003). Characterisation of heterogeneous arabinoxylans by direct imaging of individual molecules by atomic force microscopy. *Carbohydr. Res.* **338**, 771-780.
Adamcik, J, Valle, F, Witz, G, Rechendorff, K. and Dietler, G. (2008). The promotion of secondary structures in single-stranded DNA by drugs that bind to duplex DNA: an atomic force microscopy study. *Nanotechnol.* **19**, article number 384016.
Adams, E.L, Kroon, P.A, Williamson, G, Gilbert, H.J. and Morris, V.J. (2004), Inactivated enzymes as probes of the structure of arabinoxylans as observed by atomic force microscopy. *Carbohydr. Res.* **339**, 579-590.
Afrin, R, Alam, M.T. and Ikai, A. (2005). Pretransition and progressive softening of bovine carbonic anhydrase II as probed by single molecule atomic force microscopy. *Protein Sci.* **14**, 1447-1457.
Ahn, H.H, Lee, M.S, Cho, M.H, Shin, Y.N, Lee, J.H, Kim, K.S, Kim, M.S, Khang, G, Hwang, K.C, Lee, I.W, Diamond, S.L. and Lee, H.B. (2008). DNA/PEI nano-particles for gene delivery of rat bone marrow stem cells. *Colloids & Surfaces A – Physicochem. & Eng. Aspects* **313**, 116-120.
Ainavarapu, S.R, Li, L, Badilla, C.L. and Fernandez, J.M. (2005). Ligand binding modulates the mechanical stability of dihydrofolate reductase. *Biophys. J.* **89**, 3337-3344.
Akeson, M, Branton, D, Kasianowicz, J.J, Brandin, E. and Deamer, D.W. (1999). Microsecond time scale discrimination among polycytidylic acid, polyadenylic acid, and polyuredylic acid as homopolymers or as segments within single RNA molecules. *Biophys. J.* **77**, 3227-3233.
Alexandrov, B, Voulgarakis, N.K, Rasmussen, K.O, Usheva, A. and Bishop, A.R. (2009). Pre-melting dynamics of DNA and its relation to specific functions. *J. Phys.-Condensed Matter* **21**, article number 034107.
Allen, M.J, Dong, X.F, O'Neil, T.E, Yau, P, Kowalczykowski, S.C, Gatewood, J, Balhorn, R. and Bradbury, E.M. (1993a). Atomic force microscopic measurements of nucleosome cores assembled along defined DNA sequences. *Biochemistry* **32**, 8390-8396.
Allen, M.J, Lee, C, Lee, J.D, Pogany, G.C, Balooch, M, Siekhaus, W.J. and Balhorn, R. (1993b). Atomic force microscopy of mammalian sperm chromatin. *Chromosoma* **102**, 623-630.
Allen, S, Chen, X.Y, Davies, J. Davies, M.C, Dawkes, A.C, Edwards, J.C, Roberts, C.J, Sefton, J, Tendler, S.J.B. and Williams, P.M. (1997). Detection of antigen-antibody binding events with the atomic force microscope. *Biochemistry* **36**, 7457-7463.
Allison, D.P, Bottomley, L.A, Thundat, T, Brown, G.M, Woychik, R.P, Schrick, J.J, Jacobson, K.B. and Warmack, R.J. (1992a). Immobilization of DNA for scanning probe microscopy. *PNAS USA* **89**, 10129-10133.

Allison, D.P, Warmack, R.J, Bottomley, L.A, Thundat, T, Brown, G.M, Woychik, R.P, Schrick, J.J, Jacobson, K.B. and Ferrell, T.L. (1992b). Scanning tunnelling microscopy of DNA - a novel technique using radiolabelled DNA to evaluate chemically mediated attachment of DNA to surfaces. *Ultramicroscopy* **42**, 1088-1094.

Allison, D.P, Kerper, P.S, Doktycz, M.J, Thundat, T, Modrich, P, Larimer, F.W, Johnson, D.K, Hoyt, P.R, Mucenski, M.L. and Warmack, R.J. (1997). Mapping individual cosmid DNAs by direct AFM imaging. *Genomics* **41**, 379-384.

Almqvist, N, Backman, L. and Fredriksson, S. (1994). Imaging human erythrocyte spectrin with atomic force microscopy. *Micron* **25**, 227-232.

An, H.J, Guo, Y.C, Zhang, X.D. and Hu, J. (2005). Nanodissection of single- and double-stranded DNA by atomic force microscopy. *J. Nanosci. & Nanotechnol.* **5**, 1656-1659.

An, H.J, Huang, J.H, Lu, M, Li, X.L, Lu, J.H, Li, H.K, Zhang, Y, Li, M.Q. and Hu, J. (2007). Single-base resolution and long-coverage sequencing based on single-molecule nanomanipulation. *Nanotechnol.* **18**, article number 225101.

Andersson, S.B. (2007). Curve tracking for rapid imaging in AFM. *IEEE Trans. Nanosci.* **6**, 354-361.

Andrushchenko, V, Leonenko, Z, Cramb, D, van de Sande, H. and Wiesser, H. (2001). Vibrational CD (VCD) and atomic force microscopy (AFM) study of DNA interaction with Cr^{3+} ions: VCD and AFM evidence of DNA condensation. *Biopolymers* **61**, 243-260.

Antognozzi, M, Szczelkun, M.D, Round, A.N. and Miles, M.J. (2002). Comparison between shear force and tapping mode AFM–High resolution imaging of DNA. *Single Molecules* **3**, 105-110.

Aragano, I, Odetti, P, Altamura, F, Cavalleri, O. and Rolandi, R. (1995) Structure of rat tail collagen examined by atomic force microscopy. *Experientia* **51**, 1063-1067.

Aranda-Espinoza, H, Carl, P, Leterrier, J-F, Janmey, P and Discher, D.E. (2002). Domain unfolding in neurofilament sidearmsL effects of phosphorylation and ATP. *FEBS Letts.* **531**, 397-401.

Argaman, M, Golan, R, Thomson, N.H. and Hansma, H.G. (1997). Phase imaging of moving DNA molecules and DNA molecules replicated in the atomic force microscope. *Nucleic Acids Res.* **25**, 4379-4384.

Argaman, M, Bendetz-Nezer, S, Matlis, S, Segal, S. and Priel, E. (2003). Revealing the mode of action of DNA topoisomerase I and its inhibitors by atomic force microscopy. *Biochem. & Biophys. Res. Commun.* **301**, 789-797.

Ashcroft, B.A, Spadola, Q, Qamar, S, Zhang, P, Kada, G, Bension, R. and Lindsay, S. (2008). An AFM/rotaxane molecular reading head for sequence-dependent DNA structures. *Small* **4**, 1468-1475.

Avci, R, Schweitzer, M, Boyd, R.D, Wittmeyer, J, Steele, A, Toporski, J, Beech, W, Arce, F.T, Spangler, B, Cole, K.M. and McKay, D.S. (2004). Comparison of antibody-antigen interactions on collagen measured by conventional immunological techniques and atomic force microscopy. *Langmuir* **20**, 11053-11063.

Bae, A.H, Lee, S.W, Ikeda, M, Sano, M, Shinkai, S. and Sakurai, K. (2004). Polysaccharide-polynucleotide complexes, Part 18. Rod-like architecture and helicity of the poly(C)/schizophyllan complex observed byAFM and SEM. *Carbohydr. Res.* **339**, 251-258.

Baker, A.A, Helbert, W, Sugiyama, J. and Miles, M.J. (1997). High-resolution atomic force microscopy of native *Valonia* cellulose I microcrystals. *J. Struct. Biol.* **119**, 129-138.

Baker, A.A, Helbert, W, Sugiyama, J. and Miles, M.J. (1998). Surface structure of native cellulose microcrystals by AFM. *Appl. Phys. A* **66**, S559-S563.

Baker, A.A, Miles, M.J. and Helbert, W. (2001). Internal structure of the starch granule revealed by AFM. *Carbohydr. Res.* **330**, 249-256.

Baldwin, P.M, Frazier, R.A, Alder, J, Glasbey, T.O, Keane, M.P, Roberts, C.J, Tendler, S.B.J, Davies, M.C. and Melia, C.D. (1996) Surface imaging of thermally sensitive particulate and fibrous materials with the atomic force microscope: a novel sample preparation method. *J. Microscopy-Oxford* **184**, 75-80.

Baldwin, P.M, Davies, M.C. and Melia, C.D. (1997). Starch granule surface imaging using low-voltage scanning electron microscopy and atomic force microscopy. *Int. J. Biol. Macromolecules* **21**, 103-107.

Banerjee, T. and Mukhopadhyay, R. (2008). Structural effects of nogalamycin, an antibiotic agent, on DNA. *Biochem. & Biophys. Res. Commun.* **374**, 264-268.

Bartels, F.W, Baumgarth, B, Anselmettti, D, Ros, R. and Becker, A. (2003). Specific binding of the regulatory protein ExpG to promoter regions of the galactoglucan biosynthesis gene cluster of Sinorhizobium meliloti–a combined molecular biology and force spectroscopy investigation. *J. Struct. Biol.* **143**, 145-152.

Baselt, D, Revel, J. and Baldeschweiler, J. (1993). Subfibrillar structure of type I collagen observed by atomic force microscopy. *Biophys. J.* **65**, 2644-2655.

Basnar, B, Elnathan, R. and Willner, I. (2006). Following aptamer-thrombin binding by force measurements. *Anal. Chem.* **78**, 3638-3642.

Batey, S, Randles, L.G, Steward, A. and Clarke, J. (2005). Cooperative folding in a multi-domain protein. *J. Mol. Biol.* **349**, 1045-1059.

Becker, J.C, Nikroo, A, Brabletz, T. and Resfeld, R.A. (1995). DNA loops induced by cooperative binding of transcriptional activator proteins and preinitiation complexes. *PNAS USA* **92**, 9727-9731.

Benseny-Cases, N, Cocera, M. and Cladera, J. (2007). Conversion of non-fibrillar β-sheet oligomers into amyloid fibrils in Alzheimer's disease amyloid peptide aggregation. *Biochem. Biophys. Res. Commun.* **361**, 916-921.

Berg, C.H, Lindh, L. and Amebrant, T. (2004). Intraoral lubrication of Prp-1, statherin and mucin as studied by AFM. *Biofouling* **20**, 65-70.

Berge, T, Jenkins, N.S, Hopkirk, R.B, Waring, M.J, Edwardson, J.M. and Henderson, R.M. (2002). Structural perturbations in DNA caused by bis-intercainium visualised by atomic force microscopy. *Nucleic Acids Res.* **30**, 2980-2896.

Berge, T, Haken, E.L, Waring, M.J. and Henderson, R.M. (2003). The binding of the DNA bisintercalator luzopetin investigated using atomic force microscopy. *J. Struct. Biol.* **142**, 241-246.

Berry, M, McMaster, T.J, Corfield, A.P. and Miles, M.J. (20010. Exploring the molecular adhesion of ocular mucins. *Biomacromolecules* **2**, 498-503.

Best, R.B, Li, B, Steward, A, Daggett, V. and Clarke, J. (2001). Can non-mechanical proteins withstand force? Stretching barnase by atomic force microscopy and molecular dynamics simulation. *Biophys. J.* **81**, 2344-3344.

Bezanilla, M, Drake, B, Nudler, E, Kashlev, M, Hansma, P.K. and Hansma, H.G. (1994). Motion and enzymatic degradation of DNA in the atomic force microscope. *Biophys. J.* **67**, 2454-2459.

Bhatia, S.K., Shriver-Lake, L.C., Prior, K.J., Jacque H., Georger, J.H., Calvert, J.M., Bredehorst, R. and Ligler, F.S. (1989) Use of thiol terminated silanes and heterobifunctional crosslinkers for immobilisation of antibodies on silica surfaces. *Anal Biochem.* **178**, 408-413.

Bigi, A, Gandolfi, M, Roveri, N. and Valdre, G. (1997). *In vitro* calcified tendon collagen: an atomic force and scanning electron microscopy investigation. *Biomaterials* **18**, 657-665.

Binnig, G. and Rohrer, H. (1984). Scanning tunnelling microscopy. *Trends in Physics*, (eds. J. Janta and J. Pantollick). pp. 38-46.

Biochicchio, B, Pepe, A, Flamia, R, Lorusso, M. and Tamburro, A.M. (2008). Investigating the amyloidogenic nanostructured sequences of elastin: sequences encoded by Exon 28 of human Tropoelastin gene. *Biomacromolecules* **8**, 3478-3486.

Blinc, A, Magdic, J, Fric, J. and Musevic, I. (2000). Atomic force microscopy of fibrin networks and plasma clots during fibrinolysis. *Fibrinolysis & Proteolysis* **14**, 288-299.

Bogracheva, T.Y, Cairns, P, Noel, T, Hulleman, S, Wang, T.L, Morris, V.J, Ring, S.G. and Hedley, C.L. (1999). The effect of mutant genes at the *r*, *rb*, *rug3*, *rug4*, *rug5* and *lam* loci on the granular structure and physico-chemical properties of pea seed starch. *Carbohydr. Polym.* **39**, 303-314.

Bordas, J, Perez-Grau, L, Koch, M.H.J, Vega, M.C. and Nave, C. (1986). The superstructure of chromatin and its condensation mechanism. 1. Synchrotron radiation X-ray scattering results. *Eur. Biophys. J.* **13**, 157-173.

Bornschlog, T. and Rief, M. (2008). Single-molecule dynamics of mechanical coiled-coil unzipping. *Langmuir* **24**, 1338-1342.

Borovok, N, Molotsky, T, Ghabboun, J, Cohen, H, Porath, D. and Kolyar, A. (2007). Poly(dG)-poly(dC) DNA appears shorter than poly(dA)-poly(dT) and possibly adopts an A-related conformation on a mica surface under ambient conditions. *FEBS Letts.* **581**, 5843-5846.

Bouchiat, C, Wang, M.D, Allemand, J-F, Strick, T, Block, S.M. and Croquette, V. (1999). Estimating the persistence length of a worm-like chain molecule from force-extension data. *Biophys. J.* 76, 409-413.

Bonin, M, Oberstrass, J, Lukacs, N, Ewert, K, Oesterschulze, E, Kassing, R. and Nallen, W. (2000). Determination of preferential binding sites for anti-dsRNA antibodies on double-stranded RNA by scanning force microscopy. *RNA-A Publication of the RNA Society* 6, 563-570.

Borgia, A, Williams, P.M. and Clarke, J. (2008). Single-molecule studies of protein folding. *Ann. Rev. Biochem.* 77, 101-125.

Bozec, L. and Horton, M. (2005). Topography and mechanical properties of single molecules of Type I collagen using atomic force microscopy. *Biophys. J.* 88, 4223-4231.

Brant, D.A. and McIntire, T.M. (1996). Cyclic polysaccharides. In *Large Ring Molecules*, (ed. J.A. Semlyen), chapter 4, pp. 113-154. John Wiley & Sons, Oxford.

Brett, A.M.O. and Chiorcea, A.M. (2003). Effect of pH and applied potential on the adsorption of DNA on highly oriented pyrolytic graphite electrodes. Atomic force microscopy surface characterisation. *Electrochem. Commun.* 5, 178-183.

Brizzolara, R.A, Boyd, J.L. and Tate, A.E. (1997). Evidence for covalent attachment of purple membrane to a gold surface via genetic modification of bacteriorhodopsin. *J. Vac. Sci. & Technol. A* 15, 773-778.

Brockwell, D.J, Paci, E, Zinober, R, C, Beddard, G.S, Olmsted, P.D, Smith, D.A, Perham, R.N. and Radford, S. (2003) *Nature Struct. Biol.* 10, 731-737.

Brown, A.E.X, Litvinov, R.I, Discher, D.E. and Weissel, J.W. (2007). Forced unfolding of coiled-coils in fibrinogen by single-molecule AFM. *Biophys. J.* 92, L39-L41.

Brown, H.G. and Hoh, J.H. (1997). Entropic exclusion by neurofilament sidearms: a mechanism for maintaining interfilament spacing. *Biochemistry* 36, 15035-15040.

Burchard, W. (1994). Light Scattering Techniques. In *Physical Techniques for the Study of Food Biopolymers*, (ed. S.B. Ross-Murphy), chap. 4. 151-213. Blackie, London.

Bustamante, C, Vesenka, J, Tang, C.L, Rees, W, Guthold, M. and Keller, R. (1992). Circular DNA molecules imaged in air by scanning force microscopy. *Biochemistry* 31, 22-26.

Bustamante, C, Keller, D. and Yang, G. (1993). Scanning force microscopy of nucleic acids and nucleoprotein assemblies, *Current Opinion Struct. Biol.* 3, 363-372.

Bustamante, C, Eire, D.A. and Keller, D. (1994a). Biochemical and structural applications of scanning force microscopy. *Current Opinion Struct. Biol.* 4, 750-760.

Bustamante, C, Marko, J.F, Siggia, E.D. and Smith, S. (1994b). Entropic elasticity of λ-phage DNA. *Science* 265, 1599-1600.

Bustamante, C. and Rivetti, C. (1996). Visualising protein-nucleic acid interactions on a large scale with the scanning force microscope. *Ann. Rev. Biophys. Biomol. Struct.* 25, 395-429.

Butt, H-J. (1992a). Electrostatic interaction in scanning probe microscopy when imaging in electrolyte solutions. *Nanotechnol.* 3, 60-68.

Butt, H-J. (1992b). Measuring local surface charge densities in electrolyte solutions with a scanning force microscope. *Biophys. J.* 63, 578-582.

Calderon, C.P, Chen, W.H, Lin, K.J, Harris, N.C. and Kiang, C.H. (2009). Quantifying DNA melting transitions using single-molecule force spectroscopy. *J. Phys.-Condensed Matter* 21, article number 034114.

Camesano, T.A. and Abu-Lail, N.I. (2002). Heterogeneity in bacterial surface polysaccharides, probed on a single-molecule basis. *Biomacromolecules* 3, 661-667.

Cano, S, Caravaca, J.M, Martin, M. and Daban, J.R. (2006). Highly compact folding of chromatin induced by cellular cation concentrations: evidence from atomic force microscopy studies in aqueous solution. *Eur. Biophys. J. Biophys. Letts.* 35, 495-501.

Cao, Y. and Li, H. (2007). Polyprotein of GB1 is an ideal artificial elastomeric protein. *Nature Materials* 6, 109-114.

Cao, T, Wang, A.F, Liang, X.M, Tang, H.Y, Auner, G.W, Salley, S.O. and Ng, K.Y.S. (2007). Investigation of spacer length effect on immobilised *Escherichia coli* pili-antibody molecular recognition by AFM. *Biotechnol. & Bioeng.* 98, 1109-1122.

Capron, I, Alexander, S. and Muller, G. (1998). An atomic force microscopy study of the molecular organisation of xanthan. *Polymer* **39**, 5725-5730.

Cardenas, M, Elofsson, U. and Lindh, L. (2004). Salivary mucin MUC5B could be an important component of *in vitro* pellicles of human saliva: an *in situ* ellipsometry and atomic force microscopy study. *Biomacromolecules* **8**, 1149-1156.

Carrion-Vazquez, M, Oberhauser, A.F, Fowler, S.B, Marszalek, P.E, Broedel, S.E, Clarke, I. and Fernandez, J.M. (1999). Mechanical and chemical unfolding of a single protein: A comparison. *PNAS USA* **96**, 3694-3699.

Carrion-Vazquez, M, Li, L, Lu, H, Marszalek, P.E, Oberhauser, A.F. and Fernansez, J.M. (2003). The mechanical stability of ubiquitin is linkage dependent. *Nature Struct, Biol.* **10**, 738-743.

Caruso, F, Rodda, E. and Furlong, D.N. (1996). Orientational aspects of antibody immobilisation and immunological activity on quartz crystal microbalance electrodes. *J. Colloid & Interface Sci.* **178**, 104-115.

Caruso, F, Furlong, D.N. and Kingshott, P. (1997). Characterisation of ferritin adsorption onto gold. *J. Colloid & Interface Sci.* **186**, 129-140.

Cecconi, C, Shank, E.A, Bustamante, C. and Marqusee, S. (2005). Direct observation of the three-state folding of a single protein molecule *Science* **309**, 2057-2060.

Chang, Y.C, Lo, Y.H, Lee, M.H, Leng, C.H, Hu, S.M, Chang, C.S. and Wang, T.F. (2005). Molecular visualization of the yeast Dcm1 protein ring and Dcm1-ssDNA nucleoprotein complex. *Biochemistry.* **44**, 6052-6058.

Charvin, G, Allemand, J.F, Strick, T.R, Bensimon, D. and Croquette, V. (2004). Twisting DNA: single molecule studies. *Contemporary Phys.* **45**, 383-403.

Chen, A, Cao, E.H, Sun, X.G, Qin, J.F, Liu, D, Wang, C. and Bai, C.L. (2001). Direct visualization of telomeric DNA loops in cells by AFM. *Surface & Interface Anal.* **32**, 32-37.

Chen, F, Yu, J-P, Fang, X-H. and Zhang, G-Y. (2006). Characterisation of a novel AP2/EREBP transcription factor TSH1 specifically binding to GCC elements. *Progr. Biochem. & Biophys.* **33**, 627-634.

Chen, H, Fu, H.X. and Koh, C.G. (2008). Sequence-dependent unpeeling dynamics of stretched DNA double helix. *J. Comput. & Theor. Nanosci.* **5**, 1381-1386.

Cheng, G.F, Zhao, J, Tu, Y.H, He, P.G. and Fang, Y.Z. (2005). Study on the interaction between antitumor drug daunomycin and DNA. *Chin. J. Chem.* **23**, 576-580.

Chen, L.W, Haushalter, K.A, Lieber, C.M. and Verdine, G.L. (2002). Direct visualization of a DNA glycosylase searching for damage. *Chem. & Biol.* **9**, 345-350.

Cheung, J.W.C. and Walker G.C. (2008). Immuno-Atomic Force Microscopy characterization of adsorbed fibronectin. *Langmuir* **24**, 13842-13849.

Chernoff, E.A.G. and Chernoff, D.A. (1992). Atomic force microscope image of collagen fibers. *J. Vac. Sci. & Technol. A* **10**, 596-599.

Chidsey, C.E.D, Loiacono, D.N, Sleator, T. and Nakahara, S. (1988). STM study of the surface morphology of gold on mica. *Surface Sci.* **200**, 45-66.

Chim, Y.T.A, Lam, J.K.W, Ma, Y, Armes, S.P, Lewis, A.L, Roberts, C.J, Stolnik, S, Tendler, S.J.B. and Davies, M.C. (2005). Structural study of DNA condensation induced by novel phosphorylcholine-based copolymers for gene delivery and relevance to DNA protection. *Langmuir* **21**, 3591-3598.

Chtcheglova, L.A, Haeberli, A. and Dietler, G. (2008). Force spectroscopy of the fibin(ogen)-Fibrinogen interaction. *Biopolymers* **89**, 292-301.

Cieplak, M. and Sulkowska, J.I. (2005). Thermal unfolding of proteins. *J. Chem. Phys.* **123**, article number 194908.

Clausen-Schaumann, H, Rief, M, Tolksdorf, C. and Gaub, H.E. (2000). Mechanical stability of single DNA molecules. *Biophys. J.* **78**, 1997-2007.

Clemmer, C.R. and Beebe, T.P. (1991). Graphite - a mimic for DNA and other biomolecules in scanning tunnelling microscopic studies. *Science* **251**, 640-642.

Clemmer, C.R. and Beebe, T.P. (1992). A review of graphite and gold surface studies for use as substrates in biological scanning tunnelling microscopy studies. *Scanning Microscopy* **6**, 319-333.

Cocco, S, Monasson, R. and Marko, J.F. (2002). Unzipping dynamics of long DNA. *Phys. Rev. E* **66**, article number 051914.

Cocco, S, Yan, J, Leger, J.F, Chatenay, D. and Marko, J.F. (2004). Overstretching and force-driven strand separation of double-helix DNA. *Phys. Rev. E* **70**, article number 011910.

Conroy, R.S. and Danilowicz, C. (2004). Unravelling DNA. *Contemporary Phys.* **45**, 277-302.

Cowman, M.K, Li, M. and Balazs, E.A. (1998a). Tapping mode atomic force microscopy of hyaluronan: extended and intramolecularly interacting chains. *Biophys. J.* **75**, 2030-2037.

Cowman, M.K, Lui, J, Li, M, Hittner, D.M. and Kim, J.S. (1998b). Hyaluronan interactions: self, water, ions. In *The chemistry, biology and medical applications of hyaluronan and its derivatives*, (ed. T. Laurent), pp.17-24. Portland Press, London.

Crampton, N, Thomson, N.H, Kirkham, J, Gibson, C.W. and Bonass, W.A. (2006a). Imaging RNA polymerase-amelogenin gene complexes with single molecule resolution using atomic force microscopy. *Eur. J. Oral Sci.* **114**, 133-138.

Crampton, N.C, Bonass, W.A, Kirkham, J, Rivetti, C. and Thomson, N.H. (2006b). Collison events between RNA polymerases in convergent transcription studied by atomic force microscopy. *Nucleic Acids Res.* **34**, 5416-5425.

Cuerrier, C.M, Lebel, R. and Grandbois, M. (2007). Single cell transfection using plasmid decorated AFM probes. *Biochem. & Biophys. Res. Commun.* **355**, 632-636.

Cui, Y. and Bustamante, C. (2000). Pulling a single chromatin fiber reveals the force that maintain its higher-order structure. *PNAS USA* **97**, 127-132.

Cui, S.X, Albrecht, C, Kuhner, F. and Gaub, H.E. (2006). Weakly bound water molecules shorten singl stranded DNA. *J. Amer. Chem. Soc.* **128**, 6636-6639.

Dame, R.T, Wyman, C. and Goosen, N. (2000). H-NS mediated compaction of DNA visualised by atomic force microscopy. *Nucleic Acids Res.* **28**, 3504-3510.

Dammer, U, Popescu, O, Wagner, P, Anselmetti, D, Guntherodt, H-J. and Misevic, G.N. (1995). Binding strength between cell adhesion proteoglycans measured by atomic force microscopy. *Science* **267**, 1173-1175.

Danilowicz, C, Coljee, V.W, Bouzigues, C, Lubensky, D.K, Nelson, D.R. and Prentiss, M, (2003) DNA unzipped under a constant force exhibits multiple metastable intermediates. *PNAS USA* **100**, 1694-1699.

Danielson, S, Maurstad, G. and Stokke, B.T. (2005). DNA-polycation complexation and polyplex stability in the presence of competing polyanions. *Biopolymers* **77**, 86-97.

Das, C, Hizume, K, Batta, K, Kumar, B.R.P, Gadad, S.S, Ganguly, S, Lorain, S, Verreault, A, Sadhale, P.P, Takeyasu, K. and Kundu, T.K. (2006). Transcriptional coactivator PC4, a chromatin-associated protein, induces chromatin condensation. *Mol. & Cellular Biol.* **26**, 8303-8315.

Davies, J, Roberts, C.J, Dawkes, A.C, Sefton, J, Edwards, J.C, Glasbey, T.O, Haymes, A.G, Davies, M.C, Jackson, D.E, Lomas, M, Shakessheff, K.M, Tendler, S.J.B, Wilkins, M.J. and Williams, P.M. (1994). Use of scanning probe microscopy and surface plasmon resonance as analytical tools in the study of antibody-coated microtiter wells. *Langmuir* **10**, 2654-2661.

Davies, L.M. and Harris, P.J. (2003). Atomic force microscopy of microfibrils in primary cell walls. *Planta* **217**, 283-289.

de Mier-Vinue, J, Lorenzo, J, Montana, A.M, Moreno, V. and Aviles, F.X. (2008). Synthesis, DNA interaction and cytotoxicity studies of cis-{[1,2-bis(aminomethyl)cyclohexane]dihalo} platinum(II) complexes. *J. Inorg. Biochemistry.* **102**, 973-987.

del Mercato, L.L, Maruccio, G, Pompa, P.P, Lochiocchio, B, Tamburro, A.M, Cingolani, R. and Rinaaldi, R. (2008). Amyloid-like fibrils in elastin-related polypeptides: structural characterisation and elastic properties. *Biomacromolecules* **9**, 796-903.

Dean, D, Han, L, Grodzinsky, A.J. and Ortiz, C. (2006). Compressive nanomechanics on opposing aggrecan macromolecules. *J. Biomechanics* **39**, 1555-2565.

Decho, A.W. (1999). Imaging an alginate gel matrix using atomic force microscopy. *Carbohydr. Res.* **315**, 330-333.

DeGrooth, B.G. and Putman, C.A.J. (1992). High resolution imaging of chromosome-related structures by atomic force microscopy. *J. Microscopy-Oxford* **168**, 239-247.

Dessinges, M.N, Maier, B, Zhang, Y, Peliiti, M, Bensimon, D. and Croquette, V. (2002). Stretching single stranded DNA, a model polyelectrolyte. *Phys. Rev. Letts,* **29**, article number 248102.

Dettmann, W, Grandhois, M, Andre, S, Benoit, M, Wehle, A,K, Kaltner, H, Gabius, H-J. and Gaub, H.E. (2000). Differences in zero-force and force-driven kinetics of ligand dissociation from β-galactoside-specific proteins (plant and animal lectins, immunoglobulin G) monitored by plasmon resonance and dynamic single molecule microscopy. *Archives Biochem. & Biophys.* **383**, 157-176.

Dietz, H. and Rief, M. (2004). Exploring the energy landscape of GFP by single-molecule mechanical experiments. *PNAS USA* **101**, 16192-16197.

Dietz, H, Bornschloegl, T, Heym, R, Konig, F. and Rief, M. (2007). Programming protein self assembly with coiled coils. *New J. Phys.* **9** article number 424.

Ding, S.Y. and Himmel, M.E. (2006). The maize primary cell wall microfibril: A new model derived from direct observation. *J. Agric. & Food Chem.* **54**, 597-606.

Donald, A.M, Kato, K.L, Perry, P.A. and Waigh, T.A. (2001). Scattering studies of the internal structure of starch granules. *Stärke* **53**, 504-512.

Doyen, C-M, Montel, F, Gautier, T, Menoni, H, Claudet, C, Delacour-Larose, M, Angelov, D, Hamiche, A, Bednar, J, Faivre-Moskalenko, C, Bouvet, P. and Dimitrov, S. (2006). *EMBO J.* **25**, 4234-4244.

Drake, B, Prater, C.B, Weisenhorn, A.L, Gould, S.A.C, Albrecht, T.R, Quate, C.M, Cannel, D.S, Hansma, H.G. and Hansma, P.K. (1989). Imaging crystals, polymers, and processes in water with the atomic force microscope. *Science* **243**, 1586-1589.

Ebner, A, Wildling, L, Kamruzzahan, A.S.M, Rankl, C, Wruss, J, Hahn, C.D, Holzl, M, Zhu, R, Kienberger, F, Blass, D, Hinterdorfer, P. and Gruber, H.J. (2007). A new simple method for linking of antibodies to atomic force microscopy tips. *Bioconjugate J.* **18**, 1176-1184.

Eckel, R, Ros, R, Ros, A, Wilking, S.D, Sewald, N. and Anselmetti, D. (2003). Identification of binding mechanisms in single molecule-DNA complexes. *Biophys. J.* **85**, 1968-1973.

Ellis, J.S, Abdelhady, H.G, Allen, S, Davies, M.C, Roberts, C.J, Tendler, S.J.B. and Williams, P.M. (2004). Direct atomic force microscopy observations of monovalent ion induced binding of DNA to mica. *J. Microscopy-Oxford* **215**, 297-301.

Engel, A. (1991). Biological applications of scanning probe microscopes. *Ann. Rev. Biophys. Biophys. Chem.* **20**, 79-108.

Erie, D.A, Yang, G, Schultz, H.C. and Bustamante, C. (1994). DNA bending by Cro protein in specific and non-specific complexes - implications for protein site recognition and specificity. *Science* **266**, 1562-1566.

Fahlen, J. and Salmen, L. (2002). On the lamellar structure of the trachoid cell wall. *Plant Biol.* **4**, 339-345.

Fang, Y. and Hoh, J.H. (1998). Surface-directed DNA condensation in the absence of soluble multivalent cations. *Nucleic Acids Res.* **26**, 588-593.

Fang, Y, Spisz, T.S, Wiltshire, T, D'Costa, N.P, Bankman, I.N, Reeves, R.H. and Hoh, J.H. (1998). Solid-state DNA sizing by atomic force microscopy. *Anal. Chem.* **70**, 2123-2129.

Fannon, J.E, Hauber, R.J. and BeMiller, J.N. (1992). Surface pores of starch granules. *Cereal Chem.* **69**, 284-288.

Fava, J, Alzamore, S.M. and Castro, M.A. (2006). Structure and nanostructure of the outer tangential epidermal cell wall in *Vaccinium corymbosum* L. (Blueberry) fruits by blanching, freezing and ultrasound. *Food Sci. Technol. International* **12**, 241-251.

Ficarra, E, Benini, L, Macii, E. and Zucchen, G. (2005a). Automated DNA fragments recognition and sizing through AFM image processing. *IEEE Trans. Information Technol. Biomed.* **9**, 508-517.

Ficarra, E, Masotti, D, Macii, E, Benini, L, Zuccheri, G. and Samori, B. (2005b). Automatic intrinsic DNA curvature computations from AFM images. *IEEE Trans Biomed. Eng.* **52**, 2074-2086.

Finch, J.T. and Klug, A. (1976). Solenoidal model for superstructure in chromatin. *PNAS USA* **73**, 1897-1901.

Firtel, M. and Beveridge, T.J. (1995). Scanning probe microscopy in microbiology. *Micron.* **26**, 347-362.

Fishman, M, Cooke, P.H, Chau, H.K, Coffin, D.R. and Hotchkiss, A.T. (2007). Global structures of high methoxyl pectin from solution and in gels. *Biomacromolecules* **8** 573-578.

Fologea, D. (2005). Detecting single stranded DNA with a solid state nanopore. *Nano Letts.* **5**, 1905-1909.

Forman, J.R. and Clarke, J. (2007). Mechanical unfolding of proteins: insights into biology, structure and folding. *Current Opinion Struct. Biol.* **17**, 58-66.

Frank, B.P. and Belfort, G. (1997). Intermolecular forces between extracellular polysaccharides measured using the atomic force microscope. *Langmuir* **13**, 6234-6240.

Frazier, R.C, Davies, M.C, Matthijs, G, Roberts, C.J, Schacht, E, Tendler, S.J.B. and Williams, P.M. (1997a). High-resolution atomic force microscopy of dextran monolayer hydration. *Langmuir* **13**, 4795-4798.

Frazier, R.A, Davies, M.C, Matthijs, G, Roberts, C.J, Schacht, E, Tendler, S.J.B. and Williams, P.M. (1997b). *In situ* surface plasmon resonance analysis of dextran monolayer degradation by dextranase. *Langmuir* **13**, 7115-7120.

Fritz, J, Anselmetti, D, Jarchow, J. and Fernandez-Busquets, X. (1997). Probing biomolecules with atomic force microscopy. *J. Struct. Biol.* **119**, 165-171.

Fritz, M, Radmacher, M, Cleveland, J.P, Allersma, M.W, Stewart, R.J, Gieselmann, R, Janmey, P. Schmidt, C.F. and Hansma, P.K. (1995). Imaging globular and filamentous proteins in physiological buffer solutions with tapping mode atomic-force microscopy. *Langmuir* **11**, 3529-3535.

Fritzsche, W, Schaper, A. and Jovin, T.M. (1994). Probing chromatin with the scanning force microscope. *Chromosoma* **103**, 231-236.

Fritzsche, W. and Henderson, E. (1996a). Scanning force microscopy revealed ellipsoid shape of chicken erythrocyte nucleosomes. *Biophys. J.* **71**, 2222-2226.

Fritzsche, W. and Henderson, E. (1996b). Ultrastructural characterisation of chicken erythrocyte nucleosomes by scanning force microscopy. *Scanning* **18**, 138-139.

Fritzsche, W. and Henderson, E (1996c). Volume determination of human metaphase chromosomes scanning force microscopy. *Scanning Microscopy* **10**, 103-110.

Fritzsche, W, Martin, L, Dobbs, D, Jondle, D, Miller, R, Vesenka, J. and Henderson, E. (1996). Reconstruction of ribosomal subunits and rDNA chromatin by scanning force microscopy. *J. Vac. Sci. & Technol. B* **14**, 1405-1409.

Fritzsche, W, Takac, L. and Henderson, E. (1997). Application of atomic force microscopy to visualisation of DNA, chromatin, and chromosomes, *Critical Rev. Eukaryotic Gene Expression* **7**, 2311-240.

Fujita, Y, Kobayashi, K. and Hoshino, T. (1997). Atomic force microscopy of collagen molecules. Surface morphology of segment-long-spacing (SLS) crystallites of collagen. *J. Electron Microscopy* **46**, 321-326.

Funami, T, Hiroe, M, Noda, S, Asai, I, Ikeda, S. and Nishinari, K. (2007). Influence of molecular structure imaged with atomic force microscopy on the rheological behavior of carrageenan aqueous systems in the presence and absence of cations. *Food Hydrocolloids* **21**, 617-629.

Gaczynska, M, Osmulski, P.A, Jiang, Y, Lee, J-K, Bermudez, V. and Hurwitz, J. (2004). Atomic force microscopic analysis of the binding of the *Schizosaccharomyces pombe* origin recognition complex and the spOrc4 protein with origin DNA. *PNAS USA* **101**, 17952-17957.

Gale, M, Pollansen, M.S, Markiewicz, P. and Goh, M.C. (1995). Sequential assembly of collagen revealed by atomic force microscopy. *Biophys. J.* **68**, 2124-2128.

Gallant, D.J, Bouchet, B. and Baldwin, P.M. (1997). Microscopy of starch: Evidence of a new level of granule organization. *Carbohydr. Polym.* **32**, 177-191.

Gallyamov, M.Q, Tartsch, B, Khoklov, A.R, Sheiko, S.S, Borner, H.G, Matyjaszewski, K. and Moller, M. (2004a). Real-time scanning force microscopy of a macromolecular conformational transition. *Macromol. Rapid Commun.* **25**, 1703-1707.

Gallyamov, M.Q, Tartsch, B, Khoklov, A.R, Sheiko, S.S, Borner, H.G, Matyjaszewski, K. and Moller, M. (2004b). Conformational dynamics of single molecules visualized in real time by scanning force microscopy: macromolecular mobility on a substrate surface in different vapours. *J. Microscopy-Oxford* **215**, 245-256.

Giannotti, M.I. and Vancso, G.J. (2007). Interrogation of single synthetic polymer chains and polysaccharides by AFM-based force spectroscopy. *CHEMPHYSCHEM* **8**, 2290-2307.

Giardina, T, Gunning, A.P, Faulds, C.B, Juge, N, Furniss, C.S.M, Svensson, B, Morris, V.J. and Williamson, G. (2001). Influence of the two binding sites of the starch - binding domain of *Aspergillus niger* on amylose conformation. *J. Mol. Biol.* **313**, 1151-1161.

Giro, A, Bergia, A, Zuccheri, G, Bink, H.H.J, Pleij, C.W.A. and Samori, B. (2004). Single molecule studies of RNA secondary structure: AFM of TYMV viral RNA. *Microscopy Res. & Techniques* **65**, 235-245.

Golan, R, Pietrasanta, L.I, Hsieh, W. and Hansma, H.G. (1999). DNA toroids: stages in condensation. *Biochemistry.* **38**, 14069-14076.

Goldsbury, C, Kistler, J, Aebi, U, Arvinte, T. and Cooper, G.J.S. (1999). Watching amyloid fibrils grow by time lapse atomic force microscopy. *J. Mol. Biol.* **285**, 33-39.

Goldsbury, C, Frey, P, Olivieri, V, Aebi, U. and Muller, S.A. (2005). Multiple assembly pathways underlie amyloid-beta fibril polymorphisms. *J. Mol. Biol.* **352**, 282-298.

Gosal, W.S, Clark, A.H. and Ross-Murphy, S.B. (2004). Fibrillar β-lactoglobulin gels: part 1. Fibril formation and structure. *Biomacromolecules* **5**, 2408-2419.

Gosal, W.S, Myers, S.L, Radford, S.E. and Thomson, N.H. (2006). Amyloid under the atomic force microscope. *Protein & Peptide Letts.* **13**, 261-270.

Green, J, Goldbury, C, Min, T, Sunderji, S, Frey, P, Kistler, J, Cooper, G. and Aebi, U. (2003). Full-length rat amylin forms fibrils following substitution of single residues from human amylin. *J. Mol. Biol.* **326**, 1147-1156.

Greenlife, W.J, Woodside, M.T. and Block, S.M. (2007). High-resolution, single molecule measurements of biomolecular motion. *Ann. Rev. Biophys. & Biomol. Struct.* **36**, 171-190.

Guazzaroni, M.E, Krell, T, del Arroyo, P.G, Velez, M, Jimenez, M, Rivas, G. and Ramos, A. (2007). The transcriptional repressor TtgV recognizes a complex operator as a tetramer and induced convex DNA bending. *J. Mol. Biol.* **369**, 927-939.

Guillot-Nieckowski, M, Joester, D, Stohr, M, Losson, M, Adrian, M, Wagner, B, Kansay, M, Heinzelmann, H, Pugin, R, Diederich, F. and Gallani, J.L. *Langmuir* **23**, 737-746.

Gunning, A.P, Kirby, A.R, Morris, V.J, Wells, B. and Brooker, B.E. (1995). Imaging bacterial polysaccharides by AFM. *Polymer Bull.* **34**, 615-619.

Gunning, A.P, Kirby, A.R. and Morris, V.J. (1996a). Imaging xanthan gum in air by ac "tapping" mode atomic force microscopy. *Ultramicroscopy* **63**, 1-3.

Gunning, A.P, Kirby, A.R, Ridout, M.J, Brownsey, G.J. and Morris, V.J. (1996b). Investigation of gellan networks and gels by atomic force microscopy. *Macromolecules* **29**, 6791-6796.

Gunning, A.P, Morris, V.J, Al-Assaf, S. and Phillips, G.O. (1996c). Atomic force microscopic studies of hylan and hyaluronan. *Carbohydr. Polym.* **30**, 1-8.

Gunning, A.P, Kirby, A.R, Ridout, M.J, Brownsey, G.J. and Morris, V.J. (1997). Investigation of gellan networks and gels by atomic force microscopy (erratum). *Macromolecules* **30**, 163-164.

Gunning, A.P, Cairns, P, Kirby, A.R, Bixler, H.J. and Morris, V.J. (1998). Characterising semi-refined iota-carrageenan networks by atomic force microscopy. *Carbohydr. Polym.*36, 67-72.

Gunning, A.P, Giardina, T.P, Faulds, C.B, Juge, N, Ring, S.G, Williamson, G and Morris, V.J. (2003). Surfactant mediated solubilisation of amylose and visualisation by atomic force microscopy. *Carbohydr. Polym.* **51** 177-182.

Gunning A. P., Chambers S., Pin Arias C., Man A. L., Morris V. J. and Nicoletti C. (2008). Mapping specific adhesive interactions on living human intestinal epithelial cells with atomic force microscopy. *FASEB J.* **22** 2331-2339.

Gunning, A.P, Bongaerts, R.J.M. and Morris, V.J. (2009). Recognition of galactan components of pectin by galectin-3. *FASEB J.* **23**, 415-424.

Guo, C.L, Song, Y.H, Wang, L, Sun, L.L, Sun, Y.J, Peng, C.Y, Liu, Z.L, Yang, T. and Li, Z. (2008). Atomic force microscopic study of low temperature induced disassembly of RecA-dsDNA filaments. *J. Phys. Chem. B* **112**, 1022-1027.

Guthold, M, Bezanilla, M, Erie, D.A, Jenkins, B, Hansma, H.G. and Bustamante, C. (1994). Following the assembly of RNA polymerase-DNA complexes in aqueous solution with the scanning force microscope. *PNAS USA* **91**, 12927-12931.

Gutsmann, T, Fantner, G.E, Venturnoni, M, Ekani-Nkodo, A, Thompson, J.B, Kindt, J.H, Morse, D.E, Fygenson, D.K. and Hansma, P.K. (2003). Evidence that collagen fibrils in tendons are inhomogeneously structured in a tubelike manner. *Biophys. J.* **84**, 2593-2598.

Gutsmann, T, Fantner, G.E, Kindt, J.H, Venturnoni, M, Danielson, S. and Hansma, P.K. (2004). Force spectroscopy of collagen fibers to investigate their mechanical properties and structural organisation. *Biophys. J.* **86**, 3186-3193.

Guzman, C, Jeney, S, Kreplak, L, Kasas, S, Kulik, A.J, Aebi, U. and Forro, L. (2006). Exploring the mechanical properties of single vimentin intermediate filaments by atomic force microscopy. *J. Mol. Biol.* **360**, 623-630.

Habelitz, S, Balooch, M, Marshall, S.J, Balooch, G. and Marshall, G.W. (2002). *In situ* atomic force microscopy of partially demineralised human dentin collagen fibrils. *J. Struct. Biol.* **138**, 227-236.

Hagerman, P.J. (1988). Flexibility of DNA. *Ann. Rev. Biophys. Chem.* **17**, 265-286.

Hallett, P, Offer, G. and Miles, M.J. (1995). Atomic force microscopy of the myosin molecule. *Biophys. J.* **68**, 1604-1606.

Hallett, P, Tskhovrebova, L, Trinick, J, Offer, G. and Miles, M.J. (1996). Improvements in atomic force microscopy protocols for imaging fibrous proteins. *J. Vac. Sci. & Technol. B* **14**, 1444-1448.

Hamada, D, Tsumoto, K, Sawara, M, Tanaka, N, Nakahira, K, Shiraki, K. and Yanagihara, I. (2008). *Proteins* **72**, 811-821.

Hamon, L, Pastre, D, Dupaigne, P, Le Breton, C, Le Cam, E. and Pietrement, O. (2007). High-resolution AFM imaging of single-stranded DNA-binding protein-DNA complexes. *Nucleic Acids Res.* **35**, No. 8, e58 doi:10.1093/nar/gkm147

Han, L, Dean, D, Ortiz, C. and Grodzinsky, A.J. (2007a). Lateral nanomechanics of cartilage aggrecan macromolecules. *Biophys. J.* **92**, 1384-1398.

Han, L, Dean, D, Mao, P, Ortiz, C. and Grodzinsky, A.J. (2007b). Nanoscale shear deformation mechanisms of opposing cartilage aggrecan macromolecules. *Biophys. J.* **93**, L23-L25.

Han, S.W, Nakamura, C, Kotobuki, N, Obataya, I, Ohgushi, H, Nagamune, T. and Miyake, J. (2008). High-efficiencyu DNA injection into a single human mesenchymal stem cell using a nanoneedle and atomic force microscopy. *Nanomedicine – Nanotechnol. Biol. & Medicine* **4**, 215-225.

Han, W, Mou, J, Sheng, J, Yang, J. and Shao, Z. (1995). Cryo atomic force microscopy: a new approach for biological imaging at high resolution. *Biochemistry* **34**, 8215-8220.

Hanley, S.J, Giasson, J, Revol, J-F. and Gray, D (1992). Atomic force microscopy of cellulose microfibrils; comparison with transmission electron microscopy. *Polymer* **33**, 4639-4642.

Hanley, S.J, Revol, J-F, Godbout, L. and Gray, D.G. (1997). Atomic force microscopy and transmission electron microscopy of cellulose from *Micrasterias denticulata*; evidence for a chiral helical microfibril twist. *Cellulose* **4**, 209-220.

Hansma, H.G, Sinsheimer, R L, Li, M.Q. and Hansma, P.K. (1992a). Atomic force microscopy of single and double stranded DNA. *Nucleic Acids Res.* **20**, 3585-3590.

Hansma, H.G, Vesenka, J, Siegerist, C, Kelderman, G, Morrett, H, Sinsheimer, R.L, Elings, V, Bustamante, C. and Hansma, P.K. (1992b). Reproducible imaging and dissection of plasmid DNA under liquid with the atomic force microscope. *Science* **256**, 1180-1184.

Hansma, H.G. and Hansma, P.K. (1993) Potential applications of atomic force microscopy of DNA to the human genome project. *Proc. SPIE Int. Opt. Eng. (USA)* **1891**, 66-70.

Hansma, H.G, Bezanilli, M, Zenhausern, F, Adrian, M. and Sinsheimer, R.L. (1993a). Atomic force microscopy of DNA in aqueous solutions. *Nucleic Acids Res.* **21**, 505-512.

Hansma, H.G, Sinsheimer, R.L, Groppe, J, Bruice, T.C, Elings, V, Gurley, G, Bezanilla, M, Mastrangelo, I.A, Howe, P.V.C. and Hansma, P.K. (1993b). Recent advances in atomic force microscopy of DNA. *Scanning* **15**, 296-299.

Hansma, H.G. and Hoh, J.H. (1994). Biomolecular imaging with the atomic force microscope. *Ann. Rev. Biomed. Struct.* **23**, 115-139.

Hansma, H.G, Brown, K.A, Bezanilla, M. and Bruice, T.C. (1994). Bending and straightening of DNA induced by the same ligand: characterisation with the atomic force microscope. *Biochemistry* **33**, 8436-8441.

Hansma, H.G, Laney, D.E, Bezanilla, M, Sinsheimer, R. L. and Hansma, P.K. (1995). Applications for atomic force microscopy of DNA. *Biophys. J.* **68**, 1672-1677.

Hansma, H.G. (1996). Atomic force microscopy of biomolecules. *J. Vac. Sci. & Technol. B* **14**, 1390-1394.

Hansma, H.G. and Laney, D.E. (1996). DNA binding to mica correlates with cationic radius: assay by atomic force microscopy. *Biophys. J.* **70**, 1933-1939.

Hansma, H.G, Revenko, I, Kim, K. and Laney, D.E. (1996). Atomic force microscopy of long short double-stranded, single stranded and triple stranded nucleic acids. *Nucleic Acids Res*. **24**, 713-720.

Hansma, H.G, Kim, K.J, Laney, D.E, Garcia, R.A, Argaman, M, Allen, M.J. and Parsons, S.M. (1997). Properties of biomolecules measured from atomic force microscopic images: a review. *J. Struct. Biol*. **119**, 99-108.

Hansma, H.G, Golan, R, Hsieh, W, Lollo, C.P, Mullen-Ley, P. and Kwoh, D. (1998). DNA condensation for gene therapy as monitored by atomic force microscopy. *Nucleic Acids Res*. **26**, 2481-2487.

Harada, A, Yamaguchi, H. and Kamachi, M. (1997). Imaging antibody molecules at room temperature by contact mode atomic force microscope. *Chem. Letts*. part 11, 1141-1142.

Hansma, H.G, Golan, R, Hsieh, W, Daubendiek, S.L. and Kool, E.T. (1999). Polymerase activities and RNA structures in the atomic force microscope. *J. Struct. Biol*. **127**, 240-247.

Hansma, H.G, Oroudjev, E, Baudrey, S. and Jaeger, L. (2003). TectoRNA and 'kissing-loop' RNA: atomic force microscopy of self-assembling RNA structures. *J. Microscopy-Oxford* **212**, 273-279.

Hansma, H.G, Kasuya, K. and Oroudjev, E. (2004). Atomic force microscopy imaging and pulling of nucleic acids. *Current Opinion Struct. Biol*. **14**, 380-385.

Harris, S.A. (2004). The physics of DNA stretching. *Contemporary Phys*, **45**, 11-30.

Haugstad, G, Gladfelter, W.L, Keyes, M.P. and Weberg, E.B. Atomic force microscopy of AgBr crystals and adsorbed gelatin films. (1993). *Langmuir* **9**, 1594-1600.

Heckl, W.M. (1992). Scanning tunneling microscopy and atomic force microscopy on organic and biomolecules. *Thin Solid Films* **210**, 640-647.

Heddle, J.G, Mitelheiser, S, Maxwell, A. and Thomson, N.H. (2004). Nucleotide binding to DNA gyrase causes loss of DNA wrap. *J. Mol. Biol*. **337**, 597-619.

Heinemann, S, Ehrlich, H, Heinemann, C, Worch, H, Schatton, W. and Hanke, T. (2007). Ultrastructural studies on the collagen of the marine sponge *Chondrosia reniformis* nardo. *Biomacromolecules* **8**, 3452-3457.

Henderson, E. (1992). Imaging and nanodissection of individual supercoiled plasmids by atomic force microscopy. *Nucleic Acids Res*. **20**, 445-447.

Hertadi, R. and Ikai, A. (2002). Unfolding of holo- and apocalmodulin studied by the atomic force microscope. *Protein Sci*. **11**, 1532-1538.

Heslop-Harrison, J.S, Leitch, A.R, Schwarzacher, T, Smith, J.B, Atkinson, M.D. and Bennett, M.D. (1989). The volume and morphology of human chromosomes in mitotic reconstructions. *Human Genet*. **84**, 27-34.

Hirano, Y, Takahashi, H, Kumeta, M, Hizume, K, Hirrai, Y, Otsuka, S, Yoshimura, S.H. and Takeyasu, K. (2008). Nuclear architecture and chromatin dynamics revealed by atomic force microscopy in combination with biochemistry and cell biology. *Pflugers Arch –Eur. J. Physiol*. **456**, 139-153.

Hoshi, O, Owen, R, Miles, M.J. and Ushiki, T. (2004). Imaging of human metaphase chromosomes by atomic force microscopy in liquid. *Cytogenetic & Genome Res*. **107**, 28-31.

Hoshi, O, Shigeno, M. and Ushiki, T. (2006). Atomic force microscopy of native human metaphase chromosomes in a liquid. *Arch. Histol. Cytol*. **69**, 73-78.

Higgins, M.J, Sader, J.E, Mulvaney, P. and Wetherbee, R. (2003). Probing the surface of living diatoms with atomic force microscopy: the nanostructure and nanomechanical properties of the mucilage layer. *J. Phycology* **39**, 722-734.

Higgins, M.J, Sader, J.E. and Jarvis, S.P. (2006). Frequency modulation atomic force microscopy reveals individual intermediates associated with each unfolded I27 titin domain. *Biophys. J*. **90**, 640-647.

Hong, Z.N, Chasan, B, Bansil, R, Turner, B.S, Bhaskar, K.R. and Afdhal, N.H. (2005). Atomic force microscopy reveals aggregation of gastric mucin at low pH. *Biomacromolecules* **6**, 3458-3466.

Holschen, C, Bussiek, M, Langowski, J. and Diekmann, S. (2008). *Escherichia coli* low-copy number plasmid R1 centromere *parC* forms a U-shaped complex with its binding protein ParR. *Nucleic Acids Res*. **36**, 607-615.

Hou, S, Yang, K, Yao, Y, Liu, Z, Feng, X.Z, Wang, R, Yang, Y.L. and Wang, C. (2008). DNA condensation induced by a cationic polymer studied by atomic force microscopy and electrophoresis assay. *Colloids & Surfaces B-Biointerfaces* **62**, 151-156.

Hoyer, W.G, Cheny, D, Subramaniam, V. and Jovin, T.M. (2004). Rapid self-assembly of α-synuclein observed by *in situ* atomic force microscopy. *J. Mol. Biol.* **340, 127-130**.

Humphris, A.D.L, Tamayo, J. and Miles, M.J. (2000a). Active quality factor control in liquids for force spectroscopy. *Langmuir* **16**, 7891-7894.

Humphris, A.D.L, McMaster, T.J, Miles, M.J, Gilbert, S.M, Shewry, P.R. and Tatham, A.S. (2000b). Atomic force microscopy (AFM) study of interactions of HMW subunits of wheat glutenin. *Cereal Chem.* **77**, 107-110.

Humphris, A.D.L, Round, A.N. and Miles, M.J. (2001). Enhanced imaging of DNA via active quality factor control. *Surface Sci.* **491**, 468-472.

Husale, S, Grange, W. and Hegner, M. (2002). DNA mechanics affected by small DNA interacting ligands. *Single Molecules* **3**, 91-96.

Husale, S, Grange, W, Karle, M, Burgi, S. and Hegner, M. (2008). Interaction of cationic surfactants with DNA: a single-molecule study. *Nucleic Acids Res.* **36**, 1443-1449.

Iijima, M, Yoshimura, M, Tsuchiya, T, Tsukada, M, Ichikawa, H, Fuskumori, Y. and Kamiya, H. (2008). Direct measurement of interactions between stimulation-responsive drug delivery vehicles and artificial mucin layers by colloid probe microscopy. *Langmuir* **24**, 3987-3992.

Ikai, A. (1996). STM and AFM of bio/organic molecules and structures. *Surface Sci. Reports* **26**, 261-332.

Ikai, A, Afrin, R. and Sekiguchi, H. (2007). Pulling and pushing protein molecules by AFM. *Current Nanosci.* **3**, 17-29.

Ikeda, S. (20030. Heat-induced gelation of whey proteins observed by rheology, atomic force microscopy, and Raman scattering spectroscopy. *Food Hydrocolloids* **17**, 399-406.

Ikeda, S. and Morris, V.J. (2002). Fine-stranded and particulate aggregates of heat-denatured whey proteins visualized by atomic force microscopy. (2002). *Biomacromolecules* **3**, 382-389.

Ikeda, S. and Shishido, Y. (2005). Atomic force microscopy studies on heat-induced gelation of curdlan. *Agric. Food Chem.* **53**, 786-791.

Ikeda, S, Nitta, Y, Kim, B.S, Temsiripong, T, Pongsawatmananit, B. and Nishinari, K. (2004a). Single-phase mixed gels of xyloglucan and gellan. *Food Hydrocolloids* **18**, 669-675.

Ikeda, S, Nitta, Y, Temsinpong, T, Pongsawatmanit, R. and Nishinari, K. (2004b). Atomic force microscopy studies on cation-induced network formation of gellan. *Food Hydrocolloids* **18**, 727-735.

Ikeda, S, Funami, T. and Zhang, G.Y. (2005). Visualizing surface active hydrocolloids by atomic force microscopy. *Carbohydr. Polym.* **62**, 192-196.

Ill, C.R, Kievens, V.M, Hale, J.E, Nakamura, K.K, Jue, R.A, Cheng, S, Melcher, E.D, Drake, B. and Smith, M.D. (1993). A COOH-terminal peptide confers regiospecific orientation and facilitates atomic force microscopy of an IgG1. *Biophys. J.* **64**, 919-924.

Ishii, T, Murayama, Y, Katano, A, Maki, K, Kuwajima, K. and Sano, M. (2008). Probing force-induced unfolding intermediates of a single staphylococcal nuclease molecule and the effect of ligand binding. *Biochem & Biophys. Res. Commun.* **375**, 586-501.

Iwabuchi, S, Mori, T, Ogawa, K, Sato, K, Saito, M, Morita, Y, Ushiki, T. and Tamiya, E. (2002). Stomic force microscope-based dissection of human chromosomes and high resolutional imaging by carbon nanotube tips. *Arch. Histol. Cytol.* **65**, 473-479.

Iwataki, T, Kidoaki, S, Sakaue, T, Yoshikawa, K. and Abramchuk, S.S. (2004). Competition between compaction of single chains and bundling of multiple chains in giant DNA molecules. *J. Chem. Phys.* **120**, 4004-4011.

Janovjak, H, Kessler, M, Oesterhelt, D, Gaub, H. and Muller, D.J. (2003). Unfolding pathways of native bacteriorhodopsin depend on temperature. *EMBO J.* **22**, 5220-5229.

Jansen, R, Dzwolak, W. and Winter, R. (2005). Amyloidogenic self-assembly of insulin aggregates probed by high resolution atomic force microscopy. *Biophys. J.* **88**, 1344-1353.

Janus, T. and Yarus, M. (2003). Visualization of membrane RNAs. *RNA-A Publication of the RNA Society* **9**, 1353-1361.

Jeffrey, A.M, Jing, T.W, DeRose, J.A, Vaught, A, Rekesh, D, Lu, F-X. and Lindsay, S.M. (1993). *Nucleic Acids Res.* **21**, 5896-5900.

Jenko, S, Skarabot, M, Kenig, M, Guncar, G, Musevic, I, Turk, D. and Zerovnik, E. (2004). Different propensity to form amyloid fibrils by two homologous proteins-human stefins A and B: searching for an explanation. *Proteins* **55**, 417-425.

Jia, Y.X, Bi, L.J, Li, F, Chen, Y.Y, Zhang, C.G. and Zhang, X.E. (2008). Alpha-shaped DNA loops induced by MutS. *Biochem. & Biophys. Res. Commun.* **372**, 518-622.

Jiang, Y, Qin, F, Li, Y, Fang, X. and Bai, C. (2004). Measuring specific interaction of transcription factor ZmDREB1A with its DNA responsive element at the molecular level. *Nucleic Acids Res.* **32**, e101, doi:10.1093/nar/gnh100

Jiang, Y, Ke, C.H, Mieczkowski, P.A. and Marszalek, P.E. (2007). Detecting ultraviolet damage in single DNA molecules by atomic force microscopy. *Biophys. J.* **93**, 1758-1767.

Jiao, Y.K, Cherny, D.I, Heim, G, Jovin, T.M. and Schaffer, T.E. (2001). Dynamic interactions of p53 with DNA in solution by time-lapse atomic force microscopy. *J. Mol. Biol.* **314**, 233-243.

Jin, Y, Zhang, H, Yin, Y. and Nishinari, K. (2006). Comparison of curdlan and its carboymethylated derivative by means of rheology, DSC, and AFM. *Carbohydr. Res.* **341**, 90-99.

Jondle, D.M, Ambrosio, L, Vesenka, J. and Henderson, E (1995). Imaging and manipulating chromosomes with the atomic force microscope. *Chromosome Res.* **3**, 239-244.

Jørgensen, T.E, Sietmoen, M, Draget, K.I. and Stokke, B.T. (2007). Influence of oligoguluronates on alginate gelation, kinetics, and polymer organisation. *Biomacromolecules* **8**, 2388-2397.

Kaji, N, Ueda, M. and Baba, Y. (2001). Direct measurement of conformational changes on DNA molecule intercalating with a fluorescence dye in an electrophoretic buffer solution by means of atomic force microscopy. *Electrophoresis* **22**, 3357-3364.

Kalle, W.H.J, Macville, M.V.E, van der Corput, M.P.C, de Grooth, B.G, Tanke, H.J. and Raap, A.K. (1996). Imaging of RNA *in situ* hybridization by atomic force microscopy. *J. Microscopy-Oxford* **182**, 192-199.

Kaminsky, S.G.W, Susan, G.W. and Dahms, T.E.S. (2008). High spatial resolution surface imaging and analysis of fungal cells using SEM and AFM. *Micron* **39**, 349-361.

Kamruzzahan, A.S.M, Ebner, A, Wildling, L, Kienberger, F, Riener, C.K, Hahn, C.D, Pollheimer, P.D, Winklehner, P, Holzl, M, Lackner, B, Schorkl, D.M, Hinterdorfer, P. and Gruber, H.J. (2006). Antibody linking to atomic force microscope tips via disulfide bond formation. *Bioconjugate J.* **17**, 1473-1481.

Karrasch, S, Hegerl, R, Hoh, J.H, Baumeister, W. and Engel, A. (1994). Atomic force microscopy produces faithful high-resolution images of protein surfaces in an aqueous environment. *PNAS USA* **91**, 836-838.

Kasas, S, Thomson, N.H, Smith, B.L, Hansma, P.K, Miklossy, J. and Hansma, H.G (1997a). Biological applications of the AFM: from single molecules to organs. *Int. J. Imaging Syst. Technol.* **8**, 151-161.

Kasas, S, Thomson, N.H, Smith, B.L, Hansma, H.G, Zhu, X, Guthold, M, Bustamante, C, Kool, E.T, Kashlev, M. and Hansma, P.K. (1997b). *Escherichia coli* RNA polymerase activity observed using atomic force microscopy. *Biochemistry* **36**, 461-468.

Kasianowicz, J.J, Brandin, E, Branton, D. and Deamer, D.W. (1996). Characterization of individual polynucleotide molecules using membrane channels. *PNAS USA* **93**, 13770-13773.

Kaur, A.P. and Wilks, A. (2007). Heme inhibits the DNA binding properties of the cytoplasmic heme binding protein of *Shigella dysenteriae* (ShuS). *Biochemistry* **46**, 2994-3000.

Kawakami, M, Byrne, K, Khatri, B, McLeish, T.C., Radford, S.E. and Smith, D.A. (2004). Viscoelastic properties of single polysaccharide molecules determined by analysis of thermally driven oscillations of an atomic force microscope cantilever. *Langmuir* **20**, 9299-9303.

Kawakami, M, Byrene, K, Khatri, B.S, McLeish, T.C.B, Radford, S.E. and Smith, D.A. (2005). Viscoelastic measurements of single molecules on a millisecond time scale by magnetically driven oscillation of an atomic force microscope cantilever. *Langmuir* **21**, 4765-4772.

Kayed, R, Bernhagen, J, Greenfield, N, Sweimeh, K, Brunner, H, Voelter, W. and Kapurniotu, A. (1999). Conformational transitions of islet amyloid polypeptide (IAPP) in amyloid formation *in vitro*. *J. Mol. Biol.* **287**, 781-796.

Ke, C, Humeniuk, M, S-Gracz, H. and Marszalek, P.E. (2007). Direct measurements of base stacking interactions in DNA by single-molecule atomic force spectroscopy. *Phys. Rev. Letts.* **99**, article number 018302.

Kellermayer, M.S.Z, SWmith, S.B, Granzier, H.L. and Bustamante, C. (1997). Folding-unfolding transitions in single titan molecules characterized with laser tweezers. *Science* **276**, 1112-1116.

Khurana, R, Souilliac, P.O, Coats, A.C, Minert, L, Ionescu-Zanetti, D, Carter, S.A, Solomon, A. and Fink, A.L. (2003). A model for amyloid fibril formation in immunoglobulin light chains based on comparison of amyloidogenic and benign proteins and specific antibody binding. *Amyloid J. Protein Folding Disorders* **10**, 97-109.

Keinberger, F, Mueller, H, Pastushenko, V. and Hinterdorfer, P. (2004). Following single antibody binding to purple membranes in real time. *EMBO Reports* **5**, 579-583.

Kim, D.T, Blanch, H.W. and Radke, C.J. (2002). Direct imaging of lysozyme adsorption onto mica by atomic force microscopy. *Langmuir* **18**, 5841-5850.

Kim, J.M, Jung, H.S, Park, J.W, Lee, H.Y. and Kawai, T. (2004). AFM phase lag mapping for protein DNA oligonucleotide complexes. *Anal. Chim. Acta* **525**, 151-157.

Kimura, E, Hitomi, J. and Ushiki, T. (2002). Scanning near field optical/atomic force microscopy of bromodeoxyuridine-incorporated human chromosomes. *Arch. Histol. Cytol.* **65**, 435-444.

Kimura, E, Hoshi, O. and Ushiki, T. (2004). Atomic force microscopy of human metaphase chromosomes after differential staining of sister chromatids. *Arch. Histol. Cytol.* **67**, 173-177.

Kirby, A.R, Gunning, A.P. and Morris, V.J. (1995a). Imaging xanthan gum by atomic force microscopy. *Carbohydr. Res.* **267**, 161-166.

Kirby, A.R, Gunning, A.P, Morris, V.J. and Ridout, M. J. (1995b). Observation of the helical structure of the bacterial polysaccharide acetan by atomic force microscopy. *Biophys. J.* **68**, 360-363.

Kirby, A.R, Gunning, A.P. and Morris, V.J. (1996a). Imaging polysaccharides by atomic force microscopy. *Biopolymers* **38**, 355-366.

Kirby, A.R, Gunning, A.P, Waldron, K.W, Morris, V.J. and Ng, A. (1996b). Visualisation of plant cell walls by atomic force microscopy. *Biophys. J.* **70**, 1138-1143.

Kirby, A.R, MacDougall, A.J. and Morris, V.J. (2006a). Sugar Beet Pectin–Protein Complexes. *Food Biophysics* **1**, 51-56.

Kirby, A.R, Ng, A, Waldron, K.W. and Morris, V.J. (2006b). AFM investigations of cellulose fibres in Bintje potato (*Solanum tuberosum L*) cell wall fragments. *Food Biophysics* 1, 163-167.

Kirby, A.R, McDougall, A.J. and Morris, V.J. (2007). Atomic Force Microscopy of Tomato and Sugar Beet pectin molecules. *Carbohydr. Polym.* **71**, 640-647.

Kiselyova, O.I, Yaminsky, I.V, Karger, E.M, Frolova, E.M, Dorokhov, Y.L. and Atabekov, J.G. (2001). Visualization by atomic force microscopy of tobacco mosaic virus movement protein-RNA cpmplexes *in vitro*. *J. Gen. Virol.* **82**, 1503-1508.

Klinov, D.V, Martynkina, L.P, Yurchenko, V.Y, Demin, V.V, Streltsov, S.A, Gerasimov, Y.A. and Vengerov, Y.Y. (2003). Effect of supporting substrates on the structure of DNA and DNA-trivaline complexes studied by atomic force microscopy. *Russ. J. Bioorganic Chem.* **29**, 363-367.

Klinov, D, Dwir, B, Kapon, E, Borovok, N, Molotsky, T. and Kotlyar, A. (2006). A comparative study of atomic force imaging of DNA on graphite and mica surfaces. *AIP Conference Proc.* **859**, 99-106.

Kowalczyk, D, Marsault, J-P. and Slomkowski, S. (1996). Atomic force microscopy of human serum albumin (HSA) on poly(styrene/acrolein) microspheres. *Colloid & Polymer Sci.* **274**, 513-519.

Krasnoslobodtsev, A.V, Shlyakhtenko, L.S. and Lyubchenko, Y.L. (2007). Probing interactions within the synaptic DNA-Sfil complex by AFM force spectroscopy. *J. Mol. Biol.* **365**, 1407-1416.

Krautbauer, R, Clausen-Schaumann, H. and Gaub, H.E. (2000). Cisplatin changes the mechanics of single DNA molecules. *Angewandte Chemie-Internaional. Ed.* **39**, 3192-3195.

Krautbauer, R, Pope, L. H, Schrader, T.E, Allen, S. and Gaub, H.E. (2002a). Discriminating small molecule DNA binding modes by single molecule force spectroscopy. *FEBS Letts.* **510**, 154-158.

Krautbauer, R, Fischerlander, S, Allen, S. and Gaub, H.E. (2002b). Mechanical fingerprints of DNA drug complexes. *Single Molecules* **3**, 97-103.

Krautbauer, R, Rief, M. and Gaub, H.E. (2003). Unzipping DNA oligomers. *Nano Letts.* **3**, 493-496.

Kreplak, L, Bar, H, Leterrier, J.F, Hermann, H. and Aebi, U. (2005). Exploring the mechanical behaviour of single intermediate filaments. *J. Mol. Biol.* **354**, 569-577.

Kuhner, F, Costa, L.T, Bisch, P.M, Thalhammer, S, Hockl, W.M. and Gaub, H.E. (2004). LexA-DNA bond strength by single molecule force spectroscopy. *Biophys. J.* **87**, 2683-2690.

Kumar, S, Ravi, V.K. and Swaminathan, R. (2008). How do surfactants and DTT Affect the size, dynamics, activity and growth of soluble lysozyme aggregates? *Biochem. J.* **415**, 275-288.

Kuhner, F, Erdmann, M. and Gaub, H.E. (2006). Scaling exponent and Kuhn length of pinned polymers by single molecule force spectroscopy. *Phys. Rev. Letts.* **97**, issue 21, article number 218301.

Kuhner, F, Morfill, J, Neher, R.A, Blank, K. and Gaub, H.E. (2007). Force-induced DNA slippage. *Biophys. J.* **92**, 2491-2497.

Kunze, S, Lemke, K, Metze, J, Bloukas, G, Kotta, K, Panagiotidis, C.H, Sklaviadis, T. and Bodemer, W. (2008). Atomic force microscopy to characterise the molecular size of prion protein. *J. Microscopy-Oxford* **230**, 224-232.

Kusnetsov, Y.G, Daijogo, S, Zhou, J, Semier, B.L. and McPherson, A. Atomic force microscopy analysis of icosahedral virus RNA. *J. Mol. Biol.* **347**, 41-52.

Kusnetsov, Y.G. and McPherson, A. (2006) Identification of DNA and RNA from retroviruses using ribonuclease A. *Scanning* **26**, 278-281.

Kuutti, L, Peltonen, J, Pere, J. and Teleman, O. (1995). Identification and surface structure of crystalline cellulose studied by atomic force microscopy. *J. Microscopy - Oxford* **178**, 1-6.

Lal, R. and John, S.A. (1994). Biological applications of atomic force microscopy. *Amer. J. Physiol.* **266** (*Cell Physiol.* **35**), C1-C21.

Lankas, F, Sponer, J, Hobza, P. and Langowski, J. (2000). Sequence-dependent elastic properties of DNA. *J. Mol. Biol.* **299**, 695-709.

Law, R, Carl, P, Harper, S, Dalhaimer, P, Speicher, D.W. and Discher, D.E. (2003). Pathway shifts and thermal softening in temperature-coupled forced unfolding of spectrin domains. *Biophys. J.* **84**, 533-544.

Lea, A.S, Pugnor, A, Hlady, V, Andrade, J.D, Herron, J.N. and Voss Jnr, E.W. (1992). Manipulation of proteins on mica by atomic force microscopy. *Langmuir* **8**, 68-73.

Leake, M.C, Grutzner, A, Kruger, M. and Linke, W.A. (2006). Mechanical properties of cardiac titin's N2B-region by single-molecule atomic force spectroscopy. *J. Struct. Biol.* **155**, 263-272.

LeCam, E, Frechon, D, Barry, M, Fourcade, A. and Delain, E. (1994). Observation of binding and polymerization of Fur repressor onto operator-containing DNA with electron and atomic force microscopes. *PNAS USA* **91**,11816-11820.

Lee, C.K, Wang, Y.M, Huang, L.S. and Lin, S.M. (2007). Atomic force microscopy: determination of unbinding force, off rate and energy barrier for protein-ligand interaction. *Micron* **38**, 446-461.

Lee, G, Novak, W, Jaroniec, J, Zhang, G.M. and Marszalek, P.E. (2004). Molecular dynamics simulations of forced conformational transitions in 1,6-linked polysaccharides. *Biophys. J.* **87**, 1456-1465.

Lee, G, Abdi, K, Jiang, Y, Michaely, P, Bennett, V. and Marszalek. P.E. (2006). Nanospring behaviour of ankyrin repeats. *Nature* **440**, 246-249.

Lee, I. and Akiyoshi, K. (2004). Single molecular mechanics of a cholesterol-pullulan nanogel at the hydrophobic interfaces. *Biomaterials* **25**, 2911-2918.

Lee, J.H, Ahn, H.H, Shin, Y.N, Kim, M.S, Lee, B, Khang, G, Lee, I. and Lee, H.B. (2007). Gene delivery using a DNA/polyethyleneimine nanoparticle to fibroplast cells. *Tissue Eng. & Regenerative Medecine* **4**, 566-570.

Lee, I, Evans, B.R, Lane, L.M. and Woodward, J. (1996). Substrate-enzyme interactions in cellulase systems. *Bioresource Technol.* **58**, 163-169.

Leuba, S.H, Yang, G, Robert, C, Samori, B, van Holde, K, Zlatanova J. and Bustamante, C. (1994). Three-dimensional structure of extended chromatin fibers as revealed by tapping-mode scanning force microscopy. *PNAS USA* **91**, 11621-11625.

Leuba, S.H. and Zlatanova, J. (2002). Single-molecule studies of chromatin fibers: a personal report. *Arch. Histol. Cytol.* **65**, 391-403.

Li, B, Hu, J, Wang, Y, Wu, S.Y, Huang, Y.B. and Li, M.Q. (2003). Real time observation of the photocleavage of single DNA molecules. *Chin. Sci. Bull.* **48**, 673-675.

Li, B.S, Sattin, B.D. and Goh, M.C. (2006). Direct and real-time visualization of the disassembly of a single RecA-DNA-ATP gamma S complex using AFM imaging in fluid. *Nano Letts.* **6**, 1474-1478.

Li, H, Rief, M, Oesterhelt, F. and Gaub, H.E. (1998). Single-molecule force spectroscopy on xanthan by AFM. *Adv. Mater.* **3**, 316-319.

Li, H, Cao, E.H, Han, B.S. and Jin, G. (2005). Stretching short single-stranded DNA adsorbed on gold surface by atomic force microscope. *Progr. Biochem. & Biophys.* **32**, 1173-1177.

Li, H.B, Rief, M, Oesterhelt, F, Gaub, H.E, Zhang, X. and Shen, J.C. (1999). Single-molecule force spectroscopy on polysaccharides by AFM – nanomechanical fingerprint of α-(1, 4)-linked polysaccharides. *Chem. Phys. Letts.* **305**, 197-201.

Li, J.W, Tian, F, Wang, C, Bai, C.L. and Cao, E.H. (1997). Possible multistranded DNA induced by acid denaturation-renaturation. *J. Vac. Sci. & Technol. B* **15**, 1637-1640.

Li, L.Y, Chen, S.F, Oh, S.J. and Jiang, S.Y. (2002). *In situ* single-molecule detection of antibody-antigen binding by Tapping mode atomic force microscopy. *Anal. Chem.* **74**, 6017-6022.

Li, M-Q, Xu, L. and Ikai, A (1996). Atomic force microscope imaging of ribosome and chromosome. *J. Vac. Sci. & Technol. B* **14**, 1410-1412.

Li, X.J, Sun, J.L, Zhou, X.F, Li, G, He, P.G, Fang, Y.Z, Li, M.Q. and Hu, J. (2003). Height measurement of dsDNA and antibodies measured on solid substrates in air by vibrating mode scanning polarixation force microscopy. *J. Vac. Sci & Technol. B* **21**, 1070-1073.

Liac, X.M. and Wiedmann, T.S. (2006). Formation of cholesteric crystals at a mucin-coated substrate. *Pharm. Res.* **23**, 2413-2416.

Lim, B.B.C, Lee, E.H, Sotomayor, M. and Schulten, K. (2008). Molecular basis of fibrin clot elasticity. *Structure* **16**, 449-459.

Limanskaya, L.A. and Limanskii, A.P. (2006). Compaction of single supercoiled DNA molecules adsorbed onto amino mica. *Russ. J. Bioorganic Chem.* **32**, 444-459.

Limanskii, A. (2007). Compaction of single molecules of supercoiled DNA immobilized on amino mica: From duplex to minitoroidal and spheroidal conformations. *Biofizika* **52**, 252-260.

Lin, A.C. and Goh, M.C. (2002). Investigating the ultrastructure of fibrous long spacing collagen by parallel atomic force and transmission electron microscopy. *Proteins-Struct. Function & Genetics* **49**, 378-384.

Lin, H, Clegg, D.O. and Lal, R. (1999). Imaging real-time proteolysis of single collagen I molecules with an atomic force microscope. *Biochemistry* **38**, 9956-9963.

Lin, Y, Shan, X.C, Wang, J.J, Bao, L, Zhaang, Z.L. and Pang, D.W. (2007). Measuring radial Young's modulus of DNA by tapping mode AFM. *Chin. Sci. Bull,* **52**, 3189-3192.

Linke, W.A. and Grutzner, A. (2008). Pulling single molecules of titin by AFM - recent advances and physiological implications. *Pflugers Archiv.-Eur. J. Physiol.* **456**, 101-115.

Liu, R.T, McAllister, C, Lyubchenko, Y. and Sierks, M.R. (2004). Residues 17-20 and 30-35 of β-amyloid play critical roles in aggregation. *J. Neurosci. Res.* **75**, 162-171.

Liu, D, Wang, C, Lin, Z, Li, J.W, Xu, B, Wei, Z.Q, Wang, Z.G. and Bai, C.L. (2001). Visualization of the intermediates in a uniform DNA condensation system by tapping mode atomic force microscopy. *Surface & Interface Anal.* **32**, 15-19.

Liu, Z, Li, Z, Zhou, H, Wei, G, Song, Y. and Wang, L. (2005). Immobilization and condensation of DNA with 3-aminopropyltriethoxysilane studied by atomic force microscopy. *J. Microscopy-Oxford* **218**, 233-239.

Lohr, D, Bash, R, Wang, H, Yodh, J. and Lindsay, S. (2007). Using atomic force microscopy to study chromatin structure and nucleosome remodelling. *Methods* **41**, 333-341.

Lonskaya, I, Potaman, V.N, Shlyakhtenko, L.S, Oussatcheva, E.A, Lyubchenko, Y.L. and Soldatenkov, V.A. (2005). Regulation of poly (ADP-ribose) polymerase-1 by DNA structure-specific binding. *J. Biol. Chem.* **280**, 17076-17083.

Lu, J.H, An, H.J, Li, H.K, Li, X.L, Wang, Y, Li, M.Q, Zhang, Y. and Hu, J. (2006a). Nanodissection, isolation, and PCR amplification of single DNA molecules. *Surface & Interface Sci.* **38**, 1010-1013.

Lu, M, Shi, B.C, Li, X.L, Chan, R.S. and Hu, J. (2006b). A strategy for ordered sequencing based on single molecule nanomanipulation. *Progr. Biochem. & Biophys.* **33**, 660-664.

Lu, Z.Y, Nowak, W, Lee, G.R, Marszalek, P.E. and Yang, W.T. (2004) Elastic properties of single amylose chains in water: A quantum mechanical and AFM study. *J. Amer. Chem. Soc.* **126**, 9033-9041.

Luan, B. and Aksimentiev, A. (2008). Strain softening in stretched DNA. *Phys. Rev. Letts.* **101**, article number 118101.

Lubarsky, G.V, Davidson, M.R. and Bradley, R.H. (2007). Hydration-dehydration of adsorbed protein films studied by AFM and QCM-D. *Biosensors & Bioelectronics* **22**, 1275-1281.

Luginbuehl, R. and Sigrist H. (1998). Light-dependent substrate functionalization and biomacromolecule immobilization. In *Procedures in Scanning Probe Microscopy*, (eds. R.J.Colton, A. Engel, J.E. Frommer, H.E. Gaub, A.A. Gewirth, R. Guckenberger, J. Rabe, W. Heckl and B Parkinson), module 7.12.2, pp 488-492. J. Wiley & Sons, New York.

Lushnikov, A.Y, Brown, B.A, Oussatcheva, E.A, Potaman, V.N, Sinden, R.R. and Lyubchenko, Y.L (2004). Interaction of the Z α-domain of human ADAR1 with a negatively supercoiled plasmid visualized by atomic force microscopy. *Nucleic Acids Res.* **32**, 4704-4712.

Lushnikov, A.Y, Potaman, V.N, Gussatcheva, E.A, Sinden, R.R. and Lyubchenko, Y.L. (2006a). DNA strand arrangement within the SfiI-DNA complex: Atomic force microscopy analysis. *Biochemistry* **45**, 152-158.

Lushnikov, A.Y, Potaman, V.N. and Lyubchenko, Y.L. (2006b). Site-specific labelling of supercoiled DNA. *Nucleic Acids Res.* **34**, e111 doi:10.1093/nar/gkl642.

Lysetska, M, Knoll, A, Boehringer, D, Hey, T, Krauss, G. and Krausch, G. (2002). UV light-damaged DNA and its interaction with human replication protein A: an atomic force microscopy study. *Nucleic Acids Res.* **30**, 2686-2591.

Lysetska, M, Zetti, H, Oka, I, Lipps, G, Krauss, G. and Krausch, G. (2005). Site-specific binding of the 9.5 kilodalton DNA-binding protein ORF80 visualized by atomic force microscopy. *Biomacromolecules* **6**, 1252-1257.

Lyubchenko, Y.L. (2004). DNA structure and dynamics–An atomic force microscopy study. *Cell Biochem. & Biophys.* **41**, 75-98.

Lyubchenko, Y.L, Gall, A.A, Shlyakhtenko, L.S, Harrington, R.E, Oden, P.I, Jacobs, B.L. and Lindsay, S.M. (1992a). Atomic force microscopy imaging of double stranded DNA and RNA. *J. Biomol. Struct. Dynamics* **9**, 589-606.

Lyubchenko, Y.L, Jacobs, B.L. and Lindsay, S.M. (1992b). Atomic force microscopy of reovirus dsRNA: a routine technique for length measurements. *Nucleic Acids Res.* **20**, 3983-3986.

Lyubchenko, Y.L, Oden, P.I, Lampner, D, Lindsay, S.M. and Dunker, K.A. (1993). Atomic force microscopy of DNA and bacteriophage in air, water and propanol: the role of adhesion forces. *Nucleic Acids Res.* **21**, 1117-1123.

Lyubchenko, Y.L, Jacobs, B.L, Lindsay, S.M. and Stasiak, A. (1995). Atomic force microscopy of nucleoprotein complexes. *Scanning Microscopy* **9**, 705-727.

Lyubchenko, Y.L. and Shlyakhtenko, L.S. (1997). Visualization of supercoiled DNA with atomic force microscopy *in situ*. *PNAS USA* **94**, 496-501.

Lyubchenko, Y.L, Shlyakhtenko, L.S, Binus, M, Gaillard, C. and Strauss, F. (2002). Visualization of hemiknot DNA structure with an atomic force microscope. *Nucleic Acids Res.* **30**, 4902-4909.

Ma, H, Snook, L.A, Kaminsky, S.G.W. and Dahms, T.E.S. (2005). Surface ultrastructure and elasticity in growing tips and mature regions of *Aspergillus* hyphae describe wall maturation. *Microbiol.-SGM* **151**, 3579-3688.

Ma, H, Snook, L.A, Tian, C, Kaminsky, S.G.W. and Dahms, T.E.S. (2006). Fungal surface remodelling visualized by atomic force microscopy. *Mycological Res.* **110**, 679-886.

Maaloum, M, Pernodet, N. and Tinland, B. (1998). Agarose gel structure using atomic force microscopy: gel concentration and ionic strength effects. *Electrophoresis* **19**, 1606-1610.

Mackie, A.R, Gunning, A.P, Ridout, M.J. and Morris, V.J. (1998). Gelation of gelatin: observation in the bulk and at the air-water interface. *Biopolymers* **46**, 245-252.

Mann, A, Khan, M.A, Shukla, V. and Ganguli, M. (2007). Atomic force microscopy reveals the assembly of potential DNA "nanocarriers" by poly-L-orithine. *Biophys. Chem.* **129**, 126-136.

Mann, A, Richa, R and Ganguli, M. (2008). DNA condensation by poly-L-lysine at the single molecule level: Role of DNA concentration and polymer length. *J. Controlled Release* **125**, 252-262.

Marga, F, Grandbois, M, Cosgrove, D.J. and Baskin, T.I. (2005). Cell wall extension results in the coordinate separation of parallel microfibrils: evidence from scanning electron microscopy and atomic force microscopy. *Plant J.* **43**, 181-190.

Marek, P, Abedini, A, Song, B.B, Kanungo, M, Johnson, M.E, Gupta, R, Zaman, W, Wong, S.S. and Raleigh, D.P. (2007). Aromatic interactions are not required for amyloid fibril formation by islet amyloid polypeptide but do influence the rate of fibril formation and fibril morphology. *Biochemistry* **46**, 3255-3261.

Marilly, M, Sanchez-Seville, A. and Rocca-Serra, J. (2005). Fine mapping of inherent flexibility variation along DNA molecules. Validation by atomic force microscopy (AFM) in buffer. *Mol. Gen. Genomics* **274**, 658-670.

Marko, J.F. and Siggia, E.D. (1995). Stretching DNA. *Macromolecules* **28**, 8759-8770.

Marsden, S, Nardeli, M, Linder, P. and McCarthy, J.E.G. (2006). Unwinding single RNA molecules using helicases involved in eukaryotic transition initiation. *J. Mol. Biol.* **361**, 327-335.

Marszalek, P.E, Oberhauser, A.F, Pang, Y-P. and Fernandez, J.M. (1998). Polysaccharide elasticity governed by chair-boat transitions of the glucopyranose ring. *Nature* **396**, 661-664.

Maskin, L, Frankel, N, Gudesblat, G, Demergasso, M.J, Pietrasanta, L.I. and Iusem, N.D. (2007). Dimerization and DNA-binding of ASR1, a small hydrophilic protein abundant in plant tissues suffering from water loss. *Biochem. & Biophys. Res. Commun.* **353**, 831-835.

Marszalek, P.E, Lu, H, Li, H.B, Carrion-Vazquez, M, Oberhaused, M, Schulten, K. and Fernandez, J.M. (1999). Mechanical unfolding intermediates in titin molecules. *Nature* **402**, 100-103.

Marszalek, P.E, Li, H. and Fernandez, J.M. (2001). Fingerprinting polysaccharides with single-molecule atomic force microscopy. *Nature Biotechnol.* **19**, 258-262.

Marszalek, P.E, Li, H.B, Oberhauser, A.F. and Fernandez, J.M. (2002). Chair-boat transitions in single polysaccharide molecules observed with force-ramp AFM. *PNAS USA* **99**, 4278-4283.

Marszalek, P.E, Oberhauser, A.F, Li, H.B. and Fernandez, J.M. (2003). The force-driven conformations of heparin studied with single molecule force microscopy. *Biophys. J.* **85**, 2696-2704.

Marsh, T.C, Vesenka, J. and Henderson, E. (1995). A new DNA nanostructure, the G-wire, imaged by scanning probe microscopy. *Nucleic Acids Res.* **23**, 696-700.

Martin, I.D, Vesenka, J.P, Henderson, E. and Dobbs, D.L. (1995). Visualization of nucleosomal structure in native chromatin by atomic force microscopy. *Biochemistry* **34**, 4610-4616.

Maurer, S, Fritz, J, Muskhelishvilli, G. and Travers, A. (2006). RNA polymerase and an activator form discrete subcomplexes in a transcription initiation complex. *EMBO J.* **25**, 3784-3790.

Mazzola, L.T, Curtis, W.F, Fodor, S.P.A, Mosher, C, Lartius, R. and Henderson, E. (1999). Discrimination of DNA hybridization using chemical force microscopy. *Biophys. J.* **76**, 2922-2933.

McIntire, T.M, Penner, R.M. and Brant, D.A. (1995). Observations of a circular, triple helical polysaccharide using noncontact atomic force microscopy. *Macromolecules* **28**, 6375-6377.

McAllister, C, Karymov, M.A, Kawano, Y, Lushnikov, A.Y, Mikheikin, A, Uversky, V.N. and Lyubchenko, Y.L. (2005). Protein interactions and misfolding analyzed by AFM force spectroscopy. *J. Mol. Biol.* **354**, 1028-1042.

McIntire, T.M. and Brant, D.A. (1997a). Imaging carbohydrate polymers with noncontact mode atomic force microscopy. In *Techniques in Glycobiology*, (eds. R.R. Townsend and A.T. Hotchkiss, Jnr), chapter 12, 187-206. Marcel Dekker, New York.

McIntire, T.M. and Brant, D.A. (1997b). Imaging of individual biopolymers and supramolecular assemblies using noncontact atomic force microscopy. *Biopolymers* **42**, 133-146.

McIntire, T.M. and Brant, D.A. (1998). Observation of the $(1\rightarrow3)$-β-D glucan linear triple helix to macrocycle interconversion using noncontact atomic force microscopy. *J. Amer. Chem. Soc.* **120**, 6909-6919.

McKeown, T.A, Moss, S.T. and Jones, E. B. G. (1996). Atomic force and electron microscopy of sporangial wall microfibrils in *Linderina pennispora*. *Mycol. Res.* **100**, 821-826.

McMaster, T.J, Hickish, T, Min, T, Cunningham, D. and Miles, M.J. (1994). Application of scanning force microscope to chromosome analysis. *Cancer Genet. Cytogenet.* **76**, 93-95.

McMaster, T.J, Winfield, M.O, Baker, A.A, Karp, A. and Miles, M.J. (1996a). Chromosome classification by atomic force microscopy volume measurement. *J. Vac. Sci. & Technol. B* **14**, 1438-1443.

McMaster, T.J, Winfield, M.O, Karp, A. and Miles, M.J. (1996b). Analysis of cereal chromosomes by atomic force microscopy. *Genome* **39**, 439-444.

McMaster, T.J, Miles, M.J, Kasarda, D.D, Shewry, P.R. and Tatham, A.S. (2000). Atomic force microscopy of A-gliadin fibrils and *in situ* degradation. *J. Cereal Sci.* **31**, 281-286.

Medalia, O, Englander, J, Guckenberger, R. and Sperling, J. (2002). AFM imaging in solution of protein-DNA complexes formed on DNA anchored to a gold surface. *Ultramicroscopy* **90**, 103-112.

Meyer, A, Rouquet, G, Lecourtier, L. and Toulhoat, H. (1992). Characterisation by atomic force microscopy of xanthan in interaction with mica. In *Physical Chemistry of Colloids and Interfaces in Oil Production*, (eds. H. Toulhoat and J. Lecourtier), pp. 275-278. Editions Techniq., Paris.

Miller, E, Garcia, T, Hultgren, S. and Oberhauser, A.F. (2006). The mechanical properties of E-coli type 1 pili measured by atomic force microscopy techniques. *Biophys. J.* **91**, 3848-3856.

Mills, E.N.C, Huang, L, Gunning, A.P. and Morris, V.J. (2001). Formation of thermally-induced aggregates of the soya globulin β-conglycinin. *Biochem. Biophys. Acta* **1547** , 339-350.

Mine, I. and Okuda, K. (2007). Fine structure of cell wall surfaces in the giant-cellular xanthophycean alga *Vaucheria terrestris. Planta* **225**, 1135-1145.

Misovic, G.N. (20010. Atomic force microscopy measurements – measurements of binding strength between a single pair of molecules in physiological solutions. *Mol. Biotechnol.* **18**, 149-153.

Miyagi, A, Tsunaka, Y, Uchihashi, T, Mayanagi, K, Hirose, S, Morikawa, K and Ando, T. (2008). Visualization of intrinsically disordered regions of proteins by high-speed atomic force microscopy. *CHEMPHYSCHEM* **9**, 1859-1866.

Montana, V, Liu, W, Mohideen, U. and Parpure, V. (2008). Single molecule probing of exocytotic protein interactions using force spectroscopy. *Croatica Chem. Acta* **81**, 31-40.

Moreno-Herrero, F, Herrero, P, Colchero, J, Baro, A.M. and Moreno, F. (2001). Imaging and mapping protein-binding sites on DNA regulatory regions with atomic force microscopy. *Biochem. & Biophys. Res. Commun.* **280**, 151-157

Moreno-Herrero, F, Colchero, J, Gomez-Herrero, J, Baro, A.M. and Avila, J. (2004a). Jumping mode atomic force microscopy obtains reproducible images of Alzheimer paired helical filaments in liquids. *Eur. Polymer J.* **40**, 927-932.

Moreno-Herrero, F, Perez, M, Baro, A.M. and Avila, J. (2004b). Characterisation by atomic force microscopy of Alzheimer paired helical filaments under physiological conditions. *Biophys. J.* **86**, 517-528.

Mori, O. and Imae, T. (1997). AFM investigation of the adsorption process of bovine serum albumin on mica. *Colloids & Surfaces B Biointerfaces* **9**, 31-36.

Morfill, J, Kuhner, F, Blank, K, Lugmaier, R.A, Sedimair, J. and Gaub, H.E. (2007). B-S transition in short oligosaccharides. *Biophys. J.* **93**, 2400-2409.

Morris, S, Hanna, S. and Miles, M.J. (2004). The self-assembly of plant cell wall components by single-molecule force spectroscopy and Monte Carlo modelling. *Nanotechnology* **15**, 1296-1301.

Morris, V.J. (1994). Biological applications of scanning probe microscopies. *Prog. Biophys. Mol. Biol* . **61**,131-185.

Morris, V.J, Gunning, A.P, Kirby, A.R, Round, A.N, Waldron, K. and Ng, A. (1997). Atomic force microscopy of plant cell walls, plant cell wall polysaccharides and gels. *Int. J. Biol. Macromolecules.* **21**, 61-66.

Morris, V.J. (1998a). Atomic Force Microscopy. *The European Food and Drink Review.* Spring, 17-21.

Morris, V.J. (1998b). Applications of atomic force microscopy in food science. In *Gums and Stabilisers for the Food Industry 9*, (eds. P.A. Williams and G.O. Phillips), pp. 361-370. Royal Society Chemistry, Cambridge, Special Publication no. 218.

Morris, V.J. (1998c). Gelation of polysaccharides. *In Functional Properties of Food Macromolecules*, (eds. S.E. Hill, D.A. Ledward and J.R. Mitchell), pp. 143-226. Aspen Publishers, Gaithersburg, USA, chapter 4.

Morris, V.J, Mills, E.C.N, Mackie, A.R, Wilde, P, Kirby, A.R. and Gunning, A.P. (1999). Probing biopolymer functionality in foods with the atomic force microscope. *Food Industry J.* **2**, 11-35.

Morris, V.J, Gunning, A.P, Faulds, C.B, Williamson, G. and Svensson, B. (2005). AFM images of complexes between amylose and *Aspergillus niger* glucoamylase mutants, native and mutant starch binding domains: a model for the action of glucoamylase. *Stärke* **57**, 1-7.

Mosher, C, Jondle, D, Ambrosio, L, Vesenka, J. and Henderson, E. (1995). Microdissection and measurements of polytene chromosomes using the atomic force microscope. *Scanning Microscopy* **8**, 491-497.

Mou, J, Czajkowsky, D.M, Zhang, Y. and Shao, Z. (1995a). High resolution atomic-force microscopy of DNA: the pitch of the double helix. *FEBS Letts.* **371**, 279-282.

Mou, J, Yang, J. and Shao, Z. (1995b). Atomic force microscopy of cholera toxin B-oligomers bound to bilayers of biologically relevant lipids. *J. Mol. Biol.* **248**, 507-512.

Mou, J, Sheng, S.J, Ho, R. and Shao, Z. (1996a). Chaperonins GroEL and GroES: views from atomic force microscopy. *Biophys. J.* **71**, 2213-2221.

Mou, J, Czajkowsky, D.M, Sheng, S.J, Ho, R. and Shao, Z. (1996b). High resolution surface structure of *E. coli* GroES oligomer by atomic force microscopy. *FEBS Letts.* **381**, 161-164.

Moukhtar, J, Fontaine, E, Faivre-Moskalenko, C. and Arneodo, A. (2007). Probing persistence in DNA curvature properties with atomic force microscopy. *Phys. Rev. Letts.* **98**, article number 178101.

Mulhern, P.J, Blackford, B.L, Jericho, M.H, Southam. G. and Beveridge, T.J. (1992). AFM and STM studies of the interaction of antibodies with S-layer sheath of the archaeobacterium *Methanospriririllum hungatei. Ultramicroscopy* **42**, 1214-1224.

Mukhopadhyay, R, Dubey, P. and Sarker, S. (2005). Structural changes of DNA induced by mono- and binuclear cancer drugs. *J. Struct. Biol.* **150**, 277-283.

Muller, D.J. and Engel, A. (1997). The height of biomolecules measured with the atomic force microscope depends on electrostatic interactions. *Biophys. J.* **73**, 1633-1644.

Muller, D.J, Fotiaddis, D, Scheuring, S, Muller, S.A. and Engel, A. (1999). Electrostatically balanced subnanometer imaging of biological specimens by atomic force microscopy. *Biophys. J.* **76**, 1101-1111.

Murakami, M, Hirokawa, H. and Hayate, I. (2000). Analysis of radiation damage of DNA by atomic force microscopy in comparison with agarose gel electrophoresis studies. *J. Biochem. & Biophys. Methods* **44**, 31-40.

Murakami, M, Minamihisamatsu, M, Sato, K. and Hayata, I. (2001). Structural analysis of heavy ion radiation-induced chromosome aberrations by atomic force microscopy. *J. Biochem. & Biophys. Methods* **48**, 283-301.

Murakami, M, Narumi, I, Satoh, K, Furukawa, A. and Hayata, I. (2006). Analysis of interaction between DNA and *Deinococcus radiodurans* PprA protein by atomic force microscopy. *Biochem. Biophys. Acta-Proteins & Proteomics* **1754**, 20-23.

Murray, M.N, Hansma, H.G, Bezanilla, M, Sano, T, Ogletree, D.F, Kolbe, W, Smith, C.L, Cantor, C.R, Spengler, S, Hansma, P.K. and Salmeron, M. (1993). *PNAS USA* **90**, 1037-1038.

Musio, A, Mariani, T, Frediani, C, Sbrana, I. and Ascoli, C. (1994). Longitudinal patterns similar to G-banding in untreated human chromosomes: evidence from atomic force microscopy. *Chromosomes* **103**, 225-229.

Nangia-Makker, P, Conklin, J, Hogan, V. and Raz, A. (2002). Carbohydrate-binding proteins in cancer, and their ligands as therapeutic agents. *Trends Mol. Medicine* **8**, 187-192.

Nakata, S, Kido, N, Hayashi, M, Hara, M, Sasabe, H, Sugawara, T .and Matsuda, T. (1996). Chemisorption of proteins and their thiol derivatives onto gold surfaces: characterisation based on electrochemical non-linearity. *Biophys. Chem.* **62**, 63-72.

Netz, R.R. (2001). Strongly stretched semiflexible extensible polyelectrolytes and DNA. *Macromolecules* **34**, 7522-7529.

Neagu, C, van der Werf, K.O, Putman, C.A.J, Kraan, Y.M, de Grooth, B.G, van de Hulst, N.F. and Greve, J. (1994). Analysis of immunolabelled cells by atomic force microscopy, optical microscopy and flow cytometry. *J. Struct. Biol.* **112**, 32-40.

Nettikadan, S, Tokumasu, F. and Takeyasu, K. (1996). Quantitative analysis of the transcription factor Ap2 binding to DNA by atomic force microscopy. *Biochem. Biophys. Res. Commun.* **226**, 645-649.

Ng, J.D, Kuznetsov, Y.G, Malkin, A.J, Keith, G, Giege, R. and McPherson, A. (1997). Visualiazation of RNA crystal growth by atomic force microscopy. *Nucleic Acids Res.* **25**, 2582-2588.

Ng, L, Grodzinsky, A.J, Patwari, P, Sandy, J, Plaas, A. and Ortiz, C. (2003). Individual cartlage aggrecan macromolecules visualised via atomic force microscopy. *J. Struct. Biol.* **143**, 242-257.

Ng, S.P, Rounsevell, R.W.S, Steward, A, Geierhaas, C.D, Williams, P.M, Paci, E. and Clarke, J. (2005). Mechanical unfolding of TNfn2: the unfolding pathway of a fnIII domain probed by protein engineering, AFM and MD simulation. *J. Mol. Biol.* **359**, 776-789.

Ng, S.P, Randles, L.G. and Clarke, J. (2007). Single studies of protein folding using atomic force microscopy. *Methods in Mol. Biol.* **350**, 139-167.

Noda, S, Fujami, T, Nakauma, M, Asai, I, Takahashi, R, Al-Assaf, S, Ikeda, S, Nishinari, K. and Phillips, G.O. (2008). Molecular structures of gellan gum imaged with atomic force microscopy in relation to the rheological behaviour in aqueous systems. 1. Gellan gum with various acyl contents in the presence and absence of potassium. *Food Hydrocolloids* **22**, 1148-1159.

Nordgren, N, Eklof, J, Zhou, Q, Brumer, H. and Rutland, M.W. (2008). Top-down grafting of xyloglucan to gold monitored by QCM-D and AFM: enzymatic activity and interactions with cellulose. *Biomacromolecules* **9**, 942-948.

Oberhauser, A.F, Marszalek, P.E, Erickson, H.P. and Fernandez, J.M. (1998). The molecular elasticity of the extracellular matrix protein tenascin. *Nature* **393**, 181-185.

Oberhauser, A.F, Marszalek, P.E, Carrion-Vazquez, M. and Fernandez, J.M. (1999). Single misfolding events captured by atomic force microscopy. *Nature Struct. Biol,* **6**, 1025-1028.

Oberhauser, A.F. and Carrion-Vazquez, M. (2008). Mechanical biochemistry of proteins one molecule at a time. *J. Biol. Chem.* **283**, 6617-6621.

Oberleithner, H, Schneider, S. and Henderson, R.H. (1996). Viewing the renal epithelium with the atomic force microscope. *Kidney & Blood Pressure Res.* **19**, 142-147.

Oberringer, M, Englisch, A, Hainz, B, Gao, H, Martin, T. and Hartmann, U. (2003). Atomic force microscopy and scanning near-field optical microscopy studies on the characterization of human metaphase chromosomes. *Eur. Biophys. J. Biophys Letts.* **32**, 620-627.

Ohnesorge, F, Heckl, W.M, Häberle, W, Pum, D, Sara, M, Schindler, H, Schilcher, K, Kiener, A, Smith, D.P.E, Sleytr, U.B. and Binnig, G. (1992). Scanning force microscopy studies of the S layers from *Bacillus coagulans* E38-66, *Bacillus sphaericus* CCM2177 and of an antibody binding process. *Ultramicroscopy* **42**, 1236-1242.

Ohta, T, Nettikadan, S, Tokumasu, F, Ideno, H, Abe, Y, Kuroda, M, Hayashi, H and Takeyasu, K. (1996). Atomic force microscopy proposes a new model for stem-loop structure that binds a heat shock protein in the *Staphylococcus aureus* HSP70 operon. *Biochem. Biophys. Res. Commun.* **226**, 730-734.

Ohtani, T, Shichiri, M, Fukushi, D, Sugiyama, S, Yoshino, T, Kobori, T, Hagiwara. And Ushiki, T. (2002). Imaging of chromosomes at nano-meter scale resolution using near-field optical/atomic force microscopy. *Arch. Histol. Cytol.* **65**, 425-434.

Ojada, T, Sano, M, Yamamoto, Y. and Muramatsu, H. (2008). Evaluation of interaction forces between profiling and designed probes by atomic force spectroscopy. *Langmuir* **24**, 4050-4055.

Okajima, T, Tao, X.M, Azehara, H. and Tokumoto, H. (2007). Force spectroscopy on single polymer incorporated into polymer gels. *J. Nanosci. & Nanotechnol.* **7**, 790-795.

Onoa, G.B. and Moreno, V. (2002). Study of the modifications caused by cisplatin, transplatin, and Pd(II) and Pt(II) mepirizole derivatives on pBR322 DNA by atomic force microscopy. *Int. J. Pharmaceutics* **245**, 55-65.

Ortiz, C, Zhang, D.M, Ribbe, A.E, Xie, Y. and Ben-Amotz, D. (2007). Analysis of indulin amyloid fibrils by Raman spectroscopy. *Biophys. Chem.* **128**, 150-155.

Ou-Yang, Z.C, Zhou, H. and Zhang, Y. (2003). The elastic theory of a single DNA molecule. *Modern Phys. Letts.* **17**, 1-10.

Paananaen, A, Tappura, K, Tatham, A.S, Fido, R, Shewry, P.R, Miles, M.J. and McMaster, T.J. (2006). Nanomechanical force measurements of gliadin protein interactions. *Biopolymers* 83, *658-667.*

Paige, M.F, Rainey, J.K. and Goh, M.C. (1998). Fibrous long spacing collagen ultrastructure elucidated by atomic force microscopy. *Biophys. J.* **74**, 3211-3216.

Paige, M.F. and Goh, M.C. (2001). Ultrastructure and assembly of segmemted long spacing collagen studied by atomic force microscopy. *Micron* 355-361.

Paige, M.F, Rainey, J.K. and Goh, M.C. (2001). A study of fibrous long spacing collogen ultrastructure by atomic force microscopy. *Micron* **32**, 341-353.

Paige M.F, Lin, A.C. and Goh, M.C. (2002). Real-time enzymatic biodegradation of collagen fibrils monitored by atomic force microscopy. *Internat. Biodetrioration & Biodegradation* **50**, 1-10.

Parker, M.L, Kirby, A.R. and Morris, V.J. (2008). *In situ* imaging of pea starch in seeds. *Food Biophysics* **3**, 56-76.

Parkhouse, R.M, Askonas, B.A. and Dourmashkin, R. R. (1970). Electron microscopic studies of mouse immunoglobulin M structure and reconstruction following reduction. *Immunology* **18**, 575-584.

Parra, A, Casero, E, Lorenzo, E, Pariente, F. and Vazquez, L. (2007). Nanomechanical properties of globular proteins: lactate oxidase. *Langmuir* **23**, 2747-2754.

Pastre, D, Pietrement, O, Fusil, S, Landousy, F, Jeusset, J, David, M-O, Hamon, L, Le Cam, E. and Zozime, A. (2003). Adsorption of DNA to mica mediated by divalent counterions: A theoretical and experimental study. *Biophys. J.* **85**, 2507-2518.

Pastre, D, Pietrement, O, Zozime, A. and Le Cam, E. (2005). Study of the DNA/ethidium bromide interactions on mica surface by atomic force microscope: Influence of the surface friction. *Biopolymers* **77**, 53-62.

Pastushenko, V.P, Kaderabek, R, Sip, M, Borken, C, Kienberger, F. and Hinterdorfer, P. (2002). Reconstruction of DNA shape from AFM data. *Single Molecules* **3**, 111-117.

Patnaik, S, Aggarwal, A Nimesh, S, Goel, A, Ganguli, M, Saini, N, Singh, Y. and Gupta, K.C. (2006). PEI-alginate nanoparticles as efficient *in vitro* gene transfection agents. *J. Controlled Release* **114**,398-409.

Pavlicek, J.W, Lyubchanko, Y.L. and Chang, Y. (2008). Quantitative analyses of RAG-RSS interactions and conformations revealed by atomic force microscopy. *Biochemistry* **47**, 11204-11211.

Pearce, K.H, Hiskey, R.G. and Thomson, N.H. (1992). Surface binding kinetics of prothrombin fragment 1 on planar membranes measured by total internal reflection fluorescence microscopy. *Biochemistry* **31**, 5983-5995.

Perez-Jimenez, R, Garcia-Manyes, S, Ainavarapu, S.R.K. and Fernandez, J.M. (2006). Mechanical Unfolding Pathways of the Enhanced Yellow Fluorescent Protein Revealed by Single Molecule Force Spectroscopy. *J. Biol. Chem.* **281**, 40010-40014.

Perkins, S.J, Nealis, A.S, Sutton, B.J. and Feinstein, A. (1991). Solution structure of human and mouse immunoglobulin M by synchrotron X-ray scattering and molecular graphics modeling. *J. Mol. Biol.* **221**, 1345-1366.

Pernodet, N, Maaloum, M. and Tinland, B. (1997). Pore size of agarose gels by atomic force microscopy. *Electrophoresis* **18**, 55-58.

Perrin, A, Lanet, V. and Theretz, A. (1997). Quantification of specific immunological reactions by atomic force microscopy. *Langmuir* **13**, 2557-2563.,

Pietrasanta, L.I, Schaper, A. and Jovin, T.M. (1994). Probing specific molecular conformations with the scanning force microscope. Complexes of plasmid DNA and anti-Z-DNA antibodies. *Nucleic Acids Res.* **22**, 3288-3292.

Poggi, M.A, Bottomley, L.A. and Lillehei, P.T. (2002). Scanning probe microscopy. *Anal. Chem.* **74**, 2851-2862.

Poggi, M. A, Gadsby, E, Bottomley, L.A, King, W.P, Oroudjev, E. and Hansma, H.G. (2004). Scanning probe microscopy. *Anal. Chem.* **76**, 3429-3444.

Poma, A, Spano, L, Pittaluga, E, Tucci, A, Palladino, L. and Limongi, T. (2005). Interactions between saporin, a ribosome-inactivating protein, and DNA: a study by atomic force microscopy. *J. Microscopy-Oxford* **217**, 69-74.

Pope, L.H, Davies, M.C, Laughton, C.A, Roberts, C.J, Tendler, S.J.B. and Williams, P.M. (1999). Intercalation-induced changes in DNA supercoiling observed in real-time by atomic force microscopy. *Anal. Chim. Acta* **409**, 27-32.

Pope, L.H, Davies, M.C, Laughton, C.A, Tendler, S.J.B. and Williams, P.M. (2000). Atomic force microscopy studies of intercalation-induced changes in plasmid DNA tertiary structure. *J. Microscopy-Oxford* **199**, 68-78.

Pope, L.H, Davies, M.C, Laughton, C.A, Roberts, C.J, Tendler, S.B.J. and Williams, P.M. (2001). Force-induced melting of a short DNA double helix. *Eur. Biophys. J. Biophys. Letts.* **30**, 53-62.

Potaman, V.N, Lushnikov, A.Y, Sinden, R.R. and Lyubchenko, Y. L. (2002). Site-specific labeling of supercoiled DNA at the A+T rich sequences. *Biochemistry* **41**, 13199-13206.

Power, D, Larsen, I, Hartley, P, Dunstan, D. and Boger, D.V. (1998a). Molecular images of gels formed under shear using atomic force microscopy. In *Gums and Stabilisers for the Food Industry 9*, (eds. P.A. Williams and G.O. Phillips), pp. 388-394. Royal Society Chemistry, Cambridge, Special Publication no. 218.

Power, D, Larsen, I, Hartley, P, Dunstan, D. and Boger, D.V. (1998b). Atomic force microscopy studies on hydroxypropylguar gels formed under shear. *Macromolecules* **31**, 8744-8748.

Psonka, K, Brons, S, Heiss, M, Gudoeska-Nowak, E. and Taucher-Scholz, G. (2005). Induction of DNA damage by heavy ions measured by atomic force microscopy. *J. Phys.-Condensed Matter* **17**, S1443-S1446.

Puppels, G.J, Putman, C.A.J, De Grooth, B.G. and Greve, J. (1992). Raman microspectroscopy and atomic force microscopy of chromosomal banding patterns. *SPIE Proc.* **1922**, 145-155.

Putman, C.A.J, de Grooth, B.G, Hansma, P.K, van Hulst, N.F. and Greve, J. (1993a). Immunogold labels: cell-surface markers in atomic force microscopy. *Ultramicroscopy* **42**, 177-182.

Putman, C.A.J, de Grooth, B.G, Wiegrant, J, Raap, A.K, van der Werf, K.O, van Hulst, N.F. and Greve, J. (1993b). Detection of *in situ* hybridisation to human chromosomes with the atomic force microscope. *Cytometry* **14**, 356-361.

Qamar, S, Williams, P.M. and Lindsay, S.M. (2008). Can an atomic force microscope sequence DNA using a nanopore? *Biophys. J.* **94**, 1233-1240.

Qin, F, Jiang, Y.X, Ma, X.Y, Chen, F, Fang, X.H, Bai, C.L. and Li, Y.Q. (2004). A study of specific interaction of the transcription factor and the DNA element by atomic force microscopy. *Chin. Sci. Bull.* **49**, 1376-1380.

Quinto, M, Ciancio, A .and Zambonin, P.G. (1998). A molecular resolution AFM study of gold-adsorbed glucose oxidase as influenced by enzyme concentration. *J. Electroanalytical Chem.* **448**, 51-59.

Quist, A.P, Bjorck, L.P, Reimann, C.T, Oscarsson, S.O. and Sundqvist, B.U.R. (1995). A scanning force microscopy study of human serum albumin and porcine pancreas trypsin adsorption on mica surfaces. *Surface Sci.* **325**, L406-L412.

Qu, M.H, Li, H, Tian, R, Nie, C.L, Liu, Y, Han, B.S. and He, R.Q. (2004a). Neuronal tau induces DNA conformational changes observed by atomic force microscopy. *Neuroreport* **15**, 2723-2727.

Qu, M.H, Li, H, Xu, Y.J. and He, R.Q. (2004b). Interaction of human neuronal protein tau with DNA. *Progr. Biochem. & Biophys.* **31**, 918-923.

Radmacher, M, Fritz, M, Cleveland, J.P, Walters, D.R. and Hansma, P.K. (1994). Imaging adhesion forces and elasticity of lysozyme adsorbed on mica with the atomic force microscope. *Langmuir* **10**, 3809-3814.

Radmacher, M, Fritz, M. and Hansma, P.K. (1996). Imaging soft samples with the atomic force microscope-gelatin in water and propanol. *Biophys. J.* **69**, 264-270.

Raible, M, Evstigneev, M, Reimann, P, Bartels, F.W. and Ros, R. (2004). Theoretical analysis of dynamic force spectroscopy experiments on ligand receptor complexes. *J. Biotechnol.* **112**, 13-23.

Rampino, N.J. (1992.) Cisplatin induced alterations in oriented fibers of DNA studies by atomic force microscopy. *Biochem. Biophys. Res. Commun.* **182**, 201-207.

Rasch, P, Wiedemann, U, Wienberg, J. and Heckl, W.M. (1993). Analysis of banded human chromosomes and *in situ* hybridisation patterns by scanning force microscopy. *PNAS USA* **90**, 2509-2511.

Raspanti, M, Alessandrini, A, Gobbi, P. and Ruggeri, A. (1996). Collagen fibril surface: TMAFM, FEG-SEM and freeze-etching observations. *Microscopy Res. & Techniques* **35**, 87-93.

Raspanti, M, Alessandrini, A, Ottani, V. and Ruggeri, A. (1997). Direct visualisation of collagen-bound proteoglycans by Tapping-mode atomic force microscopy. *J. Struct. Biol.* **119**, 118-122.

Raspanti, M, Congiu, T. and Guizzardi, S. (2002). Structural aspects of the extracellular matrix of the tendon: an atomic force and scanning electron microscopy study. *Arch. Histol. Cytol.* **65**, 37-43.

Rees, W.A, Keller, R.W, Vesenka, J.P, Yang, G. and Bustamante, C. (1993). Scanning force microscopy imaging of transcription complexes: evidence for DNA bending in open promoter and elongation complexes. *Science* **260**, 1646-1649.

Revenko, I, Sommer, F, Minh, D.T, Garrone, R. and Franc, J-M. (1994). Atomic force microscopy of the collagen fibril. *Biol. Cell* **80**, 67-69.

Ridout, M.J, Brownsey, G.J, Gunning, A.P. and Morris, V.J. (1998). Characterisation of the polysaccharide produced by *Acetobacter xylinum* strain CR1/4 by light scattering and atomic force microscopy. *Int. J. Biol. Macromolecules* **23**, 287-293.

Ridout, M.J, Gunning, A.P, Wilson, R.H, Parker, M.L. and Morris, V.J. (2002). Using AFM to image the internal structure of starch granules. *Carbohydr. Polym.* **50**, 123-132.

Ridout, M.J, Parker, M.L, Hedley, C.L, Bogracheva, T.Y. and Morris, V.J. (2003). Atomic force microscopy of pea starch granules: Granule architecture of wild-type parent, *r*, and *rb* single mutants, and the *rrb* double mutant. *Carbohydr. Res.* **338**, 2135-2147.

Ridout, M.J, Parker, ML, Hedley, CL, Bogracheva, TY. and Morris, V.J. (2004). Atomic Force Microscopy of Pea Starch: Origins of Image Contrast. *Biomacromolecules* **5**, 1519-1527.

Ridout, M.J, Parker, M.L, Hedley, C.L, Bogracheva, T.Y. and Morris, V.J. (2006). Atomic Force Microscopy of pea starch granules: granule architecture of the *rug3*, *rug4*, *rug5*, and the *lam* mutants. *Carbohydr. Polym.* **65**, 64-74.

Rief, M, Oesterhelt, F, Heymann, B. and Gaub, H.E. (1997a). Single molecule force spectroscopy on polysaccharides by atomic force microscopy. *Science* **275**, 1295-1297.

Rief, M, Gautel, M, Oesterhelt, F, Fernandez, J.M. and Gaub, H.E. (1997b). Reversible unfolding of individual titin immunoglobulin domains by AFM. *Science* **276**, 1109-1112.

Rief, M, Gautel, M, Schemmel, A. and Gaub, H.E. (1998). The mechanical stability of immunoglobulin and fibronectin III domains in the muscle protein titin measured by atomic force microscopy. *Biophys. J.* **75**, 3008-3014.

Rief, M, Pascual, J, Saraste, M. and Gaub, H.E. (1999). Single molecule force spectroscopy of spectrin repeats: Low unfolding forces in helix bundles. *J. Mol. Biol.* **286**, 553-561.

Rivetti, C, Guthold, M. and Bustamante, C. (1999). Wrapping of DNA around the *E. coli* RNA polymerase open promoter complex. *EMBO J.* **16**, 4464-4475.

Rivetti, C. and Codeluppi, S. (2001). Accurate length determination of DNA molecules visualized by atomic force microscopy: evidence for a partial B- to A-form transition on mica. *Ultramicroscopy* **87**, 55-66.

Rivetti, C, Guthold, M. and Bustamante, C. (1996). Scanning force microscopy of DNA deposited onto mica: equilibrium versus kinetic trapping studied by statistical polymer chain analysis. *J. Mol. Biol.* **264**, 919-932.

Roberts, C.J, Williams, P.M, Davies, J, Dawkes, A.C, Sefton, J, Edwards, J.C, Haymes, A.G, Bestwick, C, Davies, M.C. and Tendler, S.J.B. (1995). Real space differentiation of IgG and IgM antibodies deposited on microtiter wells by scanning force microscopy. *Langmuir* **11**, 1822-1826.

Ros, R, Eckel, R, Bartels, F, Sischka, A, Baumgarth, B, Wilking, S.D, Puhler, A, Sewald, N, Becker, A. and Anselmetti, D. (2004). Single molecule force spectroscopy on ligand-DNA complexes: from molecular binding mechanisms to biosensor applications. *J. Biotechnol.* **112**, 5-12.

Rouillat, M.H, Dugas, V, Martin, J.R. and Phaner-Goutorbe, M. (2005). Characterization of DNA chips on the molecular scale before and after hybridization with an atomic force microscope. *Appl. Surface Sci.* **252**, 1765-1771.

Round, A. N, Kirby, A.R. and Morris, V.J. (1996) Collection and processing of AFM images of plant cell walls. *Microscopy & Analysis* **55**, 33-35.

Round, A.N, MacDougall, A.J, Ring, S.G. and Morris, V.J. (1997). Unexpected branching in pectins observed by Atomic Force Microscopy. *Carbohydr. Res.* **303**, 251-253.

Round, A.N, Rigby, N.M, Ring, S.G. and Morris, V.J. (2001). Investigating the nature of branching in pectins by atomic force microscopy and carbohydrate analysis. *Carbohydr. Res.* **331**, 337-342.

Round, A.N, Berry, M, McMaster, T.J, Stoll, S, Gowers, D, Corfield, A.P. and Miles, M.J. (2002). Heterogeneity and persistence length in human ocular mucins. *Biophys. J.* **83**, 1661-1670.

Round, A.N, Berry, M, McMaster, T.J, Corfield, A.P. and Miles, M.J. (2004). Glycopolymer charge density determines conformation in human ocular mucin gene products: an atomic force microscope study. *J. Struct. Biol.* **145**, 246-253.

Ruozi, B, Tosi, G, Leo, E. and Vandelli, M.A. (2007). Application of atomic force microscopy to characterise liposomes as drug and gene carriers. *Talanta* **73**, 12-22.

Saito, M, Kobayashi, M, Iwabuchi, S.I, Morita, Y, Takamura, Y. and Tamiya, E. (2004). DNA condensation monitoring after interaction with Hoechst 33258 by atomic force microscopy and fluorescence spectroscopy. *J. Biochem.* **136**, 813-823.

Samori, B, Nigro, C, Armentano, V, Cimieri, S, Zuccheri, G. and Quagliariello, C. (1993). DNA supercoiling imaged in 3 dimensions by scanning force microscopy. *Angew. Chem. Int. Ed. Engl.* **32**, 1461-1463.

San Paulo, A. and Garcia, R. (2000). High-resolution imaging of antibodies by tapping-mode atomic force microscopy: attractive and repulsive tip-sample interaction regimes. *Biophys. J.* **78**, 1599-1605.

Sanchez, H, Cardenas, P, Shige, H.Y, takeyasu, K. and Alonso, J.C. (2007). Dynamic structures of *Bacillus subtilis* Rec-DNA complexes. *Nucleic Acids Res.* **36**, 110-120.

Sanchez-Sevilla, A, Thimonier, J, Marilley, M, Rocca-Serra, J. and Barbet, J. (2002). Accuracy of AFM measurements of the contour length of DNA fragments adsorbed on mica in air and aqueous buffer. *Ultramicroscopy* **92**, 151-158.

Sansone, F, Dubic, M, Donofrio, G, Rivetti, C, baldini, L, Casnati, A, Cellai, S. and Ungaro, R. (2006). DNA condensation and cell transfection properties of guanidinium calixarenes: Dependence on macrocycle lipophilicity, size, and conformation. *J. Amer. Chem. Soc.* **128**, 14528-14536.

Santosh, M and Maiti, P.K. (2009). Force induced DNA melting. *J. Phys.-Condensed Matter* **21**, article number 034113.

Saoudi, B, Lacapere, J-J, Chatenay, D, Pepin, R, Derpirre, C. and Sartre, A. (1994). Imaging surface of gold-immunolabelled thin sections by atomic force microscopy. *Biol. Cell* **80**, 63-66.

Sarkar, A, Caamano, S. and fernandez, J.M. (2005). The elasticity of individual titin PEVK exons measured by single molecule atomic force microscopy *J. Biol. Chem.* **280**, 6261-6264.

Sattin, B.D. and Goh, M.C. (2004). Direct observation of the assembly of Rec/DNA complexes by atomic force microscopy. *Biophys. J.* **87**, 3430-3436.

Sattin, B.D. and Goh, M.C. (2006). Novel polymorphism of RecA fibrils revealed by atomic force microscopy. *J. Biol. Phys.* **32**, 153-168.

Schindler, H. (1980). Formation of planar bilayers from artificial or native membrane vesicles. *FEBS Letts.* **122**, 77-79.

Schotanus, M.P, Aumann, K.S. and Sinniah, K. (2002). Using force spectroscopy to investigate the binding of complementary DNA in the presence of intercalating agents. *Langmuir* **18**, 5333-5336.

Schurholz, Th. and Schindler, H. (1991). Lipid-protein surface films generated from membrane vesicles: self-assembly, composition, and film structure. *Eur. Biophys. J.* **20**, 71-78.

Schleif, R. and Hirsh, J. (1980). Electron microscopy of proteins bound to DNA. *Method Enzymol.* **65**, 885-896.

Schlierf, M, Li, H. and Fernandez, J.M. (2004). The unfolding kinetics of ubiquitin captured with single-molecule force-clamp techniques. *PNAS USA* **101**, 7299-7304.

Schneider, S.W, Larmer, J, Henderson, R.M. and Oberleithner, H. (1998). Molecular weights of individual proteins correlate with molecular volumes measured by atomic force microscopy. *Pflugers Arch-Eur. J. Physiol.* **435**, 362-367.

Schwaiger, I, Sattler, C, Hostetter, D.R. and Riej, M. (2002). The myosin coiled-coil is a truly elastic protein structure. *Nature Materials* **1**, 232-235.

Schwaiger, I, Kardinal, A, Schleicher, M, Noegel, A.A. and Rief, M. (2004). A mechanical unfolding intermediate in an actin-cross-linking protein. *Nature Struct. & Mol. Biol.* **11**, 81-85.

Seol, Y, Li, J.Y, Nelson, P.C, Perkins, T.T. and Betterton, M.D. (2007). Elasticity of short DNA molecules: Theory and experiment for contour lengths of 0.6-7 μm. *Biophys. J.* **93**, 4360-4373.

Seong, G.H, Yanagida, Y, Aizawa, M. and Kobatake, E. Atomic force microscopy identification of transcription factor NFκB bound to streptavidin-pin-holding DNA probe. *Anal. Biochem.* **309**, 241-247.

Seong, G.H, Niimi, T, Yanagide, Y, Kobatake, E. and Aizawa, M. (2000). Single-molecular AFM probing of specific DNA sequencing using RecA-promoted homologous pairing and strand exchange. *Anal. Chem.* **72**, 1288-1293.

Sewald, N, Wilking, S.D, Eckel, R, Albu, S, Wollschlager, K, Gaus, K, Becker, A, Bartels, F.W, Ros, R. and Anselmetti, D. (2006). Probing DNA-peptide interaction forces at the single-molecule level. *J. Peptide Sci.* **12**, 836-842.

Shaiu, W.L, Larson, D.D, Vesenka, J. and Henderson, E. (1993a). Atomic force microscopy of oriented linear DNA molecules labelled with 5 nm gold spheres. *Nucleic. Acids Res.* **21**, 99-103.

Shaiu, W.L, Vesenka, J, Jondle, D, Henderson, E. and Larson, D.D. (1993b). Visualization of circular DNA molecules labelled with colloidal gold spheres using atomic force microscopy. *J. Vac. Sci. & Technol. A* **11**, 820-823.

Shao, Z, Yang, J. and Somlyo, A.P. (1995). Biological atomic force microscopy: from microns to nanometers and beyond. *Ann. Rev. Cell Dev. Biol.* **11**, 241-265.

Shao, Z. and Zhang, Y. (1996) Biological cryo atomic force microscopy: a brief review. *Ultramicroscopy* **66**, 141-152.

Shao, Z, Mou, J, Czajkowski, D.M, Yang, J. and Yuan, J-Y. (1996). Biological atomic force microscopy: what is achieved and what is needed. *Adv. Phys.* **45**, 1-86.

Shao, Z, Shi, D. and Somlyo, A.V. (2000). Cryoatomic force microscopy of filamentous actin. *Biophys. J.* **78**, 950-958.

Sharma, D, Perisic, O, Peng, Q, Cao, Y, lam, C, Lu, H. and Li, H. (2007). Single-molecule force spectroscopy reveals a mechanically stable protein fold and the rational tuning of its mechanical stability. *PNAS USA* **104**, 9278-9283.

Shattuck, M.B, Gustafsson, M.G.L, Fisher, K.A, Yanagimoto, K.C, Veis, A, Bhatnagar, R.S. and Clarke, J. (1994). Monomeric collagen imaged by cryogenic force microscopy. *J. Microscopy - Oxford* **174**, RP1-RP2.

Sheng, S, Gao, Y, Khromov, A.S, Somlyo, A.V, Somlyo, A.P. and Zhang, Z. (2003). Cryo-atomic force microscopy of unphosphorylated and thiophosphorylated single smooth muscle myosin molecules. *J. Biol. Chem.* **278**, 39892-39898.

Shi, D, Somlyo, A.V, Somolyo, A.P. and Shao, Z. (2001). Visualizing filamentous actin on lipid bilayers by atomic force microscopy in solution. *J. Microscopy-Oxford* **201**, 377-382.

Shi, W.X. and Larson, R.G. (2005). Atomic force microscopic study of aggregation of RecA-DNA nucleoprotein filaments into left-handed supercoiled bundles. *Nano Letts.* **5**, 2476-2481.

Shi, W.X. and Larson, R.G. (2007). Rec-ssDNA filaments supercoil in the presence of single-stranded protein. *Biochem. & Biophys. Res. Commun.* **357**, 755-760.

Shin, M, Song, M, Rhee, J.H, Hong, Y, Kim, Y-J, Seok, Y-J, Ha, K-S, Jung, S-H. and Choy, H.E. (2005). DNA looping-mediated repression by histone-like protein H-NS: specific requirement of $E\sigma^{70}$ as a cofactor for looping. *Genes & Development* **19**, 2388-2398.

Shivji, A.P, Davies, M.C, Roberts, C.J, Tendler, S.J.B. and Wilkinson, M.J. (1996). Molecular surface morphology studies of beta-amyloid self-assembly: Effect of pH on fibril formation. *Protein & Peptide Letts.* **3**, 407-414.

Shlyakhtenko, L.S, Potaman, V.N, Sinden, R.R, Gall, A.A. and Lyubchenko, Y.L. (2000). Structure and dynamics of three-way DNA junctions: atomic force microscopy studies. *Nucleic Acids Res.* **28**, 3472-3477.

Sibley, S.P, Sosinsky, K, Gulian, L.E, Gibbs, E.J. and Pasternack, R.F. (2008). Probing the mechanism of insulin aggregation with added metalloporphyrins. *Biochemistry* **47**, 2858-2865.

Silva, L.P. (2005). Imaging proteins with atomic force microscopy: an overview. *Current Protein & Peptide Sci.* **5**, 387-395.

Sischka, A, Toensing, K, Eckel, R, Wilking, S.D, Sewald, N, Ros, R. and Anselmetti, D. (2005). Molecular mechanisms and kinetics between DNA and DNA binding ligands. *Biophys. J.* **88**, 404-411.

Sletmoen, M. and Stokke, B.T. (2005). Structural properties of polyC-scleroglucan complexes. *Biopolymers* **79**, 115-127.

Sletmoen, M, Skjåk-Braek, G. And Stokke, B.T. (2004). Single-molecular pair unbinding studies of mannuronan C-5 epimerase AlgE4 and its polymer substrate. *Biomacromolecules* **5**, 1288-1295.

Smith, B.L, Gallie, D.R, Le, H. and Hansma, P.K. (1997). Visualization of poly(A)-binding protein complex formation with poly(A) RNA using atomic force microscopy. *J. Struct. Biol.* **119**, 109-117.

Smith, S.B, Finzi, L. and Bustamante, C. (1992). Direct mechanical measurements of the elasticity of single DNA-molecules using magnetic beads. *Science* **258**, 1122-1126.

Soman, P, Rice, Z. and Siedlecki, C.A. (2008). Immunological identification of fibrinogen in dual-component protein films by AFM imaging. *Micron* **39**, 832-842.

Soon, L.L.L, Bottema, C. and Breed W.G. (1997). Atomic force microscopy and cytochemistry of chromatin from marsupial spermatozoa with special reference to *Sminthopsis crassicaudata*. *Mol. Reproduction & Development* **48**, 367-374.

Sotomayor, M. and Schulten, K. (2007). Single-molecule experiments *in vitro* and *in silico*. *Science* **316**, 1144-1148.

Stali, C, Wood, D.W. and Scoles, G. (2008). Ligand-induced structural changes in maltose binding proteins measured by atomic force microscopy. *Nano Letts.* **8**, 2503-2509.

Stanic, V, Arntz, Y, Richard, D, Affolter, C, Hguyen, I, Crucifix, C, Schultz, P, Baehr, C, Frisch, B. and Ogier, J. (2008). Filamentous condensation of DNA induced by pegylated poly-L-lysine and transfection efficiency. *Biomacromolecules* **9**, 2048-2955.

Stein, P.E, Boodhoo, A, Armstrong, G.D, Cockle, S.A, Klein, M.H. and Reid, R.J. (1994). The crystal structure of pertussis toxin. *Structure.* **2**, 45-57.

Stine Jnr, W.B, Snyder, S.W, Ladror, U.S, Wade, W.S, Miller, M.F, Perun, T.J, Holzman, T.F. and Krafft, G.A. (1996). The nanometer-scale structure of amyloid-β visualised by atomic force microscopy. *J. Protein Chem.* **15**, 193-203.

Storm, C. and Nelson, P.C. (2003). Theory of high-force DNA stretching and overstretching. *Phys. Rev. E* **67**, article number 051906.

Strick, T.R, Dessinges, M.N, Charvin, G, Dekker, N.H, Allemand, J.F, Bensimon, D. and Croquette, V. (2003). Stretching of macromolecules and proteins. *Reports Progr. Phys.* **66**, 1-45.

Strunz, T, Croszlan, K, Schafer, R. and Guntherodt, H-J. (1999). Dynamic force spectroscopy of single DNA molecules. *PNAS USA* **96**, 11277–11282.

Sui, L, Zhao, K, Ni, M.N, Guo, J.Y, Luo, H.B, Mei, J.O, Lu, X.Q. and Zhou, P. (2004). Investigation of DNA breaks induced by Li-7 and C-12 ions. *High Energy Phys. & Nuclear Phys.- Chin. Edition* **28**, 1126-1130.

Sui, L, Zhao, K, Ni, M.N, Guo, J.Y, Kong, F.Q, Cai, M.H, Lu, X.Q. and Zhou, P. (2005). Atomic force microscopy measurement of DNA fragment induced by heavy ions. *Chin. Phys. Letts.* **22**, 1010-1013.

Sun, Y.C, Arakawa, H, Osada, T. and Ikai, A. (2002). Tapping and contact mode imaging of native chromosomes and extraction of genomic DNA using AFM tips. *Appl. Surface Sci.* **118**, 499-505.

Suzuki, A, Yamazaki, M. and Kobiki, Y. (1996). Direct observation of polymer gel surfaces by atomic force microscopy. *J. Chem. Phys.* **104**, 1751-1757.

Suzuki, A, Yamazaki, M, Kobiki, Y. and Suzuki, H. (1997). Surface domains and roughness of polymer gels observed by atomic force microscopy. *Macromolecules* **30**, 2350-2354.

Suzuki, A, Yamazaki, M, Kobiki, Y. and Suzuki, H. (1998). Surface roughness of polymer gels. In *The Wiley Polymer Networks Group Review Series, Volume 1*, (eds. T. Nijenhuis and W.J. Mijs), pp. 489-503.

Szymczak, P. and Cieplak, M. (2006). Stretching of proteins in a force-clamp. *J. Phys.-Condensed Matter* 18, L21-L28.

Takashima, A, Sadamoto, H, Okuta, A. and Ito, E. (2008). Atomic force microscopic observation of nucleosomes consisting of core histones and DNA promoter regions. *Information–An International Interdisciplinary J.* **11**, 513-523.

Takeda, S, Ptaz, K, A, Nakamura, C, Miyake, J, Kageshima, M, Jarvis, S.P. and Tokumoto, H. (2001). Measurement of the length of the α helical section of a peptide directly using atomic force microscopy. *Chem. Pharm. Bull.* **49**, 1512-1516.

Tanigawa, M. and Okada, T. (1998). Atomic force microscopy of supercoiled DNA structure on mica. *Anal. Chem.* **365**, 19-25.

Tatham, A.S, Thomson, N.H, McMaster, T.J, Humphris, A.D.L, Miles, M.J. and Shewry, P.R. (1999). Scanning probe microscopy studies of cereal storage protein structures. *Scanning* **21**, 293-298.

Tasker, S, Matthijs, G, Davies, M.C, Roberts, C.J, Schacht, E.H. and Tendler, S.J.B. (1996). Molecular resolution imaging of dextran monolayers immobilized on silica by atomic force microscopy. *Langmuir* **12**, 6436-6446.

Tay, S.L, Xu, G.O. and Perera, C.O. (2005). Aggregation profile of 11S, 7S and 2S coagulated with GDL. *Food Chem.* **91**, 457-462.

Thalhammer, S, Stark, R.W, Muller, S, Wienberg, J. and Heckl W.M. (1997). The atomic force microscope as a new microdissecting tool for the generation of genetic probes. *J. Struct. Biol.* **119**, 232-237.

Thimm, J.C, Burritt, D.J, Ducker, W.A. and Melton, L.D. (2000). Celery (*Apim graveolens* L.) parenchyma cell walls examined by atomic force microscopy: effect of dehydration on cellulose microfibrils. *Planta* **212**, 25-32.

Thimonier, J, Chauvin, J.P, Barber, J. and Rocca-Serra, J. (1995). Preliminary studies of an immunoglobulin M by near field microscopies. *J. Trace & Microprobe Technol.* **13**, 353-359.

Thomson, N.H. (2005). Imaging the substructure of antibodies with Tapping-mode AFM in air: the importance of a water layer on mica. *J. Microscopy-Oxford* **217**, 193-199.

Thomson, N.H, Miles, M.J, Ring, S.G, Shewry, P.R. and Tatham, A.S. (1994). Real-time imaging of enzymatic degradation of starch granules by atomic force microscopy. *J. Vac. Sci. & Technol. B* **12**, 1565-1568.

Thomson, N.H, Kasas, S, Smith, B, Hansma, H.G. and Hansma, P.K. (1996a). Reversible binding of DNA to mica for AFM imaging. *Langmuir* **12**, 5905-5908.

Thomson, N.H, Fritz, M, Radmacher, M, Cleveland, J.P, Schmidt, C.F. and Hansma, P.K. (1996b). Protein tracking and detection of protein motion using atomic force microscopy. *Biophys. J.* **70**, 2421-2431.

Thomson, N.H, Smith, B.L, Almqvist, N, Schmitt, L, Kashlev, M, Kool, E.T. and Hansma, P.K. (1999). Oriented, active *Escherichia coli* RNA polymerase: an atomic force microscope study. *Biophys. J.* **76**, 1024-1033.

Thundat, T, Allison, D.P, Warmack, R.J, Doktycz, M.J, Jacobson, K.B. and Brown, G.M. (1993). Atomic force microscopy of single and double stranded deoxyribonucleic acid. *J. Vac. Sci. & Technol. A* **11**, 824-828.

Thundat, T, Allison, D.P. and Warmack, R.J. (1994). Stretched DNA observed with atomic force microscopy. *Nucleic Acids Res.* **22**, 4224-4228.

Tkachenko, A.V. (2004). Unfolding and unzipping of single-stranded DNA by stretching. *Phys. Rev. E* **70**, article number 051901.

Todd, B.A, Rammohan, J. and Eppall, S.J. (2003). Connecting nanoscale images of proteins with their genetic sequences. *Biophys. J.* **84**, 3982-3991.

Touhami, A, Hoffmann, B, Vasella, A, Denis, F. and Dufrene, Y.F. (2003) Aggregation of yeast cells: direct measurement of discrete lectin-carbohydrate interactions. *Microbiology* **149**, 2873–2878.

Tseng, Y.D, Ge, H.F, Wang, X.Z, Edwardson, J.M, Waring, M.J, Fitzgerald, W.J. and Henderson, R.M. (2005). Atomic force microscopy study of the structural effects induced by echinomycin binding to DNA. *J. Mol. Biol.* **345**, 745-758.

Tskhovrebova, L, Trinik, J, Sleep, J. and Simmons, R. (1997). Elasticity and unfolding of single molecules of the giant muscle protein titan. *Nature* **387**, 308-312.

Tsukamoto, K, Kuwazaki, S, Yamamoto, K, Scichiri, M, Yoshino, T, Ohtani, T. and Suglyama, S. (2006a). Nanometer-scale dissection of chromosomes by atomic force microscopy combined with heat-denaturing treatment. *Japanese J. Appl. Phys. Part 1–Regular Papers Brief Commun. & Rev. Papers* **45**, 2337-2340.

Tsukamoto, K, Kuwazaki, S, Yamamoto, K, Ohtani, T. and Suglyama, S. (2006b). Dissection and high-yield recovery of nanometre-scale chromosome fragments using an atomic-force microscope. *Nanotechnol.* **17**, 1391-1396.

Tumer, Y.T.A, Roberts, C.J. and Davies, M.C. (2007). Scanning probe microscopy in the field of drug delivery. *Advanced Drug Delivery Rev.* **59**, 1453-1473.

Uji-I, H, Foubert, P, De Schryver, F.C, De Feyter, S, Gicquel, E, Etoc, A, Moucheron, C. and Kirsch-De Mesmaeker, A. (2006). [Ru(TAP)(3)](2$^+$)-photosensitized DNA cleavage studied by atomic force microscopy and gel electrophoresis: A comparative study. *Chem. & Europ. J.* **12**, 758-762.

Umemura, K, Kometsu, J, Uchihashi, T, Choi, N, Ikawa, S, Nishinaka, T, Shibata, T, Nakayama, Y, Katsura, S, Mizuno, A, Tokumoto, H, Ishikawa, M. and Kuroda, R. (2001). Atomic force microscopy of RecA-DNA complexes using a carbon nanotube tip. *Biochem. & Biophys. Res. Commun.* **281**, 390-395.

Umemura, K, Okada, T. and Kuroda, R. (2005). Cooperativity and intermediate structures of single-stranded DNA binding-assisted RecA-single-stranded DNA complex formation studied by atomic force microscopy. *Scanning* **27**, 35-43.

Urry, D.W, and Parker, T.M. (2002). Mechanics of elastin: molecular mechanism of biological elasticity and its relationship to contraction. *J. Muscle Res. & Cell Motility* **23**, 543-559.

Ushiki, T, Hishi, G, Iwaai, K, Kimura, E. and Shigeno, M. (2002). The structure of human metaphase chromosomes: its historical perspective and new horizons by atomic force microscopy. *Arch. Histol. Cytol.* **65**, 377-390.

Valle, M, Valpuesta, J.M, Carrascosa, J.L, Tamayo, J. and Garcia, R. (1996). The interaction of DNA with bacteriophage φ29 connector: a study by AFM and TEM. *J. Struct. Biol.* **116**, 390-398.

van der Aa, B.C, Michel, R.M, Asther, M, Zamora, M.T, Rouxhel, P.G. and Dufrene, Y.F. (2001). Stretching cell surface macromolecules by atomic force microscopy. *Langmuir* **17**, 3116-3119.

van der Wel, N.H, Putman, C.A.J, Van Noort, S.J.T, de Grooth, B.G. and Emons, A.M.C. (1996). Atomic force microscopy of pollen grains, cellulose microfibrils, and protoplasts. *Protoplasma* **194**, 29-39.

van Hulst, N.F, Garcia-Parajo, M.F, Moers, M.H.P, Veerman, J-A. and Ruiter, A.G.T. (1997). Near-field fluorescence imaging of genetic material: towards the molecular limit. *J. Struct. Biol.* **119**, 222-231.

van Noort, J, van der Heijden, T, Dutta, C.F, Firman, K. and Dekker, C. (2004). Initiation of translocation by Type I restriction-modification enzymes is associated with a short DNA extrusion. *Nucleic Acids Res.* **32**, 6540-6547.

Vesenka, J, Guthold, M, Tang, C.L, Keller, D, Delanie, E. and Bustamante, C. (1992a). A substrate preparation for reliable imaging of DNA molecules with the scanning force microscope. *Ultramicroscopy* **42**, 1243-1249.

Vesenka, J, Hansma, H.G, Siegerist, C, Siligardi, G, Schabtach, E. and Bustamante, C. (1992b). Scanning force microscopy of circular DNA and chromatin in air and propanol. *SPIE Proc.* **1639**, 127-137.

Vesenka, J, Mosher, C, Schaus, S, Ambrosio, L. and Henderson, E. (1995). Combining optical and atomic force microscopy for life sciences research. *Biotechnique* **19**, 240-253.

Vieira, E.P, Hermel, H. and Mohwald, H. (2003). Change and stabilization of the amyloid-(1-40) secondaty structure by fluorocompounds. *Biochim. Biophys. Acta-Proteins & Proteomics* **1645**, 6-14.

Viglasky, V, Valle, F, Adamcik, J, Joab, I, Podhradsky, D. and Dieller, G. (2003). Anthracycline-dependent heat-induced transition from positive to negative supercoiled DNA. *Electrophoresis* **24**, 1703-1711.

Vilfan, I.D, Kamping, W, van den Hout, M, Candelli, A, Hage, S. and Dekker, N.H. (2007). An RNA toolbox for single-molecule force spectroscopy studies. *Nucleic Acids Res.* **35**, 6625-6639.

Vimik, K, Lyubchanko, Y.L, Karymov, M.A, Dahlgren, P, Tolstorukov, M.Y, Semsey, S, Zhurkin, V.B. and Adhya, S. (2003). "Antiparallel" DNA loop in gal repressosome visualizes by atomic force microscopy. *J. Mol. Biol.* **334**, 53-63.

Vinckier, A, Heyvaert, I, D'Hoore, A, McKittrick, T, Van Haesendonck, C, Engelborghs, Y. and Hellemans, L. (1995). Immobilizing and imaging microtubules by atomic force microscopy. *Ultramicroscopy* **57**, 337-343.

Volcke, C, Pirotton, S, Grandfils, C, Humbert, C, Thiry, P.A, Ydens, I, Dubois, P. and Raes, M. (2006). Influence of DNA condensation state on transfection efficiency in DNA/polymer complexes: An AFM and DLS comparative study. *J. Biotechnol.* **125**, 11-21.

von Groll, A, Levin, Y, Barbosa, M.C. and Ravazzolo, A.P. (2006). Linear DNA low efficiency transfection by liposome can be improved by the use of cationic lipid as charge neutralizer. *Biotechnol. Progr.* **22**, 1220-1224.

Voulgarakis, N.K, Redondo, A, Bishop, A.R. and Rasmussen, K.O. (2006). Probing the mechanical unzipping of DNA. *Phys. Rev. Letts.* **96**, article number 248101.

Wagner, P, Kernen, P, Hegner, M, Ungewickell, E. and Semenza, G. (1994). Covalent anchoring of proteins onto gold-directed NHS-terminated self-assembled monolayers in aqueous buffers-SFM images clathrin cages and triskelia. *FEBS Letts.* **356**, 267-271.

Wagner, P, Hegner, M, Kernen, P, Zaugg, F. and Semenza, G. (1996). Covalent immobilization of biomolecules onto Au (111) via N-hydroxysuccinimide ester functionalized self-assembled monolayers for scanning probe microscopy. *Biophys. J.* **70**, 2052-2066.

Wagner, O.I, Ascano, J, Tokito, M, Leterrier, J-F, Janmey, P.A. and Holzbaur, E.L.F. (2004). The interaction of neurofilaments with the microtubule motor cytoplamic dynein. *Mol. Biol. Cell* **15**, 5092-5100.

Wang, X.M, Jiang, X.L, Lu, Z.H. and Chen, H.Y. (2000). Evidence of DNA-ligand binding with different modes studied by spectroscopy. *Chin. Chem. Letts.* **11**, 147-148.

Wang, K, Forbes, J.G. and Jin, A.J. (2001). Single molecules measurements of titin elasticity. *Progress Biophys. & Mol. Biol.* **77**, 1-44.

Wang, H, Bash, R, Lindsay, S.M. and Lohr, D. (2005). Solution AFM studies of human Swi-Snf and its interactions with MMTV DNA and chromatin. *Biophys. J.* **89**, 3386-3398.

Wang, B, Sain, M. and Oksman, K. (2007a) Study of structural morphology of hemp fibre from the micro to the nanoscale. *Appl. Composite Mat.* **13**, 89-103.

Wang, C.Z, Li, X.F, Wettig, S.D, Badea, I, Foldvari, M. and Verrall, R.E. (2007b). Investigation of complexes formed by interaction of cationic Gemini surfactants with deoxyribonucleic acid. *Phys. Chem. Chem. Phys.* **9**, 1616-1628.

Wang, C, Huang, L.X, Wang, L.J, Hong, Y.K. and Sha, Y.L. (2007c). One-dimensional self-assembly of a rational designed β-structure peptide. *Biopolymers* **86**, 23-31.

Wei, H. and van de Ven, T.G.M. (2008). AFM-based single molecule force spectroscopy of polymer chains: Theoretical models and applications. *Appl. Spectrosc. Rev.* **43**, 111-133.

Wenner, J.R, Williams, M.C, Rouzina, I. and Bloomfield, V.A. (2002). Salt dependence of the elasticity and overstretching transition of single DNA molecules. *Biophys. J.* **82**, 3160-3169.

Whitelam, S, Pronk, S. and Geissler, P.L. (2008). Stretching chimeric DNA: A test for the putative S-form. *J. Chem. Phys.* **129**, article number 205101.

Wiegrabe, W, Monajembashi, S, Dittmar, H, Greulich, K-O, Hafner, S, Hildebrandt, M, Kittler, M, Lochner, B. and Unger, E. (1997). Scanning near-field optical microscope: a method for investigating chromosomes. *Surface & Interface Analysis* **25**, 510-513.

Wiggens, P.A. and Nelson, P.C. (2006). Generalised theory of semiflexible polymers. *Phys. Rev.* **73**, article number 031906.

Wiggens, P.A, van der Heijden, Moreno-Herrero, F, Spakowitz, A, Phillips, R, Widom, J, Dekker, C. and Nelson, P.C. (2006). High flexibility of DNA on short length scales probed by atomic force microscopy. *Nature Biotechnol.* **1**, 137-141.

Wilkins, M.J, Davies, M.C, Jackson, D.E, Mitchell, J.R, Roberts, C.J, Stokke, B.T. and Tendler, S.J.B. (1993). Comparison of scanning tunnelling microscopy and transmission electron microscopy image data of a microbial polysaccharide. *Ultramicroscopy* **48**, 197-201.

Williams M.A.K, Marshall, A, Haverkamp, R.G. and Graget, K.I. (2008). Stretching single polysaccharide molecules using AFM: a potential method for the investigation of the intermolecular urinate distribution of alginate? *Food Hydrocolloids* **22**, 18-23.

Williams, M.C, Wenner, J.R, Rouzina, L. and Bloomfield, V.A. (2001). Effect of pH on the overstretching transition of double-stranded DNA: Evidence of force-induced DNA melting. *Biophys. J.* **80**, 874-881.

Williams, M.C, Rouzina, I. and Karpel, R.I. (2008). Quantifying DNA-protein interactions by single molecule stretching. *Methods Cell Biol.* **84**, 517-540.

Winfield, M, McMaster, T.J, Karp, A. and Miles, M.J. (1996). Atomic force microscopy of plant chromosmes. *Chromosome Res.* **3**, 128-131.

Wittmar, M, Ellis, J.S, Morell, F, Unger, F, Schumacher, J.C, Roberts, C.J, Tendler, S.J.B, Davies, M.C. and Kissel, T. (2005). Biophysical and transfection studies of an amine-modified poly (vinyl alcohol) for gene delivery. *Bioconjugate Chem.* **16**, 1390-1398.

Woodcock, C.L.F, Frado, L.L.Y. and Rattner, J.B. (1984). The higher-order structure of chromatin-evidence for a helical ribbon arrangement. *J. Cell Biol.* **99**, 42-52.

Woodcock, C.L, Grigoryev, S.A, Horowitz, R.A. and Whitaker, N. (1993). Chromatin folding model that incorporates linker variability generates fibers resembling the native structures. *PNAS USA* **90**, 9021-9025.

Woodcock, C.L. (2006). Chromatin architecture. *Current Opinion Struct. Biol.* **16**, 213-220.

Woodside, M.T, Garcia-Garcia, C. and Block, S.M. (2008). Folding and unfolding single RNA molecules under tension. *Current Opinion Chem. Biol.* **12**, 640-646.

Wu, M. and Davidson, N. (1975). Use of gene 32 protein staining of single-strand polynucleotides for gene mapping by electron microscopy: application to the $\phi80d_3ilvsu+7$ system. *PNAS USA* **72**, 4506-4510.

Wu, X. and Liu, W. (1997). Secondary structure of rat ribosomal RNAs studied by atomic force microscope. *Prog. Biochem. Biophys. (China)* **24**, 430-435.

Wu, X, Liu, W, Xu, L. and Li, M. (1997). Topography of ribosomes and initiation complexes from rat liver as revealed by atomic force microscopy. *Biol. Chem.* **378**, 363-372.

Wu, Y.Z, Cal, J.Y, Cheng, L.Q, Xu, Y.F, Lin, Z.F, Wang, C.X. and Chen, Y. (2006). Atomic force microscope tracking observation of Chinese ovary cell mitosis. *Micron* **37**, 139-145.

Wyatt, H.D.M, Ashton, N.W. and Dahms, T.E.S. (2008). Cell wall architecture of *Physcomitrella patens* is revealed by atomic force microscopy. *Botany-Botanique* **86**, 385-397.

Wyman, C, Grotkoop, E, Bustamante, C. and Nelson, H.C,M. (1995). Determination of heat-shock transcription factor 2 stochiometry at looped DNA complexes using scanning force microscopy. *EMBO J.* **14**, 117-123.

Xu, Q.B, Zhang, W. and Zhang, X. (2002). Oxygen bridge inhibits conformational transition of 1, 4-linked α-D-galactose detected by single-molecule atomic force microscopy. *Macromolecules* **35**, 871-876.

Yamaguchi, H, Kubota, K. and Harada, A. (2000a). Preparation of DNA catenanes and observation of their topological structures by atomic force microscopy. *Nucleic Acids Symp. Ser. No. 44* 229-230.

Yamaguchi, H, Kubota, K. and Harada, A. (2000b). Direct observation of DNA catenanes by atomic force microscopy, *Chem, Letts, issue 4* 384-385.

Yamamoto, S, Hitomi, J, Shigeno, M, Sawaguchi, S, Abe, H. and Ushiki, T. (1997). Atomic force microscopic studies of isolated collagen fibrils of the bovine cornea and sclera. *Arch. Histol. Cytol.* **60**, 371-378.

Yamanaka, K, Saito, M, Shichiri, M, Suglyama, S, Takamura, Y, Hashiguchi, G. and Tamiya, E. (2006). AFM picking-up manipulation of the metaphase chromosome fragment by using the tweezers-type probe. *Ultramicroscopy* **108**, 647-854.

Yan, L.F. and Iwassaki, H. (2002). Thermal denaturation of plasmid DNA observed by atomic force microscopy. *Jap. J. Appl. Phys. Part 1–Regular Papers Short Notes & Review Papers* **41**, 7556-7559.

Yan, L.F. and Zhu, Q.S. (2003). Direct observation of the main cell wall components of straw by atomic force microscopy. *J. Appl. Polm. Sci.* **88**, 2055-2059.

Yan, L.F, Li, W, Yang, J.L. and Zhu, Q.S. (2004). Direct visualization of straw cell walls by AFM. *Macromol. Biosci.* **4**, 112-118.

Yang, G.L, Cecconi, C, Baase, W.A, Vetter, I.R, Breyer, W.A, Haack, J.A, Matthews, B.W, Dahlquist, F.W. and Bustamante, C. (2000). Solid-state synthesis and mechanical unfolding of polymers of T4 lysozyme. *PNAS USA* **97**, 139-144.

Yang, J, Takeyasu, K. and Shao, Z. (1992). Atomic force microscopy of DNA molecules. *FEBS Letts.* **301**, 173-176.

Yang, J, Tamm, L.K, Somlyo, A.P. and Shao, Z. (1993a). Promises and problems of biological atomic force microscopy. *J. Microscopy-Oxford* **171**, 183-198.

Yang, J, Tamm, L.K, Tillack, T.W. and Shao, Z. (1993b). New approach for atomic force microscopy of membrane proteins: the imaging of cholera toxin. *J. Mol. Biol.* **229**, 286-290.

Yang, J, Mou, J. and Shao, Z. (1994a). Molecular resolution atomic force microscopy of soluble proteins in solution. *Biochem. Biophys. Acta* **1199**, 105-114.

Yang, J, Mou, J. and Shao, Z. (1994b). Structure and stability of pertussis toxin studied by *in situ* atomic force microscopy. *FEBS Letts.* **338**, 89-92.

Yang, P.H, Gao, H.Y, Cai, J, Chui, J.F, Sun, H.Z. and He, Q.Y. (2005). The stepwise process of chromium-induced DNA breakage: Characterization by electrochemistry, atomic force microscopy, and DNA electrophoresis. *Chem. Res. Toxicol.* **18**, 1563-1566.

Yang, Y, Sass, L.E, Du, C, Hsieh, P. and Erie, D.A. (2005). Determination of protein-DNA binding constants and specificities from statistical analysis of single molecules: MutS-DNA interactions. *Nucleic Acids Res.* **33**, 4322-4334.

Yavin, E, Stemp, E.D.A, Weiner, L, Sagi, I, Arad-Yellin, R. and Shanzer, A. (2004). Direct photo-induced DNA strand scission by a ruthenium bipyridyl complex. *J. Inorganic Biochem.* **98**, 1750-1758.

Yangzhe, W, Cai, J, Cheng, L, Yun, K, Wang, C. and Chen, Y. (2006). Atomic force microscopic examination of chromosomes treated with trypsin or ethidium bromide. *Chem. Pharm. Bull.* **54**, 501-505.

Ye, Z.Y, Zhang, H.Y, Luo, H.L, Wang, S.K, Zhou, Q.H, Du, X.P, Tang, C.K, Chen, L.Y, Liu, J.P, Shi, Y.K, Zhang, E.Y, Ellis-Behnke, R. and Zhao, X.J. (2008). Temperature and pH effects on biophysical and morphological properties of self assembling peptide RADA 16-1. *J. Peptide Sci.* **14**, 152-162.

Yokokawa, M, Yoshimura, S.H, Naito, Y, Ando, T, Yagi, A, Sakai, N. and Takeyasu, K. (2006a). Fast scanning atomic force microscopy reveals the molecular mechanism of DNA cleavage by ApaI endonuclease. *IEE Proc.-Nanobiotechnol.* **153**, 60-66.

Yokokawa, M, Wada, C, Ando, T, Sakai, N, Yagi, A, Yoshimura, S.H. and Takeyasu, K. (2006b). Fast-scanning atomic force microscopy reveals the ATP/ADP conformational changes of GroEL. *EMBO J.* **25**, 4587-4576.

You, H.X. and Lowe, C. (1996). AFM studies of protein adsorption 2. Characterisation of immunoglobulin G adsorption by detergent washing. *J. Colloid & Interface Sci.* **182**, 586-601.

Yoshimura, S.H, Yoshida, C, Igarashi, K. and Takeyasu, K. (2000). Atomic force microscopy proposes a 'kiss and pull' mechanism for enhancer function. *J. Electron Microscopy* **49**, 407-413.

Yu, J.P, Sun, S, Jiang, Y.X, Ma, X.Y, Chen, F, Zhang, G.Y. and fang, X.H. (2006). Single molecule study of binding force between transcription factor TINY and its DNA response element. *Polymer* **47**, 2533-2538.

Yu, J.P, Malkova, S. and Lyubchanko, Y.L. (2008). A-synuclein misfolding: single molecule AFM force spectroscopy study. *J. Mol. Biol.* **384**, 992-1001.

Yu, H, Liu, R.G, Shen, D.W, Wu, Z.H. and Huang, Y. (2008). Arrangement of cellulose microfibrils in the wheat straw cell wall. *Carbohydr. Polym.* **72**, 122-127.

Yu, Y, Jiang, Z-H, Wang, G, Qin, D-C. and Cheng, Q. (2008). Visualization of cellulose microfibrils of Moso bamboo fibers with atomic force microscopy. *J. Beijing Forestry University* **30**, 124-127.

Zenhausern, F, Adrian, M, Heggeler-Bodrier, B, Emch, R, Jobin, M, Taborelli, M. and Descouts, P. (1992a). Imaging of DNA by scanning force microscopy. *J. Struct. Biol.* **108**, 69-73.

Zenhausern, F, Adrian, M, Heggeler-Bordier, B, Eng, L.M. and Descouts, P. (1992b). DNA and RNA polymerase/DNA complexes imaged by scanning force microscopy: influence of molecular-scale friction. *Scanning* **14**, 212-217.

Zerovnik, E, Skarabot, M, Skerget, K, Giannini, S, Stoka, V, Jenko-Kokalj, S. and Staniforth, R.A. (2007). Amyloid fibril formation by human stefin B: influence of pH and TFE on fibril growth and morphology. *Amyloid J. Protein Folding Disorders* **14**, 237-247.

Zhang, J, Wang, Y.L, Chen, X.Y., He, C.L, Cheng, C. and Cao, Y. (2004). Preliminarily investigating the polymorphism of self-organised actin filament *in vitro* by atomic force microscope. *Acta Biochim. Biophys. Sinica* **36**, 637-643.

Zhang, L, Wang, C, Cui, S.X, Wang, Z.Q. and Zhang, X. (2003). Single-molecule force spectroscopy on curdlan: unwinding helical structures and random coils. *Nano Letts.* **3**, 1119-1124.

Zhang, L, Zhong, J, Huang, L.X, Wang, L.J, Hong, Y.K. and She, Y.L. (2008). Parallel-oriented fibrogenesis of a β-sheet forming peptide on supported lipid bilayers. *J. Phys. Chem. B* **112**, 8950-8954.

Zhang, P, Bai, C, Cheng, Y, Fang, Y, Feng, L. and Pan, H. (1996). Direct observation of uncoated spectrin with atomic force microscope. *Science in China (Series B)* **39**, 378-385.

Zhang, Q.M, Lu, Z.Y, Hu, H, Yang, W.T. and Marszalek, P.E. (2006). Direct detection of the formation of V-amylose helix by single molecule force spectroscopy. *J. Amer. Chem. Soc.* **128**, 9387-9393.

Zhang, S.B, Huang, J. Zhao, H, Zhang, Y, Hou, C.H, Cheng, X.D, Jiang, C, Li, M.Q, Hu, J. and Qian, R.L. (2003). The *in vitro* reconstitution of nucleosome and its binding patterns with HMG1/2 and HMG14/17 proteins. *Cell Res.* **13**, 351-359.

Zhang, W, Barbagallo, R, Madden, C, Roberts, C.J, Woolford, A. and Allen, S. (2005). Progressing single biomolecule force spectroscopy measurements for the screening of DNA binding agents. *Nanotechnol.* **16**, 2325-2333.

Zhang, W, Machon, C, Orta, A, Phillips, N, Roberts, C.J, Allen, S. and Soultanas, P. (2008). Single-molecule atomic force spectroscopy reveals that DnaD forms scaffolds and enhances duplex formation. *J. Mol. Biol.* **377**, 705-714.

Zhang, X.W. (2005). Single-molecule RNA science. *Ann. Rev. Biophys. & Biomol. Struct.* **34**, 399-414.

Zhang, Y, Sheng, S.J. and Shao, Z. (1996). Imaging biological structures with the cryo atomic force microscope. *Biophys. J.* **71**, 2168-2176.

Zhang, Y, Shao, Z, Somlyo, A.P. and Somlyo, A.V. (1997). Cryo-atomic force microscopy of smooth muscle myosin. *Biophys. J.* **72**, 1308-1318.

Zhang, Y, Zhou, H.J. and Ou-Yang, Z.C. (2001). Stretching single-stranded DNA: Interplay of electrostatic, base-pairing, and base-pair stacking interactions. *Biophys. J.* **81**, 1133-1143.

Zhang, Y, Lu, Y.T, Hu, J, Kong, X.Y, Li, B, Zhano, G.P. and Li, M.Q. (2002). Direct detection of mutation sites on stretched DNA by atomic force microscopy. *Surface & Interface Anal.* **33**, 122-125.

Zhang, Y, Lu, J.H, Li, M.Q. and Hu, J. (2007). A strategy for ordered single molecule sequencing based on nanomanipulation (OsmSN). *Int. J. Nanotechnol.* **4**, 163-170.

Zhao, H, Zhang, Y, Zhang, S.B, Jiang, C, He, Q.Y, Li, M.Q. and Qian, R.L. (1999). The structure of the nucleosome core particle of chromatin in chicken erthrocyted visualized by using atomic force microscopy. *Cell Res.* **9**, 255-260.

Zhao, L.M, Schaefer, D, Xu, H.X, Modi, S.J, LaCourse, W.R. Marten, M.R. (2005). Elastic properties of the cell wall of *Aspergillus nidulans* studied with atomic force microscopy. *Biotechnol. Prog.* **21**, 292-299.

Zheng, H.P, Li, Z, Wu, A.G. and Zhou, H. (2003). AFM studies of DNA structures on mica in the presence of alkaline earth metal ions. *Biophys. Chem.* **104**, 37-43.

Zhu, Y, Zang, H, Xie, J.M, Ba, L, Gao, X. and Lu, Z.H. (2004). Atomic force microscopy studies on DNA structural changes induced by vincristine sulphate and aspirin. *Microscopy & Microanalysis* **10**, 286-290.

Zimmermann, T, Thommen, V, Reimann, P. and Hug, H.J. (2006). Ultrastructural appearance of embedded and polished wood cell walls as revealed by atomic force microscopy. *J. Struct. Biol.* **156**, 363-369.

Zlatanova, J, Leuba, S.H, Yang, G, Bustamante, C. and van Holde, K. (1994). Linker DNA accessibility in chromatin fibers of different conformation-a reevaluation. *PNAS USA* **91**, 5277-5280.

Zlantanova, J. and Leuba, S.H. (20030. Chromatin fibers, one-at-a-time. *J. Mol. Biol.* **331**, 1-10.

CHAPTER 5

INTERFACIAL SYSTEMS

5.1. Introduction to interfaces

The term interfacial system covers a vast array of biological samples ranging from the phospholipid bilayer membranes which encase animal cells, self-assembly of proteins into 2D crystalline arrays on bacterial surfaces, to the interfacial layers in colloidal dispersions such as emulsions or foams in foods. The major use of AFM in this area is the investigation of the structure of interfaces. However, before discussing what has been done with the AFM it is useful to introduce certain quantities used to define and characterise surfaces and interfaces.

5.1.1. Surface activity

A fundamental property of any interface between two phases is the existence of a definite quantity of free energy associated with every unit of interfacial area. Interfacial energy is related to interfacial or surface tension and sometimes these terms are used interchangeably. Interfacial tension can, in the simplest case of liquids, be understood by considering intermolecular interactions. Consider a solution of molecules at an air or oil interface (Figure 5.1). Within the bulk phase the van der Waals forces, which hold the molecules in the liquid form act uniformly on each molecule in every direction, hence the net force on each molecule is on average zero. However, this does not apply to the molecules located at the interface, as they only interact with molecules underneath them in the bulk and those alongside them on the interface. Fig. 5.1 illustrates this point schematically. These unbalanced forces at the surface can be equated to a free energy per unit area of surface. The surface free energy results in a contracting force, due to the net inward attraction from the molecules below the surface. Surface tension (γ) is defined as the amount of work required to increase the surface area by a unit amount (Fig. 5.2), therefore the surface tension (γ) and surface free energy are related. The interfacial or surface tension (γ) is defined as the force (F) per unit length required to expand the surface isothermally (Fig. 5.2). This means that the surface tension can also be described in terms of the work required to expand the surface area (A) by a unit amount.

Since reversible work and free energy (G) are equivalent then:

$$\gamma = \left(\frac{\partial G}{\partial A} \right)_{V_1 T_1}$$

(5.1)

where V is the volume and T the temperature.

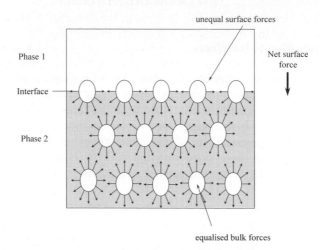

Figure 5.1. Surface tension: molecules in the bulk of phase 2 (liquid phase) have equal van der Waals forces acting on all the molecules. Molecules at the surface only interact with neighbours at the surface or in the bulk, resulting in a net inward force. This has the effect of making a liquid behave as if the surface is enclosed in an elastic skin.

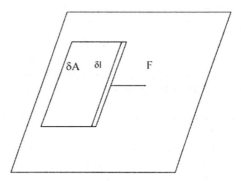

Figure 5.2. Surface tension γ is the force F per unit length δl required to expand the surface isothermally. Increasing the surface area by an amount δA leads to an increase in free energy δG. γ is the work done per unit area leading to equation 5.1.

Certain solutes, even when present in only very low concentrations, alter the surface energy of their solvents to an extreme degree. Generally the effect is to lower the surface tension, and such substances are known as *surface active* agents or *surfactants*. Although the term surfactant can be used to describe any surface active molecule it is generally used to describe only relatively small surface active molecules such as lipids and not larger molecules such as proteins. In the following pages the term surfactant will be applied according to this convention. The most familiar surfactant is soap. Surface-active species self-assemble at the interface and form a 'film'. Most surface active molecules contain hydrophilic and hydrophobic regions, i.e. they are amphiphilic. In general, the assembly at the interface is driven by expulsion of the hydrophobic regions of the amphiphilic molecules from the aqueous environment. A concentration gradient is setup at the interfacial region, and the molecules near the interface orient themselves with their hydrophobic regions toward the hydrophobic side of the interface, as illustrated in Fig. 5.3.

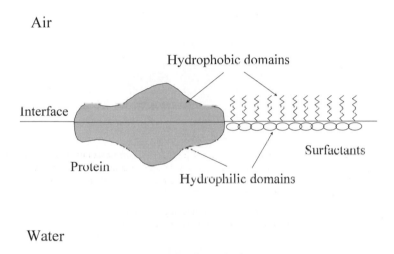

Figure 5.3. Assembly of surface active molecules at an interface by exclusion of hydrophobic regions of the molecules from the hydrophilic (aqueous) phase.

The amount of surface-active molecules present at the interface can be quantified by measuring the change in surface tension. Interfacial films are not purely of academic curiosity, they do actually confer useful physical qualities to the interface. An important property of interfaces are their rheological characteristics. To illustrate this it is instructive to consider how foams are formed and stabilised by surfactants. The formation of foams increases the interfacial area greatly and this occurs because the surfactants lower the surface free energy, meaning that the increase in surface area is energetically possible. However, lowering the interfacial energy is not the whole story since, if all that is required to produce stable foam is low interfacial energy, then pure liquids with low surface tensions such as alcohols should be able to form stable foams, which in reality they cannot. In fact it is the nature of the surfactant molecules themselves which enable the foam to be stabilised. They do this by the so-called Gibbs-Marangoni mechanism. The surfactants are highly mobile in the interfacial region; that is they can diffuse laterally with relative ease (Marangoni mechanism) and, if soluble, there is also a reasonably high rate of exchange of molecules between the bulk liquid and the surface (Gibbs mechanism). Thinning of the interfacial layer between gas bubbles in foams, if not checked would quickly lead to its collapse. In surfactant-stabilised foams such thinning causes localised depletion of the concentration of the surfactant molecules. Because of their lateral mobility within the interface, and their ability to rapidly adsorb from the bulk liquid, other surfactant molecules quickly move into the depleted region to restore equilibrium (Fig. 5.4).

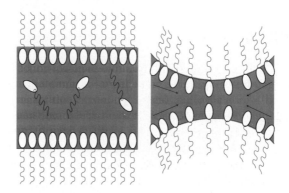

Figure 5.4. When a film is stretched, restoration of interfacial film thickness occurs by the migration of surfactant molecules with adherent water into the depleted (thin) region of the film.

As the surfactant molecules move into the depleted region they bring an attached water layer with them so that the film thickness is restored, the thin spot is 'healed' and the bubble stabilised. It is easy to see how this effect confers a real physical attribute to the interfacial film. In fact it is defined as the interfacial elasticity and the elastic modulus (E) is defined by the rate of change of surface tension with the rate of change of area, thus:

$$E = \frac{d\gamma}{d\ln A} \tag{5.2}$$

Interfacial rheology plays an important role in the biological functions of the cell membrane, for instance membrane fluidity affects cell function in ways too numerous to list.

5.1.2. AFM of interfacial systems

Interfacial regions are often inaccessible making direct imaging with the AFM problematic. For example, in an oil-in-water emulsion, oil droplets are soft, spherical and vary in size over a range of several microns. Similarly, the phospholipid bilayers which surround animal cells are, not only curved, but also are very soft, meaning that molecular resolution AFM images of cell membranes are extremely difficult to obtain *in-situ*. As a final example, the interfacial films in a protein-stabilised foam are difficult to image by AFM because the bubbles themselves are not rigid enough to scan directly. All of these examples present problems which need to be overcome in order to allow high-resolution examination of interfaces by AFM. The most convenient way to study interfacial systems is to recreate them in the controlled environment of a Langmuir trough.

5.1.3. The Langmuir trough

Langmuir troughs generally consist of a shallow rectangular PTFE bath with moveable PTFE barriers for defining the surface area, and some means of monitoring the surface tension of the liquid, usually a delicate balance which quantifies the force acting on a glass plate or metal ring dipped into the liquid surface. Perhaps most importantly they provide a relatively large planar interface which has distinct advantages for AFM imaging. Firstly, the interface can be imaged directly in the 'submarine' AFM (section 2.9.2) or the interfacial film can be transferred onto a suitable substrate without distortion of its shape, and then imaged by AFM. Interfacial films can be formed in two ways, and these are partially dependant on the solubility of the material of interest. Firstly, an interfacial film can be formed by carefully adding the sample solution dropwise to the surface of the liquid in the trough; this technique is known as spreading. Alternatively for some soluble surface active molecules such as proteins, the interfacial film can be formed by simply filling the trough with a solution of the sample, and allowing the molecules to adsorb to the interface by self-assembly; a technique known as adsorption.

So far, the basic principles of interfacial tension and elasticity have been discussed. Another parameter, directly linked to the surface or interfacial tension, and one which is particularly relevant to the Langmuir trough, is the *surface pressure* (Π). The surface pressure defines the tendency of a surface to spread out in a way analogous to the molecules in a gas, but confined to a two dimensional area rather than a three dimensional volume. The surface pressure of a given interface is defined as the difference between the surface tension of the bare interface (γ_0) and the surface tension of the interfacial film (γ_1) and is given by

$$\Pi = \gamma_0 - \gamma_1 \qquad (5.3)$$

Since there is this direct relationship between surface tension and surface pressure it can be seen that the more molecules which are present in an interfacial film the higher will be the surface pressure. This reinforces the gas analogy mentioned earlier, meaning that surface pressure is a useful concept because it can describe the state of packing of the molecules in an interfacial film. For this reason it is often quoted instead of surface tension. Fig. 5.5 illustrates how measurement of surface pressure versus interfacial area (in a graph known as a Π-A isotherm) can define the various physical states (phases) of an interfacial film.

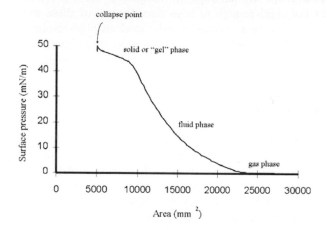

Figure 5.5. Typical Π-A isotherm of a phospholipid film showing gas, liquid, solid phases and the collapse point. Note that sometimes the units of area in a Π-A isotherm are defined in terms of molecular size rather than the absolute units shown here.

At very low surface pressure (high surface tension) the interfacial film is composed of a very low density of amphiphilic molecules, and in for example a lipid interfacial layer, they would be described as being in the expanded 'gas phase': as in a gas the molecules can diffuse freely, but remain at the surface. The packing density can be increased by adding more lipid to the surface or, more commonly, by reducing the interfacial area by moving the barriers on the Langmuir trough. In either case the molecules are forced to pack more closely and the film reaches what is termed the 'fluid phase'; the molecules can still diffuse at this stage but are more confined. If the area is reduced still further the molecules become close packed in what is known as the 'solid' or 'gel' phase. Now there is very little room for molecular diffusion. Finally if the area of the interface is reduced beyond this point then the film will quickly be forced to collapse, with the lipid molecules leaving the surface and entering the bulk or 'sub-phase'.

Now that the nature of the interfacial film has been defined, how is it transferred to a substrate for AFM imaging?

5.1.4. Langmuir-Blodgett film transfer

Generally speaking the interfacial film is very soft. Therefore it must be transferred from the Langmuir trough onto a hard substrate in order to support it mechanically so that AFM imaging may be carried out. The principle is to pull the desired substrate through the interfacial film in a controlled manner. For a planar interface this is quite straightforward, and it can be done using Langmuir-Blodgett (LB) techniques. An excellent and detailed review of LB film structures can be found elsewhere (Shwartz 1997). Measurement of surface pressure (or surface tension) can be used to monitor the transfer of films onto the substrate (Fig. 5.6).

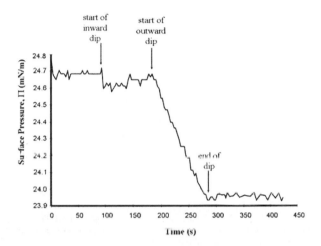

Figure 5.6. Film transfer during dipping is verified by a change in surface pressure. Data shown here is for a protein film transferring onto mica. Note that in this case film transfer, as indicated by a change in surface pressure, occurs only on the outward portion of the dip. See Fig. 5.7. for further details.

Because the surface tension of a given system is inversely related to the amount of the adsorbate at the interface a more convenient measure of film transfer is surface pressure. The principle is that, when some of the adsorbate is transferred onto a substrate during a dipping cycle, the surface pressure will be reduced. The bigger the substrate to trough ratio is, the larger the observed reduction in surface pressure will be on transfer. This change in surface pressure implies a disruption of the equilibrium state of the system and therefore a feedback loop is often employed to maintain constant surface pressure as the transfer progresses, by moving the trough barriers together and reducing the surface area of the trough; hence maintaining the packing of the interfacial layer. In this case a plot of the feedback correction signal or barrier position during the dip will confirm film transfer. Comparison of the area swept out by the moving barrier (during transfer) with that of the substrate, quantifies the transfer ratio. If dipping is performed slowly enough, the transfer ratio will be one. This ensures that the packing density of the interfacial film is not altered by the transfer process. However, since the substrate area required for an AFM sample is usually very small compared to the trough area, such feedback control is not always necessary.

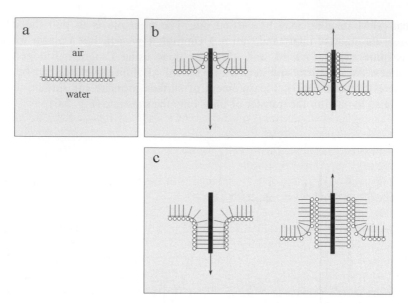

Figure 5.7. Langmuir-Blodgett film transfer from an air-water interface. (a) Surface-active molecules are oriented with their polar head groups in the aqueous phase, meaning that upon dipping a hydrophilic substrate (b) transfer occurs only on the upward stroke, producing monolayer coverage. With a hydrophobic substrate (c) transfer occurs during both strokes producing bilayer coverage.

The nature of the transferred film depends to some extent on the substrate that is dipped through the interface. It is easier for a number of reasons to start the dip with the substrate above the liquid and dip in, and then back out again. The main advantage of doing things in this order is that the substrate is exposed to the bulk solution for only the minimum length of time. This is particularly important when dealing with *adsorbed* films, where the subphase may contain a significant concentration of the sample molecules which can physisorb onto the substrate causing misleading results (for more details see section 5.7.1). For hydrophilic substrates, such as mica, only a monolayer will be transferred during the upward part of the dip, as illustrated in Fig. 5.7b and shown experimentally in Fig. 5.6. For hydrophobic substrates, such as graphite or HF-treated silicon, a bilayer will be transferred, one layer during each part of the dip (Fig. 5.7c).

5.2. Sample preparation

5.2.1. Cleaning protocols: glassware and trough

The following protocols can be used to clean glassware and the trough prior to forming interfacial films. One of the three cleaning regimes described below can be used.

1. Detergent regime:
> i. Clean thoroughly with free rinsing detergent and hot water.
> ii. Rinse with hot water.
> iii. Rinse with ethanol.
> iv. Rinse with water.

2. Acid Regime:
> i. Soak in concentrated acid (e.g. chromic acid, hydrofluoric acid or ammonium persulphate).
> ii. Rinse thoroughly with water.

3. Solvent regime:
> i. Rinse stepwise in solvents of increasing, then decreasing hydrophobicity.
> e.g. water-acetone-chloroform/methanol (2:1)-acetone water.

Note: Hydrofluoric acid (HF) is an extremely hazardous reagent and its use should not be undertaken lightly. This advice is particularly relevant when it is used to wash apparatus, since relatively large volumes of HF are required, making it especially dangerous. Also, it should be noted that, contrary to expectation, it actually becomes *more* dangerous with dilution, since dissociation liberates more F⁻ ions.

To verify that the Langmuir trough is clean and that the water being used is 'surface pure' the surface tension should be measured before spreading the sample at the air-water interface. At 20°C the surface tension value for water is 72.6 mN m^{-1}. Surface active species act to reduce the surface tension and thus, in practice, a value of surface tension lower than 72.5 mN m^{-1} (at 20°C) indicates contamination. Remember though that surface tension varies with temperature and ionic strength (corrected values can be found in most data books). Finally, a good way of ensuring that there is no surface contamination, even at very low levels, is to reduce the trough area to its smallest extent by moving the barriers together whilst measuring the surface tension. This will close-pack any surface contaminants making them easier to detect by a drop in the surface tension. If the surface tension remains constant at 72.5-72.6 mN m^{-1} then the surface is clean.

5.2.2. Substrates

Interfacial films are of course highly sensitive to any surface active species which may come into contact with them so, in addition to cleaning the Langmuir trough itself and all glassware used during sample preparation, the whole of the substrate used for film transfer must be clean, not simply the face to be imaged.

Mica

In the case of mica this means cleaving *both* sides of the sheet being used and also trimming edges which may have been handled prior to dipping through the interface. A rectangular piece of mica is cut from a larger sheet using clean scissors (rinsed with acetone and allowed to dry). The cut piece of mica is then held at the top and bottom edges only whilst both sides are cleaved using two pairs of fine pointed tweezers. This rather acrobatic task is really not that difficult with a little practice. Alternatively, the mica can be cleaved on both sides by sandwiching it between two lengths of adhesive tape and then carefully peeling them apart. This second method, although easier, does make it more difficult to verify complete cleavage of the mica, and good quality adhesive tape, i.e. tape with an even coating of adhesive, is a must. Once cleaved the mica should only be handled with clean tweezers.

Polished silicon wafers

The other useful AFM substrate for LB film transfer is polished silicon. As is the case for mica rectangular sections of substrate make the dipping process easier. These can be broken out of a wafer by scribing the back of the silicon using a diamond glass cutter, lining up the scratch with the edge of a glass slide, and then pressing gently downward to snap off the chord of silicon. Alternatively, the scribed silicon can be broken by placing it face up on a pad of compliant material (such as a Gel-Pak^{TM}, Hayward, CA. USA) and then broken by carefully placing a straight metal edge, such as the end of a steel ruler, opposite the scribed marks, and tapping sharply on the other end. The trick when breaking the silicon is that thinner wafers (thickness $\approx$ 0.4 mm) are much easier to break than thicker ones, which can be anything up to 1 mm in thickness, and often break into about twenty tiny shards! The free piece can now be carefully scribed near the end to be dipped, so that a suitable sized piece which fits into the liquid cell of the AFM can be broken off after dipping. The silicon piece should be cleaned prior to dipping in a hot mixture of Piranha solution (30% hydrogen peroxide and sulphuric acid in the ratio 3:7), followed by a water rinse. This will remove any organic contamination leaving the native hydrophilic oxide surface.

Like hydrofluoric acid, Piranha solution is another very hazardous reagent. It is prone to spontaneous detonation and should be prepared and handled with extreme caution (i.e. appropriate safety advice must be sought before its preparation. If you're a student, do yourself a favour and get an experienced technician or researcher to do it for you). A further 'processing' step in 10% hydrofluoric acid will remove the oxide layer from the silicon leaving a hydrogen-passivated hydrophobic silicon surface.

5.2.3. Performing the dip

After cleaning, a simple and convenient way to hold the substrate during dipping is to use a glass microscope slide fitted with a paper clip (Fig. 5.8). The substrate can simply be slid under the paper-clip. If using mica, the bottom edge should be trimmed off with scissors, since this may be contaminated by grease from your fingers. Now one should have a perfectly clean substrate ready to dip.

Figure 5.8. A mica substrate being dipped through an air-water interface.

The advantage of using the glass slide now becomes clear: it can easily be attached to the dipping mechanism associated with the LB trough using a crocodile clip, thus eliminating the fiddly problems encountered if one tries attach the clean substrate directly to the dipper. The substrate can then be dipped through the interface of the liquid in the trough and out again, taking care not to let the bottom of the glass slide come into contact with the liquid surface. The rate of the dip should be slow enough to ensure that minimal distortion of the interfacial film occurs during transfer. In practice we have found a rate of 8 mm per minute to be fine for protein films but for lipids a slower rate of about 2 mm per minute should be used. After this the substrate can be slid out from under the paper-clip and the dipped section cut to a suitable size to fit into the liquid cell of the AFM. For polished silicon the handling procedure is more difficult and the cutting must be done as described above (section 5.2.2).

5.3. Phospholipids

Phospholipids are a class of amphiphilic molecules of enormous importance in biology since phospholipid bilayers form the basic structure of the outer envelope, or plasma membrane, of all animal cells. Phospholipids are complex lipids consisting of polar (hydrophilic) head-groups containing glycerol with a phosphate group and two non-polar (hydrophobic) hydrocarbon tails. The tails are generally made up of fatty acids of varying length, with one tail having one or more unsaturated *cis* double bonds which introduce kinks in the tail. Variation in the length and saturation of the tails affects the packing ability of the phospholipids, and thus confers rheological properties, such as fluidity to the membranes. Fluidity of the cell membrane is biologically important; for example certain transport processes and enzyme activities are inhibited when the bilayer viscosity increases beyond a threshold level. Due to their cylindrical shape phospholipid molecules spontaneously form bilayers in aqueous solution with the non-polar tails in the middle of the 'sandwich'. The bilayers eliminate free edges where the hydrophobic tails would be in water by forming compartments, and this tendency also means that they can re-seal if torn. The phospholipid bilayer acts as a two dimensional fluid in which membrane proteins are 'dissolved' and the individual lipid molecules are free to diffuse laterally and exchange places with their neighbours. Motion of phospholipid molecules between the monolayers on either side of the bilayer, known as 'flip-flop' motion, is less common, and this confinement creates a problem for the synthesis of the bilayers, since phospholipid molecules are synthesised in only one of the monolayers of a membrane, normally the cytosolic monolayer of the endoplasmic reticulum. If transfer of the molecules to the other side of the membrane could not occur no bilayer could be formed. Transfer is in fact mediated by a special class of membrane proteins called the phospholipid translocators, which catalyse the 'flip-flop' of the phospholipid molecules (Alberts *et al* 1994).

There are a large number of different phospholipids which are defined according to the substitution on the phosphorous atom. For example two of the most abundant in the plasma membranes of mammalian cells, phosphatidylcholine, also known as lecithin, and phosphatidylethanolamine have a choline group and an ethanolamine group attached via the phosphate respectively. The nature of the headgroup confers various properties to the molecules, such as charge and solubility, which have important bearing on the reactions of proteins and enzymes which reside within the membrane, or which interact with it. This specificity of the headgroup means that very often the lipid composition of the two halves of the bilayers (also known as leaflets) are different, for reasons of functionality, and this variability is controlled by the phospholipid translocators during synthesis in the endoplasmic reticulum. For example, the enzyme kinase C, which responds to extracellular signals, binds to the cytoplasmic face of the plasma membrane which is rich in phosphatidylserine, whose negative charge is required for enzymic catalysis. As mentioned above, as well as the chemical makeup of the plasma membrane, its physical characteristics also play an important role. The bilayer can change from being in a liquid like state to a rigid crystalline or gel-like 'frozen' state as the temperature is lowered. The temperature at which this happens is determined by the length and saturation of the hydrocarbon chains in the phospholipid molecules. Longer hydrocarbon chains allow neighbouring molecules to interact more strongly and so raise the 'freezing' temperature. *Cis*-double bonds (which represent an unsaturated structure) cause kinks in the chains (tails) which makes packing of molecules more difficult, and so lowers the 'freezing' temperature. This attribute is used by the cells of primitive organisms such as yeasts, whose temperature is determined by their environment, and which increase the production of unsaturated phospholipids in their plasma membranes to preserve membrane fluidity when drops in temperature occur (Alberts *et al* 1994).

It can be seen that the detailed nature of the composition of the plasma membrane of cells, both in terms of its physical and chemical characteristics, determines functionality and so this is an area ideal for study by atomic force microscopy, offering the ability to examine both attributes at a highly localised level. As mentioned in the earlier section on sample preparation, the best way to image these materials is to prepare synthetic bilayers which can then be transferred onto suitable substrates for the AFM, since imaging of cell membranes *in-situ* produces no resolvable detail and, even under optimised imaging conditions, damages the membrane structure (Schaus and Henderson, 1997). There are two ways of forming synthetic phospholipid bilayers experimentally. The first method is to create them in bulk aqueous solution in the form of spherical vesicles known as liposomes, the size of which can vary from 25 nm to 1 μm in diameter. The liposomes can then be adsorbed onto mica for AFM imaging (Shibata-Seki *et al* 1996) and, if high resolution is required, they can be collapsed down to flat layers (Singh and Keller, 1991). The second method is to create a planar phospholipid monolayer at an air water interface on a Langmuir trough and then transfer this to the substrate by Langmuir-Blodgett dipping (Zasadzinski *et al* 1991). The first dip will produce a phospholipid monolayer on a hydrophilic substrate during the outward portion of the dip, rendering it hydrophobic. To create a bilayer the substrate is dipped for a second time through the phospholipid monolayer in the Langmuir trough. This time the transfer occurs on the inward portion of the dip producing bilayer coverage of the substrate (see section 5.1.4). In order to preserve the integrity of the newly-formed bilayer the sample must now be kept wet whilst it is transferred to the liquid cell of the AFM. Artificially prepared model cell membranes, which have been adsorbed to a planar solid substrate in either manner, are known as supported lipid bilayers (SLBs).

5.3.1. Early AFM studies of phospholipid films

In an early AFM study, membrane bilayers of dipalmitoyl phosphatidylcholine (DPPC) and dipalmitoyl phosphatidylethanolamine (DPPE) were adsorbed to a mica substrate and imaged after air-drying (Singh and Keller, 1991). Two methods were used for deposition of DPPC bilayers. In the first method the mica was incubated in a mixed solution of detergent and the lipid whilst the detergent was slowly removed by dialysis, resulting in the nucleation of small lipid islands which grew radially outwards to produce circular bilayer patches. The second method of deposition involved incubating mica in a suspension of lipid vesicles for 15-30 minutes, then blotting the excess liquid away with filter paper, producing a nearly uniform coverage of lipid bilayer with a few cracks and salt crystals. It was noted that bilayer formation was sometimes enhanced by rolling the drop of lipid suspension around over the mica surface. Despite the use of relatively high forces of around 15 nN the bilayers were reproducibly imaged, although the researchers concluded that this force was near the limit above which damage to the bilayers occurred, as shown by areas which were deliberately damaged by scanning with high forces. Thickness measurements of the DPPC bilayers at 6.3 nm were in reasonable agreement with the expected values of 4-5.5 nm (Sackmann 1983). When lipid vesicles formed from DPPE were deposited using the second method large circular discs up to 2 μm in diameter and 100-200 nm thick resulted, suggesting that, in this case, the vesicles remained intact and did not collapse. However, the samples had been stored at 4°C for several months prior to adsorption to the mica, favouring the formation of very large vesicles in the suspension. An important point arising from the study is that rinsing of the adsorbed vesicles with water removed them, leaving only 'footprints' of inconsistent height where they had been adsorbed to the mica (Singh and Keller, 1991).

5.3.2. Modification of phospholipid bilayers with the AFM

A later study utilised the ability of the AFM to modify phospholipid bilayers composed of 1,2-bis (10,12-tricosadinoyl)-*sn*-glycerol-3-phosphocholine (DC$_{8,9}$PC) (Brandow *et al* 1993). In this case lipid tubules formed by cooling the lipid-solvent dispersion slowly through its transition temperature (Yager and Schoen, 1984) were deposited onto freshly-cleaved highly oriented pyrolytic graphite (HOPG) and imaged in air after drying, whereupon they collapsed to form flat bilayers. Cutting of the lipid tubules was achieved by repeatedly scanning (some 2,500 times) over the same scan line until all of the material had been removed. A force of 12.9 ±1.0 nN was found to be the threshold needed to cut the tubules, irrespective of the angle of cut with respect to the tube axis. Above this force value the cutting process was quicker. When the cuts made to the tubules were narrow (< 80 nm) a process of self-annealing was observed over a period of 24 hours, indicating that the lipid molecules in the bilayer were mobile even in air, and that residual lipids were probably left on the graphite surface after cutting. However, for wider cuts this process could be halted. The AFM tip could, however, be used to heal such wider cuts, by shovelling lipid molecules into the cut, by scanning perpendicular to it with forces approximately 14% smaller than the threshold cutting force. The modification of the lipid tubules is shown in Fig. 5.9. Following polymerisation of the acyl chains of the $DC_{8,9}PC$ lipid tubules with UV irradiation, cutting of the tubules with the AFM tip proved to be impossible. The study was probably the first to demonstrate the ability of the AFM to image, manipulate and therefore potentially to measure and alter some fundamental properties of a lipid bilayer system, such as the intermolecular interactions which determine the viscosity in the bilayer (Brandow *et al* 1993). Following this work several studies have examined the effect of the scanning tip of the AFM on phospholipid bilayers (Hui *et al* 1995; Knapp *et al* 1995), and intact liposomes (Shibata-Seki *et al* 1996).

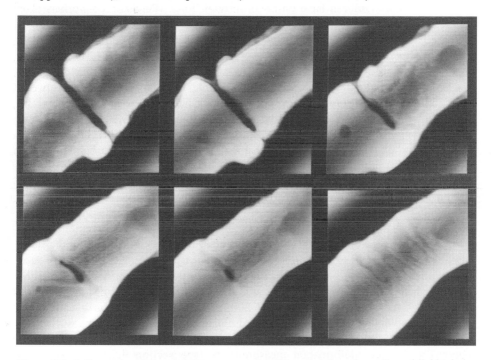

Figure 5.9. A time series of the annealing process of an 80 nm cut through a collapsed lipid tubule which occurs due to repeated scanning of the AFM tip perpendicular to the cut. Reproduced with permission from Brandow *et al* 1993 and the Biophysical Society.

Knapp and co-workers treated AFM tips by glow discharge in hexafluoropropene (HFP) to make them hydrophobic, and in air to make them hydrophilic. The interaction of the treated tips with the sample was then compared by scanning phospholipid bilayers composed of a bottom layer of DPPE and a top layer of DPPC supported on mica substrates (Knapp *et al* 1995). In addition to topography, friction and elasticity of the layers was measured, proving that with modified tips and combined imaging modes the nature of the bilayer surface, whether hydrophobic or hydrophilic, could be determined. This raised the possibility that the AFM could be used to map a mixed lipid bilayer. The study demonstrated that dissection of particular layers with the AFM tip was possible by carefully selecting an appropriate force.

The mechanical properties of phospholipid bilayers have been examined by AFM. Franz and co-workers demonstrated that the break-through force of an AFM tip pushing on a bilayer (i.e. the force at which the tip ruptures the bilayer and pushes through to the underlying mica support) can be described by a simple model (Franz *et al* 2002). Other researchers have demonstrated the effects that tip-chemistry (Pera *et al* 2004), temperature (Garcia-Manyes *et al* 2005a) and ionic strength (Garcia-Manyes *et al* 2005b) can have on break-through force. Pera and co-workers also examined the interactions between bilayers on the tip and a surface (Pera *et al* 2004). The effect of the mechanical stability of phospholipid bilayers on their tribilogical performance, when used to create biomimetic surfaces, has also been examined (Trunfio-Sfarghiu *et al* 2008). In such systems lubrication occurs due to multiple bilayers slipping past one another. A direct correlation was found between the frictional properties of the systems that they studied (DOPC and DPPC) and the force required to penetrate the bilayers. The more mechanically stable the bilayers were, the better they were at maintaining a lubricious hydration layer. However, the interpretation of AFM data on the mechanical properties of phospholipid films can be a non-trivial process. In an important contribution to the understanding of such measurements, Deleu and co-workers have shown that AFM force modulation data on phase-separated monolayers of the surface-active lipopeptide surfactin with DDPC, gave larger values for stiffness in the liquid-expanded surfactin regions of the film, than in the more closely-packed liquid condensed DPPC regions of the film, completely contrary to expectation (Deleu *et al* 2001). Furthermore, the same effect was seen by phase imaging when moderate or hard tapping regimes were used to visualise the sample. The findings suggested that the contact area with the AFM tip played the dominant role in apparent stiffness of the film and not the Young's modulus of the material (Deleu *et al* 2001). The study serves as a cautionary reminder that objectivity should remain the watchword when such measurements are carried out. Other less obvious factors than molecular packing (i.e. interfacial energy) will effect deformation measurements (see section 9.5.3 for details of deformation theory).

5.3.3. Studying intrinsic bilayer properties by AFM

Whilst the above studies concentrated mainly on the methodology in the use of AFM for imaging phospholipid bilayers, there have been several other studies which have concentrated on the nature of the bilayers themselves. In a fairly detailed study the structure and stability of bilayer films of distearoyl phosphatidylcholine (DSPC) and DPPE, transferred as LB films onto mica in the solid phase, and dilinoleoyl phosphatidylethanolamine (DLPE) transferred onto mica in the fluid phase, were examined by AFM under aqueous conditions (Hui *et al* 1995). Bilayer stability is an issue due the fact that phospholipid molecules, which are asymmetrical, tend to form curved monolayers, as illustrated schematically in Fig. 5.10. When these asymmetric lipids are incorporated into a bilayer each monolayer is constrained to a common curvature, giving rise to a bending or frustration energy which reduces the stability of the bilayer. Indeed, supplied with sufficient external energy, this stored energy can lead to the destruction of the bilayer (Seddon 1990). The bending energy in phospholipid bilayers is believed to play an active role in the function of many membrane proteins (Hui and Sen, 1989). The effect of the headgroup on the structure of the bilayer was determined by comparing high magnification AFM images of bilayers composed of DPPE/DPPE and DPPE/DSPC. The bilayer composed entirely of DPPE revealed a periodic structure of ridges 0.49 nm apart, which is close to the hexagonal pattern seen for mica of spacing 0.52 nm, but the dominant influence of the mica structure was discounted because the observed spacing agreed well with the crystallographic structure of a similar phospholipid, dilauroyl phosphatidylethanolamine.

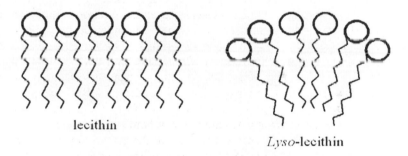

lecithin

Lyso-lecithin

Figure 5.10. Whereas symmetrical lipids such as lecithin form planar monolayers, asymmetric lipids such as *lyso*-lecithin pack in a wedge-like manner and tend to form curved monolayers.

The periodicity seen on the DPPE/DPPE bilayer was therefore attributed to rows of aligned headgroups (Hui *et al* 1995). Furthermore, high magnification AFM images of the bilayer composed of DPPE/DSPC failed to reveal any lattice details, probably confirming that the periodicity seen for the pure DPPE bilayer was not simply due to an inadvertent imaging of the underlying mica lattice. If the results are taken at face value then they represent a demonstration that AFM imaging is capable of determining the effect of the phospholipid headgroup on the packing of molecules within a bilayer, since pure DPPE bilayers would be expected to pack into an ordered crystalline arrangement when in the 'solid' phase, whereas phosphatidylcholine exists in a less ordered gel-like solid phase. Factors such as packing density, ordering of molecules and the number of layers are all known to enhance the stability of supported phospholipid layers, but the role of bending energy upon stability had not been studied. Hui and co-workers studied this effect directly by creating bilayers with a natural tendency to form highly curved surfaces under experimentally controllable conditions (Hui *et al* 1995). By varying the pH of the solution in which the samples were imaged the charge on the headgroups of the phospholipid molecules was manipulated to produce high or low bending energy systems. For example, a monolayer of the unsaturated phospholipid DLPE has a high curvature below neutral pH so a bilayer composed of DPPE/DLPE was created in order to investigate this unstable system. At high values of pH (>11) some defects were observed after the water in the liquid cell of the AFM was exchanged for buffer, but no gross changes were observed. When the imaging buffer was exchanged for one of a low pH (pH 5.0) this situation changed dramatically. The initial scan revealed many more defects than were present at pH 11, and these defects grew during the period of scanning, eventually resulting in the removal of almost the entire top layer of DLPE, correlating with the increase in bending energy which occurs at low pH. These results are reproduced in Fig. 5.11.

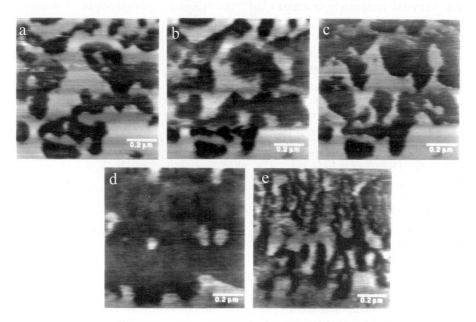

Figure 5.11 Sequence of AFM images of a DPPE/DLPE bilayer at pH5. (a-d) The top leaflet of unsaturated, fluid DLPE is systematically removed, indicating low cohesion between the leaflets at this value of pH. (e) After repeated scanning a regular pattern appears, perpendicular to the scanning direction. This is an artifact which occurs when a soft material is rubbed by a hard material: the features are known as Shallamach waves. Reproduced with permission from Hui *et al* 1995 and the Biophysical Society.

Another interesting observation, and one which should serve as a cautionary reminder of the fact that AFM can affect soft samples, was that after removal of most of the top layer of DLPE, the residual material formed a pattern of roughly parallel lines in response to repeated scanning by the tip (Hui *et al* 1995). These lines are not real structural features, but rather an artefact known as Schallamach waves, which arise when hard objects slide over soft surfaces (Schallamach 1971). The effect of surface pressure on the molecular structure of monolayers of DPPC transferred onto quartz substrates has been examined by AFM (Zhai and Kleijn, 1997). At high and intermediate surface pressures, after the onset of a condensed liquid state, LB films of DPPC on quartz gave AFM images revealing molecular order but, as expected, AFM images of films transferred from monolayers in the less ordered liquid expanded state showed no order (Zhai and Kleijn, 1997).

5.3.4. Ripple phases in phospholipid bilayers

An interesting phenomenon which can occur in phospholipid bilayers is the so-called ripple phase, which happens below the first order phase (main) transition temperature (Tardieu *et al* 1973). In stacks of bilayers it has been shown that in this ripple phase the surface of the membrane becomes wrinkled in a periodic manner (Rand *et al* 1975). Until recently the ripple phase had only been attributed to the temperature and hydration of the stacked bilayers. In an AFM study by Mou and co-workers a new ripple phase, induced by a commonly used buffer compound Tris (hydroxymethyl) aminomethane ($C_4H_{11}NO_3$), was observed in single phospholipid bilayers of phosphatidylcholine (Mou *et al* 1994). Supported phospholipid bilayers were prepared by deposition of a sonicated suspension of vesicles in 20 mM NaCl onto freshly-cleaved mica, followed by a short heating step above the main transition temperature (33 °C). The mica was then rinsed with water prior to imaging under appropriate buffer solution, with the bilayers never being exposed to air. Initially the bilayers were imaged in 20 mM NaCl solution and shown to be flat and featureless, even after temperature cycling above the main transition temperature. By scraping a patch of the phospholipid layer off the mica with the AFM tip the depth of the layer (~ 6 nm) was measured, in order to confirm that only a single bilayer was present. Upon the addition of 20 mM Tris in 50 mM NaCl the surface of the bilayer began to wrinkle and, after 16 hours, a pronounced ripple structure was observed with a periodicity of 18±2 nm, and an amplitude of 0.3 nm. The sample was then heated above its transition temperature (42-50 °C) for half an hour and a second, wider, ripple phase formed with a periodicity of 32±2 nm and an amplitude of 1.2 nm, coexisting with the narrower original phase. These results are reproduced in Fig. 5.12. Interestingly, the thickness of the bilayers changed after ripple formation; the narrow ripple phase bilayer thickness went up to about 7 nm and, for regions of the wide ripple phase, the thickness went up to about 9-10 nm.

At high magnification the thicker ripple phase was seen to be made up of two ridges with a separation of ~ 11 nm. The different phases appeared to be relatively stable; at room temperature no inter-conversion of the two phases was observed.

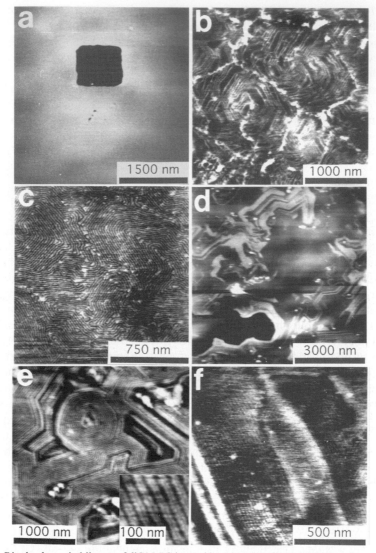

Fig 5.12. Ripple phases in bilayers of diC15-PC imaged by AFM. (a) diC15-PC bilayer imaged in pure NaCl solution (20 mM) displays no ripples. A square hole has been made in the bilayer by scanning at high speed and force so that the thickness of the bilayer could be measured (6 nm). (b) After 2 h incubation at 20 °C in 20 mM Tris and 50 mM NaCl some ripple-like features have started to form and the thickness of the bilayer has increased (7 nm). (c) By 16 h incubation in the mixed buffer, the ripple structures are more pronounced, but the bilayer thickness has not changed (7 nm). (d) Following heating of this sample above the main transition temperature (42-50 °C) two distinct ripple phases were seen; thick domains and thin domains. The thicker domains were 2-3 nm taller than the thin domains. (e) A higher magnification image of the thick domain, showing very clearly the ripple structure. At even higher magnification (inset) each ripple can be seen to consist of two ridges. (f) High magnification image of the thinner domain. Reproduced with permission from Mou *et al* 1994, copyright (1994) American Chemical Society.

When the Tris buffer was replaced with just NaCl solution however, the ripple phases began to disappear, with the wider phase initially converting to the narrow one, and then the narrow phase disappearing, a process which took about 6 hours to complete at room temperature. The whole process was fully reversible and the ripple phases could be induced again by the addition of Tris buffer to the same sample and, furthermore, this process did not appear to damage the bilayers. The AFM with its inherent high resolution was able to reveal localised subtleties in the structure of the ripples, such as domain boundaries and edges, where there were defects in the bilayer. Some examples of this are shown in Fig. 5.13. The mechanism by which the Tris molecules cause ripples in the bilayers is not understood. Nevertheless the results demonstrated that the ripple phase can occur in single bilayers, and hence inter-bilayer interaction is not required.

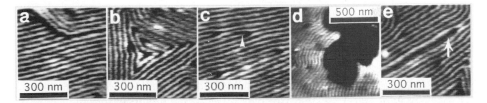

Figure 5.13. High magnification images of the boundary regions in the ripple domains. (a) At a junction one domain stops and neither connects with, nor crosses the other. (b) The triangular void in the centre of a region where three linear domains cross has been filled with a spiral ripple. (c) New ripples were sometimes observed to form in perfectly ordered domains without affecting the structure. (d) At a natural defect in the bilayer the ripple structure remained nearly flawless, both perpendicular and parallel to the edge: the ripple structure appears very stable and insensitive to defects. (e) Where two parallel domains have grown to meet, a thin gap was often seen which was too small to incorporate a complete ripple. However, in such gaps a weak line was occasionally observed (arrowed). Reproduced with permission from Mou *et al* 1994, copyright (1994) American Chemical Society.

In a later study this work was extended to examine ripple phases in *asymmetric* phospholipid bilayers composed of either DPPC or DSPC in one leaflet and 1-palmitoyl-2-oleoylphosphatidylglycerol (POPG), 1-stearoyl-2-oleoylphosphatidyl-glycerol (SOPG), or 1- palmitoyl-2-oleoylphosphatidylethanolamine (POPE) in the other leaflet (Czajkowski *et al* 1995). For these systems the ripple phase was induced by imaging in phosphate buffered saline (PBS), the sodium and phosphate ions being the necessary components. The asymmetric phospholipid bilayers were formed on mica by sequential LB film transfer of a monolayer of each phospholipid. The supported bilayer samples were then rinsed with the imaging solutions before being imaged with the AFM.

By conducting experiments with differing concentrations of the component ions in PBS (K^+, Na^+, Cl^-, PO_4^{2-}) it was established that there was a specific requirement for sodium and phosphate if the ripple phase was to occur. It was noted that the presence of 0.5 mM potassium lowered the threshold level of sodium required for the onset of ripple phase by an order of magnitude. As before more than one periodicity was observed in the ripple phase. Most of the studies were made on bilayers of DPPC/POPG but the composition of each leaflet was also studied for different systems in order to determine whether there was any specificity of the lipid for the formation of a ripple phase. Ripple phases were readily formed when the bottom layer was made up of saturated phosphatidylcholine and the top layer was made up of a mixture of saturated and unsaturated lipids with small negatively charged headgroups, such as phosphatidylglycerol at pH 7.5, and phosphatidylethanolamine at pH 11.0. For bilayers with DSPC in the bottom layer the ripple phase formation was much slower, and no ripple phase was observed for symmetric bilayers of DPPC. Thus with PBS as the initiator only the asymmetric bilayers displayed ripple phases, indicating that the nature of the inter-leaflet coupling has some bearing on the ion-induced occurrence of the ripple phase. This result is an example of an AFM study which has genuinely revealed new information on a biological system and the new observations could not be accounted for by theories of ripple formation in phospholipid bilayers (Czajkowski *et al* 1995).

The effects of temperature on the formation and disruption of ripple phases in DPPC bilayers and dual component DMPC/DSPC bilayers has been studied (Kaasgaard *et al* 2003). In DPPC the data revealed the highly anisotropic nature of the ripple phases with different phases appearing and disappearing at different temperatures. For the mixed system, onset of ripple melting was favoured at grain boundaries between different ripple types and orientations (Kaasgaard *et al* 2003). Ripple phase induction in DPPC bilayers by the addition of surfactin have been reported, and the process modelled successfully in terms of the formation of stable, curved lipid-surfactin monolayers (Brasseur *et al* 2007). The induction of the ripple phase was found to be highly dependant upon the concentrations of surfactin present in the film. In another study of biological significance, the interaction between the hydrolytic enzyme phospholipase A_2 and mixed bilayers composed of DMPC and DSPC has been shown to alter the phase behaviour of the membranes (Leidy *et al* 2004). The enzyme was found to preferentially attack DMPC in the mixed film, but the high resolution examination allowed by AFM imaging revealed that the ripple phase of the DMPC was more vulnerable to attack than the gel phase. Within the mixed film such preferential attack induced a phase transition leading to the formation of a flat gel phase lipid membrane due to the enrichment of the DSPC, whilst preserving the integrity of the membrane itself. Thus the AFM revealed subtle, but important details of the processes occurring during hydrolysis, which would otherwise have been impossible to obtain by more traditional lower resolution monitoring of either the enzyme activity or membrane integrity.

5.3.5. Mixed phospholipid films

In real cell membranes the phospholipid bilayers are not only asymmetric but generally composed of many different phospholipid molecules in each layer or leaflet. If phase separation occurs AFM has the ability to map this distribution within a single layer. In an interesting paper LB films of the non-miscible phospholipids DPPE and 1,2 Di[(*cis*)-9-octadecanoyl]-sn-glycero-3-phosphoethanolamine) (DOPE) were formed on mica and HOPG, and examined by AFM in air (Soletti *et al* 1996). The different substrates were used in order to examine the effect of the nature of the substrate on the transferred films and, in addition, the effect of surface pressure on film structure. Phase separation into distinct domains was clearly observed for the non-miscible phospholipids, since there was a difference in height of about 0.5 nm between the two species. The shape of the domains was dependent upon the molar ratio of the molecules. As the molar ratio was changed the shape of each domain varied from multi-lobed, to twin-lobed, and then to discrete circular domains of one domain within the other. The relative areas of each domain corresponded well with the expected molecular area and molar ratio allowing identification of the phases in the AFM images. Another property measured was the coefficient of friction over the mixed films utilising the by friction force imaging mode of the AFM. The friction maps produced a clear difference in contrast, with the DOPE domains having a higher coefficient of friction than the DPPE domains, and appearing liquid-like. This is a significant observation, indicating that even if there was not a size difference enabling topographic distinction between lipid phases, one may be able to differentiate between different phospholipids by their frictional characteristics. The value of the surface pressure during film transfer of mixtures with a molar ratio of 0.75/0.25 of DPPE and DOPE affected the nature of the DPPE domains. As the collapse pressure for DPPE (45 mN m⁻¹) was approached, cracks in the DPPE domains appeared, accompanied by some bilayer formation. As the surface pressure was increased to an even higher value 55 mN m⁻¹, bilayers were very clearly observed. All of the above measurements were made on films transferred onto mica. The effect of substrate on the structure seen in the transferred films was demonstrated with images of DPPE/DOPE mixtures on HOPG. The domains were no longer circular but irregularly shaped (Soletti *et al* 1996).

In a similar study mixed LB films of distearoylphosphatidylethanolamine (DSPE) and DOPE formed on mica were studied as monolayers in air and as bilayers in water (Dufrene *et al* 1997). The films were formed by LB dipping, the second leaflet of the bilayers being created on the downstroke, and the sample then kept under water by means of submerged beakers in the Langmuir trough. This allowed the LB films to be transferred to the liquid cell of the microscope without being exposed to the air, in which the formed bilayer would have been unstable. Once again phase separation was seen but, in this case, as well as in phospholipid monolayers (Fig 5.14), phase separation was observed in the phospholipid bilayers (Fig. 5.15). Also observed was good contrast in friction and adhesion maps of the lipid domains, despite the fact that both have identical headgroups. Adhesion maps were obtained by performing force-distance curves at every eighth point during imaging (to produce an array of 64 x 64 pixels) and displaying the force at which the tip detaches from the sample surface at each of these points. The differences in topographic contrast seen in all three data sets was attributed to three factors; (I) the length of the phospholipid molecules, (II) the tilt assumed by the molecules in each phase and, most importantly, (III) the mechanical response of the different layers. This last factor is responsible for the contrast seen in the friction and adhesion images. The DOPE phases being fluid-like were inelastically deformed by the AFM tip.

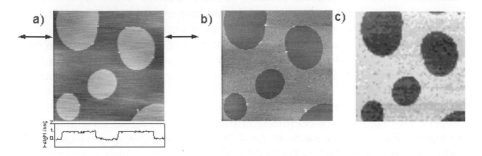

Figure 5.14. Phase separation of mixed monolayers of DSPE/DOPE on mica, imaged in air. (a) Topography image with line profile beneath the image. (b) Friction image. (c) Adhesion image (re-sampled to 512 x 512 pixels). In the images the lighter areas correspond to larger height, friction and adhesion respectively. Scan size: 15 x 15 μm. Reproduced with permission from Dufrene *et al* 1997 and the Biophysical Society.

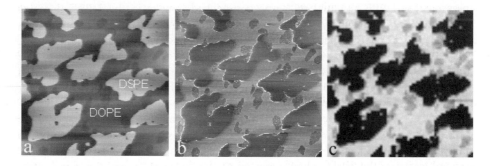

Figure 5.15. Phase separation observed in mixed phospholipid bilayers, imaged in water. The bottom leaflet of the bilayer (which is attached to the mica substrate) is composed of pure DSPE and the upper leaflet is composed of DSPE/DOPE. (a) Topography, (b) friction, (c) adhesion images (re-sampled to 512 x 512 pixels). As in Fig. 5.15 the lighter areas correspond to larger height, friction or adhesion. Scan sizes: 15 x 15 μm. Reproduced with permission from Dufrene *et al* 1997 and the Biophysical Society.

Furthermore, in the bilayers imaged under water, an additional contrast factor was observed; namely a short range repulsive force which was seen over the DSPE domains and which was speculated to arise from hydration or steric effects, since the headgroup is identical in both phospholipids. This force manifested itself as a large height difference between the domains in the topography images (4.8±0.7 nm), which could not be accounted for by molecular length, tilt, or deformation of the molecules. A large repulsive component was observed at a tip-sample separation of 3 nm in the force-distance curves recorded over the DSPE domains, suggesting that the tip hovered above the surface of these regions whilst imaging. By contrast, the force-distance curves obtained over the DOPE regions suggested that the tip actually penetrated to a depth of some 1 nm whilst imaging. In a refinement of these studies on mixed films a chemically modified AFM tip was used to differentiate between phospholipids and glycolipids in a mixed bilayer, (Dufrene *et al* 1998). Such subtle chemical sensitivity is a highly promising step for the application of AFM to the study of real biological plasma membranes.

5.3.6. Effect of supporting layers

If more realistic plasma membrane models are to be studied, such as those incorporating proteins and enzymes, then simple LB film transfer directly onto a substrate is often not straightforward, due to factors such as the insolubility of membrane proteins and instability of the bilayers themselves. A more useful method in this case is the monolayer fusion technique (Kalb *et al* 1992). The first step in this technique involves LB transfer of a lipid monolayer onto a hydrophilic substrate. After this step vesicles containing reconstituted membrane proteins (Racker, 1985) are deposited onto the coated substrate, whereupon they form a supported lipid bilayer with incorporated proteins. This has the advantages that only small amounts of protein are required, the native membrane environment of the proteins is preserved, and their orientation within the membrane will also be preserved. In addition to their application as model membranes, such lipid-protein vesicles are promising candidates for use in drug delivery systems, where they may provide a highly specific means of targeting particular cell or tissue types.

Supported phospholipid monolayers and bilayers, generated using the monolayer fusion technique, have been examined by AFM (Vikholm *et al* 1995) and a combination of AFM with the related SPM technique of scanning near field optical microscopy (SNOM - see section 8.3.) (Tamm *et al* 1996). In this study a phospholipid (either DPPC or DPPE) was transferred by LB methods onto a glass substrate to form the first supporting layer (Tamm *et al* 1996). Fluorescent lipid analogue probe molecules were mixed with these monolayers so that the structure of the layer could be characterised by SNOM, as an alternative to simple AFM examination of the surfaces. The SNOM images revealed that, under the transfer conditions used (Π = 30 mN m^{-1}), DPPC formed monolayers with numerous sub-micron sized crystallites, which could easily be visualised because the fluorescent probe molecules were forced into the grain boundaries: the crystallisation of the layer would probably not have been obvious by AFM examination alone. By contrast, the monolayers produced after transfer of DPPE in the solid condensed phase had much larger domains, the excluded fluorescent probe molecule patches were much more diffuse and no regularity was seen. AFM imaging under water of subsequently fused bilayers from vesicles of POPC, POPE, cholesterol and the disialo-ganglioside G_{D1a} revealed that the structure of the underlying supporting lipid monolayer had a profound effect upon the resultant 'biomembrane' bilayer, the results of which are shown in Fig. 5.16. The bilayer fused to the DPPC coated substrate was highly irregular and highly corrugated (Fig. 5.16a,c), whereas the bilayer fused onto the DPPE coated substrate (Fig. 5.16b,d) was relatively uniform. It appeared that the supporting lipid layer determined the structure of the adsorbed bilayer by some form of epitaxy. The studies demonstrated that the combined use of different scanning probe microscopes could be used to characterise these complex systems in biologically relevant environments.

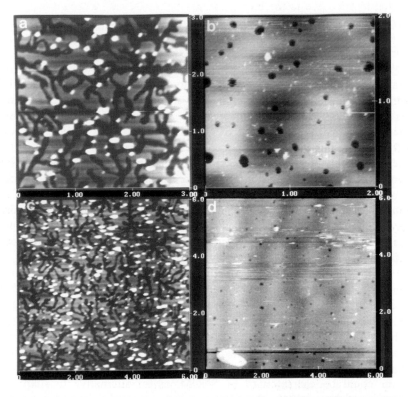

Figure 5.16. AFM images, obtained under water, showing the dramatic effect that a supporting phospholipid monolayer has on the resultant structure of an adsorbed bilayer (POPC:POPE:cholesterol:G_{D1a}). (a,c) DPPC and (b,d) DPPE. The scan sizes are indicated at the bottom of each image (units are in μm). Reproduced with permission from Tamm *et al* 1996.

Significant progress has been made in understanding the factors which determine the behaviour of supported phospholipid bilayers. In particular methods have been developed to minimise the effect of the solid support on the structure and functionality of the bilayers. It has been known for some time that even supported phospholipid bilayers can exhibit lateral fluidity of the molecules (Brain and McConnell, 1984). This has been attributed to the formation of a thin water layer which separates the bilayer from the solid surface. This layer, some 1-3 nm thick, decouples the bilayer from the solid support, allowing lateral lipid-lipid and vertical leaflet-leaflet interactions to be stronger than lipid-surface interactions (Merkel *et al* 1989; Johnson *et al* 1991; Sackman 1996). However, such a thin water layer is likely to have a significantly higher viscosity than bulk water and this can affect membrane properties (see section 6.3.8). Significantly, from an AFM point of view, the thickness and properties of this water layer can be tuned experimentally. The addition of calcium was observed to produce a large decrease in the water layer beneath DOPC and DPPC mixed bilayers (Berquand *et al* 2007). This decrease was speculated to arise due to the increased interaction between the polar head-groups of phospholipid molecules with the negatively charged mica surface. Furthermore, pre-treatment of glass surfaces prior to vesicle fusion were shown to have a marked impact on the fluidity of mixed phospholipid bilayers (Seu *et al* 2007). Another method which has been suggested to preserve membrane fluidity in supported bilayers is the use of polymer 'cushions' to bridge the gap between the bilayer and its solid support (Tanaka and Sackman, 2005). Although an intensive area of biosensor research, such polymer-supported bilayers haven't as yet been well characterised by AFM but this is a promising avenue for future research (see section 6.3.8 for more on this subject).

The progress made in supported bilayer fluidity has been coupled with the relatively recent improvements in commercially-available temperature control for the latest generation of AFMs, in order to probe the effects of heat on the phase behaviour of mixed phospholipid bilayers. By studying phase transitions in supported phospholipid bilayers, the relative strengths of lipid-lipid, leaflet-leaflet and lipid-surface interactions have been assessed for various mixed systems of the phospholipids DPPC, SOPC and DMPC (Charrier and Thibaudau, 2005; Keller *et al* 2005; Leonenko *et al* 2004). In these studies the data indicated that the lipid-surface interactions were stronger than the leaflet-leaflet interactions as peak broadening and decoupling of the phase transitions of the two leaflets were observed. However the reverse has also been seen; a study on mixed DOPC/DPPC bilayers demonstrated coupling of the phase transitions for inner and outer leaflets (Giocondi *et al* 2001). Similarly, with DMPC mica-supported bilayers, despite a broad gel-liquid crystalline transition, the melting temperature determined by AFM was only slightly raised (by 4°C) compared to the value obtained in differential scanning calorimetry (DSC) measurements (Tokamasu *et al* 2002). Furthermore, very similar values for gel - liquid transitions were found by AFM when compared with DSC studies on pure DPPC bilayers (Yarrow *et al* 2005). The likely explanation for these apparently contradictory results is the critical role that pH and ionic strength will have in determining the magnitude of the interaction between the bottom leaflet of a bilayer and the supporting surface. Additionally, the effect that sample history has on the formation of supported lipid bilayers is likely to play a role (Seantier *et al* 2008). The important point here is that apparently small variations in experimental conditions may have a large effect on the physical behaviour of supported phospholipid bilayers.

As we have seen in the preceding sections, the mechanical properties of supported lipid bilayers are also amenable to study by AFM and recent work has demonstrated the effect that both temperature and ionic strength can have on these properties (Garcia-Manyes *et al* 2005a; 2005b).

5.3.7. Dynamic processes of phospholipid layers

One of the most exciting possibilities of the use of AFM imaging is the ability to follow dynamic processes *in-situ,* and there are a few examples of real time imaging of processes on phospholipid bilayers.

In what was one of the first examples of real-time imaging in a phospholipid bilayer system, the binding of streptavidin to bilayers, composed of DPPE in the first layer and a second layer of mixed dimyristoyl phosphatidylethanolamine (DMPE) and biotinylated DPPE, was followed by AFM (Weisenhorn *et al* 1992). The study found that streptavidin bound almost exclusively to biotin which was in the fluid-phase lipid domains, but individual proteins could not be resolved due to the lipid fluidity. The structure of the few regions where streptavidin did bind to biotin in crystalline-phase lipid domains was found to be sensitive to the imaging force but, in the optimum range (< 200 pN), individual proteins were resolved. The conclusion was that at excessive force the AFM tip simply 'ploughs' through the lipid layer rendering visualisation of bound protein molecules impossible (Weisenhorn *et al* 1992).

Two studies have used AFM to examine the enzymatic modification of phospholipid layers (Turner *et al* 1996; Grandbois *et al* 1998). In the first instance AFM was combined with several other techniques (X-ray photoelectron spectroscopy, secondary-ion mass spectrometry, X-ray reflectivity and ellipsometry) to characterise the effect of free phospholipase C on a film of dimyristoyl phosphatidylcholine (DMPC). The DMPC layer was immobilised by chemical attachment to a silicon wafer, and the film examined before and after enzyme treatment. The study wasn't therefore a true dynamic study, but is worthy of mention as one of only two examples of early AFM investigations of enzymatic breakdown of phospholipid films. The results showed that the enzyme was active even against an immobilised lipid layer, removing all of the phosphate from the lipid (Turner *et al* 1996).

The second study followed the process of enzymatic degradation of a phospholipid bilayer of DPPC by phospholipase A_2 as it actually took place in the liquid cell of the AFM (Grandbois *et al* 1998). The bilayer was formed on mica by double LB dipping, with the first film allowed to dry in air for 15 minutes prior to the second dip. The surface pressure was maintained at 35 mN m^{-1} for both dips, so that the DPPC layer was in the highly packed gel-phase, and the transfer speed was 3 mm min⁻¹.

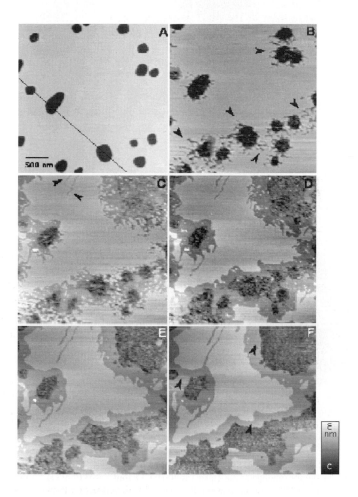

Figure 5.17. part 1. A sequence of AFM images showing the effect of the hydrolysis of a DPPC bilayer by phospholipase A_2. (A) Before addition of enzyme, and (B) 2 min. (C) 4 min. (D) 6 min. (E) 9 min. (F) 12 min. after addition of the enzyme. Scale bar 500 nm. The sequence is continued overleaf. Reproduced with permission rom Grandbois *et al* 1998 and the Biophysical Society.

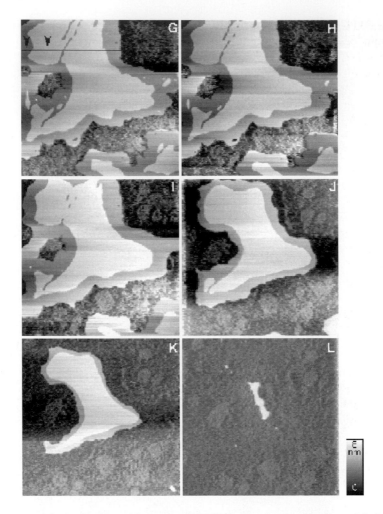

Figure 5.17. part 2. Continuation of the sequence of AFM images showing the effects of the hydrolysis of a DPPC bilayer by phospholipase A_2. (G) 14 min. (H) 17 min. (I) 30min. (J) 50 min. (K) 80 min. and, (L) 140 min. after addition of the enzyme. Reproduced with permission from Grandbois *et al* 1998 and the Biophysical Society.

Sharpened tips can be grown by the deposition of carbon using the electron beam in an EM focused onto the apex of a standard silicon nitride tip, and these are often referred to as 'e-beam tips' or 'supertips'. Such a tip was used to minimise the effects of adhesion and tip contamination by the products of hydrolysis and, in this way, the imaging force was kept below 0.5 nN. The AFM imaging was carried out under buffer solution and then bee venom phospholipase A_2 was added to a concentration of 0.26 μM. The subsequent sequence of AFM images obtained is reproduced in Fig. 5.17 and illustrates that the degradation of the DPPC bilayers by the enzyme is nucleated by defects. This is because the rate of degradation of DPPC in the gel-phase is very slow, due to the close-packing of the lipids, whereas at defects the lipids are not well packed and, indeed, no degradation was observed in defect-free regions of the bilayers. This was proven by deliberately disturbing a defect-free region of the bilayer, by scanning at slightly higher force and scan rate (5 nN and 20 Hz), in the presence of enzyme and noting that this produced a groove. When this procedure was repeated in the absence of enzyme no damage occurred. The gross pattern of degradation is strikingly clear, with the enzyme eroding the bilayer around the holes and 'eating' its way into the defect-free areas of the film (Fig. 5.17D-J) until virtually none of the film remains (Fig. 5.17L). Upon closer inspection of the image in Fig. 5.17B many small channels can be seen leading away from the holes, indicating that the erosion of the bilayer by the enzyme did not occur uniformly in every direction. At higher magnification these channels were seen to be confined, in the main, to discrete angles centred around 120°, indicating that the enzyme may be sensitive to the arrangement of the DPPC molecules since, in the gel phase, they are hexagonally packed. The AFM images of the degradation patterns were carefully analysed in terms of the channel area, to estimate the numbers of lipid molecules hydrolysed with time in order to estimate reaction rates, thus determining what level of hydrolysis was imaged in the sequence. For the narrow channels, which are only some 15 nm in width and are arrowed in Fig. 5.17B, the remarkable conclusion was that these channels represented hydrolysis by single enzyme molecules, since the figures derived from image analysis agreed well with calorimetric data obtained by others (Lichtenberg *et al* 1986). Thus the activity of single enzyme molecules may be followed by AFM and, if activity is really sensitive to the phospholipid organisation in bilayers, then it may be possible to examine the effect of factors believed to modulate bilayer organisation by AFM (Grandbois *et al* 1998).

Other dynamic studies on phospholipid bilayers of DPPC include lipid loss from mica surfaces *ex-situ* at elevated temperatures (Fang and Yang, 1997), domain motion under conditions of high relative humidity *in-situ* (Shiku and Dunn, 1998) and fillipin-induced lesions in mixed DPPE/cholesterol bilayers *ex-situ* (Santos *et al* 1998).

5.4 Liposomes and intact vesicles

All of the AFM studies described in the previous sections have been made on flat bilayers. There are some examples of AFM studies carried out on intact liposomes and vesicles. Shibata-Seki and coworkers imaged intact liposomes of DPPC/cholesterol by AFM under water after binding them to an antigen covered surface of gold on mica (Shibata-Seki *et al* 1996). The AFM images of the liposomes produced spherical features, with measured diameters which were in reasonable agreement with sizes obtained from light scattering measurements. Measurements of the heights of the liposomes indicated that the AFM tip compressed these soft structures by about 40% and, therefore, the image quality was found to depend quite strongly on the force set-point used (acceptable image quality was only obtained with forces <1 nN), and, of course, loading pressure which is determined by tip sharpness. In general the blunter standard silicon nitride tips produced better images. From rudimentary calculations it was determined that the critical loading force (for a standard tip) and pressure which the liposomes could withstand was 4.5 nN and 0.9×10^6 N m^{-2} respectively (Shibata-Seki *et al* 1996).

In a more recent and fascinating study the process of liposome collapse during adsorption to mica has been imaged by AFM (Egawa and Furusawa, 1999). Initially AFM scans were made of a freshly-cleaved mica surface under pure buffer (10 mM MgCl$_2$) in the liquid cell of the microscope. Then the buffer was replaced with a suspension of liposomes at various salt concentrations in order to assess the effect of electrolyte concentration on the interaction between the mica and liposomes. The electrostatic interactions between the liposomes and mica were quantified by measurement of the zeta potential of the systems. Phosphatidylcholine (PC) and phosphatidylethanolamine (PE) were studied and comparisons made of the the mechanisms of bilayer formation. The results obtained for PC are shown as a time sequence in Fig. 5.18.

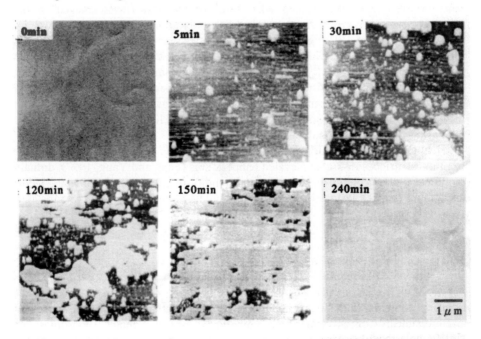

Fig 5.18. Time sequence of AFM images captured during the adsorption of a PC liposome onto mica in 10 mM MgCl$_2$ solution. Reproduced with permission from Egawa and Furusawa, 1999, copyright (1999) American Chemical Society.

Five minutes after injection of the PC suspension spherical features appear on the mica surface (Fig 5.18) which grow in number (Fig. 5.18, 30 min) until they overlap and begin to flatten (Fig 5.18, 120-150 min), and then finally form a continuous bilayer (Fig. 5.18, 240 min). In addition to the visual evidence for liposome collapse provided by the AFM images, namely spherical features flattening out, the collapse upon adsorption to mica was also monitored using a fluorimeter to measure the release of a fluorescent probe molecule encapsulated in the liposomes. The rate of PC bilayer coverage of the substrate was quantified by area measurement in the AFM images under three different ionic strengths (10^{-4} M, 10^{-3} M, 10^{-2} M $MgCl_2$) and compared to fluorescence measurements of PC liposome/mica mixtures at the same ionic strengths. The results from this experiment were in very close agreement and, when displayed graphically, the form of the curves were almost identical. The rate of adsorption and deformation of PC vesicles on mica was shown to increase with increasing electrolyte concentration due to the decreasing zeta potential of the vesicle/mica system, meaning that less electrostatic repulsion was present between them. Notably with PC only one bilayer was ever formed on the mica, no additional bilayers were seen under any of the electrolyte conditions studied. The results for PE were rather different in the respect that a second bilayer formed but the initial bilayer formation was similar to that of PC. The second bilayer, however, was relatively easy to displace with the AFM tip, indicating that the interaction between bilayers (that is between the ethanolamine headgroups) was weaker than the interaction between the bilayer and mica surface. These differences, which were observed in the absence of electrostatic repulsion between lipid and mica in both cases, were attributed to the different degrees of hydration in the headgroups of the two phospholipids, demonstrating the importance of headgroup chemistry in determining the nature of bilayer formation (Egawa and Furusawa, 1999).

5.5. Lipid-protein mixed films

In the plasma membrane of cells the phospholipid bilayer contains numerous membrane proteins and so the study of proteins incorporated into lipid bilayers is of great relevance to biological AFM. This area overlaps to a certain degree with the discussion on 2D protein crystals described in section 6.2, since most such systems are membrane proteins which naturally occur in phospholipid bilayers. There are however, a few examples of studies where proteins have been deliberately incorporated into phospholipid bilayers created on Langmuir troughs, or into lipid vesicles for reasons of improved protein stability, and therefore improved resolution by AFM imaging (Yang *et al* 1993; Mou *et al* 1995; Czajkowski *et al* 1998), to examine intrinsic interactions between the components (von Nahmen *et al* 1997), or finally, as a model system for generating material contrast in friction AFM images (Sommer *et al* 1997).

In the first example a bilayer of 1,2-dipentacosa-10,12-diynoyl-phosphatidylcholine (DAPC) containing 20 mol% ganglioside GM1 was transferred from an air-water interface onto mica using LB dipping. Both cholera toxin and its B-subunit were bound to the ganglioside in the bilayer by incubation of solutions of the toxin with the mica supported bilayer. This enabled high resolution (better than 2 nm) AFM images to be obtained under pure water and buffer as long as the imaging force was kept around 0.3 - 0.5 nN. The intact cholera toxin molecules did not pack as closely as the B-subunits, with the result that the resolution achieved for the former was lower, providing direct evidence of the importance of molecular packing for AFM imaging at high resolution. The whole cholera toxin molecules were imaged only as spherical globules but, by contrast, the B-subunits' pentameric structure was clearly resolved (Yang *et al* 1993). In an effort to increase the stability of the bilayer and protein contained within it, cross-linking of the leaflets was carried out by exposing the sample to ultraviolet light. In a later study by the same group cholera toxin B-subunits were re-examined, this time bound to bilayers of DPPE / PC, but without any cross-linking steps (Mou *et al* 1995). Bilayers were produced by both LB-dipping and the vesicle fusion technique. Slightly clearer AFM images were obtained, although the resolution was comparable to the first study (1-2 nm), and interestingly no dependence of image quality on lipid state was found; the AFM images of the oligomers were just as good whether they were bound in a gel-phase (Fig. 5.19b) or in a fluid-phase (Fig. 5.19c) lipid layer. As in previous studies the proteins were bound to ganglioside incorporated into the lipid layer and, if this step was omitted, none were bound (Fig. 5.19d). Since proteins will bind randomly and non-specifically to bare mica, this specific binding into a phospholipid bilayer provides a useful alternative method for protein deposition to a substrate which enables a far more ordered protein layer to be obtained. A further advantage of specific binding of proteins would appear to be that less tip contamination is encountered, probably because the molecules are far harder to displace. In this regard it was noted that preparation of these protein-lipid 'membranes' using the small scale lipid-coated droplet technique (Kornberg and Darst, 1991- see section 6.2.3.) was unsuitable, producing very small 'membrane' fragments which quickly contaminated the AFM tip. Finally, in this study similar results were obtained when the protein was bound to bilayers composed of several other phospholipids with differing headgroups and acyl chains (Mou *et al* 1995). *Staphylococcal* α-hemolysin (α-HL), a small (33 kD) water soluble protein, converts into a pore forming oligomer upon binding to the plasma membrane.

For some years there had been a degree of uncertainty over the structure adopted by the α-HL in membranes. Electron microscopy studies had indicated that it formed a hexamer (Gouaux *et al* 1994) but later X-ray crystallography studies demonstrated a heptameric shape (Song *et al* 1996). The crystallographic study argued that the image processing required in the EM study may have led to artifacts (Song *et al* 1996). AFM was applied in order to try to resolve this problem (Czajkowski *et al* 1998) since it has already proved capable of resolving directly the shape of several membrane proteins (chapter 6) and soluble proteins incorporated into phospholipid bilayers, without the need for image processing and in realistic environments. A lipid bilayer was formed on mica by LB dipping but, on this occasion, no ganglioside was required because oligomerisation of the α-HL occurs spontaneously upon contact with suitable amphipathic substrates. The AFM images obtained, in contact mode with oxide-sharpened silicon nitride tips under a selection of aqueous buffers, demonstrated that the α-HL could indeed form a hexamer, confirming the earlier EM study.

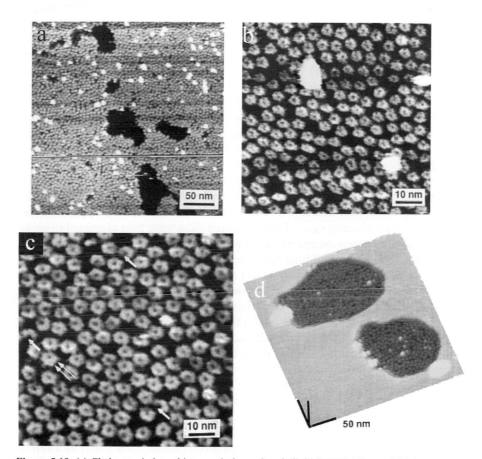

Figure 5.19. (a) Cholera toxin bound into a gel-phase phospholipid (DPPC) bilayer which incorporates a ganglioside (GM1) to bind the protein molecules. (b) At higher resolution individual toxin molecules are clearly seen, and in many the central cavity and pentameric structure is visible. (c) High resolution image of cholera toxin molecules bound to a fluid-phase phospholipid (DPPC/POPG) bilayer containing GM1. The image quality is just as good as in the gel-phase layer, indicating that membrane fluidity does not degrade AFM image resolution. The clarity of this image allows variation in the structure of individual cholera toxin molecules to be seen. In some cases (single arrows) subunits were missing, while others displayed six-fold symmetry (double arrows) instead of the expected five-fold symmetry. (d) When no ganglioside is incorporated into the bilayers cholera toxin molecules do not bind. However, many cholera toxin molecules are visible in the defect regions, where the bare substrate (mica) is exposed. Reproduced with permission from Mou *et al* 1995.

This AFM study provided proof that α-HL can exist in two different oligomeric forms, the reasons for which will require further investigation. Thus the study provided a further and encouraging example of AFM being used to tackle a real biological problem (Czajkowski *et al* 1998). In such studies AFM complements other techniques such as TEM by visualising the structures under realistic biological conditions.

The interplay between a recombinant pulmonary surfactant associated protein (SP-C) and a phospholipid bilayer composed of DPPC and dipalmitoyl phosphatidylglycerol (DPPG) was studied by AFM (von Nahmen *et al* 1997). This system was used as a model for the surface active combination of phospholipids and proteins present in lungs, which act to keep the surface tension of the fluid layer on the aveoli at low levels. Low surface tension of this fluid layer is an essential feature of the 'mechanics' of breathing, enabling the lungs to re-expand after exhalation. To model the effects found in the lungs, mixed lipid-protein films drawn at various surface pressures to simulate inhalation (the low surface pressure regime simulates the increased interfacial area of expanding lungs) and exhalation (the high surface pressure regime simulates the reduced interfacial area on emptying the lungs) were studied. LB films were formed on mica on the upstroke of the dip, which was carried out a rate of 2 mm min^{-1}. The transferred films were then examined in air by AFM. At lower surface pressures (30 mN m^{-1}), AFM images of the mixed film showed that both protein and lipid were present in the interfacial film; some regions were entirely composed of lipid, some had a mixture of lipid and protein, and others were composed almost entirely of protein. AFM images of films transferred at higher surface pressures (50 mN m^{-1}), showed, by contrast, raised lamellar islands of protein-fluidic lipid complex sitting on top of a continuous phospholipid bilayer. As the compression of the film was increased, by reducing the area on the Langmuir trough, the phase separation between the lipid and protein-lipid complex became more distinct, in that the lamellae of the complex reduced in area but increased in height. The effect was reproducible and fully reversible so that, once the surface pressure was again reduced the film structure seen by AFM reverted to a flat, partially mixed film. The conclusion was that the AFM images revealed that a protein-lipid complex moves in and out of the phospholipid bilayer depending upon the surface pressure, but never detaches from it completely. Thus it acts as a sort of reservoir of surface active material which can quickly re-spread onto the interfacial layer upon reduction of surface pressure (expansion of the lungs), ensuring no rupture of this lubricating layer occurs (von Nahmen *et al* 1997). This is an important property because the main phospholipid constituent in the lungs is DPPC which enters the gel-phase readily and will not re-spread after it has exceeded its collapse pressure. A final and important check carried out in the study was a validation of the LB film transfer procedure. By incorporating a fluorescent dye it was possible to obtain fluorescence micro-graphs of the interfacial films at the air-water interface and after transfer onto mica. These proved that no gross changes to film structure were induced by the transfer procedure (von Nahmen *et al* 1997). More recently AFM has been used to study the interaction of a lipid-binding protein (puroindolene) derived from wheat seed with phospholipid monolayers (Dubreil *et al* 2003). Comparison of monolayers of dipalmitoyl phospholipids with differing head groups (zwitterionic DPPC and anionic DPPG) revealed that incorporation of puroindolene into both films caused aggregation of the protein. The role that electrostatic interaction between the protein and lipid plays was demonstrated with the puroindolene exhibiting a greater affinity for the more charged DPPG. In DPPC films the protein was found only in the liquid expanded regions, whereas in DPPG films it was found in both the liquid-expanded and liquid-condensed regions of the monolayer. The study is notable in that the AFM observations were combined with confocal laser scanning laser microscopy (CLSM) imaging and surface-pressure versus area (Π-A) isothermal characterisation, and the study serves as a good example of the fact that when trying to understand a multi-component interfacial system the combination of several techniques is often required. One of the most biologically intriguing aspects of lipid-protein mixed films is the detail of the interaction between signalling proteins expressed at the cell surface and the phospholipid cell membrane. AFM is ideally suited to the study of such systems and a recent example is given in the work of Nicolini and co-workers (Nicolini *et al* 2006). They used a combination of AFM imaging with two-photon microscopy to follow the insertion of the important cell growth and differentiation regulatory signalling protein N-Ras into raft-forming mixtures composed of varying ratios of POPC, sphingomyelin and cholesterol. The hexadecylated and farnesylated N-Ras partitioned preferentially into the disordered liquid expanded lipid domains. The higher resolution of the AFM images revealed that the majority of the protein molecules were located at the liquid condensed-liquid expanded domain boundaries. The authors speculated that positioning at the boundaries might lead to a favourable decrease in line energy, which is associated with the rim of the demixed phases (Nicolini *et al* 2006).

Mueller and co-workers performed force-distance measurements on phospholipid bilayers, bare mica, and on bilayers in the presence of the nerve sheath protein myelin (Mueller *et al* 1999). Force-distance curves recorded during stretching of the myelin away from the mica, or the lipid coated mica surface indicated that, when in contact with the lipid layer, the protein adopts a different conformation to that observed when it is simply adsorbed to mica, a result which may not be surprising in view of the normal interfacial behaviour of proteins. In these force spectroscopy studies, the researchers reported that imaging was not possible and so the system could not be fully characterised.

5.5.1 High resolution studies of phospholipid bilayers

In what may well come to be considered as a landmark paper for biological AFM, Fukuma and Jarvis recently reported the development of an exquisitely sensitive new form of ac-mode AFM imaging in liquid (Fukuma and Jarvis, 2006). The new mode uses frequency modulation (FM) of the oscillating cantilever as the control signal for the feedback loop. Operation in liquids usually gives rise to an excessively-damped oscillation of the AFM cantilever which is unsuitable for FM detection. In fact traditional FM detection usually requires the very low damping (hence high quality factor) environment of vacuum conditions. However, through a combination of ultra-low noise laser detection circuitry and very small cantilever oscillation amplitudes (in the Ångstrom range) the authors demonstrated that the system becomes responsive enough to allow this means of control, even in liquids.

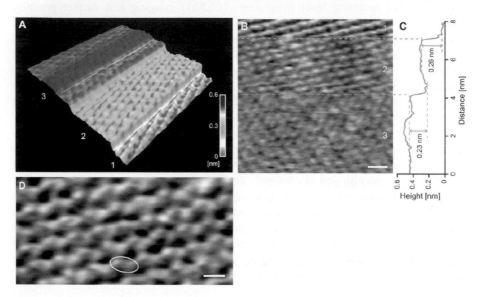

Figure 5.20 (a) FM-AFM image obtained on the surface of a DPPC bilayer in PBS showing spontaneous height jumps during imaging due to hydration layers. Image size 8 x 8 nm, height range as indicated on tonal scale at right. (b) line-flattened image of (a). Scale bar 1 nm height range 0.1 nm (black to white). (c) Line-averaged profile of (b) plotted along the slow scan direction (top to bottom). (d) Image of terrace region 1 showing submolecular resolution of the headgroup. Scale bar, 0.5nm. Height range 0.12 nm (black to white). Reproduced with permission from Fukuma *et al* 2007a and the Biophysical Society.

Such is the gain in sensitivity in this new mode that very subtle sample features can be resolved directly. The technique has been used to visualise directly the intrinsic hydration layers in gel-phase DPPC bilayers under PBS at Å resolution (Fukuma *et al* 2007a) and these results are reproduced in figure 5.20. The data reveals the presence of hydration layers which have been both imaged at the molecular scale, and detected as oscillatory jumps in force profiles of the approaching AFM tip toward the bilayer surface. The spacing of the force oscillations, seen over the final nanometer of the tip approach, was 0.28 nm, which corresponds well with the diameter of single water molecules. Geometric effects, due to confinement of the water by the AFM tip, were ruled out due to the atomic scale lateral resolution achieved in the images of the bilayer, and the relatively short range over which they acted. The data suggested that the hydration layer extended to distances of up to two water molecules, in good agreement with molecular dynamics simulations (Berkowitz *et al* 2006). Taken together the data suggested that the hydration layer observed was intrinsic to the phospholipid bilayer. Thus the study provided the first experimental conformation of the existence of an intrinsic hydration layer surrounding a phospholipid membrane which was stable enough to image. The biological significance of this finding is that it provides strong evidence that such a layer will present multiple energy barriers to approaching nanoscale objects, such as proteins and solvated ions. Indeed, the same authors went on to visualize the spatial distribution and dynamic rearrangement of ion occupancy surrounding the phospho-choline head-groups of DPPC bilayers under physiological conditions (Fukuma *et al* 2007b).

5.6. Miscellaneous lipid/surfactant films

As well as AFM studies on phospholipids there have been numerous investigations on other lipid films. These studies could justify a book in their own right but, as they are less biologically significant, they have not been included in this text. However, there are a couple of studies which are worth mentioning briefly here, since their findings may impact upon the future study of phospholipids, particularly on more complex (mixed) or dynamic systems. Firstly, chemical identification of mixed amphiphile LB films has been achieved by measurement of surface potential during scanning using a hybrid mode AFM with conducting tips (Inoue and Yokoyama, 1994; Fujihara and Kawate, 1994; Chi *et al* 1996). Secondly, the *in-situ* growth of an adsorbed film of the cationic surfactant, cetyltrimethylammonium bromide (CTAB), on mica was followed by AFM imaging (Li *et al* 1998).

5.7. Interfacial protein films

For proteins the process of assembly at an interface involves not just a simple reorientation of the molecules but also a degree of unfolding to expose hydrophobic regions. This means that unlike simple amphiphilic molecules, such as phospholipids, the assembly (or more strictly the *adsorption*) of proteins at an interface is a relatively slow process, which is influenced by many factors. As such interfacial protein films are an interesting candidate for study by AFM, although surprisingly little work has been carried out to date. In principle it should be possible to investigate many fundamental problems which have so far eluded explanation, such as how unfolding of protein molecules at an interface affects their functional behaviour and, should they interact and form networks, the characteristics of the networks formed at interfaces. For example, why do some proteins form better, more stable foams or emulsions than others, and what effect do other amphiphilic molecules, which compete for space at the interface, have on protein structure in the complex, mixed interfacial systems found in real environments?

5.7.1. Specific precautions

There are certain experimental problems which are specific to the study of interfacial protein films and which need to be considered before imaging by AFM. Obviously the objective of such studies is to examine the nature of the interfacial protein film but, because the interface is not imaged directly, it is necessary to ensure that when the film is transferred onto a substrate it is *only* the interfacial film that is deposited. For protein films this is not as straightforward as it might seem. The problem is twofold. Firstly amphiphilic proteins have a high affinity for many substrates under most conditions. A second and more specific problem is that in realistic systems the process of interfacial protein film formation is by *adsorption* of the proteins from bulk solution to the interfacial region: think of the foam produced by beating egg whites, or the foam formed as the head on a glass of beer. This means that the usual technique for creating interfacial films on a Langmuir trough, which involves spreading the amphiphiles at the air-water interface, is not the most realistic way of creating interfacial protein films. Studies of the surface rheology of protein films have demonstrated that *spread* films possess different characteristics to *adsorbed* films (Krägel *et al* 1999; Mackie *et al* 1999). If the interfacial film is created by adsorption of the protein molecules from the bulk, then as well as an interfacial layer of proteins there will also be free protein in the bulk liquid or *subphase*. During dipping the substrate is exposed to the subphase and the possibility exists of non-specific protein adsorption in addition to transfer of the interfacial film.

During a study (Mackie *et al* 1999) on the displacement of protein films from the air-water interface by surfactant, which is described in detail in the following section, it was possible to demonstrate that such 'parasitic' adsorption really does take place. Displacement of the interfacial protein films with surfactant has allowed production of an interfacial protein film with definite structural features which allow it to be distinguished from simple passive adsorption of protein onto the mica. Passive adsorption of the protein from the subphase simply produces fairly uniform monolayer coverage of the substrate. The passive adsorption can be eliminated by the following procedure. The interfacial protein film was created by filling a Langmuir trough with a 0.5 μM solution of the milk protein β-lactoglobulin and allowing it to adsorb to the air-water interface until a pseudo-equilibrium surface pressure was obtained (10 mN m^{-1} after 30 minutes). By partially displacing the interfacial β-lactoglobulin film using a surfactant (Tween-20) an interfacial protein network was produced which contains holes. Not all of the protein from the bulk solution has adsorbed to the interface, meaning that if LB dipping is done on the system as it stands parasitic adsorption of proteins from the subphase will occur as well as interfacial film transfer. This is illustrated by the AFM images in Fig. 5.21 which were obtained after LB film transfer onto mica and air-drying. Therefore the bright regions are protein and the dark regions (should be) bare mica. The two AFM images shown in Fig. 5.21 are from LB dips after different levels of perfusion of the subphase with surfactant solution to remove free protein. Fortunately interfacial protein films are immobile because they intertwine to form an elastic network, and this means that once the film has formed there is little exchange of the proteins at the interface with proteins in the bulk, as happens with films of less complex amphiphiles such as lipids and surfactants. This allows perfusion of the bulk solution or subphase with pure water or pure buffer solution without affecting the interfacial protein film, as long as the flow rate is kept low enough to prevent disruption of the surface (1-2 mL min^{-1} for a trough of 450 mL volume).

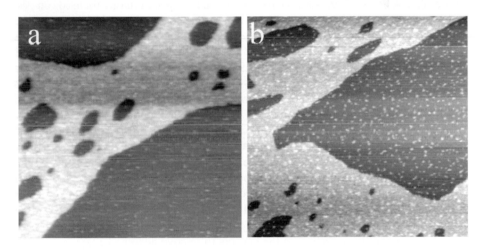

Figure 5.21. AFM images showing the effect of perfusion of the subphase prior to LB transfer, of an *adsorbed* mixed interfacial film of protein and surfactant. (a) After perfusion of the subphase of the Langmuir trough with 2L of surfactant solution most of the non-interfacial protein has been removed. This can be seen because the surfactant-rich regions from the interface have little adsorbed protein in the AFM image. (b) If perfusion of the subphase is carried out less rigorously (in this image with only 1L of surfactant solution) then more 'parasitic' adsorption of proteins to the substrate occurs-the 'dark' regions are peppered with proteins. Scan sizes: (a) 3 x 3 μm, (b) 2 x 2 μm. Imaged in dc mode under butanol.

For the situation shown in Fig. 5.21 the subphase must be perfused with the surfactant solution in order to maintain the equilibrium concentration of surfactant molecules at the interface. By transferring the interfacial films after different levels of perfusion it was possible to quantify the parasitic adsorption of protein molecules to the substrate, seen as bright specs in dark regions. Even a quick inspection of the AFM images confirms that parasitic adsorption of proteins onto the mica from the subphase is reduced as the perfusion volume increases (compare Fig. 5.21a & b which had 2L and 1L of perfusion respectively). The images prove that the protein molecules attached to the mica in the dark regions have not come from the interfacial film but from the subphase, so it is easy to see that if no perfusion is carried out when adsorbed protein films are being transferred by LB dipping, then the mica surface could potentially be completely saturated with passively-adsorbed protein before any interfacial transfer takes place, leading to highly misleading AFM images.

5.7.2. AFM studies of interfacial protein films

Early AFM studies of interfacial protein films tended to simply demonstrate the capability of AFM to image such systems (Birdi *et al* 1994; Gunning *et al* 1996). More recently however, there have been studies which have focused on the behaviour of interfacial protein films and, in particular, how they react to the presence of competing surface active species such as surfactants or lipids (Alexandre *et al* 1994, 1997; Sommer *et al* 1997; Mackie *et al* 1999). Different LB film transfer mechanisms were studied for a system of mixed fatty acid (behenic acid)/protein (glucose oxidase) which is of interest in the production of biosensors (Alexandre *et al* 1994). The study illustrated that the most commonly used vertical dipping (Langmuir-Blodgett) method created large parallel defects on the macroscopic scale and subtle differences on the microscopic scale, although the transfer speed used of 10 mm min^{-1} was rather high, and may account for these results (see section 5.2.3). In two more recent AFM studies of the same system lateral force imaging (also known as frictional force imaging, section 3.3.2.) was used together with UV-Vis and IR spectroscopy to characterise the mixed films (Alexandre *et al* 1997; Sommer *et al* 1997). The films were created by using a dilute solution (3.2 μg mL^{-1}) of glucose oxidase as the subphase in a Langmuir trough, sweeping the surface with the barriers to remove adsorbed enzyme and then spreading the fatty acid from solvent at the air-solution interface. The mixed films were allowed to equilibrate for 45 minutes before being compressed to a surface pressure of 30 mN m^{-1} and then transferred after varying time intervals onto graphite by vertical (LB) dipping. Using frictional force imaging it was possible to differentiate between the protein and fatty acid domains in the mixed films, which displayed a highly heterogeneous structure. The glucose oxidase molecules appear to adsorb to an interfacial film of behenic acid and can take up to 15 hours to reach a well ordered interfacial protein film. In essence the sample preparation mirrors the method for the production of 2D crystals of soluble proteins developed by Kornberg and Ribi (Kornberg and Ribi, 1987) and the studies therefore represent a stepwise monitoring of the events leading to crystallisation by AFM. The mixed fatty acid protein film was initially a highly heterogeneous mixture containing protein aggregates which slowly became covered by fatty acid molecules, allowing their rearrangement into ordered quasi-crystalline arrays (Alexandre *et al* 1997; Sommer *et al* 1997).

A study of the unusual fibrous protein gelatin examined the initial stages of association of the molecules, leading to the formation of gel networks, by sampling from bulk solutions and from interfacial films formed at an air-water interface (Mackie *et al* 1998). In order to achieve a high local concentration but avoid the formation of large three dimensional aggregates or gels which are difficult to image with any detail in AFM due to their softness, the process was carried out on a Langmuir trough, thereby confining the structures formed to two dimensions. At various intervals interfacial films were transferred onto mica using LB dipping and the samples imaged under butanol in the dc mode. The study highlighted the importance of the thermal history of the gelatin solutions, which determined the association of the gelatin molecules and thus the rheological characteristics of the interfacial film. The results obtained supported the model of gelation which involves molecular association via triple helix formation followed by further association of the helical structures into bundles or fibres. Occasionally these fibres displayed a collagen-like periodicity (Fig. 5.22) consistent with the proposed reformation of a 'collagen' structure during the gelation of gelatin. Gelatin solutions which had been rapidly quenched did not form any large fibres, in contrast to slowly cooled solutions, and displayed very different interfacial rheology.

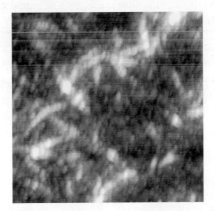

Figure 5.22. Gelatin fibres from an interfacial film occasionally display regular periodicity commonly seen in collagen fibres. Scan size: 1.6 x 1.6 µm, imaged in dc mode under butanol.

Aliquots were taken from bulk solutions at concentrations where the gelatin just gelled at room temperature (4 mg mL^{-1}). These were then diluted and examined, after drop deposition onto mica, so that the association seen in the interfacial films could be compared with the gelation process in the bulk. These bulk samples displayed the same association behaviour as the slowly cooled interfacial film samples, namely slow fibre formation over time. An important problem in interfacial protein systems is their stability in the presence of other surface active species. In isolation both proteins and surfactants can stabilise foams or emulsions but, when the two are combined, things go wrong and stability is lost. This effect is sometimes used in a beneficial manner; for example very young babies suffer from colic (trapped wind in the intestine) caused by foaming of the proteins in milk. The foam is highly stable entrapping air which prevents the baby from passing wind (at either end!) and releasing the trapped air. Anti-colic preparations contain food grade surfactants which destroy the protein foam and so release the wind. Another less welcome example is the collapse of the foam on a glass of beer. The beer foam is stabilised by interfacial adsorbed protein but surfactant molecules from a glass which hasn't been properly rinsed after washing, or lipid molecules from the lips of the drinker soon cause its destruction. Until recently the mechanisms for this effect were not understood in any detail. The competitive adsorption of proteins and surfactants to the air-water interface has been examined by AFM (Mackie *et al* 1999).

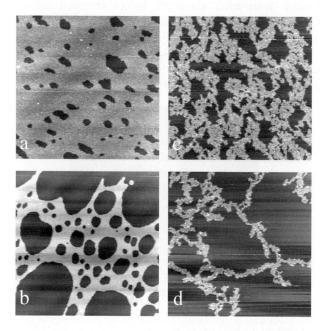

Figure 5.23. Partially displaced β-casein (a,b) and β-lactoglobulin (c,d) LB films on mica. As the surface pressure (Π) increases, due to increased adsorption of surfactant, the holes in the protein network (the areas previously occupied by surfactant) become larger. Scan sizes, surface pressure: (a) 6.4 x 6.4 μm, 16.7 mN m^{-1} (b) 6.4 x 6.4 μm, 19.2 mN m^{-1} (c) 3.2 x 3.2 μm, 21.8 mN m^{-1} (d) 3.2 x 3.2 μm, 24.6 mN m^{-1}. Imaged in dc mode under butanol.

Both spread and adsorbed interfacial protein films were created on a Langmuir trough and the experimental protocol was to progressively displace the films by the addition of surfactant to the subphase. The displacement of the protein film was monitored by periodic sampling of the film by LB film transfer onto mica. AFM imaging of the mixed protein/surfactant film was carried out under butanol which dissolved the surfactant (Tween 20) from the mica surface leaving just the protein network behind to be imaged. Examples of films of two different milk proteins, β-lactoglobulin and β-casein, partially displaced by the neutral surfactant Tween-20, are shown in Fig. 5.23. The images demonstrated clearly the formation of protein networks at the interface and provided qualitative information (in terms of the shapes of the surfactant domains) and quantitative information (in terms of the area and thickness of the protein domains) on the progression of the displacement process. One of the unique advantages of AFM is that it provides real three dimensional data. Film thickness was determined by measuring the distance between peaks in the relatively bimodal histogram of grey levels in the images (Fig. 5.24). Measurement of protein film area also utilised the essentially bimodal distribution of grey levels in the images to threshold the data to produce a binary image. Area of each phase (protein or surfactant) was then easily determined by quantifying the ratio of black to white pixels in the image.

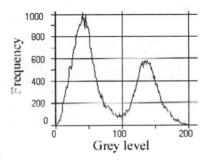

Figure 5.24. Histogram of the grey levels contained in the AFM image seen in Fig. 5.23c. The thickness of the protein film is given by the separation between the peaks, because the left hand peak represents the surfactant rich (dark) region and the right hand peak represents the protein rich (bright) region in the image.

The protein film responds to the increased surface pressure by thickening or buckling, and hence the process was termed 'orogenic' displacement, after the geological term for the creation of mountains by colliding plates in the earth's crust. At the final stages of displacement the protein network is fractured and the AFM images show islands of protein formed within a continuous surfactant phase (Fig. 5.25). Only at this final stage is the protein displaced from the interface. Because the proteins interact and form a network, individual proteins can not be displaced into the bulk until the network structure is broken. In this case AFM studies have produced a completely new and unexpected model for protein displacement, which could not, at present, have been deduced from other surface techniques which either lack the required resolution, or spatially average the interfacial structure. The 'orogenic' mechanism has also been demonstrated *in-situ* in a 'real time' AFM study of the displacement (cleaning) of a protein film from a graphite surface (Gunning *et al* 1999). The orogenic displacement mechanism has proved to be a generic process which is valid for all proteins studied to date, at air-water and oil-water, for oil-soluble, and water-soluble charged and uncharged surfactants. The early studies on single protein systems have been extended to characterise the displacement of complex mixed protein systems as models for real protein isolates. The generic nature of the process suggests that the stability of all protein-stabilised foams or emulsions can be enhanced by strengthening the protein network (Morris and Gunning, 2008).

Figure 5.25. At the final stage of collapse of the interfacial protein film islands of protein are seen surrounded by a continuous surfactant phase. Scan size: 10 x 10 μm, surface pressure Π=27.1 mN m⁻¹.

References

Alberts, B.A, Bray, D, Lewis, J, Raff, M, Roberts, K. and Watson, J. (1994). Membrane Structure. In *Molecular biology of the cell*, 3rd ed. pp. 477-484. Garland Publishing Inc., New York & London

Alexandre, S, Dubreuil, N, Fiol, C, Malandain, J-J, Sommer, F. and Valleton, J-M. (1994). Comparison at the microscopic scale of mixed fatty acid-protein Langmuir-Blodgett films resulting from vertical or horizontal transfer. *Microsc. Microanal. Microstruct.* 5, 359-371.

Alexandre, S, Dubreuil, N, Fiol, C, Lair, D, Sommer, F, Duc, T.M. and Valleton, J-M. (1997). Analysis of the dynamic organization of mixed protein/fatty acid Langmuir films. *Thin Solid Films* 293, 295-298.

Berkowitz, M.L, Bostick, D.L. and Pandit, S. (2006). Aqueous solutions next to phospholipid membrane surfaces: insights from simulations. *Chem. Rev.* 106, 1527-1539.

Berquand A., Levy D, Gubellini F, Le Grimellec C. and Milhiet P.E. (2007). Influence of calcium on direct incorporation of membrane proteins into in-plane lipid bilayer. *Ultramicroscopy* 107, 928-933.

Birdi, K.S, Vu, D.T, Moesby, L, Andersen, K.B. and Kristensen, D. (1994). Structures of lipid and biopolymer monolayers investigated as Langmuir-Blodgett films by atomic force microscopy. *Surface & Coatings Technol.* 67, 183-191.

Brian A.A. and Mcconnell H.M. (1984). Allogeneic stimulation of cyto-toxic t-cells by supported planar membranes. *PNAS USA* 81, 6159-6163.

Brandow, S.L, Turner, D.C, Ratna, B.R. and Gbaer, B.P. (1993). Modification of supported lipid membranes by atomic force microscopy. *Biophys. J.* 64, 898-902.

Brasseur, R, Braun, N, El Kirat, K, Deleu, M, Mingcot-Leclercq, M-P. and Dufrene, Y.F. (2007). The biologically important surfactin lipopeptide induces nanoripples in supported lipid bilayers. *Langmuir* 23, 9769-9772.

Charrier A. and Thibaudau F. (2005). Main phase transitions in supported lipid single-bilayer. *Biophys. J.* 89, 1094-1101.

Chi, L.F, Jacobi, S. and Fuchs, H. (1996). Chemical identification of differing amphiphiles in mixed Langmuir-Blodgett films by scanning surface potential microscopy. *Thin Solid Films* 284-285, 403-407.

Czajkowski, D.M, Huang, C. and Hao, Z. (1995). Ripple phase in asymmetric unilamellar bilayers with saturated and unsaturated phospholipids. *Biochemistry* 34, 12501-12505.

Czajkowski, D.M, Sheng, S. and Shao, Z. (1998). Staphlococcal α-hemolysin can form hexamers in phospholipid bilayers. *J. Mol. Biol.* 276, 325-330.

Delcu, M, Nott, K, Brasseur, R, Jacques, P, Thonart, P. and Dufrene, Y.F. (2001). Imaging mixed lipid monolayers by dynamic atomic force microscopy. *Biochimic.t Biophys. Acta* 1513, 55-62.

Dubreil, L, Vie´, V, Beaufils, S, Marion, D. and Renault, A. (2003). Aggregation of Puroindoline in Phospholipid Monolayers Spread at the Air-Liquid Interface. *Biophys. J.* 85, 2650-2660.

Dufrene, Y.F, Barger, W.R, Green, J.B.D. and Lee, G.U. (1997). Nanometer-scale surface properties of mixed phospholipid monolayers and bilayers. *Langmuir* 13, 4779-4784.

Dufrene, Y.F, Boland, T, Schneider, J.W, Barger, W.R. and Lee. G.U. (1998). Characterization of the physical properties of model biomembranes at the nanometer scale with the atomic force microscope. *Faraday Discussions* 111, 79-94.

Egawa, H. and Furusawa, K. (1999). Liposome adhesion on mica surface studied by atomic force microscopy. *Langmuir* 15, 1660-1666.

Fang, Y. and Yang, J. (1997). The growth of bilayer defects and the induction of interdigitated domains in the lipid loss process of supported phospholipid bilayers. *Biochim. Biophys. Acta* 1324, 309-319.

Franz, V, Loi, S, Muller, H, Bamberg, E. and Butt, H.H. (2002). Tip penetration through lipid bilayers in atomic force microscopy. *Colloids & Surf. B* 23, 191-200.

Fujihira, M. and Kawate, H. (1994). Scanning surface potential microscope for characterisation of Langmuir-Blodgett films. *Thin Solid Films*, 242, 163-169.

Fukuma, T, and Jarvis, S.P. (2006). Development of liquid-environment frequency modulation atomic force microscope with low noise deflection sensor for cantilevers of various dimensions. *Rev. Sci. Inst.* **77**, article number 043701.

Fukuma, T, Higgins, M.J. and Jarvis, S.P. (2007a). Direct Imaging of Individual Intrinsic Hydration Layers on Lipid Bilayers at Ångstrom Resolution. *Biophys. J.* **92**, 3603-3609.

Fukuma, T, Higgins, M.J. and Jarvis, S.P. (2007b). Direct Imaging of Lipid-Ion Network Formation under Physiological Conditions by Frequency Modulation Atomic Force Microscopy. *Phys. Rev. Letts.* **98**, article number 106101.

Garcia-Manyes, S, Oncins, G. and Sanz, F. (2005a). Effect of temperature on the nanomechanics of lipid bilayers studied by force spectroscopy. *Biophys. J.* **89**, 4261–4274.

Garcia-Manyes, S, Oncins, G. and Sanz, F. (2005b). Effect of ion-binding and chemical phospholipid structure on the nanomechanics of lipid bilayers studied by force spectroscopy. *Biophys. J.* **89**, 1812–1826.

Giocondi, M.C, Besson, F, Dosset, P, Milhiet P.E. and Le Grimellec, C. (2001). Remodelling of Ordered Membrane Domains by GPI-anchored Intestinal Alkaline Phosphatase. *Langmuir* **23**, 9358–64.

Gouaux, J.E, Braha, O, Hobaugh, M.R, Song, L, Cheley, S, Shustak, C. and Bayley, H.(1994). Subunit stoichiometry of staphylococcal alpha-hemolysin in crystals and on membranes: a heptameric transmembrane pore. *PNAS USA* **91**, 12828-12831.

Grandbois, M, Clausen-Schaumann, H. and Gaub, H. (1998). Atomic force microscope imaging of phospholipid bilayer degradation by phospholipase A_2. *Biophys. J.* **74**, 2398-2404.

Gunning, A.P, Wilde, P.J, Clark, D.C, Morris, V.J, Parker, M.L. and Gunning, P.A. (1996). Atomic force microscopy of interfacial protein films. *J. Colloid & Interface Sci.* **183**, 600-602.

Gunning, A.P, Mackie, A.R, Wilde, P.J. and Morris, V.J. (1999) *In-situ* observation of the surfactant induced displacement of protein from a graphite surface by atomic force microscopy. *Langmuir* **15**, 4636-4640.

Hui, S.K. and Sen, A. (1989). Effects of lipid packing on polymorphic phase behaviour and membrane properties. *PNAS USA* **86**, 5825-5829.

Hui, S.W, Viswanathan, R, Zasadzinski, J.A. and Israelachvili, J.N. (1995). The structure and stability of phospholipid bilayers by atomic force microscopy. *Biophys. J.* **68**, 171-178.

Inoue, T. and Yokoyama, H. (1994). Surface potential imaging of phase separated Langmuir-Blodgett monolayers by scanning Maxwell stress microscopy. *Thin Solid Films* **243**, 399-402.

Johnson, S.J, Bayerl, T.M, McDermott, D.C, Adam, G.W, Rennie, A.R, Thomas, R.K. and Sackmann, E. (1991). Structure of an adsorbed dimyristoylphosphatidylcholine bilayer measured with specular reflection of neutrons. *Biophys. J.* 59:289–294.

Kaasgaard, T, Leidy, C, Crowe, J.H, Mouritsen, O.G. and Jørgensen, K. (2003). Temperature-controlled structure and kinetics of ripple phases in one- and two-component supported lipid bilayers. *Biophys. J.* **85**, 350-360.

Kalb, E, Frey, S. and Tamm, L.K. (1992). Formation of supported planar bilayers by fusion of vesicles to supported phospholipid monolayers. *Biochim. Biophys. Acta* **1103**, 307-316.

Keller D, Larsen NB, Moller IM and Mouritsen OG. (2005). Decoupled phase transitions and grain-boundary melting in supported phospholipid bilayers. *Phys. Rev. Lett.* **94**, Article number 025701.

Knapp, H.F, Wiegrabe, W, Heim, M, Eschrich, R. and Guckenberger, R. (1995). Atomic Force Microscope Measurements and Manipulation of Langmuir-Blodgett Films with Modified Tips. *Biophys. J.* **69**, 708-715.

Krägel, J, Grigoriev, D.O, Makievski, A.V, Miller, R, Fainerman, V.B, Wilde, P.J. and Wüstneck, R. (1999). Consistency of surface mechanical properties of spread protein layers at the liquid-air interface at different spreading conditions. *Colloids & Surfaces B: Biointerfaces* **12**, 391-397.

Kornberg, R.D. and Ribi, H.O. (1987). Formation of two dimensional crystals of proteins on lipid layers. In *Protein Structure, folding and design 2*. (ed. D.L. Oxender), pp. 175-186, Alan R. Liss: New York.

Leidy, C, Mouritsen, O.G, Jørgensen, K. and Peters, G.H. (2004). Evolution of a rippled membrane during phospholipase A_2 hydrolysis studied by time-resolved AFM. *Biophys. J.* **87**, 408–418.

Leonenko Z.V, Finot E., Ma H., Dahms T.E.S. and Cramb D.T. (2004). Investigation of temperature-induced phase transitions in DOPC and DPPC phospholipid bilayers using temperature-controlled scanning force microscopy. *Biophys. J.* **86**, 3783-3793.

Li, B, Fuji, M, Fukakada, K, Kato, T. and Seimiya, T. (1998). In situ AFM observation of heterogeneous growth of adsorbed on cleaved mica surface. *Thin Solid Films* **312**, 20-23.

Lichtenberg, D.G, Romero, M, Menashe, M. and Biltonen, R.L. (1986). Hydrolysis of dipalmitoylphosphatidylcholine large unilamellar vesicles by porcine pancreatic phospholipase A₂. *J. Biol. Chem.* **261**, 5334-5340.

Mackie, A.R, Gunning, A.P, Ridout, M.J. and Morris, V.J. (1998). Gelation of gelatin: observation in the bulk and at the air-water interface. *Biopolymers*, **46**, 245-252.

Mackie, A.R, Gunning, A.P, Wilde, P.J. and Morris, V.J. (1999). Orogenic displacement of protein from the air/water interface by competitive adsorption. *J. Colloid & Interface Sci.* **210**, 157-166.

Merkel, R, Sackmann, E. and Evans. E. (1989). Molecular friction and epistatic coupling between monolayers in supported monolayers. *J. Physique France.* **50**, 1535-1555.

Morris, V.J. and Gunning, A.P. **(2008).** Microscopy, microstructure and displacement of proteins from interfaces: implications for food quality and digestion. *Soft Matter* **4**, 943-951

Mou, J, Yang, J. and Shao, Z. (1994). Tris(hydroxymethyl)aminomethane (C₄H₁₁NO₃) induced a ripple phase in supported unilamellar phospholipid bilayers. *Biochemistry* **33**, 4439-4443.

Mou, J, Yang, J. and Shao, Z. (1995). Atomic force microscopy of cholera toxin B-oligomers bound to bilayers of biologically relevant lipids. *J. Mol. Biol.* **248**, 507-512.

Mueller, H, Butt, H-J. and Bamberg, E. (1999). Force Measurements on Myelin Basic Protein Adsorbed to Mica and Lipid Bilayer Surfaces Done with the Atomic Force Microscope. *Biphys. J.* **76**, 1072-1079.

Nicolini, C, Baranski, J, Schlummer, S, Palomo, J, Lumbierres-Burgues, M, Kahms M, Kuhlmann, J, Sanchez, S, Gratton, E, Waldmann, H. and Winter, R. (2006). Visualizing association of N-Ras in lipid microdomains: Influence of domain structure and interfacial adsorption. *J. Amer. Chem. Soc.* **128**, 192-201.

Pera, I, Stark, R, Kappl, M, Butt H-J. and Benfenati, F. (2004). Using the Atomic Force Microscope to Study the Interaction between Two Solid Supported Lipid Bilayers and the Influence of Synapsin I. *Biophys. J.* **87**, 2446-2455.

Racker (1985). In *Reconstitution of transporters, receptors and pathological states*, Academic Press, Orlando.

Rand, R.P, Capman, D. and Larrson, K. (1975). Tilted hydrocarbon chains of dipalmitoyl lecithin become perpendicular to the bilayer before melting. *Biophys. J.* **15**, 1117-1124.

Sackmann, E. (1983). Physical foundations of the molecular organisation and dynamics of membranes. In *Biophysics*, (eds. W. Hoppe, W. Lohmann, H. Markl & H. Zeigler), pp. 425-457. Springer-Verlag, New York.

Sackmann, E. (1996). Supported membranes: scientific and practical applications. *Science.* **271**:43-48.

Santos, N.C, Ter-Ovanesyan, E, Zasadzinski, J.A, Prieto, M. and Castanho, M.A.R.B. (1998). Filipin-induced lesions in planar phospholipid bilayers imaged by atomic force microscopy. *Biophys. J.* **75**, 1869-1873.

Schallamach, A. (1971). How does rubber slide? *Wear* **17**, 301-312.

Schaus, S.S. and Henderson, E.R, (1997). Cell viability and probe-cell membrane interactions of XR1 glial cells imaged by atomic force microscopy. *Biophys. J.* **73**, 1205-1214.

Seantier, B, Giocondi, M-C, Le Grimellec, C, Milhiet, P-E. (2008). Probing supported model and native membranes using AFM. *Current Opinion Colloid & Interface Sci.* **13**, 326-337.

Seddon, J.M. (1990). Structure of the inverted hexagonal (HII) phase, and non-lamellar phase transitions of lipids. *Biochim. Biophys. Acta* **1031**, 1-69.

Seu K.J., Pandey A.P, Haque F, Proctor E.A, Ribbe A.E. And Hovis J.S. (2007). Effect of surface treatment on diffusion and domain formation in supported lipid bilayers. *Biophys. J.* **92**, 2445-2450.

Shibata-Seki, T, Masai, J, Tagawa, T, Sorin, T. and Kondo, S. (1996). In-situ atomic force microscopy study of lipid vesicles adsorbed on a substrate. *Thin Solid Films* **273**, 297-303.

Shiku, H. and Dunn, R. (1998). Direct observation of DPPC phase domain motion on mica surfaces under conditions of high relative humidity. *J. Phys. Chem. B* **102**, 3791-3797.

Shwartz, D.K. (1997). Langmuir-Blodgett film structure. *Surface Science Reports* **27**, 241-334.

Singh, S. and Keller, D.J. (1991). Atomic force microscopy of supported planar membrane bilayers. *Biophys. J.* **60**, 1401-1410.

Solleti, J.M, Boteau, M, Sommer, F, Duc, T.M. and Celio, M.R, (1996). Characterisation of mixed miscible and nonmiscible phospholipid Langmuir-Blodgett films by atomic force microscopy. *J. Vac. Sci. & Technol. B*, **14**, 1492-1497.

Sommer, F, Alexandre, S, Dubreuil, N, Lair, D, Duc, T.M. and Valleton, J-M. (1997). Contribution of lateral force and "Tapping Mode" microscopies to the study of mixed protein Langmuir-Blodgett films. *Langmuir* **13**, 791-795.

Song, L, Hobaugh, M, Shustak, C, Cheley, S, Bayley, H. and Gouaux, J.E. (1996). Structure of staphylococcal alpha-hemolysin, a heptameric transmembrane pore. *Science* **274**, 1859-1856.

Tamm, L.K, Böhm, C, Yang, J, Shao, Z, Hwang, J, Edidin, M. and Betzig, E. (1996). Nanostructure of supported phospholipid monolayers and bilayers by scanning probe microscopy. *Thin Solid Films* **284-285**, 813-816.

Tanaka, M. and Sackmann, E. (2005). Polymer-supported membranes as models of the cell surface. *Nature* **437**, 656-663.

Tardieu, A, Luzzati, V. and Reman, FC. (1973). Structure and polymorphism of the hydrocarbon chains of lipids: a study of lecithin-water phases. *J. Mol. Biol.* **75**, 711-733.

Tokumasu, F, Jin, A.J. and Dvorak, JA. (2002). Lipid membrane phase behaviour elucidated in real time by controlled environment atomic force microscopy. *J. Electron Microscopy* **51**, 1–9.

Trunfio-Sfarghiu A.M, Berthier Y., Meurisse M.H. and Rieu J.P. (2008) Role of nanomechanical properties in the tribological performance of phospholipid biomimetic surfaces. *Langmuir* **24**, 8765-8771.

Turner, D.C, Peek, B.M, Wertz, T.E, Archibald, D.D, Geer, R.E. and Gaber, B.P. (1996). Enzymatic modification of a chemisorbed lipid monolayer. *Langmuir* **12**, 4411-4416.

Von Nahmen, A, Schenk, M, Sieber, M. and Amrein, M. (1997). The structure of a model pulmonary surfactant as revealed by scanning force microscopy. *Biophys. J.* **72**, 463-469.

Vikholm, I, Peltonen, J. and Telleman, O. (1995). Atomic force microscope images of lipid layers spread from vesicle suspensions. *Biochim. Biophys. Acta* **1233**, 111-117.

Weisenhorn, A.L, Schmitt, F-J, Knoll, W. and Hansma, P.K. (1992). Streptavidin binding observed with an atomic force microscope. *Ultramicroscopy* **42-44**, 1125-1132.

Yager, P. and Schoen, P.E. (1984). Formation of tubules by a polymerizable surfactant. *Mol. Cryst. Liq. Cryst.* **106**, 371-381.

Yang, J, Tamm, L.K, Tillack, T.W. and Shao, Z. (1993.) New approach for Atomic force microscopy of membrane proteins. *J. Mol. Biol.* **229**, 286-290.

Yarrow F, Vlugt TJH, Van der Eerden JPJM, Snel MME. (2005) Melting of a DPPC lipid bilayer observed with atomic force microscopy and computer simulation. J Crystal Growth **275**, e1417–21.

Zasadzinski, J.A.N, Helm, C.A, Longo, M.L, Weisenhorn, A.L, Gould, S.A.C. and Hansma, P.K. (1991). Atomic force microscopy of hydrated phosphatidylethanolamine bilayers. *Biophys. J.* **59**, 755-760.

Zhai, X. and Kleijn, J.M. (1997). Molecular structure of dipalmitoyl phosphatidylcholine Langmuir-Blodgett monolayers studied by atomic force microscopy. *Thin Solid Films* **304**, 327-332.

CHAPTER 6

ORDERED MACROMOLECULES

6.1. Three dimensional crystals

The study of the surfaces of 3D crystals by AFM offers the prospect of achieving the highest resolution on biological samples. The periodicity of the surface will minimise probe broadening effects and allow image reconstruction to remove noise. The structure at the surface may differ from that in the bulk and it may be possible to identify features which help in the subsequent analysis by X-ray diffraction. If the crystals are small then AFM offers an alternative to electron microscopy, with the possible advantages of acquiring images under aqueous or buffered conditions. If the crystalline structures occur naturally, then AFM allows these faces to be examined under natural conditions, with the prospect of observing biological processes such as enzymatic reactions at the surfaces. Finally the AFM permits crystal growth to be examined, and an investigation of growth kinetics and mechanisms. As a microscopic technique it is able to visualise the effects of foreign particles and the incorporation of defects into the crystal lattice. This type of information provides a basis for more systematic optimisation of crystal growth for X ray diffraction work.

6.1.1. Crystalline cellulose

There are a number of AFM studies on isolated crystalline cellulose fibres which report high resolution images and analysis of the surface structure (Hanley *et al* 1992; Baker *et al* 1997; 1998; Kuutti *et al* 1995): research on isolated cellulose is described in more detail in chapter 4. The most detailed studies are on cellulose isolated from *Valonia*, which is regarded as the source of the most crystalline form of cellulose I. *Valonia* cellulose contains two distinct allomorphs I_α (triclinic) and I_β (monoclinic). Recent high resolution AFM studies have been used to probe the detailed surface structure of *Valonia* cellulose. AFM images of the purified cellulose from *V. macrophysa* were enhanced by Fourier processing and compared with model Connolly surfaces generated from electron diffraction data for the I_α and I_β forms. The surface structures observed were attributed to the monoclinic phase (Kutti *et al* 1995). More recent AFM studies on *V. ventricosa* cellulose have produced images of the surface in which it has been possible to image the repeating cellobiose unit along the cellulose molecules (Fig. 6.1), through identification of the bulky hydroxymethyl group, and thus to identify a triclinic phase directly from the images (Baker *et al* 1997; 1998). The distinction between the two crystal phases required the detection of differences in the displacement of the cellulose chains along their axes by 0.26 nm. This was achieved without filtering or averaging and is believed to represent the highest resolution achieved to date on a biological sample. At present AFM has not been used to investigate possible coexistence of I_α and I_β on the crystal surfaces. The cellulose structures could be imaged in air (Kutti *et al* 1995), or under propanol (Baker *et al* 1997; 1998) or water (Baker *et al* 1997; 1998). The ability to image under water means that it is possible to image the surface at which biological processes occur, and that it may also be possible in the future to follow processes such as enzymatic breakdown.

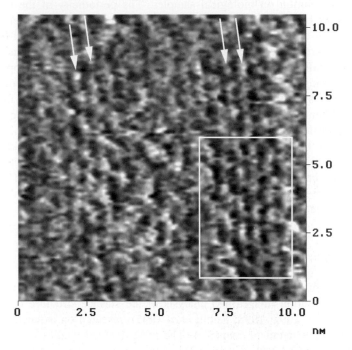

Figure 6.1. Contact error signal mode AFM images of a *Valonia* surface under water. The white arrows indicate cellulose molecules which are running almost vertically down the page. The white box highlights a region where bright spots can be seen along the length of the molecules, separated by a distance closely matching the cellobiose repeat unit. The angle of the spots within the box to the molecular axis is $64 \pm 2°$. Data provided by the authors and based on Baker *et al* 1997.

6.1.2. Protein crystals

The first AFM study of 3D protein crystals was on crystals of the membrane protein Ca-ATPase, the calcium pump from the sarcoplasmic reticulum (Lacapere *et al* 1992). For small dried crystals it was possible to measure step changes in height allowing the unit cell dimension normal to the substrate to be determined. This value was found to increase on hydration, presumably due to hydration and swelling of extramembranous domains. The in-plane resolution for both dried and hydrated crystals was poor: it was not possible to resolve individual proteins or characteristic periodicities of the crystal lattice. The difficulties in imaging were attributed to displacement of layers within the crystal by the probe during scanning. A major problem identified in these early studies was the difficulty of immobilising small crystals on the substrate, particularly for studies on hydrated crystals or investigations of crystal growth in mother liquor. Approaches to resolving this problem include optimising physical adsorption to substrates (Lacapere *et al* 1992), nucleation of crystal growth directly on the substrate (Malkin *et al* 1996a, 1999b; Kuznetsov *et al* 1997), and sticking the crystals to the substrate (Konnert *et al* 1994; Yip and Ward, 1996) with an adhesive. If the crystals are firmly attached then the second limiting factor is damage or 'etching' of the surface with the probe. Provided the crystals are resilient enough, and that the imaging force is minimised, then dc contact mode imaging can give molecular resolution (Kuznetsov *et al* 1997). It was shown for soft crystals such as insulin that if the crystal surface is damaged in contact mode by shear forces then the use of Tapping can improve the images (Yip and Ward, 1996).

Figure 6.2. An AFM image of a canavalin crystal surface showing growth from multiple hillocks. Scan size 100 x 100 μm. Data reproduced from Land *et al* 1997 with permission.

AFM studies have been made on a number of protein crystals including insulin (Yip and Ward, 1996), canavilin (Land *et al* 1995; 1997; Malkin *et al* 1996a), thaumatin (Malkin *et al* 1995a; 1996b), catalase (Malkin *et al* 1995a), lysozyme (Durbin and Carlson, 1992; Durbin *et al* 1993; Malkin *et al* 1995a; Konnert *et al* 1994;

Kuznetsov *et al* 1997), apoferritin (Malkin *et al* 1995a), bacteriorhodopsin (Kouyama *et al* 1994) and lipase (Kuznetsov *et al* 1997). In many of these studies it has been possible to obtain molecular resolution, or to observe periodicities characteristic of the crystal lattice (Konnert *et al* 1994; Land *et al* 1995; Yip and Ward, 1996; Kuznetsov *et al* 1997). The information obtained from high resolution AFM images of 3D protein crystals is unlikely to rival that obtainable from X-ray diffraction but it may complement images or image reconstruction obtainable by electron microscopy. However, AFM data can be used to obtain information useful for crystal structural analysis by X-ray diffraction: the data can be used to resolve questions on space group enantiomers, about packing of molecules within unit cells, the number of molecules per asymmetric units, or the disposition of multiple molecules within asymmetric units (Kuznetsov *et al* 1997). AFM studies of the surfaces of bacteriorhodopsin revealed a novel assemblage of bacteriorhodopsin in its 3D crystal: the crystal consists of hexagonal close-packed arrays of spherical protein clusters. Both AFM and EM have advantages for studying small crystals. AFM provides a means of observing differences in the structure of the surface when compared to the bulk; but to date no such differences   have been reported. The major advantage of the AFM is the ability to investigate mechanisms of crystallisation and to follow crystal growth. The first studies on protein crystal growth were by Durbin and coworkers (Durbin and Carlson, 1992; Durbin *et al* 1993). Although largely non-invasive there are reports that growth rates may be modified by the scanning process (Land *et al* 1995; Yip and Ward, 1996) and the tip may displace loosely bound proteins, or protein aggregates, which might otherwise influence growth. Thus Tapping mode images of insulin crystals revealed aggregates attached to the crystal surface which were not seen in dc contact mode images (Yip and Ward, 1996). Studies on a range of protein crystals suggest that the main method of crystal growth is by the nucleation and growth of 2D islands on the surface of the crystal. This has been observed to be the dominant mechanism for lysozyme (Durbin and Carlson, 1992; Durbin *et al* 1993; Malkin *et al* 1995a; Konnert *et al* 1994), catalase (Malkin *et al* 1995a) and thaumatin (Malkin *et al* 1995a; 1996). At low supersaturations screw dislocations have been observed for lysozyme (Markin *et al* 1995a; Konnert *et al* 1994) and thaumatin (Markin *et al* 1995a; 1996). At high supersaturation 3D clusters are deposited onto thaumitin surfaces and their 2D tangential growth results in multilayer stacks (Malkin *et al* 1996). The mechanisms of crystal growth, surface morphology and growth kinetics have been studied in detail for canavilin (Malkin *et al* 1995a; Land *et al* 1995; 1997). Depending on the level of supersaturation growth occurs on steps generated either by simple or complex screw dislocation sources (Fig. 6.2), 2D nucleating islands or deposited 3D clusters (Fig. 6.3). Apoferritin and occasionally trigonal catalase and tetragonal lysozyme were observed to develop very rough growing surfaces by intensive random nucleation; a rare process for conventional crystal growth from solution known as normal crystal growth (Malkin *et al* 1995a).

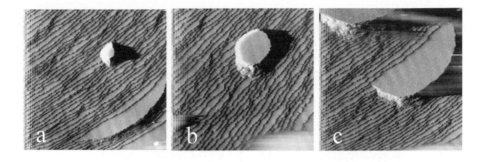

Figure 6.3. A series of AFM images (scan size 40 x 40 μm) collected at 30 second intervals showing the growth of a macrocluster which has landed on the growing crystal surface. The nucleus grows out radially, forming macrosteps at the edge and a large plateau on top. Note that the nucleus merges flawlessly into the existing crystal. Data reproduced from Land *et al* 1997 with permission.

For the range of proteins studied, all mechanisms of crystal growth have been observed and lysozyme has been shown to exhibit all of these mechanisms under appropriate conditions. Adsorption of foreign particles results in holes which remain within the crystal upon growth of additional layers. This has been observed for canavilin (Land *et al* 1995) and lysozyme (Konnert *et al* 1994) crystals.

6.1.3. Nucleic acid crystals

When compared to proteins very few nucleic acids have been crystallised. The crystallisation process is still largely empirical because virtually nothing is known about the growth mechanisms, growth kinetics, development of surface morphology, or the effect of defects on crystallisation. Recently there has been an AFM study [Ng *et al* 1997) of tRNA crystal growth aimed at providing a basis for optimising crystallisation of RNA. Crystal growth was observed to occur in steps on steep vicinal hillocks generated by screw dislocations. Different growth mechanisms were observed in different temperature ranges which were used to vary supersaturation (Fig 6.4). At low supersaturation a 2D nucleation process occurs in which the step edges of the hillocks grow tangentially, and small islands nucleate and grow tangentially on the plateaus of the hillocks. As observed for proteins, at higher supersaturation a 3D mechanism occurs in which multilayer stacks appear and grow both normally and tangentially on the crystal surface. Adsorption of foreign particles results in holes which remain upon growth of additional layers: these types of holes have also been observed in the growth of protein (Konnert *et al* 1994; Land *et al* 1995) and virus (Malkin *et al* 1995a; 1996a) crystals. Lattice resolution images of the crystal surface were obtained but no molecular features which are characteristic of tRNA were observed. The study has been used to suggest ways of improving the crystallisation of RNA: it is proposed that after nucleation supersaturation is reduced to allow growth by regular and unique mechanisms (Ng *et al* 1997).

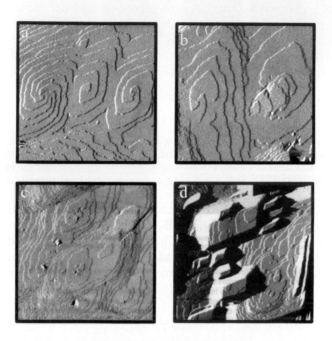

Figure 6.4. AFM images showing different surface morphologies of yeast tRNA crystals seen at different temperature ranges. (a) Dislocation hillocks formed by right handed, single left handed and double right handed screw dislocations (15° C). (b) Development of double and single screw dislocations (14°C). (c) Growth by 2D nucleation at 13°C. Also seen is the formation of a hole caused by the incorporation of a foreign particle during growth. (d) Dominant mechanism of growth is by 3D nucleation with multilayer stack macrostep production at 12°C. Scan sizes (a, b) 23 x 23 μm, (c) 20 x 20 μm and (d) 34 x 34 μm. Reproduced with permission of the authors and Oxford University Press from Ng *et al* 1997.

6.1.4. Viruses and virus crystals

Viruses consist of a nucleic acid packaged within a protein coat. There is interest in the detailed molecular structure, assembly and mechanisms of viral infection. Usually it is sufficient to passively adsorb the virus particles to substrates such as mica or silica wafers. If necessary, substrates such as mica can be coated with poly-L-lysine to prevent the viruses particles being displaced on the surface by the probe. Images can be obtained in air, under propanol or in aqueous media. A range of types of viruses have been imaged by AFM including bacteriophages (Kolbe *et al* 1992; Thundat *et al* 1992; Imai *et al* 1993; Ikai *et al* 1994), plant viruses (Thundat *et al* 1992; Zenhausern *et al* 1992; Imai *et al* 1993; Bushell *et al* 1995; Kirby *et al*

1996; Falvo *et al* 1997) and animal viruses (Ohnesorge *et al* 1997). They have been considered as standards for height measurements or for tip deconvolution. The bacteriophage T4 polyheads were used as a standard material for appraising image processing of STM images (Engel 1991). If the virus particles are assembled into arrays then probe broadening effects are minimised and the measured widths approach the true diameters of the virus particles (Thundat *et al* 1992; Kirby *et al* 1996). Most studies are low resolution revealing size and shape, or in the case of bacteriophages features such as heads, tails and tail fibres (Ikai *et al* 1994). There are few studies at present of the assembly of the protein coats (see for example section 6.3.7). Submolecular resolution has been obtained by AFM on the crystalline arrays of bacteriophage T4 polyheads (Karrasch *et al* 1993). Images of pox viruses under water and buffer have revealed tubular 'protein' structures consistent with those observed previously by electron microscopy (Ohnesorge *et al* 1997). As well as intact viruses it is also possible to observe damaged virus particles from which the nucleic acid has spilled out (Fig. 6.5) onto the substrate (Kolbe *et al* 1992: Shao and Zhang, 1996).

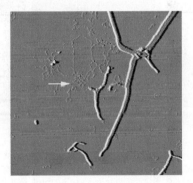

Figure 6.5. AFM image of papaya mosaic virus air dried onto mica and imaged under butanol. The arrow indicates RNA released from a damaged virus particle. Scan size 3 x 3 μm.

It has been suggested (Ikai *et al* 1994) that the heights of bacteriophage heads can be used to distinguish between normal heads and ghosts devoid of nucleic acid: the ghosts being either deformed more by the probe or collapsing down onto the substrate. Virus particles can be damaged or dissected with the probe tip (Ikai *et al* 1994; Bushell *et al* 1995; Falvo *et al* 1997): for example partial removal of the protein coat of TMV revealed the central channel (Bushell *et al* 1995). Controlled manipulation of TMV particles deposited onto graphite was carried out using a modified AFM (Fig. 6.6) and the data obtained was interpreted in terms of a mechanical model for the virus particles (Falvo *et al* 1997).

In a unique set of experiments it has been possible to study living cells infected with virus particles. Living cultured monkey kidney cells were imaged using a 'pipette-AFM' (Häberle *et al* 1991; 1992; Hörber *et al* 1992). For cells infected with pox virus it was possible to image adsorbed viruses on the cell surface

(Ohnesorge *et al* 1997) and to generate a time sequence of images believed to demonstrate release of new viral particles from the cell surface (Häberle *et al* 1992; Hörber *et al* 1992; Ohnesorge *et al* 1997).

There are several reported studies on the growth of satellite tobacco mosaic virus crystals (Malkin *et al* 1995a, 1995b; 1996a; Kuznetsov *et al* 1997). The growth mechanism differs from that seen for protein crystals (section 6.1.2), in that 2D nucleation and growth only made a major contribution to crystal growth at very low supersaturation, and no screw dislocations were observed. The dominant growth mechanism appears to be by addition of 3D clusters or microcrystals, and their subsequent expansion. If the 3D additions are misaligned with respect to the larger crystal lattice then their incorporation generates a defect structure in the crystal. Other sources of defects are adsorbed foreign particles which are also incorporated in the growing crystal. High resolution images of the crystal surfaces permit resolution of individual virus particles (Kuznetsov *et al* 1997; Malkin *et al* 1995b).

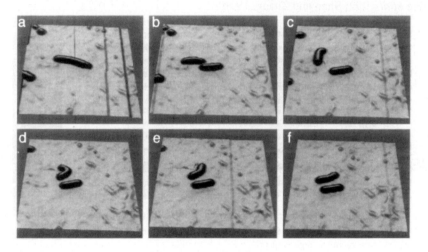

Figure 6.6. AFM images showing controlled dissection and manipulation of a TMV virus particle after deposition onto graphite. The sequence of images shows the TMV particle in its original position and orientation (a), after dissection (b), rotated (c), translated (d), and straightened so that it is parallel with the remaining segment of the virus particle (e and f). Scan size 560 x 560 nm. Reproduced with permission from Falvo *et al* 1997 and the Biophysical Society.

More recently there have been significant advances in the field of viral imaging with the AFM (for a useful review see Kuznetsov *et al* 2001), with methods being developed to allow surface as well as internal detail to be resolved for several important human viruses, namely *Herpes simplex virus-1,* intracellular mature vaccinia virus and HIV (Plomp *et al* 2002; Malkin *et al* 2003; Kuznetzov *et al* 2003). In a rather elegant study, the gradual degradation of *herpes simplex virus-1* virions by means of a detergent treatment was followed by AFM (Plomp *et al* 2002). This allowed visualisation of the outer envelope, the underlying capsid, the caposomeres comprising the capsid and their surface arrangement, damaged and partially degraded capsids with missing capsomeres and finally DNA which had leaked out of damaged particles. Data from the first two stages are shown in figure 6.7, which illustrates the stripping away of the outer membrane to reveal the exquisite detail of the underlying capsid, quite an achievement on such a highly curved surface.

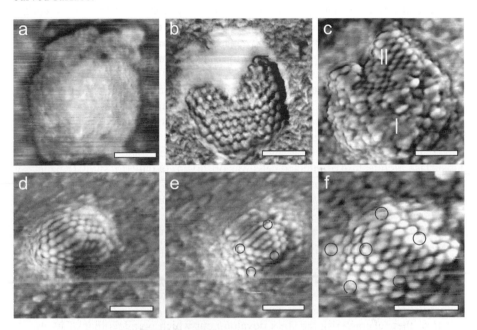

Figure 6.7 Herpes HSV-1 virions in different confirmations adsorbed on mica and imaged with tapping-mode AFM in air. (a) The capsid is fully covered in the lipid envelope. (b-f) Addition of 0.2% Triton X-100 removes the lipid envelope. (b) The capsid is still partly covered by the lipid envelope. (c) Most of this capsid is covered by an irregular collection of particles of approximately 10 nm in size (area I), which correspond to tegument proteins. In a smaller region the underlying, highly regular capsid is exposed (area II). (d-f) Three capsids with the lipid bilayer and tegument completely removed, which fully exposes the regular capsomere structure consisting predominantly of hexons and pentons. In (e) and (f), pentons are denoted by black circles. Scale bars are 100 nm. Image kindly supplied by M. Plomp and A.J. Malkin.

The resolution obtained by the AFM allowed direct visualisation of variations in the structure of individual capsids, something which is not available from electron tomography data where averaging is required. Additionally the data were obtained in much more biologically-relevant conditions, at ambient temperatures and pressures and without the need for staining nor coating to induce contrast into the images. In a similar way the structural elements of intracellular mature vaccinia virus (IMV) particles were visualised using a series of detergent, disulphide reducing compounds and enzymatic attack (Malkin *et al* 2003). Vaccinia from the *Poxviridae* family is of interest due to its role in the elimination of smallpox, and its study has received new impetus because of the potential threat of deliberate release in the post 9/11 era. This study was probably the first to compare high-resolution images of virus in a fully hydrated state under buffer with images obtained following air drying, demonstrating that reversible shrinkage (up to 2.5 times in height) and other structural changes occur upon dehydration of the particles. A key to the success of liquid imaging was to achieve a close-packed array of virus particles on the mica, which allowed them to resist tip forces which might otherwise sweep away discrete particles under liquid (Kuznetsov *et al* 2001, see also section 7.2.1). This can be achieved by slow drying during the deposition stage to encourage particle-jamming to occur as the liquid film thins out, leaving behind islands of close-packed virions.

Another area where AFM had been brought to bear is the viral invasion of cells, with the effect that HIV has upon the surface structures expressed by Lymphocytes and the effect of West-Nile virus upon Vero cells from African green monkey kidney cells being two examples (Kuznetzov *et al* 2003; Lee and Ng, 2004). In both cases structural alteration of the surface and in the latter case, subsurface architecture of infected cells was observed. In the HIV study (Kuznetzov *et al* 2003) high-resolution images of viral particles on the cell surface were obtained, these displayed a larger degree of size variation than when seen following deposition onto glass, a fact tentatively attributed to the budding process. Furthermore, detergent treatment allowed measurements to be obtained of the thickness of the enveloping protein-membrane-matrix, which forms the outer shell of the virion.

Because of their small and fixed size a new application of viruses which has emerged in the last few years is their use as scaffolds for the engineering of nanostructures (Smith *et al* 2003; Nam *et al* 2004), with AFM imaging being ideally suited to their characterisation. In a development which parallels that of its use in bacterial and mammalian cells (see section 7.2) AFM has been used to probe the mechanical properties of viruses (Carrasco *et al* 2006; Kol *et al* 2006; 2007) and, in one case, viral adhesion properties (Negishi *et al* 2004). Notable amongst these is the demonstration of a 'stiffness-switch' in HIV, with newly emergent budding particles being some 14-fold stiffer than mature particles, primarily mediated by the HIV envelope cytoplasmic tail domain (Kol 2007). A striking correlation was observed between the maturation-induced softening of the viral particles and their ability to enter cells. Thus the authors have shown how HIV regulates its mechanical properties at different stages of its life cycle (stiff during budding and soft during cell entry) which may be important for efficient infectivity (Kol *et al* 2007).

6.2. Two dimensional protein crystals: an introduction

Two dimensional (2D) protein crystals are a ubiquitous class of proteins occurring naturally on bacterial and algal cell surfaces. The interest in these proteins from a microscopy point of view was sparked off by Henderson and Unwin (1975) who demonstrated that regular two dimensional crystalline arrays enabled elegant image processing (including electron diffraction) to be carried out *on-line* using transmission electron microscopy. In further refined studies they were able to produce a 3D map of bacteriorhodopsin (Baldwin and Henderson, 1984, Henderson *et al* 1986; 1990, Baldwin *et al* 1988) from which an atomic model of the protein has been proposed (Henderson *et al* 1990). Since then many other systems have been examined using TEM, notable examples including 2D crystals of the bacterial porins OmpF (Sass *et al* 1989) and PhoE (Jap 1988) and of the plant light harvesting complex (LHC-II) (Kühlbrandt and Downing, 1989, Kühlbrandt and Wang, 1991). The justification for the study of 2D protein crystals is the structural determination of the proteins at atomic or near atomic resolution. Normally this would of course be done by X-ray crystallography but membrane proteins in particular are notoriously difficult to crystallise in three dimensions so 2D crystals provide an attractive alternative route involving microscopy rather than X-ray crystallography (they are typically only 5-20 nm thick which is ideal for electron diffraction but too thin for X-ray crystallography). To date most work has been carried out on naturally abundant membrane proteins which are easy to isolate and purify in large amounts. However, there are a vast array of other interesting candidates for study which are naturally rare such as receptors, channels and transporters and, with recent advances in genetics, these proteins can now be produced in reasonable quantities by over-expression of genes in cell culture (Kühlbrandt, 1992; Müller *et al* 1997a).

6.2.1. What does AFM have to offer?

AFM has, not surprisingly, been brought to bear on 2D protein crystals offering the inherent advantages of being able to work in physiological buffers, not exposing the sample to any radiation and, in terms of raw data or image contrast, achieving resolution better than TEM (Müller *et al* 1995b). There are limitations to the use of AFM though; compared to TEM the image processing is more difficult since the pixel density of an AFM image cannot possibly compete with an 'on-line' electron beam which effectively has no pixel limit as such. This problem is partly obviated by the fact that AFM images are of far higher contrast than TEM images, and so the need for image processing is much less. Generally, the effect of tip convolution on AFM images of specimens presents a significant problem. Image de-convolution is a non-trivial process due to factors such as compression and distortion of molecules by the AFM tip (Shao *et al* 1996; Yang *et al* 1996) and inherently unpredictable events such as changes in the shape of the tip during scanning due to wear or, more likely with biological samples, contamination (Schabert and Engel, 1994b). However for samples which are close packed and have very low surface roughness, tip convolution becomes negligible, since only the apex of the tip takes part in the imaging process as illustrated in Fig. 6.8. Incidentally this demonstrates one of the unique features of images generated by probe microscopes, namely that image resolution is enhanced as specimen roughness decreases (Bustamante *et al* 1997). This is contrary to all other forms of microscopy where such flatness leads to poor contrast. TEM, with its ability to perform electron diffraction, offers the advantage of being able to generate electron density maps and, from these maps atomic scale models of the protein molecules.

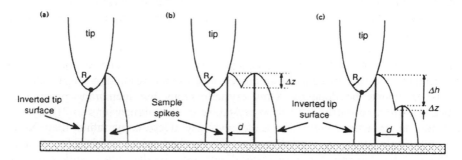

Figure 6.8 Image formation in the AFM. (a) Image of a spike obtained with a parabolic tip. (b) Image of two spikes of similar height close together. (c) Image of two spikes of different height. The ability of the AFM to resolve two spikes separated by a distance *d* depends on the size of the dimple ΔZ in the image and decreases with *increasing* height difference Δh between the spikes. Thus flatter samples can lead to higher resolution images. *R* is the radius of curvature of the tip. Reproduced from Bustamante *et al* 1997 with kind permission from Elsevier Science Ltd, London, UK.

At present AFM cannot offer this capability and it is difficult to envisage how this could ever be achieved with AFM. However, even without the capability to generate atomic models of the proteins, comparison of predicted surface topography data derived from such models with directly acquired AFM images of the surface topography, which can be obtained in realistic environments, can assess the reliability and biological relevance of the TEM data. TEM suffers particularly because the images need to be processed using averaging techniques to produce acceptable contrast, and this means that only regular repeating structures can be resolved, defects in the crystals are lost along with the noise, reducing the resolution in the TEM data, and other local variations in structure such as twinning can actually produce incorrect data. AFM with its ability to generate high contrast images directly does not suffer from this drawback, and is in fact exceptionally good at visualising point defects in crystals (Devaud *et al* 1992). The conclusion from a structural determination point of view is that AFM provides a powerful complementary technique to be used alongside rather than instead of TEM. There are other areas though where AFM can provide unique information, for example by working in physiological buffers the AFM can image dynamic events *in-situ* (Müller *et al* 1996b).

6.2.2. Sample preparation: membrane proteins

Some membrane proteins such as bacteriorhodopsin occur naturally as 2D crystalline arrays and sample preparation in this case simply involves retrieving them from the cell wall and then applying various treatment steps which depend upon the system being examined. These are many and varied and beyond the scope of this book, but are described in detail in various comprehensive review articles (Boekema 1990; Kühlbrandt 1992; Jap 1992) and the book, *Crystallization of membrane proteins* (Hartmut 1991).

There is, however, a more general method which, although the precise details vary for different proteins, can be applied to all membrane proteins. The best 2D crystals produced *in-vitro* have been grown from detergent solubilised and purified material. Conditions can be controlled more easily, and consequently the results are more reproducible (Kühlbrandt 1992). The principle features of this method are as follows. Membrane proteins are insoluble in water and so the first step to forming 2D crystals is to solubilise the proteins using detergents. They are then added in the detergent solution to a suspension of lipid-detergent micelles and finally the detergent is removed by dialysis or adsorption onto latex beads (Kühlbrandt 1992). This leaves the proteins in a continuous lipid bilayer just as they would be in the biological membrane, resembling the situation *in-vivo*. Because they are much smaller than 3D crystals a relatively small amount of the protein is required and, furthermore, 2D protein crystals grow much more quickly, meaning that the protein is exposed to high levels of detergent for only a short time, favouring protein stability.

The Kühlbrandt method has been adapted recently to directly incorporate membrane proteins into supported phospholipid bilayers (Milhiet *et al* 2006). In the revision, which is ideally suited to AFM imaging, a supported phospholipid bilayer is formed on mica by vesicle fusion (see section 5.3.1). The supported bilayer is then partially disrupted by the addition of a sugar-based detergent solution (dodecyl-β-D-maltoside or dodecyl-β-D-thiomaltoside) and then detergent-solubilised membrane proteins are added. Within a matter of minutes the proteins are unidirectionally-incorporated into the supported bilayers. The method requires only picomolar amounts of the membrane protein. Thus very small quantities of freshly-purified membrane proteins can be used without the usual sample concentration steps required in previous methods, to produce patches of locally self-assembled close-packed 2D crystalline arrays in the supported phospholipid bilayer (Milhiet *et al* 2006). The small quantities required should allow the study of much rarer or difficult to purify membrane proteins and so holds great promise.

Reconstitution of protein membranes can lead to several different arrangements of 2D membrane protein crystals. In native membranes and in vesicle crystals all of the protein molecules face the same way in the lipid bilayer (Fig. 6.9a, b). Tubular crystals can also form: these are basically vesicle crystals with the proteins arranged helically on the surface of a cylinder (Fig. 6.9c). Finally crystals with protein molecules facing alternately up and down can form when the crystals are grown from isotropic detergent solutions (Fig. 6.9d).

6.2.3. Sample preparation: soluble proteins

Although membrane proteins form 2D crystals more or less spontaneously when mixed with detergents and lipids, most soluble proteins need to be coerced into forming crystals. The general principle of 2D crystallisation for these proteins has four basic requirements. The first is that the molecules should be fixed in a plane, the second that they should have sufficient mobility within the plane to permit reordering, allowing the third requirement to be satisfied, namely identical orientation of all of the molecules.

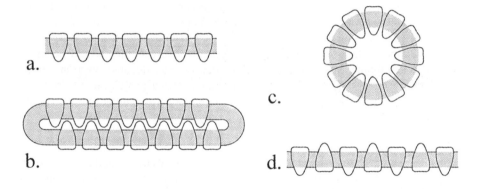

Figure 6.9. Different types of 2D protein crystals. Reproduced in part from Kühlbrandt 1992, with permission of the author and Cambridge University Press.

The fourth and final requirement is a high concentration of the molecules in the plane so that crystallisation will be favoured over the formation of two dimensional solutions. The molecules can be fixed in two dimensions by adsorbing them onto an air-water interface but pure protein monolayers lack the necessary mobility to allow molecular rearrangement. This is overcome by adsorbing the proteins to a lipid monolayer (Uzgiris and Kornberg, 1983) either electrostatically to a charged lipid monolayer or by binding the proteins to ligands attached to the polar head groups of the lipids (Fig. 6.10). The mobility of the lipid monolayer can be controlled by adjusting the packing density and/or the hydrocarbon chain length of the lipid molecules, which turns out to be an important parameter for achieving 2D crystallisation of different proteins (Kornberg and Ribi, 1987). Most proteins require a high lipid density in the *fluid* phase. Lipids compete with protein for sites on the air-water interface and it is speculated that a high level of lipid prevents denaturation of the proteins, which would otherwise partially unfold given free access to the air-water interface, and unfolded or denatured protein will not easily crystallise (Kornberg and Ribi, 1987).

There are two options for preparing 2D crystals. The first method is to use the traditional means of constructing the lipid monolayer on a Langmuir trough and the protein is then added to the subphase (Darst *et al* 1990). Film transfer is achieved by performing an LB-dip (as described in section 5.1.4.) using a hydrophobic substrate. The lipid layer binds to the substrate on the way into the liquid and the surface should be swept with the barriers before pulling the substrate back out again in order to avoid a bilayer being transferred, in which case a second layer would be transferred with lipid molecules uppermost on the substrate surface.

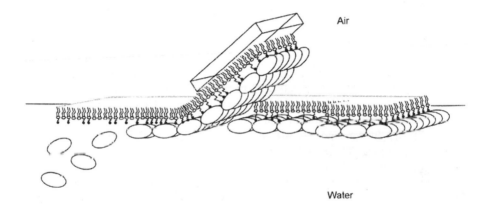

Air

Water

Figure 6.10. The production of 2D crystals of soluble proteins is achieved by binding the proteins to an interfacial monolayer of lipid molecules either electrostatically or chemically to allow the rearrangement and ordering needed for crystallisation. Reprinted from Kornberg and Darst, 1991 with kind permission from Elsevier Science Ltd, London, UK.

This method suffers from the drawback that most Langmuir troughs require a significant volume of liquid to fill them and, because protein concentration in the subphase needs to be relatively high (typically 1-3 mg mL^{-1}), large amounts of protein are needed, negating one of the advantages of two dimensional crystals. For small scale work the second method is to simply place the substrate directly onto the surface of protein droplets (10-20 μL) coated with lipid and then withdraw the substrate, which in this case was a carbon coated hydrophobic TEM grid (Kornberg and Darst, 1991). Some of the lipid layer and attached protein adhere to the substrate as a result of hydrophobic interactions between the hydrocarbon chains of the lipid and the carbon film. For AFM imaging highly oriented pyrolitic graphite (HOPG) could be used as an alternative, flatter hydrophobic substrate. Although this method has the advantage of requiring only very small amounts of sample it is a little more 'hit and miss', lacking the control of the Langmuir trough method.

An alternative method requiring only small volumes of protein solution (~ 40 µL) is to use a Teflon cell well as a mini Langmuir trough (Czajkowsky *et al* 1998). Recently a means of monitoring, and so improving the 2D crystallisation process, has been proposed based upon ellipsometry and shear rheology measurement of the interfacial layers on the Langmuir trough (Vénien-Bryan *et al* 1998). Ellipsometry monitors the attachment of the proteins to the lipid monolayer. Shear rheology, which measures the resistance of the interfacial film to flow, monitors the crystallisation of the protein molecules.

Another rather simpler, though less controlled method for the formation of 2D soluble protein crystals (Harris 1992) is an adaptation of the mica spreading, negative staining carbon films procedure originally developed for forming 2D ordered arrays of viral particles for TEM (Horne and Pasquali-Ronchetti, 1974). The crucial extra step in the adapted procedure is the inclusion of polyethylene glycol (PEG). A solution of highly purified protein, around 1mg mL^{-1} in low concentration buffer, is mixed with a solution of 2% ammonium molybdate containing 0.1-0.2% PEG (molecular weight 1000-10000 D). This mixture is spread over freshly-cleaved mica and excess fluid removed to leave a thin, even layer of liquid on the mica. This is then allowed to dry at room temperature, which takes approximately 5 to 10 minutes. Protein crystallisation occurs during drying, but depends upon the optimisation of conditions such as protein concentration, ionic strength of the buffer and its pH, the PEG grade and concentration, drying temperature and time. These all have to be determined by trial and error making the process potentially somewhat time consuming. Another limitation of this technique is of course the drying step which, although fine for EM, discards one of the advantages of AFM.

6.3. AFM Studies of 2D membrane protein crystals

6.3.1. Purple membrane (bacteriorhodopsin)

Probably the first 2D membrane protein crystals to be examined by AFM were the purple membranes of *Halobacterium halobium*. They are 2D crystals composed of 75% of the protein bacteriorhodopsin and 25% lipid, and function as light powered proton pumps. The bacteriorhodopsin molecules are densely packed into trigonal lattices. Structural analysis by cryo-electron microscopy has revealed a retinal unit embedded in seven closely packed α helices (Henderson *et al* 1990). Early AFM studies of purple membrane met with only limited success, which was almost certainly due to instrumental inadequacies (Worcester *et al* 1988; 1990). These investigations were made in the days before optical sensing methods for AFM had been developed and, more importantly with home-made AFM tips, *and* in air where the factors of poor tip quality and high imaging forces combined to make resolution of any fine detail impossible. Things improved considerably with the advent of commercially manufactured AFM tips and AFMs with the ability to operate in liquids so that forces could be reduced, and images containing some structural information were obtained. Nevertheless, the contrast in the images was poor and the structural information was derived from Fourier analysis (Butt *et al* 1990). Much more recently, however, a drastic improvement in image quality was obtained by Müller and colleagues who demonstrated the absolute importance of controlling the imaging forces in an AFM study of purple membrane, by recording a force-induced conformational change of the protein molecules on the cytoplasmic surface (Müller *et al* 1995a). The conformational change is clear and can be seen in Fig. 6.11. This represents an important step along the road to assessing the potential for sample distortion by the scanning tip of the AFM on soft biological systems at very high resolution. In addition to demonstrating the effect that the imaging force can have on such delicate systems, the conformational change itself revealed a hint about the importance of the choice of imaging environment. The conformational change was noticed to be fully reproducible in high pH buffer but less so in acidic buffer.

Figure 6.11. Force dependent topography of the cytoplasmic surface of purple membrane. The force was adjusted manually during the scan from 300 pN at the bottom of the image to 100 pN at the top. A distinct conformational change can be seen: doughnut-shaped bacteriorhodopsin trimers transform into units with three pronounced protrusions at their edges. Scale bars (left) 10 nm (right) 4 nm. This image has been Fourier filtered to improve its clarity. Reproduced with permission from Müller *et al* 1995a.

This observation strengthened the interpretation that what was actually being seen in the AFM images was the protrusion of a loop, predicted from the sequence of the protein, which connects two helical regions in bacteriorhodopsin, since this would pop out of the cytoplasmic surface at basic pH due to electrostatic forces. The conclusions drawn from this study were twofold. Firstly and perhaps not surprisingly, choice of imaging environment is particularly important if imaging is carried out in aqueous buffers, and secondly, conformational changes due to buffer conditions may be visualised directly with AFM. Indeed this was confirmed in a later study of HPI layers by the same group (Müller *et al* 1997b). In a related study (Müller *et al* 1995b) an attempt to determine the orientation of the purple membranes was made by adsorbing them to poly-L-lysine coated mica at different values of pH, which favour attachment by either the cytoplasmic surface (pH 7.4), or the extracellular surface (pH 3). Compared with samples adsorbed to bare mica, which produced diffraction patterns with reflections extending beyond a lateral resolution of 0.7 nm, the samples attached to poly-L-lysine treated mica were rougher, due to buckling of the membrane sheets, and generated diffraction orders to (only!) 1.18 nm, partly because of the deviations of individual units in the images from the lattice symmetry.

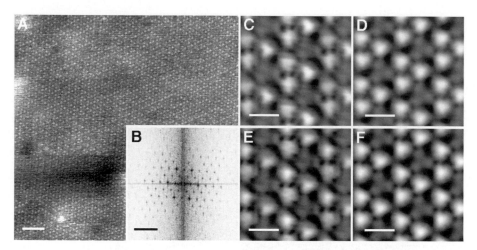

Figure 6.12. Extremely high resolution AFM images of the extracellular surface of purple membrane. (A) Image recorded in the trace direction (left to right). (B) The peaks in the power spectrum extend to a resolution of 0.7 nm. Correlation averages derived from images scanned in trace and retrace direction (right to left) calculated from 770 unit cells can be seen in C and E respectively. Symmetrized averages are shown in D and F which come from the average topography of C and E, the height in these images is 0.2 nm. Scale bars: (A) 20 nm, (B) 2 nm^{-1} and (C-F) 4 nm. Reproduced with permission from Müller *et al* 1996a and the Biophysical Society.

Identification of the different sides of the purple membrane sheets were tentatively made but, more recently, immunolabelling has been used to confirm this and to produce 0.7 nm lateral resolution on the extracellular face (Müller *et al* 1996a); the improvement presumably being due to attachment to bare mica, which avoids the problems with poly-L-lysine coated mica. The AFM images obtained are reproduced above in Fig. 6.12. Several interesting observations were made in the immunolabelling study. Firstly, whilst the antibody molecules could be swept off the mica surface at forces of 0.2-0.3 nN, those attached the purple membrane required the higher forces of around 1nN to remove them, suggesting greater interaction between the antibody and antigen, than between antibody and mica. Secondly, after removal of the antibody molecules from the membranes the underlying crystalline topography could then be recorded. Thirdly, areas on the membranes cleared of antibodies by the AFM tip were rapidly re-labelled indicating that the surface retained its native antigenicity, despite contact with the AFM tip. Fourthly, despite imaging with very low forces (100 pN), no high resolution images of the antibody molecules themselves were obtained - only 'blobs' of the appropriate dimensions, illustrating the advantages of forming 2D crystalline arrays of proteins if one wishes to image structural detail. A final point worth mentioning is that this study is one of only a few at the time in which the use of antibodies was checked with a proper control sample (Müller *et al* 1996a).

6.3.2. Gap junctions

After purple membranes, the first AFM study on a protein membrane system which demonstrated good contrast of ordered arrays on a membrane surface was that carried out on gap junctions (Hoh *et al* 1991). Gap junctions are regions in the cell membranes of vertebrates which allow the free passage of small molecules (< 1 kD) and, having a low electrical resistance, provide a signalling pathway (Flagg-Newton *et al* 1979). They are made up of two plasma membranes sandwiched together with cell to cell channels bundled into an array as the 'filling' (Revel and Karnovsky, 1967). They were a good sample to pick for an early AFM study since they had already been extensively studied by X-ray diffraction (XRD) and EM, and a model for their structure proposed (Makowski *et al* 1977; Makowski 1985; Unwin and Zampighi, 1980). In the model the channels were aligned head to head across a 2-3 nm gap between the membrane sheets. In the most regular samples the channels were arranged hexagonally with lattice constants of 8 to 10 nm, but this varied according to the preparation conditions. The channel itself was composed of two connexons, one from each membrane. The connexons were proposed to be cylindrically shaped, 7.5 nm high and 7 nm wide. Each had a pore running through its centre, some 1.5-2.0 nm in diameter. Finally, the connexon was believed to consist of six identical protein subunits, since it has a six-fold symmetry. The gap junction dispersions in PBS were allowed to adsorb to glass coverslips for 10-20 minutes and then the substrate was rinsed with an excess of buffer prior to imaging. Samples revealing the hexagonal arrays were trypsinised and fixed with glutaraldehyde. AFM images, collected with forces around 1 nN, of the gap junction at low magnification under PBS buffer revealed them to be irregularly shaped flat sheets of 14 nm thickness and typically 0.5 microns in size. At higher magnification the top surface was simply seen to be undulating and featureless. Fixation of the gap junctions did not appear to change their overall morphology; they all exhibited approximately the same height, although the fixed membranes were more easily scraped off the glass surface because they are less positively charged after fixation, reducing their interaction with the negatively charged glass. When the fixed samples were imaged at higher forces (up to 15 nN) the top membrane sheet was removed.

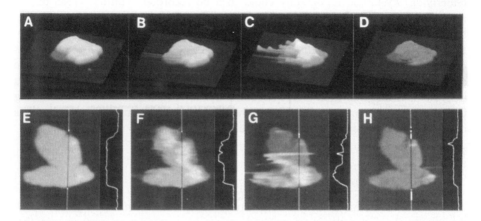

Figure 6.13. AFM tip induced dissection of gap junction membranes. Increasing the force set-point from (A) 0.8 nN (B) 3.6 nN (C) 6.1 nN and (D) 9.6 nN is shown to 'shovel' off the upper membrane from right to left on a single gap junction plaque. The bottom row of images shows a top view of another gap junction plaque which has again been subjected to increasing imaging force; (E) 0.8 nN (F) 3.1 nN (G) 10.1 nN and (H) a repeat scan at 10.1 nN. Line profiles are shown beside each image illustrating that the height of the structure is reduced in successive images to finally half the original value (from ≈ 15 nm to ≈ 7 nm). Scan sizes are 1.5 x 1.5 μm. Reprinted with permission from: Hoh, J.H, Lal, R, John, S.A, Revel, J-P. and Arnsdorf, M.F. (1991). Atomic force microscopy and dissection of gap junctions. *Science* **253**, 1405-1408. Copyright 1991 American Association for the Advancement of Science.

The process is shown in Fig. 6.13 and the changes in height can be seen in the line profiles beside each image. Once the top layer had been 'dissected', high magnification imaging clearly revealed the hexagonally packed connexons. The centre to centre spacing of the connexons was determined by Fourier analysis and yielded a value of 9.1 nm, in excellent agreement with the previous XRD and EM studies (Makowski *et al* 1977; Makowski, 1985; Unwin and Zampighi, 1980). The diameter of the connexons appeared smaller than the models at 4-6 nm and they protruded 0.4-0.5 nm from the membrane surface. Hoh and coworkers followed up this study with a more detailed re-examination of gap junctions with better control of imaging forces and other instrumental factors (Hoh *et al* 1993). The study also contained a useful discussion of potential problems associated with AFM, which at that time was a very new and underdeveloped technique (Hoh *et al* 1993). Damage induced by the AFM tip during both routine imaging and during membrane dissection was considered, as was image distortion due to instrumental factors such as tip shape and feedback-loop characteristics. The images obtained were much clearer and are reproduced in Fig. 6.14a, allowing a central pore to be discerned in the connexons (Fig. 6.14b). The AFM images were the first to reveal substructure on a membrane protein, namely details of the pore running through the connexons, but it was acknowledged that concerns regarding artifacts due to the use of AFM needed to be addressed before the full significance of the images could be assigned (Hoh *et al* 1993).

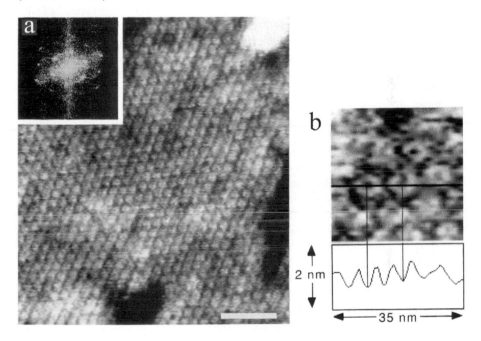

Fig. 6.14 (a) Connexon structure of the gap junction is revealed after removal of the top layer (see above Fig. 6.12). The power spectrum (inset) demonstrates the hexagonal packing. Scale bar 50 nm. (b) Cross section through selected connexons revealing pores, with a depth of approximately 0.7 nm suggesting that the AFM tip is too broad to penetrate fully, or that there is an obstruction. Reproduced with permission from Hoh *et al* 1993 and the Biophysical Society.

In hindsight it would appear that the researchers were overcautious since many of the safeguards, such as scanning in different directions, force minimisation, feedback optimisation and thorough analysis of the images using Fourier and correlation methods have been both adopted and validated subsequently by other workers.

6.3.3. Photosynthetic protein membranes

Another early AFM study which achieved clear resolution of single molecules of a membrane protein was carried out on the photosynthetic protein membrane from the bacteria *Rhodopseudomonas viridis* (Yamada *et al* 1994). A suspension of the membranes, which had been extracted from bacterial cells was spread onto an air-water interface in a Langmuir trough and then transferred onto glass coverslips by LB dipping. In contrast to most of the other studies on membrane protein, mica was found be an unsuitable substrate, causing wrinkling and aggregation of the membranes. Because no details are given of the imaging forces used, it must be assumed that the samples were imaged in air, since there is no mention of working under liquid. Despite this the protein molecules were imaged quite clearly as spherical features found to be hexagonally packed in most regions of the membranes. In some regions the protein molecules displayed less regular packing, and these differences were tentatively ascribed to the different faces of the membrane sheet which have different, so-called C and H subunits protruding outwards. Force modulation by oscillation of the cantilever during imaging was carried out on the samples yielding information on the local mechanical properties of the sample, although it simply confirmed that the protein membrane is softer than the glass substrate (Yamada *et al* 1994). More recently studies have been made on the effect of lighting conditions during growth on the detailed structural arrangement of light harvesting membrane proteins from *Rhodospirillum photometricum* (Scheuring & Sturgis, 2005). Large changes in the ratio of the light harvesting component proteins were observed for low-light level and high-light level adapted samples. The local environment of the core complexes appeared the same for both samples but the number of light-harvesting antennae domains was higher in the low-light adapted samples, ensuring sufficient coupling of the reaction centre is maintained in low light conditions. Thus a submolecular adaption to the photosynthetic machinery of a plant, induced by environmental stress was demonstrated for the system; a significant milestone in the application of AFM imaging in biology.

6.3.4. ATPase in kidney membranes

In another early example, a combined TEM/AFM study of membrane proteins examined the structures formed by sodium and potassium ATPase in purified canine kidney membranes. The AFM imaging was performed in air and the forces were kept to reasonably low levels through the use of Tapping, allowing molecular resolution to be obtained (Paul *et al* 1994). By comparing the TEM and AFM images it was concluded that the cytoplasmic face of the membranes was imaged by AFM and that the images (AFM) demonstrated that the ATPase forms channel-like structures, with a distinct pore of internal diameter 0.6-2.0 nm. An interesting observation in the study was that the dimensions of the protein molecules varied when uranyl acetate (negative stain) was applied, with some 50% shrinkage resulting from the staining procedure, demonstrating even in this early study that AFM could be used to assess the effect of common EM preparation procedures on such samples (Paul *et al* 1994).

6.3.5. OmpF porin

The outer membranes of gram-negative bacteria protect the cells from harmful agents such as proteases, bile salts, antibiotics, toxins and 'phages, and against drastic changes in osmotic pressure (Cowan *et al* 1992). The barrier contains a family of proteins known as porins which act as channels to mediate the transfer of nutrients, metabolites and waste products (Nikaido and Vaara, 1985). A major species of porin in the bacterium *Escherichia coli* is the trimeric matrix porin OmpF whose atomic structure has been resolved by X-ray crystallography (Cowen *et al* 1992). In the case of OmpF porin the channels or pores are formed by strands of β-barrel, connected by short turns on the periplasmic face, and by loops of variable length on the extracellular face. Thus the faces are smooth and rough respectively. OmpF can be reconstituted in the presence of phospholipids into 2D crystals of various forms. The first AFM study of reconstituted crystals of OmpF porin from *E. coli* was reported by Lal and coworkers in 1993, who observed a mixed pattern of rectangular and hexagonal motifs when high magnification scans were performed (Lal *et al* 1993). The researchers attributed this to the extracellular face of the porins, because of the thickness of the vesicles and the protrusion height of the features. The centre to centre spacings of the arrays were measured at 8.4 x 9.8 nm and 7.2 nm respectively, in good agreement with previously published EM data (Sass *et al* 1989). No fine detail of the actual channels was achieved, which with the benefit of hindsight, was probably due to two factors, namely, imaging in air which may induce collapse of the structure, and imaging with relatively high forces ~ 1 nN. Improvements in resolution were demonstrated the following year when AFM micrographs exhibiting 2 nm lateral resolution (Schabert *et al* 1994a) and then 1 nm lateral resolution (Schabert and Engel, 1994b) were obtained by imaging under aqueous buffers, and eliminating the drying step of the earlier study. The improved resolution allowed both extracellular and periplasmic faces to be identified in the images. The rougher extracellular face showed two distinct conformations related to the imaging force, whereas the smooth periplasmic face appeared the same over a range of forces from 200-600 pN. Furthermore, on the periplasmic face of OmpF porin a novel hexagonal crystal packing arrangement, and its transition to the more common rectangular arrangement were observed, demonstrating once again that crystal defects, (Devaud *et al* 1992) and different crystalline packing arrangements are within the grasp of AFM (Schabert and Engel, 1994b). The paper by Schabert and Engel also details the range of image analysis techniques which are possible on such highly ordered systems, many of which have been 'borrowed' from electron microscopy. Techniques such as correlation averaging (Saxton and Baumeister,

1982) and Fourier analysis were used to allow accurate measurement of unit cell parameters, and to assess the accuracy of the measurements respectively (Schabert and Engel, 1994b). Before applying such treatment to the images scans were performed in forward and reverse direction, and the images added to compensate for frictional effects. Reconstitution of the membranes generally produced a double layer where the extracellular sides were sandwiched between the periplasmic sides and so not accessible to the AFM tip. This was overcome by imaging regions where the overlap was not perfect, and also by 'teasing away' edges of the upper layers with the AFM tip to expose sections of lower extracellular layer, although this was only possible on bilayers which lacked the two loosely bound lipopolysaccharide molecules per OmpF porin trimer (Schabert and Engel, 1994b). This 'nanodissection' was repeated in a later study, and an example of how this technique reveals both faces of the 2D OmpF porin crystal is shown in Fig. 6.15 (Schabert *et al* 1995). Although AFM cannot produce atomic models of 2D protein crystals it can be used to validate such models. The protein-protein and protein-lipid interactions in OmpF porin were modelled and AFM topographs used to verify which model was the most accurate. AFM images also revealed details about the flexibility of protein loops in the structure which were impossible to obtain from X-ray crystallography studies (Schabert *et al* 1995).

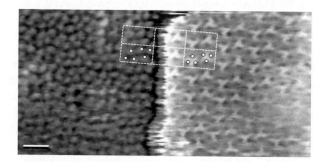

Fig 6.15. Visualisation of both faces of OmpF porin was possible after removal of the upper layer. The corrugated extracellular surface is visible on the left side of the AFM image, and the smoother periplasmic face to the right. Rectangular unit cells (a=13.5 nm, b=8.2 nm) which comprise two trimers are framed, the dots mark the positions of the trimers on both faces. Scale bar 10 nm. Reprinted with permission from: Schabert, F.A, Henn, C. and Engel, A. (1995). Native *Eschericia coli* OmpF porin surfaces probed by atomic force microscopy. *Science*, **268**, 92-94. Copyright 1995 of the American Association for the Advancement of Science.

6.3.6. Bacterial S layers

Virtually all archaebacteria have cell envelopes which incorporate a regular surface layer (S layer) of proteins or glycoproteins (König 1988). S layers make up some

7-12% of the total protein content of the cells, and provide an important interface between the cell and its surroundings (Beveridge 1981). The first S layer to be examined by AFM was from the bacterium *Sulfolobus acidocaldarius* (Devaud *et al* 1992). Indeed it was probably the first AFM study on 2D protein crystals to attain lattice resolution without resorting to Fourier processing of the images. The reason for this success was almost certainly due to the coating of the sample with titanium by electron beam evaporation, making it able to withstand the forces encountered when imaging in air. The resolution, low by present standards at around 10 nm, was nevertheless sufficient to distinguish between the extracellular and cytoplasmic faces of the crystals. Perhaps the most important finding in this study was that twin boundaries in 2D crystals, which are not easily visible in TEM images, are highly structured areas with the two S-layer crystal domains merging in such a way as to preserve the overall pattern of pores. Thus, even in this early study, the role that AFM can to play as a rather powerful complementary technique to TEM was demonstrated (Devaud *et al* 1992).

Another bacterial S-layer, and one which has been extensively studied, is the hexagonally packed intermediate (HPI) layer of *Deinococcus radiodurans* (Baumeister *et al* 1982). It is extracted from the outer membrane of whole cells with detergent (Baumeister *et al* 1982). Assembled from a 107 kD protein it forms hexamers of 655 kD, producing a lattice of unit cell size of 18 nm. Each hexamer is composed of a massive core from which spokes that connect adjacent hexamers emanate. Modelling of electron microscopy data indicates that the core encloses a pore which is surrounded by six relatively large openings, centred around the three-fold line of symmetry (Baumeister *et al* 1986). In what was an important step for the application of AFM to biological molecules at high magnifications, an early study of HPI layers quantitatively compared AFM images (obtained under buffer) with TEM data (obtained *in vacuo*) in order to try to verify the reliability of the AFM in the study of protein structure (Karrasch *et al* 1994). A resolution of 1 nm laterally and 0.1 nm vertically was achieved and, furthermore, measurements taken from the AFM images were in excellent agreement with previous electron microscopy data. It was also clear that regions of differing rigidity in the structure could be discriminated in the AFM by evaluating the variations between images of many molecules. The authors (Karrasch *et al* 1994) tentatively concluded that this might allow observations of function related structural changes in such systems using AFM. Within two years, such structural changes were actually observed on HPI layers; the opening and closing of pores (Müller *et al* 1996b). An important instrumental prerequisite to such delicate work was the use of very low imaging forces of around 100 pN in order to eliminate probe force as the dominant mechanism of structural alteration. In a further effort to eliminate tip-induced effects, both hydrophilic oxide-sharpened Si_3N_4 tips and electron beam deposited carbon hydrophobic tips were used, and produced the same results. Fig. 6.16a, b shows the unprocessed AFM images that were obtained (Müller *et al* 1996b). It can be seen that the hydrophobic inner surface of the HPI layer exhibits two conformations in the AFM images. These have been attributed to the opening and closing of the central pore of the hexameric core because, in some images, this region is unobstructed (round circles) and in others it is plugged (square boxes). Most importantly, upon re-scanning of the same region some five minutes later, different hexamers were unobstructed (square boxes-Fig. 6.16b) eliminating the possibility that the original image simply represented a natural heterogeneity in the structure of the hexamers. To quantify the observed structural changes, 330 units from ten different images were aligned by adjusting their translational and rotational position (or put another way, jiggling them around until the best fit was found) and then performing multivariate statistical analysis. The results of this can be seen in Fig. 6.16c-g. The difference in terms of height between the obstructed and empty pore was 0.8 ± 0.5 nm, only just above the uncertainty level, but probably significant since this figure is derived from the average of many images. When one considers the small size of the pore, electron microscopy data suggests it is about 2.2 nm in diameter, it is easy to see that this will limit the penetration of the AFM tip, which is not infinitely sharp, meaning that the depth measured by AFM is bound to be underestimated.

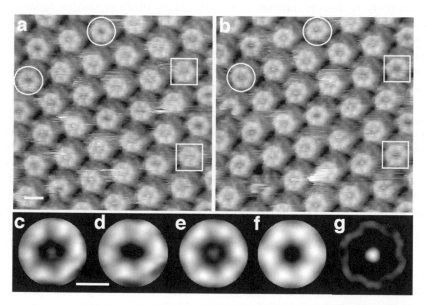

Figure 6.16. Conformation changes of the inner surface of HPI layer captured in successive AFM images of the identical area, taken some 5 minutes apart. (a) Open pores in the image are marked with circles and closed or plugged pores are marked with squares. (b) Pores which were open have now closed (circles) whilst some that were closed have opened (squares). (c-f) Images resulting from various mathematical operations (details in text) on the raw images demonstrate that there are real and measurable differences between the images of the two states of the pores. Scale bars / grey level range (a-b) 10 nm / 6 nm (c-g) 6 nm / 3 nm. Reproduced with permission from Müller *et al* 1996b.

The only "fly in the ointment" is that it is not known what induces the observed conformational change, since no changes to the imaging or buffer conditions occurred between scans, leaving the somewhat tentative conclusion that random switching between open and closed states is a specific property of the HPI layer (Müller *et al* 1996b; 1997b). Nevertheless, the very high resolution achieved in this study (approximately 0.8 nm laterally and 0.1 nm vertically) proved that dynamic events can be followed with AFM, even at the sub-molecular level, and this surely represents a very significant milestone in its application to biology.

6.3.7. Bacteriophage φ29 head-tail connector

The assembly of bacteriophage particles is a complex process involving interaction between unassembled components into various intermediates of the virus. Typical 'phages consist of a head and a tail which are connected by proteins, known as connector or portal proteins. These are implicated in the translocation of DNA inside the viral head. Necks from the bacteriophage φ29, made up of connector proteins p10 and p11, have a 12 fold rotational symmetry (Carrascosa *et al* 1982). Electron microscopy data from 2D crystals of the φ29 connector revealed an open channel in the system (Carazo *et al* 1986). However this channel was found to be closed in necks extracted from native viral particles (Carazo *et al* 1985). This suggested that the different conformational states may play a role in the packing of DNA within the viral particles (Carrascosa *et al* 1990). Although this system had been studied fairly extensively there was still relatively little information on the actual topography of the inner and outer surfaces of the connector and so AFM was applied to tackle this problem (Müller *et al* 1997a). For the purposes of 2D crystal formation and AFM imaging, relatively large amounts of φ29 connector protein were obtained by the over-expression of the gene encoding it in *E. coli*. In contrast to the usual methods, 2D crystals were formed by very slowing raising the ionic strength of a 3-4 mg mL^{-1} solution of the connector proteins up to 2M NaCl. As well as 2D crystals, 3D crystals were also prepared and imaged, and interestingly these failed to produce sub-molecular resolution, this being attributed to their relative lack of stability to the shear forces of the AFM tip. It seems that the mica substrate plays an important role in supporting the sample, and indeed this conclusion is probably applicable to other interfacial systems if high resolution AFM images are required, justifying LB film transfer.

As for bacteriorhodopsin, a force-induced conformational change was observed in the AFM images of the φ29 connectors and this is shown in Fig. 6.17. At the top of the image in Fig. 6.17, which was scanned at low force ~50-100 pN, the narrow domains of the connectors dominate the image contrast (regions a-b), but, as the force between AFM tip and sample is increased (between regions b-c, force ~150-200 pN), the narrow part was squashed down onto the surface of the crystal, allowing the wider central channel beneath to be more clearly resolved. When the force was reduced again the extended conformation of the narrow, and apparently flexible, end of the connectors once more became visible (regions c-d). To prove the reproducibility of the conformational changes the force was again raised and the wide connector domain became clear (region d-e). Finally, at the bottom of the image the probe force was increased to 250-300 pN, resulting in the AFM tip pushing right through the sample to the mica underneath (region e), demonstrating the delicate nature of the sample, and the need to control the imaging force accurately. The process of disruption is probably actually due to the shearing force of the tip in contact mode, rather than simple piercing of the crystal sheet by the loading (normal) force of the tip. When the force was increased to image the wider connector domains even the raw data showed, with reasonable clarity, the presence of twelve subunits in good agreement with previous EM studies (Carazo *et al* 1986). Additionally, however, the AFM data revealed that the wide end of the connector had a right handed orientation vorticity. This may have some bearing on the DNA packing mechanism of the connectors.

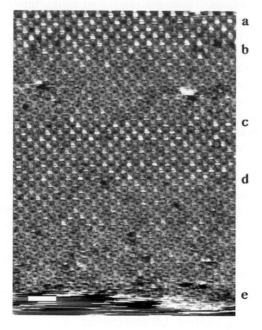

Fig 6.17. Force-induced conformational change of φ29 connectors. Extended protrusions from the narrow connector end which are visible at low force (50-100 pN, region a-b) were pushed down by the AFM tip with only a moderate increase in the force (150 pN, region b-c). This process was fully reversible as can be seen when the force was lowered again (region c-d) and raised (d-e). Finally if the force was increased to too high a value (300 pN) the sample was damaged by the tip (region e). Scale bar 5 nm, Grey range 5 nm. Reproduced with permission from Müller *et al* 1997a.

The AFM images clearly show that there is a difference between the channel on each side of the connector; one wide with a diameter of 3.7 nm, and one narrow with a diameter of 1.7 nm. Furthermore, the AFM data indicated that the shape of the channel is not cylindrical but conical, an important observation, since knowledge of the precise nature of the channel enables verification of models for the packing of DNA (Carazo *et al* 1985; Carrascosa *et al* 1990). The data obtained by AFM is consistent with the model for channel opening and closing involving small concerted movements of the subunits, and closing of the channel has been linked to the final step of DNA packaging (Carrascosa *et al* 1990) by the $\phi29$ connectors.

6.3.8. AFM imaging of membrane dynamics

Recently there has been a vast increase in the number of membrane protein systems studied by AFM. One area which has received particular attention is the observation of membrane proteins at work. This aspect plays to the major strength of AFM over other forms of high resolution microscopy, namely its ability to image and manipulate unlabeled samples in physiological, or near physiological environments (more of which later), at submolecular resolution, where the samples have the possibility at least of retaining their biological activity. Even with the significant advances in electron and fluorescence microscopy, neither can compete with the AFM on these terms. Although apparently random conformational changes had been observed in HPI layers discussed in section 6.3.6, the field has now moved forward to report changes attributable to specific stimuli. Probably the first example published was the closing of the channel entrance in OmpF porin induced by the application of an electrical potential across the membrane (Müller and Engel, 1999). Closure of the pore occurred through collapse of large extracellular loops of the porin onto the channel entrance. This effect could also be seen by changing the pH of the imaging solution. Thus environmental stimuli were shown to control the gating function of the transmembrane channel. Gating using the same mechanism was later demonstrated for other porins of the same family (Andersen *et al* 2002; Yildiz *et al* 2006). In a rare example of the observation of the functional behaviour of a membrane protein from a vertebrate, connexons in the gap junctions from rat liver cells were observed closing their transmembrane channel by two different mechanisms. Addition of Calcium ions caused the connexon to move radially inwards to close the channel entrance (Müller *et al* 2002). Gap junctions were also found to close at acidic pH in the presence of aminosulphonate compounds, which are found naturally in the cytoplasmic environment, but the molecular mechanism is not understood. Nevertheless, AFM images revealed that pH-induced closure caused the connexon subunits to rotate and close the channel like the iris in a camera.

Although the studies cited above have successfully observed membrane protein function directly, a major limiting factor for such experiments with the AFM is the effect that the supporting surface has on the mobility of the various membrane constituents. Many membrane proteins are believed to undergo relatively large conformational transitions in order to carry out their function and, in the plasma membrane of their native aqueous environment, are limited only by the fluidity of the surrounding lipid molecules. For a supported system this is not the case; the underlying solid substrate can present a significantly more formidable barrier, which in *extremis* may prove insurmountable to the particular function. For example measurements of the diffusion of single membrane proteins within a supported lipid bilayer have been made, and the rates of diffusion were found to be one order of magnitude slower than in free membranes (Müller *et al* 2003). This rather thorny issue is essentially the same as that discussed in chapter 5 (section 5.3.6) for AFM characterisation of supported phospholipid bilayers, and the approaches to solving these problems are similar in both cases. The approaches currently being investigated for membrane proteins are summarised in figure 6.18 (for an excellent review of this area see Müller and Engel, 2008). One of the more promising of these is the use of regenerated cellulose as a polymer cushion (Goennenwein *et al* 2003). Probing of the interaction between membranes containing human platelet integrin supported on cellulose layers, and vesicles containing integrin-specific ligands, gave larger adhesion values than equivalent experiments using solid supported membranes. These larger adhesion values were closer to those inferred from the integrin-ligand dissociation constant.

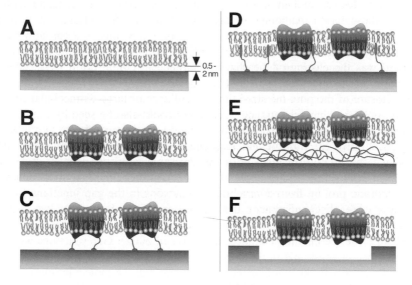

Fig 6.18 Strategies for retaining functionality of supported membranes. (A) If the membrane is directly supported by mica the water layer beneath is too thin to allow biological function since such a confined layer is highly viscous. (B) In such a case the protruding membrane proteins may even be directly adsorbed to the underlying mica. To overcome this limitation polymer spacers can be incorporated into either the proteins (C) or the phospholipid layer (D). Two alternative strategies are the use of a polymer cushion (E) or the micro-patterning of holes in the supporting surface, over which the membrane sheets are draped (F). Reproduced with permission from Müller and Engel 2008 and Elsevier.

Another approach to solving the problems associated with solid supports is to drape the membrane over holes in the underlying support. This has been achieved by depositing the S-layers from *Corynebacterium glutamicum* onto a silicon surface patterned with nanoscopic wells (Goncalves *et al* 2006). The mechanical properties of the non-supported regions of the membranes were probed giving access to the lateral interaction energy between proteins. These regions were rigid enough to resolve the 1.5 nm wide protein pore in the AFM images, although whether more typical softer membrane proteins would allow such resolution is still not clear. The functionality of membrane proteins supported in this way was demonstrated by observation of proton-pumping in purple membranes from *Halobacterium salinarium* by parallel fluorescence microscopic imaging of the resultant induced pH gradient (Goncalves *et al* 2006).

6.3.9 Force spectroscopy of membrane proteins

As mentioned already the truly unique aspect that sets AFM apart from the other high-resolution microscopes is its ability to not only image at extremely high resolution but also to directly manipulate samples. In the study of membrane proteins this has developed from fairly simple mechanical measurements of the rigidity of the membranes themselves (Xu *et al* 1996), through extremely detailed spectroscopic description and quantification of the extraction of single protein molecules from membranes (Janovjak *et al* 2005), all of the way to dramatic new developments in measuring the free energy involved in the unfolding of membrane proteins in the context of their lipidic environment (Preiner *et al* 2007). Such studies have now reached the stage where they can quantify directly the structural stability of membrane proteins and, perhaps most excitingly, provide a means of 'fingerprinting' their functional states in response to external stimuli (Kedrov *et al* 2005; 2007; Park *et al* 2007). For example, Kedrov and co-workers have shown that single molecule force spectroscopy measurements can detect interactions occurring upon the activation of a transmembrane protein due to the binding of a single sodium ion in the sodium/proton driven antiporter NhaA from *E. coli* (Kedrov *et al* 2005). The impact that such new abilities will bring to cell biology is hard to overstate, and this area provides an example of an application of AFM which is genuinely likely to push forward the frontiers of the science.

6.3.10. Gas vesicle protein

Gas vesicles are hollow tubular structures composed entirely of proteins which provide buoyancy for aquatic micro-organisms. The gas vesicle from the cynaobacterium *Anabaena flos-aquae* was examined recently under propanol after spray deposition onto mica by AFM (McMaster *et al* 1996). The study found that the major protein of the vesicles, GvpA, was packed into ordered arrays of rib-like structures with a periodicity of 4.6 nm. In contrast to all previous studies reported in this section, AFM imaging was carried out in the error-signal mode at high scan speeds. It was speculated that the high scan speed reduced instrumental drift during the scan resulting in less image distortion. Furthermore, working in error signal mode rendered the instrument more sensitive to the small deviations of the AFM tip on this flat specimen, since the data collection bandwidth is not limited by the control loop, as is the case in constant force mode. In error signal mode the only limiting factor on bandwidth is the inertia of the tip-cantilever assembly, hence a very low spring constant lever was used (k = 0.06 N m^{-1}). Resolution of the rib-structure formed by the proteins was clearly observed and, in some areas, an even finer level of detail, namely the β-sheet secondary structure of the protein molecules which had a periodicity of 0.57 nm, was seen. An advantage of using AFM was that by imaging the surface of the vesicles directly the packing sense of the molecules was determined, something which is impossible with X-ray diffraction, since the diffraction pattern is a projection map which has contributions from both sides of the vesicle superimposed. The study achieved a level of resolution believed to be amongst the highest to be obtained to date on a biological system (0.57 nm laterally). The downside of the error signal mode images was that only lateral dimensions could be determined from the images, since no height information is available in this imaging mode. The study nevertheless suggests an interesting alternative technique for achieving very high resolution on ordered arrays of proteins.

6.4. AFM studies of 2D crystals of soluble proteins

The soluble proteins ferritin and catalase have been imaged by AFM as 2D crystalline arrays (Ohnishi *et al* 1993; 1996; Furono *et al* 1998). Ferritin concentrates insoluble Fe (III) ions into a soluble protein-mineral complex. The iron is transferred through the protein into an internal cavity in the molecule, achieving concentrations of iron some one thousand times greater than for the free ion. Ferritin is found in all animals, plants and many bacteria, and provides iron for the proteins involved in respiration, nitrogen fixation, cell division and biosynthesis. Catalase is one of the most abundant enzymes found in nature, and protects cells from the damaging effects of hydrogen peroxide by catalysing its breakdown into water and free oxygen. Beef liver catalase is commonly used as a calibration standard in electron microscopy, forming a tetrameric structure composed of four identical subunits.

Both proteins were crystallised in two dimensions by binding them to a charged lipid monolayer of poly (1-benzyl-L-histidine) (PBLH) which had been spread at an air-water interface, using the method described above for soluble proteins (Kornberg *et al* 1987; 1991). The charged lipid, PBLH, played a double role in the studies. Firstly, it bound the opposite charged protein molecules allowing crystallisation to occur, and secondly, once the crystals were transferred onto the substrate for AFM imaging in pure water, it acted to screen the charges on the protein itself, as demonstrated by the absence of hysteresis in force-distance curves upon tip approach and withdrawal from the sample. This allowed low force imaging to be performed (Ohnishi *et al* 1993). In common with the studies on 2D membrane protein crystals described in the preceding sections, low forces (less than 100 pN) were a prerequisite to successful imaging. In the earliest study, which was carried out on ferritin, regular hexagonal packed arrays of the proteins were clearly seen, and there was also a hint of sub-molecular detail, paradoxically in areas where the crystalline packing was imperfect (Ohnishi *et al* 1993).

These studies were extended to catalase where a strong dependence of the image quality on buffer pH during the crystallisation step was demonstrated (Ohnishi *et al* 1996). Catalase is negatively charged in buffers above its isoelectric point at pH 5.7. This meant that in order to cancel the charge on the catalase the PBLH layer needed to be positively charged. In the range pH 6-7 PBLH bears only a slight positive charge, which is not enough to counteract the negative charge contribution from the catalase. The AFM tip is also negatively charged in pure water, in which the 2D catalase sheets were imaged, and the net result was repulsion between the AFM tip and sample preventing proper tracking of the tip, and hence poor images were obtained. This effect can be likened to a record-player stylus skipping or jumping when trying to play a record which has an excessive amount of static charge on its surface. When the 2D crystals were prepared at pH 6.0 however, the force-distance curves obtained suggested that the positive charge on the PBLH was slightly greater than the reduced negative charge on the catalase at this pH, being as it is close to the isoelectric point of the protein (pH 5.7), since there was a small amount of adhesion present between tip and sample. This led to optimum imaging by the AFM, since now the tip could track the sample surface accurately. This optimum pH value sits within the range of the isoelectric point of the protein and the pK_a of the histidyl residues in the PBLH. At lower values of pH during crystal formation (pH <5.0), both the PBLH and catalase are positively charged, so of course the net charge on the 2D crystals was also positive. This produced excessive adhesion between the negatively charged AFM tip and the 2D crystals, rendering AFM imaging impossible due to destruction of the sample. The results obtained by AFM were in general agreement, in terms of unit cell dimensions, with electron microscopy data, but the tetrameric subunits of the protein were not resolved (Ohnishi *et al* 1996). Nevertheless, the study clearly demonstrated the need to control the electrostatic interactions between the AFM tip and sample which arise when imaging is carried out in aqueous solutions (Ohnishi *et al* 1996), and this point is discussed further in the next section.

Not surprisingly, the electrostatics of the catalase-PBLH system also affected the actual formation of the 2D protein crystals, since they determine the extent of interaction between the two molecules, as confirmed by TEM studies (Sato *et al* 1993). It can now be appreciated that the formation and subsequent AFM imaging of 2D crystals of *soluble* proteins is a little more complicated than that for membrane protein crystals since the interaction between the protein and the lipid layer, needed to induce crystallisation, can affect AFM image quality. In the most recent study, using a specially prepared electron beam-deposited 'supertip', 2D crystals of ferritin and catalase were re-examined and higher resolution was achieved (Furuno *et al* 1998). Images were obtained both in water and in air at molecular resolution. Because of the capillary forces present when working in air (2-8 nN) the samples required fixation with negative stain (methylamine tungstate) to prevent probe damage. Under water much lower forces were achieved and better images were obtained with no need for negative staining. The AFM images are reproduced in Fig.6.19.

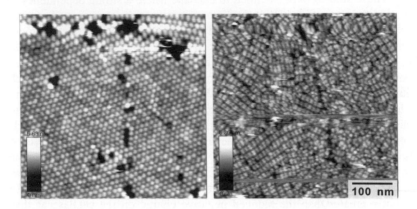

Figure 6.19. AFM images of 2D crystals of the soluble proteins (a) ferritin and (b) catalase in water. Reproduced with permission from Furuno *et al* 1998.

It was noted that the crystal sheets remained flatter in water than when imaged in air, perhaps due to buckling of the film during the drying step. In both environments the catalase molecules appeared to pack into a rectangular unit in some regions of the images and into a different unit form in others. The 'supertip' used to image the proteins was grown by focusing the electron beam from a field-emission scanning electron microscope (FESEM) onto the apex of a standard Si_3N_4 tip. FESEMs have a much smaller spot size compared to conventional SEMs enabling sharper tips to be grown, typically with radii of curvature of around 2.5-3.0 nm.

6.4.1. Imaging conditions

It is worth mentioning at this point some details of the imaging conditions under which such high resolution studies were accomplished. For these, and indeed virtually all of the highest resolution studies of 2D protein crystals, the AFM was operated in constant force (dc) mode and under aqueous buffers, but there are two notable exceptions. In one case high resolution images were obtained in air, but on a metal coated sample (Devaud *et al* 1992) and in the other case high resolution images were obtained by AFM imaging under propanol in error-signal mode at high scan speeds (McMaster *et al* 1996). Because dc mode is prone to thermal drift, the samples contained in the liquid cell were often left for several hours to reach stability before imaging (Müller *et al* 1995b). In principle the problem of thermal drift could be overcome by using an ac mode of operation, such as Tapping in liquids, since the amplitude of oscillation upon which the set-point is controlled is less affected by thermal fluctuation. However, in a study of specimen heights using Tapping mode AFM, HPI layers were imaged by both contact mode and Tapping mode in liquid, and the following pertinent observations were made (Schabert and Rabe, 1996). The edges of the membrane sheets appeared to be damaged in successive scans using contact mode, whereas Tapping mode eliminated this effect, indicating that Tapping mode in liquid may be less damaging than contact mode. However, when small areas on the surface of the membrane patches were examined at high resolution, contact mode produced no obvious damage, and revealed much greater detail than Tapping mode images of the same area using the same AFM tip, proving that for ultimate resolution it appears to be better to use contact mode imaging. Another point to note is that generally short cantilevers (~100 μm long) were used, these having greater angular sensitivity than their longer counterparts and greater stability against flexure. When it comes to the actual AFM tip itself, it is reported that for 2D crystals, oxide-sharpened silicon nitride probes are the best, although they did suffer more frequently from multiple tip artifacts and tip astigmatism (Schabert and Engel, 1994). The quality of the AFM tip was checked before imaging 2D crystals in two ways. First, contamination of the tip was checked for by the presence of hysteresis in the force-distance curve upon retraction of the tip from a bare mica surface. Only tips with no hysteresis were used. Secondly, the stick and slip effect, which occurs at the end of each scan line as the tip changes its fast scan direction, seemed to be correlated with tip quality, and so only tips which displayed the minimum amount of this behaviour were selected (Schabert and Engel, 1994).

Scan speed, or tip velocity, was also found to be an important factor. There is a critical limit of 2 μm s^{-1} which cannot be exceeded for high resolution imaging (Butt *et al* 1993) and speeds of around 1 μm s^{-1} were typical for most of the 2D crystals cited here. A useful trick when engaging the AFM tip was to set the scan size to zero in order to avoid sample damage and tip contamination (Müller *et al* 1995b). This last step may not be necessary depending upon the nature of the approach mechanism for the microscope. Deformation of the sample was monitored by comparing height profiles obtained in forward and reverse scan directions and at differing scan angles. A very important point noted in the work cited here (Müller *et al* 1995b) is that the applied force was adjusted manually to compensate for thermal drift of the cantilever. This advice is good for any sample, one should never assume that the force will not vary as a scan progresses. Despite the term 'constant force' imaging, the instrument only controls the deflection of the cantilever relative to a predetermined null point (where the tip is not in contact with the surface). This null point can vary with time and temperature, and the instrument has no way of detecting such variation during image acquisition, other than de-coupling the tip from the sample which can, of course, ruin the image. Choice of substrates is another important factor to successful imaging at very high resolution. Generally mica is the best. The charged nature of its surface in aqueous buffers allows control over the binding of the 2D crystals through choice of ions and ionic strength (Schabert and Engel, 1994b). A general feature of these studies was that membrane sheets which were not tightly bound to the substrate, after an incubation of the protein solutions on the mica, made imaging conditions unstable and so were best removed by gentle rinsing of the substrate with pure buffer solution. Functionalised glass coverslips were also tried as a substrate but resulted in buckling of the sheets, which increased their apparent roughness to the detriment of image resolution (Schabert and Engel, 1994b), although glass coverslips were used successfully to deposit photosynthetic membranes (Yamada *et al* 1994) and gap junctions (Hoh *et al* 1991). The studies on soluble proteins successfully employed silicon wafers as the substrate (Ohnishi *et al* 1993; 1996; Furono *et al* 1998).

Another interesting point to note is that in many of the studies, low magnification images of entire sheets were obtained with forces of ~ 500 pN with no apparent damage to the specimen. This illustrates an important point in AFM studies of soft samples: as the magnification increases the density of scan lines increases and so the damage induced by the AFM tip increases (and of course the converse is also true accounting for the higher forces allowed at low magnifications). So it is that, very often, optimum image resolution is not obtained by simply increasing the magnification endlessly, but rather scanning at slightly lower magnification, where sample damage is minimised.

6.4.2. Electrostatic considerations

When AFM imaging is performed under aqueous buffers electrostatic forces play a pivotal role in determining how the tip interacts with the sample. Since the tip must accurately track the surface of the sample to produce a faithful image of its topography this means that particular attention needs to be paid to this interaction. Some background on the origin of electrostatic forces is discussed in chapter 3 (sections 3.1.2 & 3.1.4). In practice the two most important force contributions which need to be considered when working in aqueous liquids are the double layer forces between tip and sample, which are affected by the ionic strength and pH of the buffer solution, and the van der Waals force between the tip and sample surface. The van der Waals force is a long range force which is always attractive, but becomes significant only over very short distances (<1 nm), when working in buffered aqueous liquids where it is often opposed by repulsive double layer forces. Biological molecules and AFM tips both exhibit a surface charge in aqueous environments, since acidic and functional groups at their surfaces dissociate according to their pK values. The magnitude and sign of the charge depends on the pH and temperature of the surrounding buffer solution. As explained in chapter 3, the result of this surface charge is to create a form of ionic 'atmosphere' of opposite charged ions surrounding the surfaces and this is known as a *double layer*. The double layer *force* comes about because an electrostatic interaction occurs when the electrical double layers of the tip and the sample overlap. The distance over which the double layer force acts is known as the Debye length and this is affected by the ionic strength of the solution, higher ionic strength reducing the Debye length. The trick to successful imaging at high resolution is to try to balance the attractive van der Waals force with a repulsive double layer force of very short Debye length so that the AFM tip can track the sample surface intimately. Since the van der Waals forces are more or less fixed, this can best be achieved by manipulation of the double layer force by variation of ionic strength and, if need be, the pH of the buffer solution in order to make the double layer force repulsive.

Müller and coworkers set out experimental protocols for achieving this situation (Müller *et al* 1999). They demonstrate that what may intuitively seem the best situation, where perfect balance between attractive van der Waals forces and repulsive double layer forces is achieved, isn't actually ideal. This is because the AFM controls the force interaction between the sample and tip in a proper sense only in repulsive situations, where there is sufficient gradient in the force versus distance characteristics of the system to allow it to detect positive deflections of the cantilever. Remember that the AFM cannot determine the force directly, it can only estimate it by measuring cantilever deflection, and because AFM control loops cannot work with negative force set-points this limits stable control to positive deflections of the lever. When perfect balance is achieved between the repulsive double layer force and the attractive van der Waals forces the repulsive force versus distance gradient is very small, making it impossible for the AFM control loop to work properly. When such conditions prevail one can appreciate that the result will be that the attractive van der Waals force will pull the AFM tip onto the sample and may deform it, even if the applied external force due to bending of the cantilever is low (in fact a *negative* bend on the lever would be required to counteract this adhesion rendering the instrument unstable, since any small reduction in the adhesive force would cause the tip to detach from the surface). Indeed this attractive force is shown to be large enough to generate frictional effects in AFM images at very high resolution, such that forward and reverse scans of the same surface of purple membranes produce different topographies (Müller *et al* 1999). The best situation, particularly when a realistic tip shape is considered - this being one where a blunt hemispherical tip has a small sharp asperity on the end - is to have a small but measurable repulsive double layer force of magnitude approximately 0.1 nN counteracting the attractive van der Waals forces. In this way the tip can image in a stable manner in constant force mode, with the bulk of the applied force being distributed by the longer range double layer force acting over the blunter portion of the tip, but the sharp asperity on the apex will nevertheless track the surface intimately and with very low net force (Müller *et al* 1999). For the samples studied, HPI layers, purple membranes and OmpF porin, the optimum buffer conditions were between 100-200 mM for monovalent cations and around 50 mM for divalent cations at neutral pH (pH 7.6). These figures vary according to the charge density of the sample and the AFM tips themselves, which are not always identical even from the same wafer (Müller *et al* 1999).

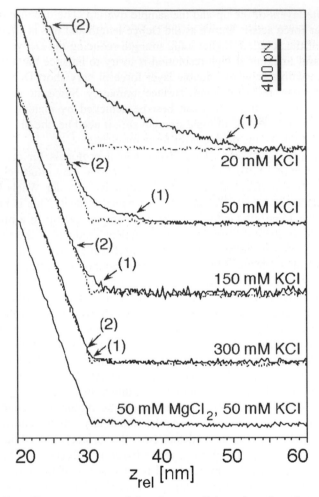

Figure 6.20 Force-distance curves recorded on the extracellular surface of purple membrane. The ionic strength of the imaging liquid was varied at constant pH, in order to alter the thickness of the electrostatic double layer. In order to show the effect of the distance dependence of the electrostatic double layer and its effect on the repulsive force between sample and tip, the dotted lines show an approach curve for a sample where there is no repulsion (i.e. after electrostatic balancing). The sharp turning point on the dotted lines marks the position of tip-sample contact. Taking this into consideration the electrostatic repulsion in the top curve for 20 mM KCl is seen to extend some 20 nm from the sample surface (arrow 1) and prevents the tip from making contact with the sample surface until point 2, where the force due to cantilever flexure is high enough to overcome the repulsion (>400 pN). Such a large force will damage the sample and prevent high resolution imaging. Optimum imaging conditions require most of the electrostatic repulsion to be removed, and this is achieved by raising the ionic strength which reduces the thickness of the double layer. Perfect balance between the attractive van der Waals force and repulsive electrostatic double layer force is achieved in 300 mM KCl (and in 50 mM $MgCl_2$, 50 mM KCl). However the optimum images were obtained when there was a small repulsive force (100 pN) between sample and tip as seen in the curve for 150 mM KCl (arrow 1). Reproduced with permission from Müller *et al* 1999 and the Biophysical Society.

The take-home message is that happily there is no need to go through long-winded calculations to achieve electrostatic balance, rather the electrolyte concentrations under which the AFM works can simply be tuned empirically, with reference to force-distance curves for each sample and tip, in order to achieve optimum imaging conditions. A series of force-distance curves taken in buffer of varying ionic strength is shown in Fig. 6.20 to illustrate this principle.

The fact that the tracking of an AFM tip over the surface of a membrane protein is influenced by the ionic strength of the imaging buffer has been used to good effect to actually *visualize* the electrostatic potential generated by the protein trimers in *OmpF* porin (Philippsen *et al* 2002). By comparing AFM images obtained in various electrolyte concentrations a potential *map* was generated. This map agreed qualitatively with continuum electrostatic calculations based on the atomic OmpF porin embedded in a lipid bilayer at the same electrolyte concentrations. When the nature of the AFM probe was included in calculations, numerical simulation was able to quantitatively reproduce the experimental measurements. Thus the method provides a means of determining the electrostatic potential of native membrane protein surfaces at submolecular resolution.

References

Andersen, C, Schiffler, B, Charbit, A. and Benz, R. (2002). pH-induced collapse of the extracellular loops closes Escherichia coli maltoporin and allows the study of asymmetric sugar binding. *J. Biol. Chem.* **277**, 41318–41325.

Baker, A.A, Helbert, W, Sugiyama, J. and Miles, M.J. (1997). High-resolution atomic force microscopy of native *Valonia* cellulose I microcrystals. *J. Struct. Biol.* **119**, 129-138.

Baker, A.A, Helbert, W, Sugiyama, J. and Miles, M.J. (1998). Surface structure of native cellulose microcrystals by AFM. *Appl. Phys. A* **66**, S559-S563.

Baldwin, J. and Henderson, R. (1984). Measurement and evaluation of electron diffraction patterns from two-dimensional crystals. *Ultramicroscopy* **14**, 319-336.

Baldwin, J.M, Henderson, R, Beckmann, R. and Zemlin, F. (1988). Images of purple membrane at 2.5 Å resolution obtained by cryo-electron microscopy. *J. Mol. Biol.* **202**, 585-591.

Baumeister, W, Barth, M, Hegerl, R, Guckenberger, R, Hahn, M. and Saxton, W.O. (1986). Three dimensional structure of the regular surface layer (HPI layer) of *Deinococcus radiodurans*. *J. Mol. Biol.* **187**, 241-253.

Baumeister, W, Karrenberg, F, Rachel, R, Engel, A, Ten-Heggler, B. and Saxton, W.O. (1982). The major cell envelope protein of *Micrococcus radiodurans* (R1). *Eur. J. Biochem.* **125**, 535-544.

Beveridge, T.J. (1981). Ultrastructure, chemistry and function of the bacterial wall. *Int. Rev. Cytol.* **72**, 229-317.

Boekema, E.J. (1990). The present state of two-dimensional crystallisation of membrane proteins. *Electron Microscopy Rev.* **3**, 87-96.

Bushell, G.R, Watson, G.S, Holt, S.A. and Myhra, S. (1995). Imaging and nano-dissection of tobacco mosaic virus by atomic force microscopy. *J. Microscopy-Oxford* **180**, 174-181.

Bustamante, C, Rivetti, C. and Keller, D.J. (1997). Scanning force microscopy under aqueous solutions. *Current Opinion Struct. Biol.* **7**, 709-716.

Butt, H-J, Downing, K.H. and Hansma, P.K. (1990). Imaging the membrane protein bacteriorhodopsin with the atomic force microscope. *Biophys. J.* **58**, 1473-1480.

Butt, H-J, Siedle, P, Seifert, K, Fendler, K, Seeger, E, Bamberg, E, Weisenhorn, A.L, Goldie, K. and Engel, A. (1993). Scan speed limit in atomic force microscopy. *J. Microscopy-Oxford* **169**, 75-84.

Carrasco, C, Carreira, A, Schaap, I.A.T, Serena, P.A, Gomez-Herrero, J, Mateu, M.G. and de Pablo, P.J. (2006). DNA-mediated anisotropic mechanical reinforcement of a virus. *PNAS USA* **103**, 13706–13711.

Carazo, J.M, Santisteban, A. and Corrascosa, J.L. (1985). Three dimensional reconstruction of the bacteriophage φ29 neck particles at 2.2nm resolution. *J. Mol. Biol.* **183**, 79-88.

Carazo, J.M, Donate, L.E, Herranz, L, Secilla, J.P. and Corrascosa, J.L. (1986). Three dimensional reconstruction of the connector of bacteriophage φ29 at 1.8nm resolution. *J. Mol. Biol.* **192**, 853-867.

Carrascosa, J.L, Vinuela, E, Garcia, N. and Santisteban, A. (1982). Structure of the head-tail connector of bacteriophage φ29. *J. Mol. Biol.* **154**, 311-324.

Carrascosa, J.L, Carazo, J.M, Herranz, L, Donate, L.E. and Secilla, J.P. (1990). Study of two related configurations of the neck of bacteriophage φ29. *Computers Math. Appl.* **20**, 57-65.

Cowan, S.W, Schirmer, T, Rummel, G, Steiert, M, Ghosh, R, Pauptit, R.A, Jansonius, J.N. and Rosenbusch, J.P. (1992). Crystal structures explain functional properties of two *E. coli* porins. *Nature* **358**, 727-733.

Czajkowsky, D.M, Sheng, S. and Shao, Z. (1998). Staphylococcal α-Hemolysin can form hexamers in phospholipid bilayers. *J. Mol. Biol.* **276**, 325-330.

Darst, S.A, Ahlers, M, Kubalek, E.W, Meller, P, Blankenburg, R, Ribi, H.O, Ringsdorf, H. and Kornberg, R.D. (1990). Two dimensional crystals of streptavidin on biotinylated lipid layers and their interactions with biotinylated macromolecules. *Biophys. J.* **59**, 387-396.

Devaud, G, Furcinitti, P.S, Fleming, J.C, Lyon, M.K. and Douglas, K. (1992). Direct observation of defect structure in protein crystals by atomic force and transmission electron microscopy. *Biophys. J.* **63**, 630-638.

Durbin, S.D. and Carlson, W.E. (1992). Lysozyme crystal growth studied by atomic force microscopy. *J. Cryst. Growth* **122**, 71-79.

Durbin, S.D, Carlson, W.E. and Saros, M.T. (1993). *In situ* studies of protein crystal growth by atomic force microscopy. *J. Phys. D* **50**, B128-B132.

Engel, A. (1991). Biological applications of scanning probe microscopes. *Ann. Rev. Biophys. Biophys. Chem.* **20**, 79-108.

Falvo, M.R, Washburn, S, Superfine, R, Finch, M, Brooks Jnr, F.P, Chi, V. and Taylor, R.M. (1997). Manipulation of individual viruses: friction and mechanical properties. *Biophys. J.* **72**, 1396-1403.

Flagg-Newton, J, Simpson, I. and Loewenstein, W.R. (1979). Permeability of the cell to cell membrane channels in mammalian cell junctions. *Science* **205**, 404-407.

Furuno, T, Sasabe, H. and Ikegami, A. (1998). Imaging two dimensional arrays of soluble proteins by atomic force microscopy in contact mode using a sharp supertip. *Ultramicroscopy* **70**, 125-131.

Goennenwein, S, Tanaka, M, Hu, B, Moroder, L. and Sackmann, E. (2003). Functional incorporation of integrins into solid supported membranes on ultrathin films of cellulose: impact on adhesion. *Biophys J.* **85**, 646–55.

Goncalves, R.P, Agnus, G, Sens, P, Houssin, C, Bartenlian, B. and Scheuring, S. (2006). Two-chamber AFM: probing membrane proteins separating two aqueous compartments. *Nature Methods* **3**, 1007–12.

Häberle, W, Hörber, J.K.H. and Binnig, G. (1991). Force microscopy on living cells. *J. Vac. Sci. & Technol. B* **9**, 1210-1213.

Häberle, W, Hörber, J.K.H, Ohnesorge, F, Smith, D.P.E. and Binnig, G. (1992). *In situ* investigations of single living cells infected by viruses. *Ultramicroscopy* **42-44**, 1161-1167.

Hanley, S.J, Giasson, J, Revol, J-F. and Gray, D. (1992). Atomic force microscopy of cellulose microfibrils; comparison with transmission electron microscopy. *Polymer* **33**, 4639-4642.

Harris, J.R. (1992). 2D crystallisation of soluble protein molecules for TEM: the negative staining carbon film procedure. *Microscopy & Analysis* **30**, 13-16.

Hartmut, M. (1991). General and practical aspects of membrane protein crystallisation. In *Crystallization of membrane proteins* (ed. Hartmut Michel), pp. 73-87. CRC press, London.

Henderson, R. and Unwin, P.N.T. (1975). Three-dimensional model of purple membrane obtained by electron microscopy. *Nature* **257**, 28-32.

Henderson, R, Baldwin, J.M, Downing, K.H, Lepault, J. and Zemlin, F. (1986). Structure of purple membrane from *Halobacterium halobium:* recording, measurement and evaluation of electron micrographs at 3.5Å resolution. *Ultramicroscopy* **19**, 147-178.

Henderson, R, Baldwin, J.M, Ceska, T.A, Zemlin, F, Beckmann, E. and Downing, K.H, (1990). Model for the structure of bacteriorhodopsin based on high resolution electron cryo-microscopy. *J. Mol. Biol.* **213**, 899-929.

Hoh, J.H, Lal, R, John, S.A, Revel, J-P. and Arnsdorf, M.F. (1991). Atomic force microscopy and dissection of gap junctions. *Science* **253**, 1405-1408.

Hoh, J.H, Sosinsky, G.E, Revel, J-P. and Hansma, P.K. (1993). Structure of the extracellular surface of the gap junction by atomic force microscopy. *Biophys. J.* **65**, 149-163.

Hörber, J.K.H, Häberle, W, Ohnesorge, F, Binnig, G, Liebich, H.G, Czerny, C.P, Mahnel, H. and Mayr, A. (1992). Investigation of living cells in the nanometer regime with the atomic force microscope. *Scanning Microscopy* **6**, 919-930.

Horne, R.W. and Pasquali-Ronchetti, I. (1974). A negative staining-carbon film technique for studying viruses in the electron microscope. I. Preparative procedure for examining icosahedral and filamentous viruses. *J. Ultrstruct. Res.* **47**, 361-383.

Ikai, A, Imai, K, Yoshimura, K, Tomitori, M, Nishikawa, O, Kokawa, R, Kobayashi, M. and Yamamoto, M. (1994). Scanning tunneling microscopy/atomic force microscopy studies of bacteriophage T4 and its tail fibres. *J. Vac. Sci. & Technol. B* **12**, 1478-1481.

Imai, K, Yoshimura, K, Tomitori, M, Nishikawa, O, Kokawa, R, Yamamoto, M, Kobayashi, M. and Ikai, A. (1993). Scanning tunnelling and atomic force microscopy of T4 bacteriophage and tobacco mosaic virus. *Japan. J. Appl. Phys.* **32**, 2962-2964.

Janovjak, H, Müller, D.J. and Humphris, A.D.L. (2005). Molecular Force Modulation Spectroscopy Revealing the Dynamic Response of Single Bacteriorhodopsins. *Biophys. J.* **88**, 1423–1431

Jap, B.K, (1988). High resolution electron diffraction of reconstituted PhoE porin. *J. Mol. Biol.* **199**, 229-231.

Jap, B.K, Zulauf, M, Scheybani, T, Hefti, A, Baumeister, W, Aebi, U. and Engel, A. (1992). 2D crystallisation: from art to science. *Ultramicroscopy*, **46**, 45-84.

Karrasch, S, Dolder, M, Schabert, F, Ramsden, J. and Engel, A. (1993). Covalent binding of biological samples to solid supports for scanning probe microscopy in buffer solution. *Biophys. J.* **65**, 2437-2446.

Karrasch, S, Hegerl, R, Hoh, J.H, Baumeister, W. and Engel, A. (1994). Atomic force microscopy produces faithful high resolution images of protein surfaces in an aqueous environment. *PNAS USA* **91**, 836-838.

Kedrov, A, Krieg, M, Ziegler, C, Kuhlbrandt, W, Müller, D.J. (2005). Locating ligand binding and activation of a single antiporter. *EMBO Reports* **6**, 668–74.

Kedrov, A, Janovjak, H, Sapra, K.T and Müller, D.J. (2007). Deciphering molecular interactions of native membrane proteins by single-molecule force spectroscopy. *Ann. Rev. Biophys. Biomol. Struct.* **36**, 233–60.

Kirby, A.R, Gunning, A.P. and Morris, V.J. (1996). Imaging polysaccharides by atomic force microscopy. *Biopolymers* **38**, 355-366.

Kol, N, Gladnikoff, M, Barlam, D, Shneck, R.Z, Rein, A. and Rousso, I. (2006). Mechanical Properties of Murine Leukemia Virus Particles: Effect of Maturation. *Biophys. J.* **91**, 767-774.

Kol, N, Shi, Y, Tsvitov, M, Barlam, D, Shneck, R.Z, Kay, M.S. and Rousso, I. (2007). A Stiffness Switch in Human Immunodeficiency Virus. *Biophys. J.* **92**, 1777–1783.

Kolbe, W.F, Ogletree, D.F. and Salmeron, M.B. (1992). Atomic force microscopy imaging of T4 bacteriophages on silicon substrates. *Ultramicroscopy* **42**, 1113-1117.

König, H. (1988). Archaebacterial cell envelopes. *Can J. Microbiol.* **34**, 395-406.

Konnert, J.H, D'Antonio, P. and Ward, K.B. (1994). Observation of growth steps, spiral dislocations and molecular packing on the surface of lysozyme crystals with the atomic force microscope. *Acta Cryst.* **D50**, 603-613.

Kornberg, R.D. and Ribi, H.O. (1987). Formation of two dimensional crystals of proteins on lipid layers. In *Protein Structure, folding and design 2.* (ed. D.L. Oxender), pp. 175-186, Alan R. Liss: New York.

Kornberg, R.D. and Darst, S.A. (1991). Two dimensional crystals of proteins on lipid layers. *Current Opinion Struct. Biol.* **1**, 642-646.

Kouyama, T, Yamamoto, M, Kamiya, Iwasaki, H, Ueki, T. and Sakurai, I. (1994). Polyhedral assembly of a membrane protein in its three-dimensional crystal. *J. Mol. Biol.* **236**, 990-994.

Kühlbrandt, W. and Downing, K.H. (1989). Two-dimensional structure of plant light harvesting complex at 3.7 Å resolution by electron crystallography. *J. Mol. Biol.* **207**, 823-828.

Kühlbrandt, W. and Wang, D.N. (1991). Three-dimensional structure of plant light harvesting complex determined by electron crystallography. *Nature* **350**, 130-134.

Kühlbrandt, W. (1992). Two-dimensional crystallisation of membrane proteins. *Quart. Rev. Biophys.* **25**, 1-49.

Kuutti, L, Peltonen, J, Pere, J. and Teleman, O. (1995). Identification and surface structure of crystalline cellulose studied by atomic force microscopy. *J. Microscopy* **178**, 1-6.

Kuznetsov, Y.G, Malkin, A.J, Lucas, R.W, Plomp M. and McPherson, A. (2001). Imaging of viruses by atomic force microscopy. *J. General Virology* **82**, 2025–2034.

Kuznetsov, Y.G, Victoria, J.G, Robinson Jr, W.E. and McPherson, A. (2003). Atomic Force Microscopy Investigation of Human Immunodeficiency Virus (HIV) and HIV-Infected Lymphocytes. *J. Virology* **77**, 11896–11909.

Kuznetsov, Yu, G, Malkin, A.J, Land, T.A, DeYoreo, J.J, Barba, A.P, Konnert, J. and McPherson, A. (1997). Molecular resolution imaging of macromolecular crystals by atomic force microscopy. *Biophys. J.* **72**, 2357-2364.

Lacapere, J-J, Stokes, D.L. and Chateny, D. (1992). Atomic force microscopy of three-dimensional membrane protein crystals. *Biophys. J.* **63**, 303-308.

Lal, R, Kim, H, Garavito, M. and Arnsdorf, M.F. (1993). Imaging of reconstituted biological channels at molecular resolution by atomic force microscopy. *Amer. J. Physiology* **265** (*Cell Physiol. 34*), C851-C856.

Land, T.A, DeYoreo, J.J. and Lee, J.D. (1995). An *in-situ* AFM investigation of canavalin crystallisation kinetics. *Surface Sci.* **384**, 136-155.

Land, T.A, Malkin, A.J, Kuznetsov, Yu.G, McPherson, A. and DeYoreo, J.J. (1997). Mechanisms of protein crystal growth: an atomic force microscopic study of canavalin crystallisation. *Phys. Rev. Letts.* **75**, 2774-2777.

Lee, J.W.M. and Ng, M-L. (2004). A nano-view of West Nile virus-induced cellular changes during infection. *J. Nanobiotechnology* **2**, article number 6.

Makowski, L, Casper, D.L.D, Philips, D.A. and Goodenough, J. (1977*).* Gap junction structures II. Analysis of the X-ray diffraction data. *J. Cell Biol.* **74**, 629-645.

Makowski, L. (1985). In *Gap Junctions*, (ed. M.V.L. Bennett, D.C. Spray) pp. 5-12. Cold Spring Harbor: New York.

Malkin, A.J, Kuznetsov, Yu.G, DeYoreo, J.J. and McPherson, A. (1995a). Mechanisms of growth for protein and virus crystals. *Nature Struct. Biol.* **2**, 956-959.

Malkin, A.J, Land, T.A, Kuznetsov, Yu.G, McPherson, A. and DeYoreo, J.J. (1995b). Investigation of virus crystal growth mechanisms by *in situ* atomic force microscopy. *Phys. Rev. Letts.* **75**, 2778-2781.

Malkin, A.J, Kuznetsov, Yu.G. and McPherson, A. (1996a). Incorporation of microcrystals by growing protein and virus crystals. *Proteins: Structure, Function, Genetics* **24**, 247-252.

Malkin, A.J, Kuznetsov, Yu.G, Glantz, W. and McPherson, A. (1996b). Atomic force microscopy studies of surface morphology and growth kinetics in thaumatin crystallisation. *J. Phys. Chem.* **100**, 11736-11743.

Malkin, A.J, McPherson, A. and Gershon, P.D. (2003). Structure of Intracellular Mature Vaccinia Virus Visualized by In Situ Atomic Force Microscopy. *J. Virology* **77**, 6332–6340.

McMaster, T.J, Miles, M.J. and Walsby, A.E. (1996). Direct observation of protein secondary structure in gas vesicles by atomic force microscopy. *Biophys. J.* **70**, 2432-2436.

Milhiet, P.E, Gubellini, F, Berquand, A, Dosset, P, Rigaud, J.L, LeGrimellec, C. and Lévy, D. (2006). High-resolution AFM of membrane proteins directly incorporated at high density in planar lipid bilayer. *Biophys J*, **91**, 3268–75.

Müller, D.J, Büldt, G. and Engel, A. (1995a). Force induced conformational change of bacteriorhodopsin. *J. Mol. Biol.* **249**, 239-243.

Müller, D.J, Schabert, F.A, Büldt, G. and Engel, A. (1995b). Imaging purple membranes in aqueous solutions at sub-nanometer resolution by atomic force microscopy. *Biophys. J.* **68**, 1681-1686.

Müller, D.J, Schoenenberger, C, Büldt, G. and Engel, A. (1996a). Immuno-atomic force microscopy of purple membrane. *Biophys. J.* **70**, 1796-1802.

Müller, D.J, Baumeister, W. and Engel, A. (1996b). Conformational change of the hexagonally packed intermediate layer of *Deinococcus radiodurans* imaged by atomic force microscopy. *J. Bacteriol.* **178**, 3025-3030.

Müller, D.J, Engel, A, Carrascosa, J.L. and Velez, M. (1997a). The bacteriophage φ29 head-tail connector imaged at high resolution with the atomic force microscope in buffer solution. *EMBO J.* **16**, 2547-2553.

Müller, D.J, Schoenenberger, C-A, Schabert, F. and Engel, A. (1997b). Structural changes in native membrane proteins monitored at subnanometer resolution with the atomic force microscope. *J. Struct. Biol.* **119**, 149-157.

Müller, D.J, Fotiadis, D, Scheuring, S, Müller, S. A. and Engel, A. (1999). Electrostatically balanced subnanometer imaging of biological specimens by atomic force microscope. *Biophys. J.* **76**, 1101-1111.

Müller, D.J. and Engel, A. (1999). Voltage and pH-induced channel closure of porin OmpF visualised by atomic force microscopy. *J. Mol. Biol.* **285**, 1347-1351.

Müller, D.J, Hand, G.M, Engel, A. and Sosinsky, G. (2002). Conformational changes in surface structures of isolated Connexin26 gap junctions. *EMBO J.* **21**, 3598–607.

Müller, D.J, Engel, A, Matthey, U, Meier, T, Dimroth, P, Suda, K. (2003). Observing membrane protein diffusion at subnanometer resolution. *J Mol Biol.* **327**, 925–30.

Müller, D.J. and Engel, A. (2008). Strategies to prepare and characterize native membrane proteins and protein membranes by AFM. *Current Opinion Colloid & Interface Sci.* **13**, 338–350.

Nam, K.T, Peelle, B.R, Lee, S-W. and Belcher, A.M. (2004). Genetically Driven Assembly of Nanorings Based on the M13 Virus. *Nano Letters* **4**, 23-27.

Negishi, A, Chen, J, McCarty, D.M, Samulski, R.J, Liu, J. and Superfine, R. (2004). *Glycobiology* **14**, 969-977.

Ng, J.D, Kuznetsov, Yu.G, Malkin, A.J, Keith, G, Giege, R. and McPherson, A. (1997). Visualisation of RNA crystal growth by atomic force microscopy. *Nucleic Acids Res.* **25**, 2582-2588.

Nikaido, H. and Vaara, M. (1985). Molecular basis of bacterial outer membrane permeability. *Microbiol. Rev.* **49**, 1-32.

Ohnesorge, F.M, Hörber, J.K.H, Häberle, W, Czerny, C-P, Smith, D.P.E. and Binnig, G. (1997). AFM review study on pox viruses and living cells. *Biophys. J.* **73**, 2183-2194.

Ohnishi, S, Hara, M, Furuno, T, Knoll, W. and Sasabe, H. (1993). AFM imaging of ferritin molecules bound to LB films of poly-1-benzyl-1-histidine - imaging the ordered arrays of water-soluble protein ferritin with the atomic-force microscope. *Atomic-Scale Imaging of Surface and Interfaces* **295**, 145-150.

Ohnishi, S, Hara, M, Furuno, T. and Sasabe, H. (1996). Imaging two dimensional crystals of catalase by atomic force microscopy. *Japan. J. Appl. Phys.* **35**, 6233-6238.

Park, P.S, Sapra, K.T, Kolinski, M, Filipek, S, Palczewski, K. and Müller, D.J. (2007). Stabilizing effect of Zn2+ in native bovine rhodopsin. *J. Biol. Chem.* **282**, 11377–85.

Paul, J.K, Nettikadan, S.J, Ganjeizadeh, M, Yamaguchi, M. and Takeyasu, K. (1994). Molecular imaging of Na^+, K^+-ATPase in purified kidney membranes. *FEBS Letts*, **346**, 289-294.

Philippsen, A, Im, W, Engel, A, Schirmer, T, Roux, B. and Müller, D.J. (2002). Imaging the Electrostatic Potential of Transmembrane Channels: Atomic Probe Microscopy of OmpF Porin. *Biophys. J.* **82,** 1667–1676.

Plomp, M, Rice, M.K, Wagner, E.K, McPherson, A. and Malkin, A.J. (2002). Rapid Visualization at High Resolution of Pathogens by Atomic Force Microscopy: Structural Studies of Herpes Simplex Virus-1. *Amer. J. Pathology* **160**, 1959-1966.

Preiner,J, Janovjak, H, Rankl, C. Knaus, H, Cisneros, D.A, Kedrov, A, Kienberger, F, Müller, D.J. and Hinterdorfer, P. (2007). Free Energy of Membrane Protein Unfolding Derived from Single-Molecule Force Measurements. *Biophys. J.* **93,** 930–937.

Revel, J.P. and Karnowski, M. (1967). Hexagonal array of subunits in intercellular junctions of the mouse heart and liver. *J. Cell Biol.* **33**, C7-C12.

Sass, H.J, Beckmann, R, Zemlin, F, van Heel, M, Zeitler, E, Rosenbusch, J.P, Dorset, D.L. and Massalski, A. (1989). Densely packed ß-structure at the protein lipid interface of porin is revealed by high-resolution cryo-electron microscopy. *J. Mol. Biol.* **209**, 171-175.

Sato, A, Furuno, T, Toyoshima, C. and Sasabe, H. (1993). 2-dimensional crystallisation of catalase on a monolayer film of poly (1-benzyl-L-histidine) spread at the air-water interface. *Biochim. Biophys. Acta* **1162**, 54-60.

Saxton, W.O. and Baumeister, W. (1982). Image averaging for biological specimens - the limits imposed by imperfect crystallinity. *Institute of Physics Conference Series* **61**, 333-336.

Schabert, F.A, Hoh, J.H, Karrasch, S, Hefti, A. and Engel, A. (1994a). Scanning force micrscopy of *E coli* OmpF porin in buffer solution. *J. Vac. Sci. & Technol. B*, **12**, 1504-1507.

Schabert, F.A. and Engel, A. (1994b). Reproducible acquisition of *Eschericia coli* porin surface topographs by atomic force microscopy. *Biophys. J.* **67**, 2394-2403.

Schabert, F.A, Henn, C. and Engel, A. (1995). Native *Eschericia coli* OmpF porin surfaces probed by atomic force microscopy. *Science*, **268,** 92-94.

Schabert, F.A. and Rabe, J.P. (1996). Vertical dimension of hydrated biological samples in tapping mode scanning force microscopy. *Biophys. J.* **70**, 1514-1520.

Scheuring, S. and Sturgis, J.N. (2005). Chromatic adaptation of photosynthetic membranes. *Science* **309**, 484–487.

Shao, Z. and Zhang, Y. (1996). Biological cryo atomic force microscopy: a brief review. *Ultramicroscopy* **66**, 141-152.

Shao, Z, Mou, J, Czajkowsky, D.M, Yang, J. and Yuan, J-Y. (1996). Biological atomic force microscopy: what is achieved and what is needed. *Adv. Phys.* **45**, 1-86.

Smith, J.C, Lee, K-B, Wang, Q, Finn, M.G, Johnson, J.E, Mrksich, M. and Mirkin, C.A. (2003). Nanopatterning the Chemospecific Immobilization of Cowpea Mosaic Virus Capsid. *Nano Letters* **3**, 883-886.

Tanaka, M. and Sackmann, E. (2005). Polymer-supported membranes as models of the cell surface. *Nature*, **437**, 656–63.

Thundat, T, Zheng, X-Y, Sharp, S.L, Allison, D.P, Warmack, R.J, Joy, D.C. and Ferrel, T.L. (1992). Calibration of atomic force microscope tips using biomolecules. *Scanning Microscopy* **6**, 903-910.

Unwin, P.N.T. and Henderson, R. (1975). Molecular structure determination by electron microscopy of unstained crystalline specimens. *J. Mol. Biol.* **94**, 425-440.

Unwin, P.N.T. and Zampighi, G. (1980). Structure of the junction between communicating cells. *Nature* **283**, 545-549.

Uzgiris, E.E. and Kornberg, R.D. (1983). Two dimensional crystallisation technique for imaging macromolecules, with an application to antigen-antibody-complement complexes. *Nature* **301**, 125-129.

Vénien-Bryan, C, Lenne, P-F, Zakri, C, Renault, A, Brisson, A, Legrand, J-F and Berge, B, (1998). Characterisation of the growth of 2D protein crystals on a lipid monolayer by ellipsometry and rigidity measurements coupled to electron microscopy. *Biophys. J.* **74**, 2649-2657.

Worcester, D.L, Miller, R.G. and Bryant, P.J. (1988). Atomic force microscopy of purple membranes. *J. Microscopy-Oxford* **152**, 817-821.

Worcester, D.L, Kim, H.S, Miller, R.G. and Bryant, P.J. (1990). Imaging bacteriorhodopsin lattices in purple membranes with atomic force microscopy. *J. Vac. Sci. & Technol.* **A8**, 403-405.

Xu, W, Mulhern, P.J, Blackford, B.L, Jericho, M.H, Firtel, M. and Beveridge, T.J. (1996). Modeling and measuring the elastic properties of an archeal surface, the sheath of *Methanospirillum hungatei*, and the implication for methane production. *J. Bacteriol.* **178**, 3106-3112.

Yamada, H, Hirata, Y, Hara, M. and Miyake, J. (1994). Atomic force microscopy studies of photosynthetic protein membrane Langmuir-Blodgett films. *Thin Solid Films* **243**, 455-458.

Yang, J, Mou, J, Yuan, J-Y and Shao, Z. (1996). The effect of deformation on the lateral resolution of atomic force microscopy. *J. Microscopy-Oxford* **182**, 106 113.

Yıldız, Ö, Vinothkumar, K.R, Goswami, P. and Kuhlbrandt, W. (2006). Structure of the monomeric outer-membrane porin OmpG in the open and closed conformation. *EMBO J.* **25**, 3702–13.

Yip, C.M. and Ward, M.D. (1996). Atomic force microscopy of insulin single crystals: direct visualisation of molecules and crystal growth. *Biophys. J.* **71**, 1071-1078.

Zenhausern, F, Adrian, M, Emch, R, Taborelli, M, Jobin, M. and Descout, P. (1992). Scanning force microscopy and cryo-electron microscopy of tobacco mosaic virus as a test specimen. *Ultramicroscopy* **42-44**, 1168-1172.

CHAPTER 7

CELLS, TISSUE AND BIOMINERALS

7.1. Imaging methods

The AFM is designed to produce high resolution images on hard, flat surfaces. Very few biological systems approximate to that ideal. Biominerals such as teeth, bone or shells are hard, and often partially crystalline. However, the major problem lies in the production of sufficiently flat surfaces. As will be seen later in section 7.12 this can often be achieved by cutting and polishing surfaces. Cells are normally large with respect to the size of the tip and highly deformable. This has led to novel methods of sample presentation (Fig. 7.1) and the analysis of the deformation and indentation of the cells.

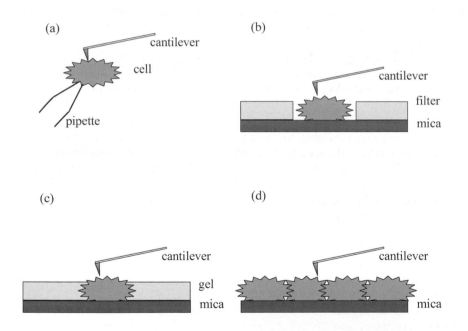

Figure 7.1. Schematic diagram showing different methods for immobilising and imaging cells. (a) Pipette method. (b) Use of porous media. (c) Trapping in a gel. (d) Confluent monolayer.

Tissues again present unique problems. In general, tissues can be examined using modified procedures developed for light and electron microscopy, with a major problem being the need to produce surfaces which are flat enough to allow imaging with the AFM.

7.1.1. Sample preparation

A range of methods have been developed for immobilising cells. Ideally such methods should allow imaging of living cells under physiological conditions. An elegant solution to this problem is to pinion individual cells on the end of a micropipette (Fig. 7.1a). The cells retain their natural shape, they can be immersed in physiological media, and the pipette can be used to scan the surface of the cell beneath the AFM probe tip (Haberle *et al* 1991; 1992; Horber *et al* 1992; Ohnesorge *et al* 1997). If individual cells are deposited onto flat substrates several difficulties occur. If the cells are fairly rigid and retain their shape then they present a rough surface, which is difficult to image without damaging or displacing the cells, resulting in distorted images heavily convoluted by the shape of the probe, or an inability to image the entire cell because the height of the cell exceeds the available z displacement of the piezoelectric scanner. The solution to this problem is to smooth the surface by effectively embedding the cells. Two approaches have been tried: entrapping the cells in porous media (Fig. 7.1b) (Holstein *et al* 1994; Kasa and Ikai, 1996) or completely enclosing the cells in a medium such as agar (Fig. 7.1c) (Gad and Akai, 1996). In the former case the pores restrict cell growth. In media such as agar the cells can grow but, once they grow out onto the surface of the gel, the imaging difficulties return.

Growth of confluent monolayers provides a smoother sample surface allowing the surfaces of the cells to be imaged (Fig. 7.1d). For individual cells it is possible to immobilise them by air drying onto the substrate, although drying is undesirable if living cells are to be imaged. In this case it is possible to employ the use of non-specific surface coatings such as poly-L-lysine (Butt *et al* 1990) collagen, gelatin (Doktycz *et al* 2003), Cell-Tak (Schilcher *et al* 1997) or specific coatings such as bound antibodies (Prater *et al* 1990) or lectins (Gad *et al* 1997). Individual cells grown on flat substrates tend to be flattened and spread out on the surface. These structures are easier to image although, in the vicinity of large organelles such as nuclei, the structure may be too high to image. By far the most limiting factor though is the fact that cells are soft and will deform during scanning. As will be described later in section 7.1.2 such deformation increases the contact area between the tip and the sample reducing the achievable resolution. Resolution can be enhanced by stiffening the structure by fixation but this will preclude functional mapping of the cell surfaces in most instances. Deformation of the outer surface of the cell results in an unexpected bonus: the cell moulds itself over the stiffer internal organelles or cytoskeleton allowing visualisation of cellular substructure even in living cells (Fig. 7.2).

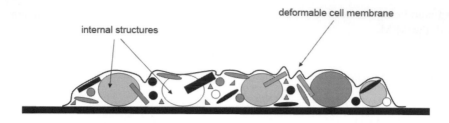

Figure 7.2. Schematic diagram showing a pliable outer cell surface moulding itself down onto more rigid intracellular components.

For tissue samples it is usually necessary to employ methodology developed for light or electron microscopy. Samples normally need to be fixed, embedded and sectioned or fractured. The final surface needs to be sufficiently smooth to allow imaging. Generally steps such as dehydration or metal coating can be avoided although dried, metal-coated samples, or metal replicas, can be imaged with the AFM. Biomineral surfaces can generally be cut and polished. Powdered materials can be dried down onto substrates, or imaged after embedding in media such as KBr discs.

7.1.2. Force mapping and mechanical measurements

Force-distance curves

In the common mode of operation the feedback loop of the AFM operates to maintain a constant cantilever deflection, and images are nominally acquired at constant force. The assumption is that locally the force-distance curve is the same and hence, at constant cantilever deflection (assumed constant force), the image is determined solely by the topography of the sample surface. If the force-distance curve varies between the sample and substrate, or locally across the sample, then the image contrast is not simply a reflection of the sample topography, but also depends on material properties. Chapter 9 provides a detailed description of force measurements with the AFM but the following figure (Fig. 7.3) illustrates a variety of types of force-distance curves which may be observed for biological systems pertinent to this chapter.

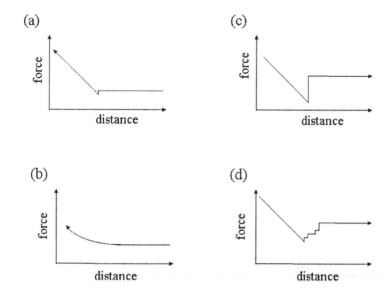

Figure 7.3. Schematic diagram illustrating the types of curves of force-distance curves observed for biological samples. Curves (a, b) are approach curves and curves (c, d) are retraction curves, as indicated by the arrows. (a) Hard, homogeneous sample and van der Waals forces. (b) Soft, and/or charged sample surface, or surface containing attached polymer which leads to an entropic repulsive force. (c) Adhesive interactions or capillary forces. (d) Binding of the tip to molecules on the surface.

Different types of interaction can all lead to differences in contrast. Important factors for cellular systems will be surface charge, elasticity, mobile surface layers, adhesion and localised molecular binding. Several, or all of these factors, can contribute to image contrast, or can be emphasised to select surface features or properties. At present the most important factor influencing contrast in studies on cells is sample deformation, although with the advent of ever more sophisticated molecular recognition imaging strategies this situation is starting to change. (Hinterdorfer and Dufrêne, 2006)

Force mapping

Because the image contrast is sensitive to the details of the force-distance curve there has been increasing interest in its measurement and interpretation for soft biological systems. In a method called force mapping force-distance curves are recorded at each sample point of an image. An increasingly accepted way of displaying such data is as force-volume plots, and an excellent account of the acquisition of force-volume data has been written by Hoh and coworkers (Hoh *et al*

1997) who have also reviewed the general interpretation of such data for mapping biological surfaces (Heinz and Hoh, 1999a). The force volume plot consists of a topographical image of an area of the sample together with an array of force-distance curves recorded over the same area. The data can be displayed as slices of the volume plot showing force at constant height, slices corresponding to constant force images, or as individual force-distance curves. Data is recorded for both approach and retraction from the surface. In studies of soft biological systems the force-distance data can contain information on surface charge, sample viscoelasticity, adhesion or other factors modifying the force-distance curve such as specific binding. Thus the force volume data can be used to generate different types of force maps which can be used to generate different types of contrast for comparison with normal topography images (Heinz and Hoh, 1999a). A major application of this approach has been the analysis of the mechanical properties of cellular surfaces.

Elasticity and elasticity mapping

As mentioned earlier, the major factor which has been considered in depth is the effect of sample deformation during imaging. When the AFM is used to image hard samples the force-distance curve is of the form shown in Fig. 7.3a. As the tip approaches the sample there is flat region where the tip is not in contact with the surface. Once contact is achieved then the force (or cantilever deflection) increases linearly as sample and tip are driven together, and a plot of cantilever deflection against distance would have a slope of 1. For soft samples the increase in cantilever deflection following contact is more gradual reflecting deformation of the sample (Fig. 7.3b). In this case the cantilever deflection will be dependent on the viscoelastic properties of the sample and this will contribute to contrast in the image. On a positive note this effect permits the mapping of the mechanical properties of the sample (Fig. 7.4f). However, a consequence of the sample deformation is that the contact area between the tip and the sample increases and this reduces the practical resolution which can be achieved. In effect the sample deformation on scanning blurs the image by effectively increasing the size of the tip. It is possible to estimate the expected resolution if the tip size and geometry and the elastic properties of the sample are known. This type of analysis has been made for soft gelatin films where the modulus was varied by imaging the sample in different propanol/water mixtures (Radmacher *et al* 1995). Another, and perhaps originally unexpected, consequence of the deformation of the sample surface is the ability to image cellular substructure using AFM. As mentioned in the last section the pliable outer coating of the cell can be folded down onto more rigid internal organelles or cytoskeleton revealing these structures (Fig. 7.2).

Topography   Surface map

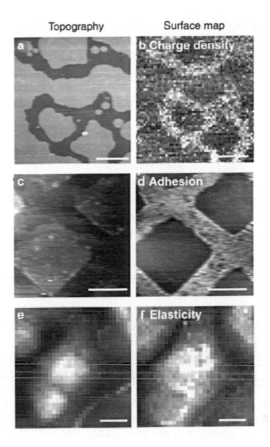

Figure 7.4. Surface maps. Topographical (a) and charge (b) maps (scale bars 1 μm) of a phospholipid bilayer on mica. The more highly charged mica surface appears bright in the charge map whereas the higher bilayer region appears bright in the topographic image. Topographical (c) and adhesion (d) maps (scale bars 500 nm) of a patterned streptavidin surface. The streptavidin is bound on the rows between the squares and appears bright in the adhesion map made with a biotinylated tip. Topographical (e) and elasticity (f) maps (scale bars 10 μm) of mitotic epithelial cells produced by FIEL mapping. This figure is reprinted from: Heinz, W.F. and Hoh, J.H. (1999a). Spatially resolved force spectroscopy of biological surfaces using the atomic force microscope. *Trends Biotechnol.* **17**, 143-150. (1999), Copyright with permission from Elsevier Science. The original experimental data maps are (a, b) from Heinz and Hoh (1999b), (b, c) from Ludwig *et al* 1997 and (e, f) from A-Hassan *et al* 1998.

Different geometries can be used to model the tip-sample interaction. Thus the Young's modulus of platelets was determined by treating the tip as a cone with an opening angle of about 30^0 and the platelet as flat plane (Radmacher *et al* 1995; 1996). In this case:-

$$F_{cp} = 0.5\ \pi\delta^2\ E\ (1\text{-}v)^{-1}.\ \tan(\alpha) \tag{7.1}$$

where α is the opening angle of the cone, E Young's modulus and ν the Poisson ratio of the soft sample. For studies on vesicles where the radius of the sample is similar to that of the probe tip a more appropriate model is a spherical tip and a spherical particle (Laney *et al* 1997) which gives the relationship:-

$$F_{ss} = 1.335\delta^{1.5} E (1-\nu)^{-1}.[R_T R_V/(R_T + R_V)]^{0.5} \tag{7.2}$$

where R_T and R_V are respectively the radii of the tip and the sample. Different expressions of this kind are obtained for different tip and sample geometries (see section 9.5.3). Estimates of tip shape and size can be made by direct measurement (e.g. scanning electron microscopy) or by deconvolution of images of biological or non-biological standards. The model assumes that the sample is homogeneous. Under these limiting conditions the analysis yields reasonable values for Young's modulus. In practice any limitations imposed by the use of the Hertz model are not too severe because, in general, the main interest is in the variation of mechanical properties, rather than measurement of absolute modulus values. Examples of the use of force-distance curves to measure elastic properties of biological material are studies on cartilage (Jurvelin *et al* 1996), glial cells (Haydon *et al* 1996) and epithelial cells (Putman *et al* 1994; Hoh and Schoenenberger, 1994). Good examples of the application of mechanical mapping of biological systems are work on chromosomes (Fritzsche and Henderson, 1997), cholinergic synaptic vesicles (Laney *et al* 1997), MDCK cells (Hoh and Scoenenberger, 1994), platelets (Radmacher *et al* 1996), bone (Tao *et al* 1992) and cardiocytes (Hofmann *et al* 1997; Domke *et al* 1999).

A-Hassan and coworkers (A-Hassan *et al* 1999) describe a refined mapping method called force integration to equal limits (FIEL) mapping which eliminates difficulties involved in the measurement of the tip-sample contact point and the spring constant of the cantilever. In this approach mechanical properties are determined from measurements of the work done by the cantilever, which is calculated from the area under the force-distance curve. In this approach it is necessary to isolate the purely elastic contributions to elasticity and this is done by separating time dependent (viscous) components from time independent (elastic) components of the deformation. The FIEL mapping method is designed to monitor and display spatial or time dependent variations in mechanical properties (A-Hassan *et al* 1999). In practice the force-distance curve may contain elastic (reversible) and inelastic or plastic (irreversible) components. The elastic components may also be time dependent. Thus the force volume data will be sensitive to imaging conditions such as scan rates, or whether contact or non-contact modes are used. For example, ac modes, such as Tapping, minimise or eliminate frictional effects, but also indent the sample at high frequency.

An alternative way for measuring elastic behaviour is to use the AFM as a microrheometer. This approach has been used to study the elastic properties of bacterial sheaths (Xu *et al* 1996). The sheaths were suspended between bars on a GaAs grating and compressed with the AFM tip (Fig. 7.5). Equations developed to describe the force-indentation curves were first tested on model plastic films and then used to evaluate the elasticity of the bacterial sheaths.

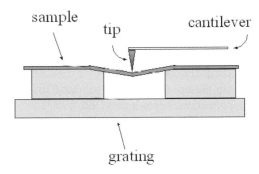

Figure 7.5. Schematic diagram illustrating the use of an AFM as a microrheometer to study the elasticity of bacterial sheaths. Based on a diagram in Xu *et al* 1996.

Charge, adhesion and other maps

Most cellular systems will be charged and surface charge will influence image contrast. If the charge varies over very short distances on the surface then surface deformation may smooth the charge distribution eliminating this effect. However, if charge differences occur in large patches, or are localised at specific surface structures, then this may provide additional contrast (Fig. 7.4b). Examples of charge maps for biological samples include bacteriorhodopsin membrane patches (Butt, 1992; Rotsch and Radmacher, 1997; Heinz and Hoh, 1999b) and phospholipid bilayer patches (Heinz and Hoh, 1999b) on hard substrates. From the known surface charge density of the substrate it was possible to calculate a reasonable value for the surface charge density of the membrane (Butt, 1992). By producing relative maps of surface charge density (or surface potential) it is possible to avoid the need to measure tip charge and radius, the cantilever spring constant and the tip-sample contact position, necessary to evaluate absolute tip-sample separations. This is achieved by subtracting iso-force surfaces collected at different ionic strengths; a method known as D-D mapping (Heinz and Hoh, 1999b). Charge mapping has been used to illustrate phase separation in mixed surfactant films (Yuan and Lenhoff, 1999).

Adhesion data can be used to map tip-sample binding and hence, if functionalised tips are used, specific structures can be identified on sample surfaces. Adhesion maps can be generated by selecting the most negative force on the retraction curve and plotting iso-force maps as a function of position on the surface (Fig. 7.4d). This approach overcomes one of the difficulties with AFM imaging which is identification of surface features. The feasibility of this approach was shown by using biotinylated tips to map distributions of streptavidin on a patterned surface (Ludwig *et al* 1997). The method was termed affinity imaging and was used to compare topographical, elasticity and adhesion maps of the sample. Antibody coated tips can be used for selective imaging and have been used to image monolayers of intracellular adhesion molecules (Willemson *et al* 1998). Mapping of structures on the surfaces of living cells was first demonstrated by (Gad *et al* 1997). These authors used concanavillin A coated tips to detect mannan polysaccharides on the surface of living yeast cells. The mapping showed a non-uniform distribution of the polysaccharide on the cell surface. As described in section 4.5.3, neurofilaments are branched and the branches give rise to long range repulsive (entropic) interactions with the tip. Brown and Hoh have used iso-force difference mapping to reveal such interactions (Brown and Hoh, 1997).

Practical issues in AFM adhesion mapping of live cells

Although adhesion mapping of cells to reveal the presence of biologically relevant molecules on the cell surfaces is still a very demanding application, two factors are opening this once highly complex process up to the more general AFM user. The first is the vastly improved tip functionalisation protocols which have been developed and refined in the last decade or so (Hinterdorfer *et al* 1996). The second factor is the new generation of environmentally-controlled liquid cells, provided by most of the major manufacturers, which offer fine control over temperature and flow of liquids. These allow many of the conditions of the cell-culture cabinet to be reproduced within the confines of the AFM. However several problems still exist which present the AFM experimentalist with challenges. An appreciation of the potential difficulties and limitations in live cell imaging/mapping allows a more rational assessment to be made of whether the AFM is likely to provide useful information on a particular system. The first and most dogged problem for the application of AFM to live cell imaging is that old chestnut, sample deformability. This imposes quite severe practical limits on image resolution, which correspondingly limit the spatial resolution of adhesion mapping data. Additionally, there is the problem of the relatively large topography of most cell surfaces; this greatly enhances the effects of probe-broadening (section 2.8.2), which once again acts to the detriment of spatial resolution. Problems relating to the strength of attachment of cells to the underlying substrate are another source of experimental difficulty. Although adhesion-promoting culture dishes are available, cell monolayers which are weakly bound to the underlying substrate are often more representative of their *in-vivo* environment. The problem is that weakly attached cells can give rise to cells budding out of the monolayer and into the media, producing cell debris which can easily foul a functionalised tip. The removal of such debris before AFM imaging is not always trivial, as weakly bound cells can withstand only gentle rinsing. If budding occurs during the AFM measurements there is no way to remove the debris. Tip contamination during a scan is likely to eliminate any biological specificity by blocking the attached probe molecule(s). If the fouling is transient then highly misleading data can result, and repeatability will be poor.

Poor repeatability can stem from tip contamination, but there is another potential source; the very fact that the cells are *alive* means that they are likely to respond to external stimuli. The expression and position of receptors on a cell surface is a dynamic process, which aside from any biological factors, might be influenced by the cell being repeatedly prodded with the equivalent of a sharpened stick during the course of a typical force-volume scan! The fact that the location and presence of receptor molecules on cell surfaces may be dynamic, presents AFM adhesion mapping with a problem. A typical scan must be performed very slowly on live cells to avoid tip-induced damage, since as we already know they are both rough and so difficult to track with a tip, and also delicate structures. If cell membrane dynamics occur on timescales faster than the AFM measurement then repeatability becomes a rather 'moot point'. The extent to which the temporal resolution of AFM adhesion mapping of live cells might be a limiting factor in repeatability is something that will only be fully resolved with the advent of faster scanning modes. Although much higher scan rates have been achieved for relatively flat samples (Yokokawa *et al* 2006; Picco *et al* 2007) the prospect of achieving such scan speeds for very rough, yet very soft samples such as cells *and* combining this with force measurement is still some way off. However, in an important and fascinating study Almqvist and colleagues demonstrated that even current force-volume methodology had sufficient time resolution to observe the clustering of VEGF receptors toward the boundaries of bovine aortic endothelial cells following stimulation with VEGF (Almqvist *et al* 2004). The data also revealed a remarkable reduction in elasticity of the regions under the clusters, and a concomitant increase in membrane fluidity. These combined effects were considered to provide a mechanism for cell growth and angiogenesis (Almqvist *et al* 2004).

At present the slow data acquisition rates mean that AFM adhesion measurements are also often plagued by thermal drift of the AFM cantilever null-point (particularly in force-volume mode which is much slower than standard imaging). If drift is bad enough it can lead to loss of feedback control, and/or loss of portions of the deflection data as the reflected laser light from the AFM cantilever ends up missing some (or in extreme cases all) of the photodetector. However, the advances in environmentally-controlled stages for the latest generation of instrument have helped this enormously. One very useful method, which goes some way towards overcoming the lack of temporal resolution for AFM adhesion mapping studies, is the dynamic molecular recognition protocol (Raab *et al* 1999; Ebner *et al* 2005, see section 9.3.2). By eliminating the need to perform force curves at each imaging point, this approach speeds up the scanning process significantly when compared to the conventional force-volume method. However, the procedure produces only qualitative information (Stroh *et al* 2004a; 2004b), but methods to quantify adhesion in this mode are beginning to be developed (Chtcheglova *et al* 2004; Gabai *et al* 2005) which point the way forward for affinity mapping of live cells with the AFM.

Occasionally the sample can present fundamental limitations which will require a modification of approach. For example, certain cell types have hydrolytic and proteolytic enzymes associated with their apical surfaces (e.g. the brush-border enzymes in intestinal epithelial cells) which can preclude the use of single molecule recognition strategies if the probe molecule on the AFM tip is susceptible to attack. In such cases multiple, rather than single, molecular functionalisation of the AFM tip may become necessary.

Finally assuming that adhesion data *can* be obtained, there are specific problems with its subsequent analysis. Interpreting and analysing the force curves which result from live-cell mapping is typically much more complex (Almqvist *et al* 2004) than interpreting force spectra obtained in the more controlled environment of experiments performed on functionalised mica surfaces, which are both flat and rigid. The deformable nature of cells means that approach and retract data frequently do not overlay, or conform to the linear baseline ideal obtained on rigid surfaces, making estimation of adhesion tricky. This is further complicated by the relative lack of control over what may bind to the probe molecule as it scans across the highly heterogeneous surface of a cell, increasing the possibility of non-specific interaction between sample and tip. The result is that multiple unbinding events are commonly seen in force spectra from live cells, producing potentially confusing and apparently "noisy" data, making discrimination and analysis difficult for automatic routines. However, manual analysis is generally unrealistic because, in order to *actually map* a cell surface, a reasonable number of points have to be covered to enable correlation of the adhesion and topography data. A typical force-volume run might consist of a 50 x 50 array of sampling points generating 2,500 force curves! The most sensible compromise within these constraints is the use of a routine which allows the user to manually inspect particular curves from the adhesion map to ensure that the curve fitting and event discrimination algorithms are working sensibly.

Despite the problems listed above, many studies have been successfully carried out where functionally-relevant molecules have been spatially mapped on cell surfaces, and specificity demonstrated by the addition of free ligands to provide control data for comparison. These have been achieved on cell types ranging from bacterial and fungal cells (for a review see Dupres *et al* 2007) to mammalian cells (Grandbois *et al* 2000; Lehenkari *et al* 2000; Horton *et al* 2002; Almqvist *et al* 2004; Gunning *et al* 2008).

Using the AFM to measure or induce cellular processes

In early studies AFM was the only form of microscopy with the potential for obtaining ultra-high resolution images of cell surfaces in physiological conditions, and thus much of the early emphasis and excitement was focussed on this objective, and indeed on methodology for overcoming the problems associated with tracking the relatively rough yet soft surfaces of living cells. However, in the past decade two factors have changed this situation. The first is the growing acceptance that very rough and very soft is a combination that may always elude high resolution imaging by AFM (although scanning ion conductance microscopy (SICM) appears to provide an attractive alternative probe microscopy for high resolution cell surface imaging, see section 8.4). The second and more fundamental factor is the revolution that has occurred in fluorescence microscopy in recent years, with methodologies being developed to beat the diffraction limitation in far-field optics (Hell 2007; Betzig *et* al 2006; Patterson *et al* 2007). Put simply, the dramatic improvements in resolution which have resulted mean that the AFM may be overtaken as the method of choice for high-resolution cellular imaging. The emphasis in AFM studies has now, therefore, shifted away from the death-or-glory quest for ever higher resolution of cellular systems (which in any case is much better done on supported, reconstituted or extracted membranes, see chapter 6) to concentrate on the area where AFM still holds a trump card; namely the mechanical monitoring or manipulation of samples with incredible precision. In the context of cell biology this gives the AFM a uniquely powerful ability, since many cellular processes either generate mechanical responses or are induced by mechanical stimuli, and so are ideally suited to study by this aspect of AFM. An excellent review of the many applications to date has been written recently (Lamontagne *et al* 2008) and a brief summary of examples is given in the following section.

Mechanical manipulation of cells

The mechanical behaviour of cells is an area of significant biological interest in the various dynamic processes which cells undergo such as tissue restructuring and transduction of mechanical stimuli. Cell types such as those involved in hearing, pressure sensing in vascular systems, bone restructuring and cell motility have been studied using AFM (Radmacher 2007; Charras *et al* 2001; Li *et al* 2005; Prass *et al* 2006). For example, the mechanical stimulation of osteoblasts by an AFM tip was shown to propagate to neighbouring cells through measurements of intra-cellular calcium migration (Charras and Horton, 2002). Even cells which are not 'mechanically' active in terms of their biological roles require the ability to preserve their integrity in response to environmental changes. These processes are governed by modulation of the cytoskeletal architecture of the cell and also the composition of its plasma membrane, each of which has been studied by AFM (Kasas *et al* 2005; Wu *et al* 1998). In an example of a mechano-sensitive cell system, characterisation of endothelial cell stiffening, caused by stimulation of the aldesterone receptor, has been monitored in real time by AFM (Oberleithner *et al* 2006), a process which has direct relevance to blood pressure regulation. Such real-time monitoring of membrane and cytosolic fluctuations in living cells has also been applied to a number of other systems (Rotsch *et al* 1999; Prass *et al* 2006; Szabo *et al* 2002). The principle of this method is to place the AFM tip gently onto the surface of a cell and then monitor the cantilever deflection over time.

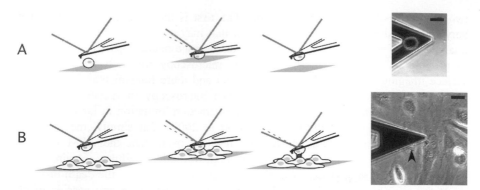

Figure 7.6 Schematic of cell capture. (A) The lectin-decorated cantilever is positioned over a cell in suspension, at close proximity to the surface. Then the cantilever is gently pushed for a few seconds on the cell. After this, the cantilever-bound cell and support are vertically separated by ~100 μm. This allows the cell to establish a firm adhesion to the cantilever. The last panel presents an optical image of a cell captured on a tipless cantilever. (B) Schematic of an adhesion assay of a single captured melanoma cell on an endothelial cell layer. The probe cell is positioned over a zone of interest, which has been selected by phase-contrast microscopy. A given force (usually of several hundred pN) is applied for a given time (usually in the range of seconds) on the cell layer. The probe cell is subsequently separated from the surface, and the de-adhesion events are recorded. The last panel shows a phase-contrast image of a cantilever-bound melanoma cell (arrow) in contact with an HUVEC layer (scale bars, 20 μm). Reproduced with permission from (Puech *et al* 2006) and Elsevier.

Because this method requires neither scanning nor active feedback control of the tip (the feedback loop is typically frozen at a certain z piezo position where the tip is in contact with the cell) it can achieve spatial resolution of nanometres for the height fluctuations of the plasma membrane, and temporal resolution on the order of tens of microseconds (Shroff *et al* 1995). This has been shown to be sensitive enough to detect fluctuations due to cellular processes as subtle as the metabolic activity of yeast cells (Pelling *et al* 2004). Localised membrane fluctuations were also observed for human foreskin fibroblasts which could be linked to the physiological state and cytoskeletal dynamics through distinct sets of correlation time constants in human cells (Pelling *et al* 2007).

Measuring adhesion between cells

In addition to mapping adhesion due to specific biological receptors sites on cell surfaces, new developments in AFM instrumentation have allowed adhesive interactions *between* cells to be quantified directly (Puech *et al* 2006). The AFM tip (actually often a tip-less cantilever) is functionalised to allow live cells to be attached to it, usually with a cell-surface specific lectin which glues the cell whilst preserving its viability. The cantilever is gently placed on the surface of a weakly bound cell and held in position for a several seconds (Fig. 7.6). When the lever is retracted the cell comes with it (i.e. it is pulled off the substrate) and the tip-bound cell can be repositioned over another so that cell-cell adhesion measurements can be performed.

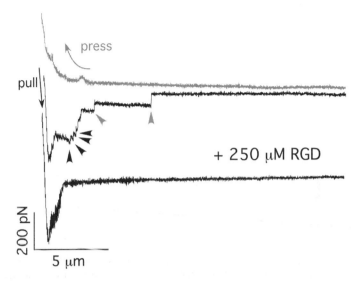

Figure 7.7 Unbinding of WM115 Melanoma cell from a fibronectin (FN) coated surface. The approach and retract speeds were set to 10 µm/s, cell-surface contact was maintained for 10 s, and pulling distance was set to 20 µm. A subset of small adhesion events, 44+8 pN were preceded by a force plateau (grey arrowheads). Black arrowheads indicate where small adhesion events were not preceded by a force plateau. All small adhesion events disappeared upon the addition of 250 mM of RGD peptide to the medium, which competes with FN to bind to the active site of integrins,. Reproduced with permission from (Puech *et al* 2006) and Elsevier.

Extended *z* range piezo devices are often incorporated into the scanner, allowing force curves to be obtained over distances up to 100 µm. Such long (by AFM standards) distances are needed due to the large size and deformability of cells; cell-cell interactions can occur over much larger ranges than the inter-molecular interactions probed in traditional single molecule force spectroscopy. Despite the large scanner range, Peuch and co-workers (Puech *et al* 2006) demonstrated sufficient sensitivity to detect single molecule rupture events in force-distance curves extending over 60 µm (Fig. 7.7).

Biological manipulation of cells

A clutch of very exciting recent studies have demonstrated that the AFM presents the biologist with a tool to not only mechanically manipulate living cells, but also to carry out biological intervention on a cell by cell level (Osada *et al* 2003; Cuerrier *et al* 2007; Uehara *et al* 2004; Chen *et al* 2007). For example, direct interaction with living cells using the AFM tip has allowed the insertion (Cuerrier *et al* 2007) and removal (Osada *et al* 2003) of genetic material with the AFM. Osada and co-workers used the silicon nitride tip of a standard AFM cantilever to pick up messenger-RNA after penetration of a cell membrane. The tips were inserted into the cytoplasm region around the nucleus and the mRNA allowed to physisorb onto the tip. The captured mRNA was then quantified by reverse PCR of the β-actin transcript. Thus removal of endogenous molecules from the cytosol of a living cell was demonstrated. By using quantitative PCR the authors demonstrated that time dependant gene expression patterns in single cells do not always mirror those in large populations (Osada *et al* 2003). Thus, this technique has the potential to provide valuable insights into the biology of gene expression at the single cell level and quantify cell-cell variability. Importantly the cells were found to tolerate this procedure well, showing no signs of subsequent damage. Indeed several other studies have found that such intrusions into live cells appear to be non-lethal for mammalian (Uehara *et al* 2004; Chen *et al* 2007; Cuerrier *et al* 2007) and very recently also for bacterial cells (Suo *et al* 2009). Although the retrieval of mRNA in Osada's study was by simple passive physisorption onto the AFM tip, presumably removal of specific cytosolic components might be possible with a biologically functionalised tip. This would be a very exciting prospect indeed.

In an example of the opposite process, Cuerrier and co-workers have shown that genetic material can be successfully *inserted* through the plasma membrane of living cells (Cuerrier *et al* 2007). Plasmid DNA encoding green fluorescent protein (EGFP) was physisorbed to an AFM tip which was then pushed (controllably) through the plasma membrane of living human embryonic kidney 293 (HEK 293) cells (which were grown as a monolayer in tissue culture). The process of penetrating the cell membrane was monitored in real-time by reference to force-distance data from the approaching tip. As the tip pressed against the cell membrane the force rose relatively linearly (due to cantilever deflection) until a critical break-through force was reached, characterised by an inflection in the curve. Once this had been achieved the tip was held in position (i.e. in the cytosol) for a few minutes before being withdrawn again. Fluorescence micrographs were taken of the treated cells after 24 h to quantify how many of them expressed EGFP, and thus also gauge the success of the transfection. The success rate was some 30% and, furthermore, the transfected cells continued to divide and grow normally. The fluorescence intensity in the progeny cells (images taken 48 and 72 h post treatment) appeared similar to the parents, indicating that the level of EGFP expression was similar, and that stable transfection had occurred (Cuerrier *et al* 2007).

7.2. Microbial cells: bacteria, spores and yeasts

7.2.1. Bacteria

Bacterial surface layers

A number of bacterial surface layers have been studied extensively by AFM. These structures are generally rigid, well ordered and can be isolated as large sheets.

Examples include the HPI layer of *Deinococcus radiodurans* (Schabert *et al* 1992), *Halobacterium* purple membrane (Muller *et al* 1995), S-layers from *Bacillus coagulans* and *Bacillus sphaericus* (Ohnesorge *et al* 1992) and *Escherichia coli* porin surfaces (Schabert and Engel, 1994; Schabert *et al* 1995). Other more complex structures include the sheath, hoops and plugs of the surface structures of *Methanospirillum hungatei* (Southam *et al* 1993) and the gas vesicles of the cyanobacterium *Anabaena flos-aquae* (McMaster *et al* 1996a). The highest resolution images reported are observations of β-sheet protein secondary structure in *A. flos-aquae* gas vesicles (McMaster *et al* 1996a). Atomic force microscopy studies on these ordered naturally occurring structures have already been discussed in section 6.3. In addition to obtaining high resolution images of cell surface structures it is also possible to measure and model their elastic properties using the AFM. This type of study is well illustrated by investigations on the sheath of the methanogen *Methanospirillum hungatei* (Xu *et al* 1996). The sheaths were air-dried onto grooved GaAs plates and the AFM tip used to compress the sheaths into the grooves (Fig. 7.5). The measured Young's modulus suggested that these surface features were more than adequate to maintain structural integrity of the cells, and may therefore play an additional role as a pressure regulator controlling release of methane generated by these bacteria (Xu *et al* 1996).

Bacterial cells

Isolated bacteria deposited on substrates can be imaged in air (Fig. 7.8a) or under aqueous conditions if they adhere sufficiently strongly to the substrate (Fig. 7.8b).

Immobilising bacterial cells

The two most commonly used polymers for immobilising bacterial cells are poly-L-lysine or polyethyleneimide (PEI), which effectively generate a positively charged surface which promotes adhesion (Velegol and Logan, 2002). Doktycz and co-workers compared the efficacy of several commonly used non-specific polyelectrolyte polymers for immobilising bacterial cells under aqueous liquids and found gelatin to be better than the more commonly used poly-L-lysine (Doktycz *et al* 2003). Whichever polyelectrolyte is used for binding, we find in our laboratory that the key to successful AFM imaging and/or force mapping of bacterial cells in liquids is to achieve a confluent monolayer on the substrate. Replacement of the fermentation broth with phosphate buffered saline (PBS) is the first step to remove other charged species which compete for binding sites on the polyectrolyte layer (although see figure 7.9 and associated text in the following section for discussion on the effects of buffer choice for bacteria). The 'washing' step can involve simply centrifugation and then replacement of the supernatant with PBS; this needs to be repeated a few times. The second step is to ensure that a very large number of cells are available for attachment. This requires incubating the highest possible concentrated cell suspension with the coated glass or mica for around 1 hour or so, followed by gentle rinsing to remove non-adherent cells. To achieve this high concentration, following the final centrifugation step, the pelleted cells can be re-dispersed in a smaller volume of liquid than the original one to give around 10^{10} cells.mL^{-1}. As a rule of thumb the cell suspension should have a similar turbidity to skimmed milk – if it's not noticeably turbid you are way too low in numbers. If confluence can be achieved on incubation with the cell suspension it brings two significant advantages. Firstly, it locks the cells into the attached layer giving them much greater resistance to tip-induced displacement, even in contact mode imaging (figure 7.8b). Secondly, it reduces the potential problems of contaminating the AFM tip through contact with the sticky, underlying adhesive polyelectrolyte film (this is a particular issue with the gelatin method) which can lead to highly misleading data if force spectroscopy is the objective of your experiment. Recently a more specific immobilisation method, utilising antibody functionalisation of the substrate, has been demonstrated to be effective for *Salmonella enterica* (Suo *et al* 2008). An antibody to the fimbrae protein CFA-I was covalently attached to silicon and gold substrates in a patterned manner, and these functionalised substrates were then incubated for 3 h in the bacterial broths. Anti CFA-I was chosen because *Salmonella enterica* express fimbrae in abundance. Furthermore, the hair-like fimbrae protrude some distance away from the body of the cells, thus maximising the chances of potential attachment, by placing the adhesive site well outside the extracellular polysaccharide layer (which would interfere with the antibody-antigen binding). The *Salmonella* cells were shown to stick preferentially to the antibody patterned regions and to remain viable, since cell growth and division of the immobilised cells was observed. This work is encouraging because it provides a method which maintains the cells in their native environment by eliminating the need for washing and centrifugation (both of which will of course subject them to stress) and might be applicable to other bacterial cell lines containing other types of secretion-system appendages such as pili. However, this strategy will not work with flagellae, which are too long and too mobile to provide a useful anchor point (Suo *et al* 2008). Finally, a third and less 'chemically-intrusive' method is to mechanically immobilise bacterial cells onto Isopore (Millipore) polycarbonate filters, as shown schematically in figure 7.1b (Dufrêne 2004). This approach allows imaging of the cells in aqueous conditions, while minimizing denaturation of the specimen (Dufrêne 2004).

Imaging bacterial cells

The cell wall structures of bacterial cells are rigid and, for air-dried specimens that partially collapse upon dehydration, it is possible to reveal the roughness of the surface (Gunning *et al* 1996; Braga and Ricci, 1998) as shown in figure 7.8a. Images obtained under liquid reveal most bacterial surfaces to be smooth and relatively featureless (Fig. 7.8b) and the cells appear to be 2-3 fold taller than when imaged dry.

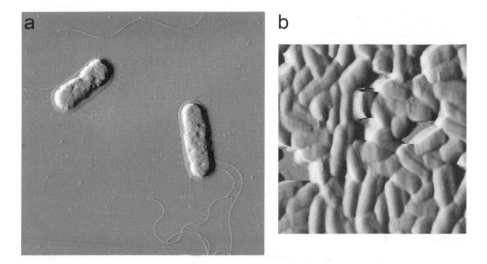

Figure 7.8. Error-signal AFM images of *Salmonella typhimurium* bacteria, (a) images obtained in air reveal cell surface wrinkling, flagellae and fimbrae, scan size 8.5 x 8.5 μm (b) Imaged in water the cell surfaces are smooth and relatively featureless, scan size 10 x 10 μm.

The lack of surface detail seen when bacteria are imaged in aqueous liquids is almost certainly due to the soft and deformable nature of the hydrated extracellular polysaccharides which coat many bacterial cells. Bacterial flagella can be imaged in either the dc (Jaschke *et al* 1994; Beech *et al* 1996) or ac (Gunning *et al* 1996) modes (Fig. 7.11d), and substructure and flagella motors have been observed (Jaschke *et al* 1994). AFM has been used to 'read' photoresists generated by X-ray microscopy of *E. coli*. The images revealed the outer Gram-negative envelope and internal structure attributed to chromosomal DNA (Rajyaguru *et al* 1997). The action of antibiotics on bacteria has been investigated by AFM. Studies include observation of the action of penicillin on *Bacillus subtilis* (Kasas *et al* 1994) and the use of AFM to examine changes in surface structure of *E. coli* due to exposure to the β-lactam antibiotic cefodizime (Braga and Ricci, 1998). The major advantages of AFM in such studies are the ability to image under natural conditions, and the simpler sample preparation, when compared to electron microscopy methods. The effect that sample preparation has upon bacterial cells is an area that has received little attention to date but a very interesting recent study has revealed the unexpected role that the apparently simple process of rinsing can have upon the extracellular polysaccharide coats (EPS) produced by *Salmonella typhimurium* cells (Suo *et al* 2007). Rinsing with HEPES buffer was found to stabilise and promote capsule formation around the cells. This was attributed to the piperazine moiety of the HEPES cross-linking the acidic exopolysaccharide by means of electrostatic attraction. The resultant AFM images, which are reproduced in figure 7.9, showed clear evidence for encapsulated cells. The topography images (fig. 7.9a) show a relatively large, amorphous and apparently unpromising 'blob' on the mica. Had the experiment been performed in contact mode, then that may have been the end of the story. However, the use of Tapping mode in air allowed simultaneous phase (fig. 7.9c) and amplitude images (7.9d) to be obtained of the blob-like features, and what a revelation they produce! Phase imaging generates contrast based upon dissipative losses between sample and tip and is therefore sensitive to the local mechanical properties of the sample (see section 3.3.3). The dramatic result is the image shown in figure 7.9c, which easily reveals the presence of a bacterial cell and associated flagellae enveloped within the 'blob', which can now be firmly assigned as capsular extracellular polysaccharide (EPS). It was noted that image contrast in the phase images was related to the level of residual hydration of the sample, with samples which had been dried under ambient conditions for around 12 hours appearing optimal. Rinsing of the bacteria with other buffers containing potential EPS cross-linking moieties (calcium and a second piperazine containing compound) also yielded encapsulated cells. In contrast, rinsing of the cells with buffers having no EPS cross-linking potential was found to remove most (PBS, Tris, MOPS or glycine) or virtually all (Na_2SO_3 or Na_2SO_4) of the capsular EPS from around the cells.

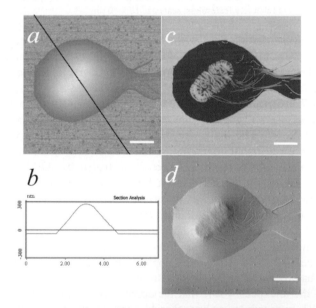

Figure 7.9 *Salmonella typhimurium* rinsed with HEPES buffer (pH 7.4). (a) Topography (b) A line profile taken along the black line in (a) to illustrate height of the capsule in nanometres, (c) Phase, and (d) Amplitude images. Scale bar 1μm. Reproduced with permission from Suo *et al* 2007 and the American Chemical Society.

The study provides direct evidence for the need to carefully consider each and every preparatory step when studying complex living samples such as bacterial cells. Indeed, a nagging problem which is still something of an unknown quantity in AFM measurements of bacterial cells is the effect that immobilisation may have on their behaviour, since it is well known that bacteria adapt to the environment in which they find themselves. Stuck to the abiotic surface of a glass slide or the tip of an AFM cantilever may not be the most biologically representative scenario for certain bacteria (Wright and Armstrong, 2006).

Adhesion behaviour and functional mapping of bacterial cells

The progress of AFM applications to bacterial systems mirrors in many aspects that for mammalian cells, and has come mainly in the fields of affinity mapping (Dupre *et al* 2005) and the study of the adhesion of bacteria (Emerson and Camasano, 2004). Bacterial adhesion is an extremely important phenomenon with huge industrial relevance, being the first stage in the colonisation of abiotic surfaces. Bacterial cell attachment is also generally the first step in invasion of tissue. As such this is an area ripe for study by AFM since it allows the direct measurement of adhesive interactions between bacteria and surfaces right down to the single cell level, and in some cases even at the single molecule level. Traditional methods for determining bacterial adhesion have limited potential for uncovering new molecular aspects of the biological processes, since they require an *ensemble* approach which prevents detailed information being obtained about individual molecular events. Conventional studies of bacterial adhesion are also hideously laborious. However, it would be naïve to assume that AFM measurements of bacterial adhesion are plain sailing. There are challenges in both sample presentation and data analysis. A general approach to the analysis of AFM data of bacterial interaction with surfaces is to model the interactions via non-specific effects such as colloidal interaction and DLVO theory (see section 9.5.7). This approach is not adequate for many systems due to the presence of extracellular polysaccharide coats: the analysis needs to include description of the steric effects of these coats on adhesion (Camesano and Logan, 2000; Emerson and Camesano, 2004). Comparison of DLVO models with interaction force boundary layer theory showed the latter to produce more realistic results (Cail and Hochella, 2005). Retraction data obtained upon withdrawing AFM tips from bacterial surfaces frequently show adhesive events due to the formation of weak physical attachments between cell surface molecules and the tip. Histograms and statistical analysis of the data can be used to demonstrate differences between the adhesion forces between different bacterial strains or bacteria grown or measured under different conditions, but it is difficult to model the forces more explicitly in the absence of knowledge of the number and type of interactions which have occurred (Abu-Lail and Camesano, 2006). Furthermore, it is likely that several simultaneous interactions will occur when probing intact cells. Given this complex scenario *Poisson* statistical analysis has been shown to be best suited to try to separate the biologically specific interactions involving hydrogen bonding from the other sources of non-specific interaction, such as electrostatic and van der Waals forces (Abu-Lail and Camesano, 2006). The work found a typical hydrogen bonding force of 125 pN for the system studied (*E.coli* JM109). Hydrogen bonding is claimed to play a biologically significant role in bacterial interactions with surfaces and antibiotics (Loll and Axelsen, 2000; Walsh *et al* 1996).

In an important contribution to the understanding of AFM quantification of bacterial adhesion to surfaces, Li & Logan used a glass bead colloid-probe AFM tip (see section 9.4.1) to investigate the origins for previously reported differences in the stickiness of three different strains of *Eschericia coli* toward glass (Li & Logan 2004). The study included a revealing exploration of different features seen in the force-distance curves obtained on the bacteria. Gradient analysis of the data obtained upon approach of the colloid tip to the cell surface revealed four distinct regions. These were defined as a non-interaction region, a non-contact phase, a contact phase and a constant compliance region (Fig. 7.10). After the initial non-interaction region, seen at large tip-sample distances, the tip reached a 'non-contact' region characterised by a subtle deflection of the cantilever. This region extended some 28-59 nm from the cell surface; this was suggested to arise due to steric repulsion of the colloid probe by extra-cellular polysaccharide coating the bacterial surfaces (electrostatic repulsion was ruled out as the source of this effect because the Debye length is around 1 nm for the ionic strength of the buffer solution used). As the tip was pushed in further it reached the 'contact phase', which spanned the next 59-113 nm, and was believed to arise from the initial pressure of the colloid tip on the outer membrane of the cell. Finally with further pushing, the 'constant compliance' region was reached; this was attributed to the response of the colloid probe cantilever to the stiff peptidoglycan layers which confer strength and rigidity to gram negative bacteria.

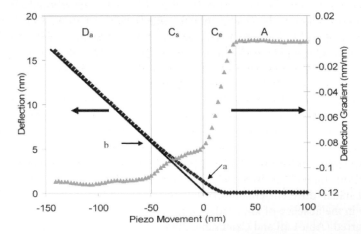

Figure 7.10 AFM approach curve (diamonds) and the corresponding gradient analysis curve (triangles) for *E. coli* strain JM109. The four phases are defined on the basis of the linear regions observed in the gradient analysis data: A, non interaction region; Ce, non contact phase; Cs, contact phase; Da, constant compliance. (The force curve shown is an average of 10 measurements.). Reproduced with permission from Li and Logan 2004 and the American Chemical Society.

These regimes are subtle and difficult to distinguish in the raw deflection-distance data (Fig. 7.10 black diamonds), but they become quite distinct following gradient analysis of the data (fig. 7.10 grey triangles). Armed with a quantitative knowledge of the extent of the different regions in the approaching deflection data, the authors were able to demonstrate that the sticking coefficients of the three strains were correlated with only one aspect of data, namely the width of the non-contact phase (Li and Logan, 2004). Remarkably, the seemingly more obvious candidate parameters determined from the retraction curves, such as the pull-off distances and separation energies, exhibited no correlation with bacterial stickiness. This study provides an excellent demonstration of the fact that bacterial cells are a highly complex multi-component system, and that the interplay between components may not always be straightforward. As such the traditional *ensemble* approaches are highly unlikely to produce meaningful information about the details of the adhesion behaviour, other than defining the degree of adhesion. By the same token it also serves as a useful indication that measuring bacterial adhesion with the AFM is not always as simple as quantifying pull-off forces.

In addition to the measurement of the adhesion of bacteria to surfaces, the technique of molecular recognition spectroscopy (see section 9.3.2) allows the AFM to actually map the spatial locations of potential adhesion molecules on the surfaces of live bacterial cells (Dupre *et al* 2007). In a rather elegant example of the power of this technique, heparin-functionalised AFM tips were used to map the location of heparin-binding haemagglutinin adhesin (HBHA) molecules on the surfaces of a living mycobacterium, *Mycobacterium bovis* (Dupre *et al* 2005). Rather than being randomly distributed, the adhesins were found to be concentrated into nanodomains on the cell surface. The observed clustering of the HBHA molecules on the cell surfaces may be biologically significant, as heparin-sulphate proteoglycan adhesins such as HBHA may induce oligomerisation of receptors upon binding (Bernfield *et* al 1999) and recruitment of these oligomerised receptors within membrane rafts (Tkachenko and Simons, 2002). Thus Dupre and colleagues speculated that the clustering of HBHA observed in *M. bovis* may promote adhesion to target cells. Because of their potential biological significance, the nanodomains of HBHA were termed 'adherosomes' (Dupre *et al* 2005). Control measurements on a HBHA deletion mutant of *M. bovis* exhibited no significant adhesion between the heparin coated tip and the bacterial surfaces. Furthermore, prior checks of the functionality and kinetics of the HBHA-heparin interaction were performed using adhesin-functionalised tips and heparin-coated glass surfaces. These studies revealed that the HBHA-heparin complex is formed via multiple intermolecular bridges.

Bacterial biofilms

In addition to imaging bacteria deposited onto solid supports it is also possible to use AFM to investigate the formation and structure of biofilms formed at interfaces or on solid surfaces. There are a range of microscopic methods which can be applied to study biofilms (Surman *et al* 1996; Beech *et al* 1996) and AFM complements the use of optical and electron microscopy methods. The major advantages of the use of AFM are the ability to cover the magnification range of both optical and electron microscopy, but under natural imaging conditions with minimal sample preparation, and the production of quantifiable 3D images of the surfaces. The major limitation in the use of AFM is the curvature, or roughness of the substrates, or the biofilms formed on them. As the curvature or roughness of the sample increases the AFM images become more distorted and, eventually, when the surface curvature or roughness becomes comparable to the height of the probe, it is no longer possible to obtain images by AFM.

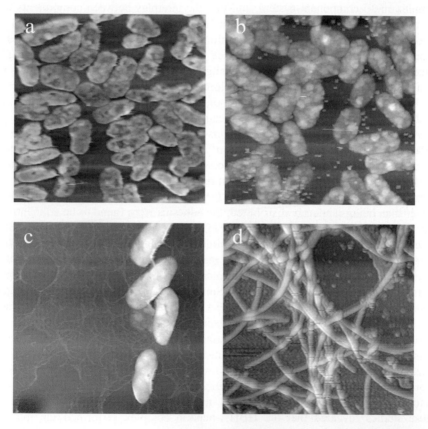

Figure 7.11. AFM images of a bacterial *P. putida* biofilm prepared at an oil-water interface, transferred onto mica and imaged in air. Images (a-c) are contact mode images. Topographic (a) and antibody labelled topographic (b) images, both having scan sizes of 10 x 10 μm. The antibody labels have been enhanced by gold labelled secondary antibodies and silver enhancement. The bright labels are seen attached to the bacterial cells and regions between the cells. The image (c) shows bacterial flagella trapped in an extracellular polysaccharide layer secreted by the bacteria. Scan size 10 x 10 μm. The antibodies are labelling this polysaccharide film. (d) Non-contact ac images of the bacterial flagella. Scan size 2 x 2 μm.

At the present time AFM has been used to study biofilms formed on glass (Surman *et al* 1996), metal surfaces (Bremer *et al* 1992; Steele *et al* 1994; Beech, 1996; Beech *et al* 1996), hydrous Fe(III)-oxides (Maurice *et al* 1996) and at oil-water interfaces (Gunning *et al* 1996). Studies of hydrated bacterial biofilms formed on copper surfaces have revealed that the bacteria are observed bound adjacent to pits with extracellular polymer extending into the pits (Bremer *et al* 1992). This is important because metal ion binding by extracellular polysaccharides has been suggested as a basis for pit corrosion of copper surfaces (Geesey *et al* 1986; Jolley *et al* 1988). More recent studies of the formation of *Pseudomonas* species biofilms on copper also illustrated the importance of extracellular polysaccharide in film formation (Beech *et al* 1996). Related studies of bacterial biofilms on steel surfaces have highlighted the presence of extracellular polysaccharide, and have been used to examine pitting of the surface caused by biofilm formation (Steele *et al* 1994; Beech, 1996; Beech *et al* 1996). AFM has also demonstrated the importance of extracellular polysaccharide in the *Pseudomonas putida* bacterial biofilms formed at oil-water interfaces in emulsions (Gunning *et al* 1996). These biofilms are stiff, can be isolated from emulsions, and then imaged by light and electron microscopy (Parker *et al* 1995). Such isolates are rough and the resolution of the AFM images is poor, revealing only the packing of the bacterial cells (Gunning *et al* 1996). If, however, biofilm formation is modelled by growing such biofilms at flat oil-water interfaces, then high resolution AFM images can be obtained for samples of these flat biofilms pulled from the interface onto mica substrates by LB techniques (Gunning *et al* 1996). In addition to visualising the packing of the bacteria it is also possible to identify remnants of bacterial flagella trapped within the extracellular polysaccharide matrix (Fig. 7.11d). AFM images of flat films labelled with polyclonal antibodies, gold labelled secondary antibodies and silver enhancement suggest that the antibodies bind to the bacterial surface, the extracellular polysaccharide and the flagella. There is probably further scope for the use of AFM to study the early stages of biofilm formation and/or the efficiency of present, or novel cleaning methods for removing such biofilms.

More recent studies include the works of Bolshakova and co-workers and Steinberger and co-workers on the use of AFM to characterise biofilms (Bolshakova *et al* 2004, Steinberger *et al* 2002). Some studies have been made of interaction and attachment to surfaces of biofilms under aqueous conditions (Beech *et al* 2002; Kolari *et al* 2002) and a novel method has been developed to quantify the cohesiveness of biofilms (Ahimou *et al* 2007). Biofilms were grown from activated sludge taken from a wastewater treatment plant which contained a diverse community of bacteria. The biofilms were grown on microporous polyolefin flat sheet membranes that had been treated by cross-linking with a fluorocarbon polyurethane coating. In order to determine the biofilm cohesion energy, selected areas were scraped off the membrane supports, and images of the sample before and after removal were recorded. By subtracting the two images the volume of the material removed could be quantified. This data was combined with the frictional work done by the AFM cantilever to determine the energy expended per unit volume. The resultant data revealed that the cohesive energy of the biofilms increased proportionately with their depth. This result was shown to be reproducible, with four different biofilms exhibiting the same behaviour. Thus the method provides a direct means for obtaining an important parameter, even in the absence of detailed knowledge of the biofilm constituents (Ahimou *et al* 2007).

Fungal Spores

There have been only a limited number of AFM studies on fungal spores: rare examples of such studies include adhesion measurements on *Aspergillus niger* in air (Bowen *et al* 2000a) and in liquid (Bowen *et al* 2000b), and studies on *Aspergillus oryzae* (van der Aa *et al* 2001) and *Phanerochaete chrysosporium* (Dufrêne *et al* 1999). The surface ultra-structure and adhesive properties of *P. chrysosporium* were found to be dramatically different for germinating and dormant spores (Dufrêne *et al* 1999). In *A. niger* comparison of force measurements obtained using colloid probes and fungal or vegetative cell probes revealed that multiple bonds are broken as the cell surface deforms while being retracted from the surface (Bowen *et al* 2000c). The deformation of vegetative fungal cell surfaces was found to be greater than that of fungal spores, reflecting the different physiological role that each plays in the lifecycle of the fungi and its efforts to survive the physical and chemical extremes that they encounter during their existence (Bowen *et al* 2000c).

7.2.2. Yeasts

A few AFM studies have been made on yeasts, all of which are on *Saccharomyces cerviceae* (Henderson, 1994; Pereira *et al* 1996; Kasas and Ikai, 1996; Gad and Ikai, 1996). Yeast cells are large and malleable, making dynamic studies of living cells in natural environments difficult: the cells are easily deformed or displaced, and large rough samples are often impossible to image due to the limited vertical motion of the tip. Air drying yeasts onto glass substrates immobilises them and allows surface morphology to be imaged in air. Characteristic features such as bud scars can be recognised and it is possible to demonstrate differences in surface morphology for different strains (Henderson, 1994; Pereira *et al* 1996). Methods have been developed to allow imaging of live yeasts under natural conditions. The first approach involved immobilising the yeasts in Millipore filters (section 7.1) with pore sizes similar to the dimensions of the yeast cells (Kasa and Ikai, 1996). This is similar to the approach used by Holstein and coworkers (Holstein *et al* 1994) to immobilise and dissect fixed *Hydra vulgaris* polyps. This reduced the roughness of the sample, allowing imaging in a natural environment; namely liquid culture medium. Although morphological features, such as bud scars, could be identified no growth activity was observed, possibly due to the restrictive effects of the pores of the filter. A refinement of the above procedure is to immobilise yeasts in agar gel (section 7.1 & Fig. 7.12) again permitting imaging in liquid culture medium (Gad and Akai, 1996).

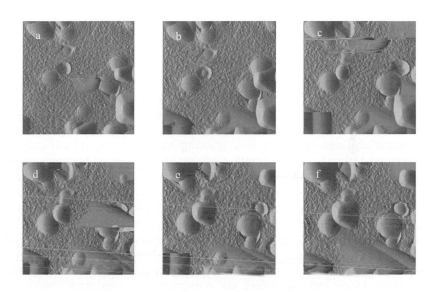

Figure 7.12. AFM deflection mode real-time images of yeast cell growth and budding. The yeast cells have been embedded in and on an agar gel. The scan size is 20 x 20 µm and the time lapse between successive frames is 6 minutes. Data described by Gad and Ikai, 1996 and reproduced with permission of the authors and the Biophysical Society.

High resolution images revealed birth and bud scars and, in addition, it was possible dynamically to follow growth and budding of the yeast cells (Fig. 7.12). As the cells grow on the gel surface the images gradually lose their spherical shape adopting a pyramidal character; an artifact caused by the finite size and shape of the probe tip. Eventually the cell size makes the surface impossible to image. The use of Con A labelled tips has allowed the mapping of the distribution of mannan polysaccharide on the surface of living yeast cells (Gad *et al* 1997). Enzymatic digestion by protease of the surface of living *Saccharomyces cerviceae* cells has been imaged in real-time using AFM. Protease caused a progressive increase of surface roughness. Large depressions surrounded by protruding edges, 50 nm in height, were formed and attributed to the erosion of the mannoprotein outer layer (Ahimou *et al* 2003).

7.3. Blood cells

AFMs were originally designed for high resolution studies on surfaces. Thus the earliest AFMs had sub micron scan ranges which did not permit studies of intact cells. Once the scan range had been enhanced then cellular systems began to be studied, and blood cells were amongst the earliest samples investigated.

7.3.1. Erythrocytes

Red blood cells are readily available, easy to recognise, and were ideal for the early assessment of applications of AFM to study cellular systems. The early studies were concerned with imaging intact cells and in determining the level of resolution at which surface morphology could be imaged. The shape of red cells is characteristic of particular diseases and there is a growing interest in the identification of ultrastructural features associated with diseases, or infection of erythrocytes with parasites. The earliest images of red blood cells were on fixed cells either in air (Gould *et al* 1990) or in buffer (Butt *et al* 1990). Fixation prevented deformation of the cells by the probe and the images revealed the characteristic doughnut shape of the cells. Higher resolution scans revealed surface detail but the origin of these features remained obscure. Spectrin is the most abundant protein of the red blood cell membrane and isolated spectrin preparations have been imaged by AFM (Almqvist *et al* 1994; Zhang *et al* 1996) revealing structures of various oligomeric forms whose shape and dimensions are consistent with structures observed by TEM. Using Tapping mode in air Zhang and coworkers (Zhang *et al* 1995) have mapped the surface structure of fixed red blood cell surfaces down to nanometre resolution. They observed close packed arrays of particles of varying shape and in size from nanometres to several hundred nanometres. Cryo-AFM studies (Zhang *et al* 1996) of glutaraldehyde-fixed erythrocytes revealed the presence of domains with closed boundaries, having lateral dimensions in the range of several hundred nanometres in size, similar to those revealed by cryo-AFM on red blood cell ghosts (Han *et al* 1995). Takeuchi and coworkers (Takeuchi *et al* 1998) describe the optimisation of methods for imaging the skeletal network of red blood cell ghosts by conventional AFM. These authors compared various fixation, drying and freezing methods and recommend rapid freezing in a liquid cryogen, followed by freeze drying, as the best method for preparing ghost specimens on glass cover slips for imaging by AFM (Takeuchi *et al* 1998). Their studies clearly showed that air drying is not suitable for preserving the intact membrane skeletal structure even after fixation in glutaraldehyde. Labelling with gold particles coated with anti-spectrin antibodies, and the effects on images of partial extraction of spectrin molecules, were used to confirm that the networks observed were spectrin networks. Images of the membrane structure on the cytoplasmic and extracellular surfaces have been used to discuss the 3D folded structure of the spectrin network and its possible influence on cell deformation during circulation, or on how abnormalities in the spectrin network can lead to loss of mechanical strength and/or deformability of the red blood cell membrane. Specific labels can be used to locate and identify specific sites on the surface of red blood cells. Neagu and coworkers (Neagu *et al* 1994) used superparamagnetic beads coupled to antibodies to locate the transferrin receptors on the surface of erythrocytes.

It has been shown that AFM can be used to compare the structures of normal and pathological red blood cells (Zachee *et al* 1992; 1994; 1996) and reveal differences in the cytoskeleton networks of normal and *Plasmodium falciparum* infected erythrocyte ghosts (Garcia *et al* 1997). AFM has been used to image erythrocytes from patients with hereditary sperocytosis revealing the abnormal surface pseudopodia (Zachee *et al* 1992). These cells become normal on removal of the spleen. The peripheral blood plasma of uremic patients contain echinocytic erythrocytes. AFM has been used to image type 3 echinocytes showing the smaller more rounded ovoid shape with an even distribution of spicules (needle-like projections) across the cell surface (Fig. 7.13). Unlike the normal discoid shaped cells the type 3 echinocytes show only a small central crater. The echinocytic erythrocytes revert to a normal discoid shape when the uremic plasma is washed away and replaced by normal blood plasma. There is clearly scope for both low and high resolution studies of the interconversion of pathogenic and normal erythrocytes due to the influence of pharmacological or pathogenic factors. The malarial parasite *P. falciparum* alters the morphology of the surfaces of the erythrocytes. AFM studies of normal and infected red blood cell ghosts show that the surface of the infected cells is smoother, contains identifiable parasites and exhibits large (0.2 - 0.7 μm) particulate protrusions. Higher resolution images are claimed to show differences in the density of the spectrin networks for normal and infected cells (Garcia *et al* 1997).

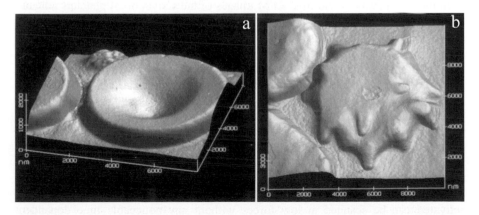

Figure 7.13. Comparative AFM images of (a) a normal erythrocyte and (b) an uremic echinocyte. The uremic echinocyte shows spicules and a smaller central cavity. Data reproduced from Zachee *et al* 1994.

7.3.2. Leukocytes and lymphocytes

Fixed white blood cells can also be imaged in phosphate buffered saline. If diluted blood is applied to a silanised glass cover slip and then washed, the red blood cells are removed leaving the white cells adhered to the derivatised glass surface (Butt *et al* 1990). The white cells have a more irregular spread-eagled shape and a rougher surface. The use of combination optical and AFM microscopes (Putman *et al* 1992) and/or immunological labelling methods (Putman *et al* 1993) are of obvious importance in identifying cell types and in the subsequent use of AFM for high resolution studies under natural conditions. Gold labelled antibodies, with or without silver enhancement, can be observed bound to the surface of air dried lymphocytes and offer the prospect of marking specific surface features of the cells (Putman *et al* 1993; Neagu *et al* 1994). Comparative imaging of cells before and after labelling provides a means of identifying specific surface features and imaging them at high resolution. Specific labels can be used to isolate and bind particular types of cells. Thus antibody coated glass slides have been used to preferentially bind B-lymphocytes for imaging by AFM (Prater *et al* 1990).

7.3.3. Platelets

Human platelets play an important role in blood clotting and wound healing. Normally they exist in the bloodstream in a 'resting state' in which the cells are discoid in shape. Injury to a blood vessel promotes activation of the cells: this involves a drastic change in the cytoskeletal structure leading to a marked change in cell shape. Activation can be induced by contact with wettable surfaces and it has been possible to obtain real time AFM images of the activation of platelets adhered to glass surfaces under physiological conditions (Fritz *et al* 1993; 1994). The resting platelets adhere poorly to the glass substrate and are difficult to image. However, the time dependent changes in the cells following activation can be visualised: initially thin filopodia (spike-like protrusions) are seen to protrude from the interior of the cell which become filled with granules transported from the centre of the cell eventually forming a flat lamelliporadium. The AFM studies provide support for the view that, during activation, the granules fuse directly with the plasma membrane. Details of the cytoskeleton and changes during activation could be resolved in unstained living cells. In unfixed platelets the tip does deform the cells during scanning to different extents depending on the height of the cell, structural differences within different regions within the cell, and on the magnitude of the applied force. Platelets which have bound to the surface but have not been activated can be scanned at low forces without any noticeable time dependent changes in shape. However, scanning at higher forces has been reported to promote activation (Fritz *et al* 1993). Deformation of the cells permits information to be obtained on the elasticity of cell structure (Fritz *et al* 1993; 1994; Radmacher *et al* 1996). Using force mapping techniques (section 7.1.2) it has been possible to generate simultaneous topographic and elasticity maps (Fig. 7.14) of the living activated cells (Radmacher *et al* 1996). The variation of elastic properties across the cell can be correlated with standard features of the cell such as the pseudonucleus, the inner and outer filamentous zones and the cortex, and then interpreted in terms of the ultrastructure of these regions. Derivatised tips can also be used for selected mapping of cell surfaces and one such study has been reported on platelets (Siedlecki and Marchant, 1998). Identification of specific features on platelet surfaces can be achieved through the use of specific labels. 14 nm gold particles conjugated to fibrinogen receptors have been used to map the distribution of these sites on the cell surface. The AFM studies were validated through complementary low voltage, high resolution scanning electron microscopy (LVHRSEM) (Eppell *et al* 1995).

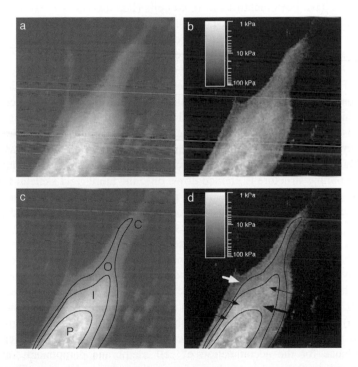

Figure 7.14. Topography (a) and elasticity (b-d) maps of human platelets. Scan size 4.3 x 4.3 μm. The range of grey scales in image 'a' is 2 μm. In image 'b' the elastic modulus is encoded logarithmically so that black corresponds to 100 kPa and white to 1 kPa (as shown in the grey scale bar in the diagram). Different parts of the platelet are identified in 'c': P-pseudonucleus, I-inner web, O-outer web, and C-cortex. In 'd' it can be seen that the pseudonucleus is softest (1.5- 4 kPa), the inner web has a stiffness about 4 kPa, and the outer web ranges about 10-40 kPa. Some areas in the cortex (white arrow) are stiffest (50 kPa). There are some areas that deviate from the general pattern; the area indicated by the thick black arrows is 10 kPa and the area indicated by the thin, black arrows is 4 kPa. Data reproduced from Radmacher *et al* 1996 with permission of the authors and the Biophysical Society.

7.4. Neurons and glial cells

Glial cells provide a system in which it has been possible to probe the structure and dynamics of intercellular components in living cells using the AFM. Biologically glial cells play an important role in the development, maintenance and regeneration of the nervous system. There is an interest in the growth and development of these cells and their interactions with surfaces or neurons. Although AFM is a technique for probing the structures of surfaces, early studies on glial cells demonstrated the possibility of imaging intracellular components. The cells used (XR1) are from a standard cell line derived from *Xenopus laevis* retinal neuroepithelium, or from rat hippocampi, and were plated onto glass coverslips and imaged in growth medium. The nucleus of the cells can easily be identified in the images because of its height, suggesting that the nucleus is less deformable than the rest of the cell. Although mitochondria have been identified in fixed cells (Papura *et al* 1993a) they have not been seen in living cells by AFM, possibly because these structures may easily deform during scanning. At the extremities of the spreading cells the dynamics of internal filamentous structures can be observed (Henderson *et al* 1992; Papura *et al* 1993a; 1993b). Inactivation of actin polymerisation and labelling studies revealed that the AFM was observing actin filaments but not microtubules within the cells (Henderson *et al* 1992). The microtubules are believed to be obscured because isolated microtubules have been imaged by AFM (Vinckier *et al* 1995). These studies raised the question as to how the AFM imaged internal structures. It has been shown that imaging of filaments requires a minimum threshold force (Papura *et al* 1993a). Two suggestions for visualising internal structure were made: firstly, a pliable membrane structure is pressed down onto the more rigid skeletal framework during imaging or, secondly, the probe penetrates the 'fluid' membrane directly imaging the underlying filamentous network. In this context it was found that holes created in the surface by imaging at high force would reseal as shown by subsequent imaging at lower forces (Henderson *et al* 1992). At present the evidence suggests that, at least with the use of standard tips, the membrane deforms around the underlying rigid filaments: this has been deduced from force-distance studies, evidence that AFM imaging does not release intracellular trapped fluorescent dyes or impair physiological functions such as signal transduction mechanisms (Haydon *et al* 1996). Recent investigations have shown that repeated AFM imaging of live cells does not reduce cell viability or increase cell death rates. However, there is clear evidence for the accumulation of cell membrane components on the tip, indicating substantial probe-membrane interactions during imaging. These effects were found to be more pronounced in contact rather than tapping mode imaging (Schaus and Henderson, 1997).

In studies of mixed cultures of neurons and glial cells it was possible to observe neuron-glial cell interactions in both fixed and live cells (Papura *et al* 1993a). In fixed cells high resolution images could be obtained of the growth cones of the neurons, and the actin-based filopodia extending from the body of the growth cones. By controlling the imaging force it was possible to manipulate the biological system; e.g. cut neurites, remove growth cones or selectively displace the weaker bound neurons (Papura *et al* 1993a). For glial cells alone, if the imaging force is gradually increased, then it is possible to observe selective removable of different features of the cells, providing relative indications of the levels of adhesion to the surface (Papura *et al* 1993b). A later paper compares AFM images of glial cells with images acquired using a surface plasmon resonance microscope (SPRM) (Giebel *et al* 1999). Contrast or height in SPRM images are related to cell-substrate contact distance and the microscope provides a new method for probing cell adhesion.

AFM has been used to investigate the localisation of individual calcium channels on the release face of a presynaptic nerve terminal. The channels were located by binding of biotinylated neurotoxin ω-conotoxin, fixation and then enhancement by tagging with avidin-gold. The resolution obtained was a 10-fold improvement over the use of light microscopy, and the detection of a 40 nm interchannel spacing, which was found to be independent of channel density, has been taken to imply anchoring of the channels on the release face of the transmitter (Haydon *et al* 1994). Synaptic vesicles play an important role in the transmission of nerve signals. The vesicles release the neurotransmitter acetylcholine into the synaptic cleft, where it triggers an action potential in postsynaptic cells. AFM has been used to visualise synaptic vesicles in fluid environments and to characterise changes in size and shape (Papura *et al* 1995). Force mapping methods (section 7.1.2) have been used to show that the vesicles possess a substructure consisting a stiffer central core region, the elasticity of which varies in different buffers, becoming harder in the presence of calcium ions (Laney *et al* 1997), surrounded by a more elastic, less rigid outer region. It is suggested that the core, which may correspond to electron dense particles seen by TEM, may in fact be proteoglycan.

7.5. Epithelial cells

Cell monolayers provide useful model systems for studying cell-cell interactions and cell differentiation. They are also attractive systems for AFM investigations of soft cells. The formation of a monolayer effectively reduces sample roughness: the scanning probe no longer has to climb over the entire height of individual cells but rather scans along the surface of the monolayer, without the need to penetrate deep down onto the substrate. A number of investigations have been made on the Martin Darby canine kidney (MDCK) cell line. AFM has been used to examine the morphology of the membrane surface of polarised renal epithelial cells (MDCK cells) (Grimellec *et al* 1994). The most detail was observed on air-dried cells. At the lowest magnification entire cells and their junction zones within the monolayer were visible. With increasing magnification it was possible to visualise microvilli and surface pits, and at the highest magnification, globular structures possibly attributable to proteins or protein aggregates. Such fine structure is also observable in fixed cells imaged under phosphate saline buffer. Treatment with 'Pronase' smoothes the surface suggesting that the protruding particles are proteins. Images of the surfaces of living cells under buffer were fuzzy or blurred (Grimellec *et al*1994; Hoh and Schoenenberger, 1994) although in some cases the nucleus of cells could be identified (Hoh and Schoenenberger, 1994). Treatment with enzymes, such as neuraminidase which partially degrade the glycocalix, improved imaging allowing visualisation of globular protrusions which were then susceptible to treatment with 'Pronase', suggesting that they are surface glycoproteins (Grimellec *et al* 1994). Surface microvilli observed in SEM images are not seen on living cells by AFM, although progressive fixation does reveal surface corrugation, possibly due to mechanical stabilisation of microvilli, allowing them to be seen by AFM (Hoh and Schoenenberger, 1994). Fixation also allowed visualisation of microvilli on the surface of rat carcinoma cells RCMD cells (Pietrasanta *et al* 1994) and CC531 cells (Braet *et al* 1998a) in confluent monolayers. An alternative model for studying cell polarisation and differentiation is the human adenocarcinoma cell line HT29. As with MDCK monolayers the AFM images of HT29 monolyers are fuzzy and blurred. However, if the cells are air-dried and then imaged under butanol, then the microvilli are clearly visible (Fig. 7.15) and globular features on the surface of the microvilli, attributed to glycoproteins can be seen (Kirby *et al* 1998). In the normal hydrated state where the microvilli extend out from the cell the probe can permeate between the microvilli deforming the structure on scanning. Air-drying flattens the microvilli down onto the surface and butanol appears to maintain this state allowing imaging (Kirby *et al* 1998). If the probe deforms the microvilli during scanning then, at low applied forces, it may be possible to reveal these structures. Such studies have been made by Lesniewska and coworkers (Lesniewska *et al* 1998) on living MDCK cell monolayers. At normal imaging forces the cell surfaces are fuzzy whereas for forces < 300 pN a close packed array of the tips of the extended microvilli could be seen. On living MDCK cells it was possible to follow dynamic events such as the formation of bulges and spikes (Hoh and Schoenenberger, 1994; Schoenenberger and Hoh, 1994), and these events appear to be intrinsic properties of the cells and not effects triggered, or stimulated by the action of the probe. Measurement of force-distance curves allowed the elastic properties of the cells to be monitored and it was possible to follow the mechanical stiffening of the cells during fixation (Hoh and Schoenenberger, 1994). A more complete force mapping of living MDCK cells provided unexpected contrast variations within the elastic profiles revealing structural features of the cells. The inter-cell contact boundaries were found to be stiffer than the centres of the cells. Intercellular features highlighted included a boundary between the nucleus and the cytoplasm (A-Hassan *et al* 1998). An intriguing use of AFM to measure changes in the mechanical properties of MDCK cells is an attempt to develop a biosensor based on micromechanical interrogation of living cells (Antonik *et al* 1997). The biosensor consists of a tip-less cantilever coated on one side with a cultured layer of MDCK cells. Exposure of the biosensor to the lytic bee venom melittin and the respiratory inhibitor sodium azide caused a deflection of the cantilever, detected in an AFM, and attributed to changes in the rigidity of the cells.

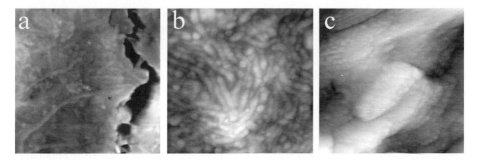

Figure 7.15. AFM contact mode images of an HT29 confluent monolayer imaged under butanol. (a) Boundary between two cells showing microvilli and possibly filopdia, scan size 30 x 30 µm. (b) Magnified image of a cell surface clearly showing microvilli, scan size 4.8 x 4.8 µm. (c) High resolution image of the surface of a microvillus showing globular 'glycoproteins', scan size 924 x 924 nm. Data based upon Kirby *et al* 1998.

It is suggested that such sensors could be developed for studying cellular responses to chemicals in areas such as the screening of pharmaceuticals, or toxicity testing (Antonik *et al* 1997). More recently Zhang and co-workers used AFM cell-cell adhesion measurements to demonstrate that leukaemia cells preferentially bind to the border of adjacent epithelial cells (Zhang *et al* 2004). The toxicity of inhaled nanoparticles entering the body through the lung is thought to be initially defined by the electrostatic and adhesive interaction of the nanoparticles with the lung wall. Thus, in an effort to understand this phenomena, AFM adhesion measurements have been used to probe the interaction between an AFM tip (which served as a model nanoparticle) and lung epithelial cells (Leonenko *et al* 2007). The interaction was monitored over time and found to increase strongly for the first 100 s of contact and thereafter level out. Interestingly the work of adhesion and its progression over time were not dependant on the loading force of the tip on the cells. The penetration behaviour of the AFM tip into the cells indicated that the process did not occur simply by passive penetration of the viscous medium of the cell. Rather it appeared that the tip was either actively taken-up by the cell, or that the cell rearranged its plasma membrane cytoskeletal elements to accommodate the tip. The authors concluded that detailed investigations of the initial thermodynamic aspects and time course of the uptake of nanoparticles by lung epithelia should be possible using AFM (Leonenko *et al* 2007).

7.6. Non-confluent renal cells

AFM images of the inner cytoplasmic leaflet of the plasma membrane of MDCK cells in air or aqueous media reveal a distribution of globular structures believed to be surface proteins. In air this layer is partially decorated with cytoskeleton features such as actin fibres but these decorative features disappear when imaged in aqueous media (Le Grimellic *et al* 1995). Treatment with enzymes such as neuraminidase is unnecessary in order to obtain high resolution images of the 'protein' structure under aqueous conditions, indicating that glycosylation is confined to the outer layer of the membrane.

Certain derivatives of MDCK cells do not form homogeneous polarised monolayers. R5 cells spread out on glass substrates and, in their highly flattened extremities, it is possible to image features such as cytoskeletol fibres and vesicles, which cannot be imaged by SEM (Hoh and Schoenenberger, 1994; Schoenenberger and Hoh, 1994). Time-resolved events can be observed by AFM: these include dynamics of stress fibres, wave-like rearrangements of the cytoplasm and even motion of vesicles along fibres (Schoenenberger and Hoh, 1994). Dynamic processes have also been observed by AFM in migrating MDCK-F cells (Oberleithner et al 1993; 1994; 1995; 1996; 1997a). The AFM has been used to excise membrane patches from the surface of MDCK cells (Oberleithner et al 1996) and to measure changes in the height of cloned and isolated ROMK 1 potassium channels, arising from addition of ATP (Henderson et al 1996a,b; Oberleithner et al 1997b). The latter studies used the 'molecular sandwich' technique: the protein is bound to the tip and then touched down onto, and interacts with the mica substrate. Changes in shape of the sandwiched protein can be detected by variations in the deflection of the cantilever.

Studies on monkey kidney cells by both contact and tapping mode AFM suggest that it may be possible to simultaneously acquire images of the surface structure of the cell membrane and the underlying cytoskeleton network. This has been achieved either by comparing topographic images (Putman et al 1994) or deflection (error signal mode) images (Le Grimellec et al 1997) in contact and tapping mode. Both static (Le Grimellec et al 1997) and dynamic (Putman et al 1994) studies have been made of the morphology of the cells. Phase imaging provides an alternative method for probing the viscoelasticity of cell surfaces: phase images of cultured CV-1 kidney cells have recently been reported by Lesniewska and coworkers (Lesniewska et al 1998).

Finally, time dependent AFM studies on individual monkey kidney cells have been used to follow the release of virus particles from cells infected with pox virus (Haberle et al 1992; Horber et al 1992; Ohnesorge et al 1997). Individual cells were held on a pipette and scanned beneath an AFM tip in physiological medium. By imaging the surface of the cell as a function of time it was possible to observe what is believed to be the exocytosis of a virus through the cell wall. The images can be combined to produce a video of this biological process. Measurements of the elastic properties of the cell surface have been used in an attempt to identify changes in the structure of the cell surfaces during infection or expulsion of virus particles (also see section 6.1.4 for more on viral invasion of cells).

7.7. Endothelial cells

The endothelium acts as a mechanical transducer between blood and the vascular walls of blood vessels, and the effects of flow can induce a range of cellular responses. The earliest AFM studies on living endothelium cells investigated the effects of shear stress on cellular structure. Comparative studies were made on unsheared and sheared confluent monolayers (Barbee *et al* 1994) of cultured bovine aortic endothelial cells in phosphate buffered saline. Whereas the unsheared cells were polygonal with well defined cell boundaries, the sheared cells were elongated and aligned in the flow direction, their boundaries were less distinct, and new aligned fibrous surface ridges were seen. Comparative fluorescent studies on fixed cells suggested that the changes observed were due to shear-induced rearrangement of the cytoskeleton structure, with the surface ridges corresponding to aligned bundles of actin fibres (Barbee *et al* 1994). AFM has also been used to probe changes in surface structure with time after confluence (Barbee 1995) in order to assess structural factors which may influence the mechanical response of the cells to flow. The response of HEK-293 cells to the stimulant angiotensin II (AngII) has been monitored by AFM (Auger-Meisser *et al* 2005). AngII is a hormone involved in vascular regulation which activates AT_1 receptors on the surface of endothelial and smooth muscle cells, leading to reorganisation of actin. The AFM tip was placed onto the surface of the HEK-293 cells and height fluctuations monitored as AngII was added to the buffer. The data revealed an initial relatively large upward displacement of the cell membrane (~500 nm) followed by a series of nano-scaled height fluctuations which would have been impossible to record using traditional optical methods. These were considered to reflect subtle structural changes occurring within the cell following AngII stimulation (Auger-Meisser *et al* 2005).

Sinusoidal liver endothelial cells (LEC) possess fenestrae which act as sieves controlling the exchange of fluids, solutes and particles between blood and the microvilli coats of parenchyma cells. AFM has provided information on the structure and dynamic response of fenestrae to different stimulants (Braet *et al* 1995; 1996a,b; 1997ba,b; 1998b). Comparative SEM and AFM images on dried, gold coated cells revealed similar features: fenestrae arranged in sieve plates surrounded by an elevated border corresponding to an underlying tubular structure. Height measurements suggest that the sieve plates are about 200 nm below the surface of the nearby cytoplasm (Braet *et al* 1996a; 1996b). AFM images could also be obtained on dried, uncoated cells and wet, glutaraldehyde-fixed cells, with the latter yielding the best images (Braet *et al* 1996a; 1997b). Comparative studies of the elasticity of fixed and unfixed cells confirmed that fixation improves image quality by stiffening the cells and inhibiting distortion on scanning (Braet *et al* 1998b). Use of non-contact imaging modes improved the quality of the images of wet cells by further eliminating deformation artifacts, such as distortion of the fenestrae in the scan direction (Braet *et al* 1997b). Imaging wet cells showed that dehydration, critical point drying or drying by evaporation of hexamethyldisilazane resulted in a considerable shrinkage of fenestrae (Braet *et al* 1996a; 1996b; 1997a).

This is a clear demonstration of the advantages of the AFM in allowing imaging without drying the samples. It was also confirmed that treatment with ethanol and serotonin led respectively to enlargement and shrinkage of the fenestrae diameter, and that treatment with cytochalasin increased the number of fenestrae (Braet *et al* 1996a) (Fig. 7.16).

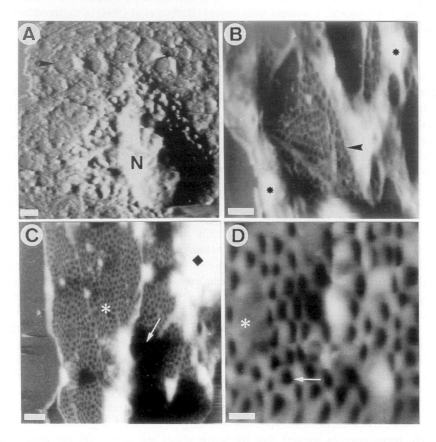

Figure 7.16. AFM images of wet fixed endothelial cells. (A) Low magnification (scale bar 2.5 μm) image showing the central nucleus (N) and sieve plates (arrowhead). (B) Higher magnification (scale bar 1 μm) image showing a sieve plate (arrowhead) in surrounding cytoplasm. The asterisks show white bumps. (C) Liver endothelial cells treated with 10 μg mL^{-1} of cytochalasin B for two hours, inducing a highly fenestrated cytoplasm (asterisk). White bumps (black diamond) and shadowing of structures (arrow) are also present; scale bar 1 μm. (D) Higher magnification (scale bar 500 nm) of the fenestrated cytoplasm after cytochalasin B treatment, which show fenestrae (arrow). Notice also the presence of typical small unfenestrated dots (asterisk) that occur after treatment with the microfilament-inhibiting drug. Data reproduced from Braet *et al* 1996a with permission. Copyright of the Royal Microscopical Society.

7.8. Cardiocytes

The ability of the AFM to image structure in living cells at molecular resolution, plus the ability to map mechanical properties of cells at high spatial and temporal resolution, provides a means of studying mechanisms of beating in contractile cells. AFM images of living rat atrial cardiomyocytes revealed the centrally located nucleus plus the submembraneous fibrous cytoskeleton structure concentrated at the extremities of the cells (Shroff *et al* 1995). Fixation enhanced the images of the cytoskeleton. Light microscopy of cells stained for F-actin showed that the fibrous structures seen by AFM are actin bundles. Force-distance measurements on quiescent cells showed that the cells are softer above the nuclear region, and become stiffer towards the periphery, where the fibrous structures are visible. Increases in cell stiffness were observed on the addition of calcium, or on fixation with formalin. By following changes in cantilever deflection, due to expansion or contraction of the cells, it was possible to monitor the localised activity of beating cells at high spatial (1-3 nm) and temporal (60-100 μs) resolution, and to probe changes in stiffness during single contractions (Shroff *et al* 1995).

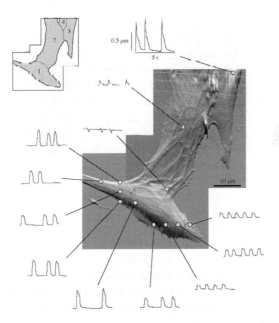

Figure 7.17. Use of the AFM to pulse map a group of active cardiomyocytes. Although the cells were beating it was still possible to image them. In the figure the deflection images of two scans were superimposed. The inset shows a sketch of the margins of the cells. Several time sequences were recorded at different locations on the cells. The presented sequences are are scaled identically; their locations on top of the cell are marked with white spots. No pulses were found that showed negative amplitudes. Only the series between cells '1' & '2' showed a biphasic pulse shape. Data reproduced from Domke *et al* 1999 with permission of the authors and Springer-Verlag GmbH & Co., KG who own the copyright.

Force mapping methods (section 7.1.2) have been used to map the elastic properties of living chicken cardiocytes (Hofmann et al 1997). Using a simple geometrical model (Hertz, 1882) for the tip indenting the cell the force-distance curves have been interpreted to map the Young's modulus across the cell. The fibres appear as stiffer structures embedded in the softer cell. The theoretical model matches the force-distance curves well in the softer regions, but fails in the vicinity of the fibres, probably because the model assumes a homogeneous structure for the cell surface. Imaging at increasing force enhanced the submembraneous cytoskeleton. Topographic and elastic maps have been used to follow the breakdown of the actin network due to the addition of the drug cytochalasin B (Hofmann et al 1997). Most recently AFM has been used to analyse the mechanical pulsing of individual active single chicken cardiocytes and the synchronised beating of cells in a confluent layer (Domke et al 1999). With the AFM it was possible to map the localised beating across complete cells (Fig. 7.17).

7.9. Other mammalian cells

There are a number of other studies on cells which do not fit easily into the above classifications. In some cases these studies represent new methodology or new types of investigation which are worth recording. A few such studies are collected in this section. Novel methods for improving the imaging and identification of subcellular structures are described by Pietrasanta and coworkers (Pietrasanka et al 1994). In fixed and dried rat mammary carcinoma cells (RMCD cells) it was possible to recognise subcellular structures such as the cell nucleus and nucleoli, plus microfilaments of the cytoskeleton, and also to visualise surface features such as microvilli and microspikes. Treatment of the fixed cells with Triton X-100 to remove lipids and soluble protein allowed visualisation of mitochondria, the identification of which was confirmed by rhodamine 123 staining and fluorescence microscopy. The detergent resistant sample surface was then dry-cleaved with cellotape to remove the dorsal part of the cell, and presumably the membrane associated microfilament layer, in order to reveal the stress fibres and an underlying network of finer filamentous structures, some associated with granular structures. Fluorescence labelling revealed the presence of both actin and microtubules although individual components could only be tentatively assigned, and not specifically recognised from the AFM data alone. This type of dissection approach, coupled with labelling methods may provide a route to assigning features observed in intact living cells.

The visualisation of submembraneous cytoskeleton structure makes it possible to image time dependent structural changes and/or the associated changes in mechanical properties of the cell. A number of such studies have been reported including work on skin fibroblasts (Braet et al 1998a), cardiocytes (Hofmann et al 1997), and MDCK epithelial cells (A-Hassan et al 1998). Time lapse images have been obtained showing the disruption of the cytoskeleton network due to the action of drugs such as latrunculin (Fig. 7.18) on skin fibroblasts (Braet *et al* 1998a), colchicine on RBL-2H3 rat leukemic mast cells (Chang *et al* 1993), and cytochalasin B on cardiocytes (Hofmann *et al* 1997). Of perhaps more interest is the observation of events resulting from specific stimulation of cells. AFM images of living unstimulated RBL-2H3 mast cells are featureless. However, if IgE antibodies are bound to the cells and they are then stimulated by the addition of a multivalent antigen, then the cells show a well defined cytoskeleton structure (Chang *et al* 1993). Time dependent changes in quiescent and activated cells can be imaged by AFM (Braunstein and Spudich, 1994). In quiescent cells granular motion along cytoskeleton filaments and cytoplasmic surface waves were imaged.

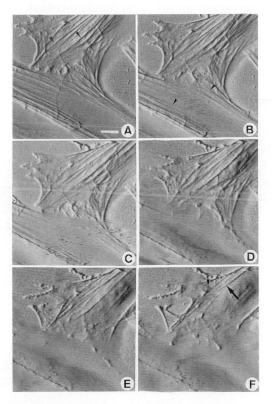

Figure 7.18. Time lapse series of AFM images of latrunculin A induced disruption of microfilaments in skin fibroblasts. (A) Untreated fibroblasts showing a parallel fibre orientation indicated by the black arrowheads. Latrunculin was added and the images B-F were recorded sequentially afterwards. Image acquistion time is about 10 minutes. (B) First signs of microfilament disruption indicated by black arrowheads. (C-E) General loss of latrunculin A sensitive fibres can be seen; peripheral filaments disappear first, followed later by the central filaments. (F) After 45-50 minutes treatment, only a few latrunculin A insensitive fibres remain (indicated by black arrow and arrowhead). Scan bar 10 μm. Data reproduced from Braet *et al* 1998a with permission. Copyright of the Royal Microscopical Society.

Granular motion has also been observed in R5 cells (Schoenenberger and Hoh, 1994), and cytoplasmic surface waves have been seen in R5 (Schoenenberger and Hoh, 1994) and lung cells (Kasas *et al* 1993). Following activation it was possible to observe an increase in the number of cell associated granules, granular motion and changes in cytoskeleton at the cell borders (Kasas *et al* 1993). The detergent extraction method (Pietrasanka *et al* 1994) was used on fixed, dried cells to reveal details of the changes in submembraneous structures between unactivated and activated cells (Kasas *et al* 1993). Such images demonstrated the spreading of the filamentous networks after activation, the fact that granules and/or vesicles sit on top of the cytoskeleton network, and also provided very detailed images of nuclear pore complexes on the nuclear surface of the activated cells (Kasas *et al* 1993). The role played by the softness of the outer cell surface in allowing imaging of intracellular structures is dramatically illustrated in images of ingested latex beads (Beckman *et al* 1994; Braet *et al* 1998a) and zymosan particles (Beckman *et al* 1994) in macrophages after phagocytosis.

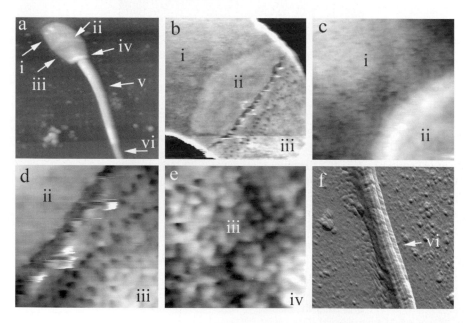

Figure 7.19. AFM images of live ram sperm. The sperm have been treated with sodium azide to reduce their motility. (a) Tapping mode image in air of an intact sperm showing different structural features: (i) acrosome, (ii) equatorial region, (iii) post acrosome, (iv) neck region, (v) mid-piece, and (vi) principal piece. Scan size 20 x 20 μm. (b) Topographical contact mode image of a sperm head in buffer using an elongated tip of height 7 μm. Structural detail has been enhanced by 'unsharp' masking (Adobe Photoshop). Scan size 5 x 5 μm. Contact mode imaging in buffer of regions (c) 'i' & 'ii', scan size 3 x 3 μm, (d) 'ii' & 'iii', scan size 2 x 2 μm, and (e) 'iii' & 'iv', scan size 1.5 μm. (f) Error signal mode image in air of the principal piece of the tail, scan size 20 x 20 μm. Data obtained in collaboration with P James, R Jones (IAH, Babraham UK) and C Wolf (IFR, Norwich UK).

Once again use of the detergent extraction method enhanced imaging of intracellular components allowing, for example, investigation of particle-filament interactions and the elimination of the role of filaments in phagocytosis (Beckman *et al* 1994). AFM has been used to image sperm and to investigate the structure and hydration of sperm chromatin. Combined AFM and X-ray microscopy of rat sperm suggested that a significant fraction of the volume of air-dried nucleus must be hydrated (Da Silva *et al* 1992). AFM has been used to compare the volume of bull and mouse sperm heads and amembraneous nuclei in both a fully hydrated and dehydrated state, and to follow the effect on single cells of the addition of increasing concentrations of propanol (Allen *et al* 1996). Previous estimates of volume, and their implications for the packing of sperm DNA, were based on EM data. The AFM studies showed that the EM preparation resulted in substantial dehydration: they suggested that the chromatin is extensively hydrated with water which comprises at least half the volume occupied by the chromatin in the nuclei. AFM has been used to investigate the structural organisation of chromatin and some of this work has already been discussed in section 4.2.3. Studies of marsupial spermatozoa by AFM revealed differences in the organisation of chromatin packaged by histones or protamines (Soon *et al* 1997): the nucleoprotamine particles appearing as tighter bundles than the nucleohistones particles. Fig. 7.19 shows AFM images of ram sperm showing how different imaging methods and analysis can be used to reveal structural detail on living sperm.

7.10. Plant cells

Plant cells were one of the first living systems to be imaged using AFM (Butt *et al* 1990). Small sections from plant leaves were cut out and stuck to a stainless steel disc and then imaged by AFM under water. Images of the underside of leaves of *Lagerstroemia subcostata*, a small Indian tree, showed cellular features. High resolution images failed to reveal structural features below 200 nm in size and this was attributed to the presence of a thick cuticle. The leaves of the water lily *Nymphaea odorata* are believed to possess thinner cuticles and AFM images revealed more detail. In addition to features resembling cells it was possible to identify fibrous structures. At large applied force the sample surface was scraped and damaged but, provided the imaging force was kept below 2 nN, features down to 12 nm could be resolved (Butt *et al* 1990).

   More recently AFM studies have been reported on isolated ivy leaf cuticles (Canet *et al* 1996). The cuticles were extracted enzymatically from plant leaves. Transverse sections were cut with a microtome after embedding in Epon. Images of the inner and outer faces of the isolated cuticles were obtained after binding them down with double sided cellotape. In AFM images of sections the stacked lamella could be seen in the outer lamella zone. The inner reticulate region was largely amorphous with evidence for fibrous inclusions at regions which would have been close to the epidermal cell wall. The outer surface of the cuticle appeared featureless and difficult to image because the probe tip tended to stick to the surface. Low resolution images of the internal surface of the cuticles showed imprints of the epidermal cells surrounded by high walls. Higher resolution images of the imprints revealed a helicoidal stacking of fibrous structures. These fibrous structures were removed by acid hydrolysis suggesting that they are polysaccharide fibres emanating out from the epidermal cell wall into the wax cuticle. The fibrous structures seen on the inner face by AFM are consistent with the fibres seen in transverse sections near the cell wall by AFM and TEM (Canet *et al* 1996).

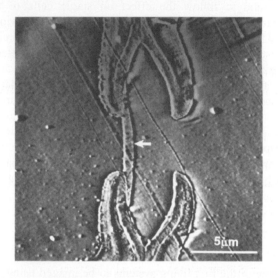

Figure 7.20. AFM deflection mode image of the surface of a transverse section of black spruce (*Picea mariana*), showing a bordered pit joining a pair of tracheids. The white arrow indicates the torus. It is possible to distinguish the middle lamella, primary cell walls and secondary cell walls. The diagonal lines arise from imperfections in the diamond knife and indicate the knife direction. Data reproduced from Hanley and Gray, 1994 with permission of the authors.

The surfaces of mechanically pulped fibres, and transverse and radial sections of black spruce (*Picea marianna)* wood, have been examined by AFM (Hanley and Gray, 1994). Thin sections were microtomed from Epon embedded pieces of wood. In sections AFM revealed details of the cell wall and of characteristic features such as bordered pits (Fig. 7.20). Within images of sections it is possible to resolve the middle lamella and different regions of the secondary cell wall (Fig. 7.21). These 'textural' differences revealed by AFM are thought to arise on cutting the sections. The orientation of microfibrils within the cell wall, relative to the cut direction, will influence the roughness of the cell wall, and possibly the extent to which it is deformed during scanning, thus influencing the contrast as imaged by AFM. Woody tissues are lignified and phase images of wood pulp fibres reveal bright patches, attributed to residual lignin, which are invisible in the normal topographical images

(Hansma *et al* 1997). The lignin is considered to be more hydrophobic than the cellulose thus giving rise to the difference in contrast.

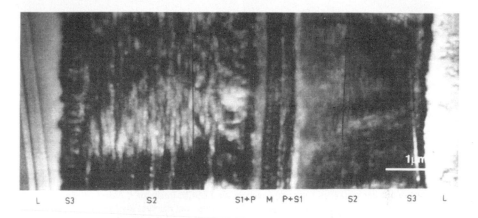

Figure 7.21. A composite of three AFM images of adjacent fibre walls of black spruce. The sections have been taken radially through the tangential wall. The middle lamella (M), primary cell wall (P), and the secondary cell wall (S1, S2 & S3) can be easily distinguished. The resin filled lumen (L) is also shown in the image. The knife marks in the lumen on the left-hand side of the image indicate the knife direction. The difference in contrast between adjacent cell walls arises due to differences in surface roughness on cutting and is attributed to different orientations of microfibrils in the cell wall (e.g. the transition regions S1- S3). Data reproduced from Hanley and Gray, 1994 with permission of the authors.

There are a few studies of pollen grains (Rowley *et al* 1995; van der Wel *et al* 1996; Demanet and Sankar, 1996). In general the images are comparable to SEM data although the sample preparation is easier, and AFM can reveal higher resolution data on exine surface substructure. Rowley and coworkers studied sections cut from grains embedded in resin (Rowley *et al* 1995). Pollen grains from *Kalanchoe blossfeldiana* and *Zea mays* were held in place on the substrate with double-sided cellotape and then imaged in the contact mode. The surface detail seen in error signal mode images (Fig. 7.22) is similar to that seen with field emission scanning electron microscopy (FESEM) (van der Wel *et al* 1996) but higher resolution images could be obtained by AFM: for example, both AFM and FESEM images of the exine surface of *Z mays* pollen grains reveal globular projections interspersed with small holes, but AFM shows substructure of the protrusions not seen by FSEM. Demanet and Sakar used the simplest sample preparation: the pollen grains were deposited directly onto stainless steel discs and then imaged in the non-contact mode (Demanet and Sakar, 1996). The detailed imaging of isolated plant cell walls and the extracted cellulose microfibres is discussed in detail in section 4.4.4. There is a least one reported study on plant cell protoplasts (van der Wel *et al* 1996).

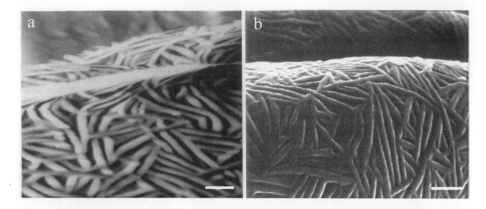

Figure 7.22. Surface structure of fresh *Kalanchoe blossfeldiana* pollen grain. (a) AFM error signal mode image in air, bar marker is 1 μm. (b) Comparative FESEM image of *K blossfeldiana* pollen after coating with Au/Pt, bar marker is 1 μm. Data reproduced from van der Wel *et al* 1996 with permission.

Images were obtained for fixed samples in air after drying onto glass slides, in water for cells attached to poly-L-lysine coated slides, or for unfixed protoplasts in air, buffer or incubation medium. The best images were obtained for fixed, air-dried protoplasts. No attempt was made to identify features revealed in the images (van der Wel *et al* 1996).

Methodology to visualise the ultrastructure of starch granules *in-situ* (see section 4.4.4) whilst still in the plant tissue has been developed (Parker *et al* 2004). The method involves cutting into a seed using a microtome in order to generate a surface which is flat enough to allow AFM imaging. The seed is held in the chuck of the microtome and flatted off at the both ends to facilitate gluing of the cut seed onto a small mica sheet in a manner such that the face being imaged by the AFM is square with respect to the scanner. AFM imaging of the upper surface is then a relatively straightforward procedure and can be carried out in contact or Tapping mode in air. The resulting AFM images reveal all of the major components of the seed, namely protein bodies, cell walls and starch granules, without the need for staining of the tissue (Fig. 7.23). The only treatment required to visualise the detailed ultrastructure of the starch granules is controlling the hydration of the cut face of the granule. Hydration softens and swells the amorphous regions within the granules, generating contrast between the amorphous and crystalline regions. The advantage of such *in-situ* imaging is that it delivers the best images of the native ultrastructure of the starch and allows variations in granule structure within seeds to be described, reflecting changes in structure during development and growth. The level of detail obtained on starch granules using this method suggests that it might find application as a screening tool for identifying natural mutants with interesting starch structures or for the direct determination of the effects of genetic manipulation/variation or selective breeding on functionally important aspects of plant tissue (Parker *et al* 2004).

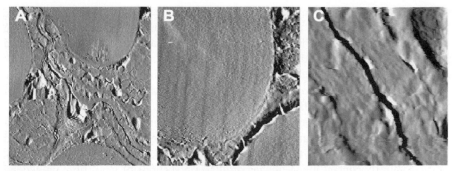

Figure 7.23. AFM image of the cut surface of a Pea seed (A) shows an overall view revealing starch granules, cell walls and protein bodies in the tissue. Close-up views of (B) starch granules and (C) cell wall boundary. Scan sizes: 18 x 18 μm, 10 x 10 μm, 3 x 3 μm respectively. Image taken by Mike Ridout.

7.11. Tissue

Intact tissue is too large and rough to be directly imaged by AFM. In addition the AFM would only probe the outermost structure of the sample. Standard preparative methods in histology and cytology convert soft biological material into rigid specimens and replicas. Furthermore there are standard techniques for fracturing or sectioning specimens to reveal internal structures. Are there advantages in combining such methods with the use of AFM imaging? AFM should be easier to carry out on such specimens because they are more rigid, and are thus less deformable than untreated biological material. Because the AFM 'feels' the surface structure then, if fracture or sectioning results in excessively rough surfaces, it will not be possible to obtain images at all. Provided the surfaces are not too rough then the AFM offers a potential resolution better than that of the light microscope and the SEM, and comparable with that of the TEM. At the highest magnifications it is possible that the AFM may, in some circumstances, be limited with respect to TEM by surface roughness. However, if sample roughness is small and results from specimen related properties then this may provide unique image contrast in the AFM.

7.11.1. Embedded sections

Early AFM studies of thin sections of embedded biological material revealed the sensitivity of AFM to the roughness of the surface of the cut section (Amako *et al* 1993). In TEM images of thin sections of Epon embedded *Vibrio cholerae*, cut either with a diamond or glass knife, the bacterial cells are clearly visible. Whereas bacterial cells are identifiable in AFM images of diamond cut sections, the additional roughness of the glass cut sections obscures the bacterial cells in the AFM image. This clearly illustrates the different contrast mechanisms, and the additional sensitivity of AFM to sample preparation. Later studies (Yamashina and Shigeno, 1995) confirmed the advantage of the use of a diamond knife but did not support suggestions (Yamamoto and Tashiro, 1994) that embedding in LR white resin was preferable to Epon resin. Tapping mode images of ultrathin sections of tissue samples collected from rats revealed details of cytoplasmic organelles (mitochondria or secretory granules), nuclear components (nucleolus and chromatin), rough endoplasmic recticulum and associated ribosomes, microvilli and cilia, and basal bodies in the tracheal epithelial cell (Yamashino and Shigeno, 1995). Similarly contact mode images of ultrathin sections of wool fibre embedded in LR white resin revealed details of the cellular composition; ortho- and para-cortical cells, cell membrane complexes, macrofibrils and nuclear remnants (Titcombe *et al* 1997). Gross surface roughness will limit use of AFM to the study of thin sections. However, detailed differences in surface roughness related to sample structure may provide novel contrast in the images. This type of effect has already been mentioned in section 7.10. When sections were cut from embedded pieces of wood the orientation of fibres with respect to the cut direction caused changes in surface texture, which manifested themselves in the AFM images (Fig. 7.21) as changes in contrast across sections of cell walls (Hanley and Gray, 1994).

7.11.2. Embedment-free sections

An interesting approach has been the use of AFM to study embedment-free sections of cells and tissue (Ushiki *et al* 1994; 1996). This uses methodology originally introduced by Wolosewick (Wolosewick, 1980; Kondo, 1984). Basically fixed samples are embedded in polyethylene glycol, solidified with liquid nitrogen, sectioned with a microtome, deposited on poly-L-lysine coated glass slides, dehydrated through a graded ethanol series, and critical point dried. Tapping mode has been used to obtain an impressive range of images of kidney, liver, pancreas and small intestine tissue from mice (Ushiki *et al* 1994). Sections of liver tissue revealed an erythrocyte in a sinusoid, chromatin fibres in the nucleus of a hepatocyte, glycogen granules in hepatocyte cytoplasm, mitochondria in liver hepatocytes and the collecting duct cell of the kidney. Microvilli were clearly resolved in samples of small intestine and higher resolution images suggested globular structures, possibly glycoproteins, on the surface of the microvilli (Fig. 7.24). The embedment free method is regarded as an acceptable preservation technique for electron microscopy and may provide a routine methodology for AFM.

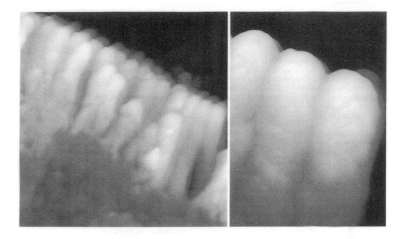

Figure 7.24. AFM images of embedment-free sections of the upper portion of the absorptive cell of the small intestine. (a) Images of microvilli, magnification about 45,000. (b) Higher magnification image showing granular structure which coats the surface of the microvilli, magnification about 192,000. Data reproduced from Ushiki *et al* 1994 with permission.

7.11.3. Hydrated sections

One of the potential benefits of AFM is imaging samples in their native state. An example of the use of freshly cut, wet sections is a study of bovine cornea and sclera (Fullwood *et al* 1994). Sections, 100-200 microns thick, were cut by hand and then glued onto flat steel discs. The samples were imaged in air and, under these conditions, the level of hydration was considered to fall from about 70% to between 20-40%. The major experimental difficulty was locating sufficiently flat regions which remained stable during imaging. However, once such regions were found it proved possible on occasions to resolve features down to 2-3 nm in size. In sections of both the cornea and the sclera the collagen network was visualised, showing details of the periodicity of the collagen fibrils, their surface morphology and high resolution images of cross-bridges believed to involve proteoglycan structures (Fig. 7.25). Because the collagen fibres are packed in arrays their measured dimensions are relatively unaffected by probe broadening effects: the measured values correlate better with data from X-ray diffraction rather than TEM, in accordance with evidence from X-ray diffraction that fibre shrinkage only becomes marked when the water content drops below 30%. The resolution achieved by AFM on wet, unstained samples close to their native state approaches that obtainable by conventional TEM studies. Similar high resolution AFM images have been obtained for collagen networks mechanically dissociated from fixed human corneal and scleral tissue (Meller *et al* 1997).

Figure 7.25. AFM error signal mode image of bovine scleral fibres. Cross bridges between the fibrils can clearly be seen. Scan size 369 x 369 nm. Data reproduced from Fullwood *et al* 1994 Current Eye Research 14, 529-535, copyright Swete and Zeitlinger Publishers with permission.

7.11.4. Freeze-fracture replicas

Freeze-fracture replicas have been used successfully to obtain detailed information on cellular and subcellular architecture. Kordylewski and coworkers have reported an AFM study on freeze-fracture replicas of rat atrial tissue (Kordylewski *et al* 1994) which have been well characterised by TEM (Kordylewski *et al* 1993). The AFM can provide direct information on the topography of the replica and both sides of the replica can be imaged. At low magnification the TEM and AFM images reveal similar features. At higher magnification the AFM was able to resolve details not apparent in the TEM data. It was most useful to compare TEM images with error signal mode AFM images, but to use topography images for measurement of dimensions. As with TEM it is possible to generate stereo pairs for viewing the topography of the surface. Features observed by both TEM and AFM included atrial granules, mitochondria with cristae, membrane patches, nuclei and both open and closed pores on the surface of nuclei (Kordylewski *et al* 1994). The major advantages of the use of AFM were direct visualisation of surface geometry, and the ease with which the dimensions of surface features could be determined, although the latter needs to be weighed against the difficulties in correcting for probe broadening effects.

7.11.5. Immunolabelling

A major problem with AFM of cells and tissue is recognition or identification of structural features. Combined AFM and optical microscopy, coupled with specific staining or fluorescent labelling, is a powerful solution to this problem and examples of this approach have already been described. Immunolabelling techniques are well established in both light and electron microscopy. Antibodies have been used to label specific antigens on individual molecules (e.g. DNA section 4.2.4), macromolecular complexes such as chromosomes (section 4.2.5), bacterial biofilms (section 7.2.1), cells (Putman *et al* 1993; Takeuchi *et al* 1998; Neagu *et al* 1994; Eppell *et al* 1995) or tissue (Saoudi *et al* 1994; Yamashina and Shigeno, 1995). If the surface features are individual molecules, or for flat layered structures, such as bacterial S layers (Ohnesorge *et al* 1992), it may be possible to recognise the antibody-antigen complex directly, although care must be taken to discriminate between specifically bound and passively adsorbed labels. When the surface is rough enhancement of the labels is necessary. Gold labelled antibodies may be used alone or to locate primary antibody-antigen complexes (Mulhern *et al* 1992; Putman *et al* 1993; Saoudi *et al* 1994). The procedure may not be as straightforward for AFM as it is for EM studies. Gold labels may be confused with surface protrusions, or even compressed into the surface by the probe, making them difficult to spot (Mulhern *et al* 1992). Whereas large gold labels may be easy to identify, smaller labels can be difficult to locate by AFM alone (Eppell *et al* 1995). Location of labels can be improved by generating larger particulate deposits: examples used with AFM include silver enhancement (Neagu *et al* 1994; Putman *et al* 1993), peroxidase labelled antibodies and their reaction with DAB (sections 4.2.5 & 4.3), or even fluorescently labelled complexes (McMaster *et al* 1996b; 1996c) (section 4.2.5). With AFM it is possible to modify the label or tip in order to improve sensitivity. Magnetic labels can be detected with magnetic tips and should give enhanced sensitivity. There is at least one report of the use of superparamagnetic beads as labels (Neagu *et al* 1994). Yamashina and Shingo (Yamashina and Shingo, 1995) used Kelvin force probe microscopy to enhance imaging of gold labels: the surface potential is raised in the vicinity of the gold label. Antibody coated tips could be used to locate antigens on sample surfaces. Immunolabelling methods can be used with AFM and there is scope for improving the sensitivity of the technique and in the use of controls.

7.12. Biominerals

Biominerals are generally complex but the mineral component may mean that they are rigid, possibly well ordered, and hence can be imaged by AFM. A few examples of AFM studies on such materials are collected below.

7.12.1. Bone, tendon and cartilage

The use of AFM to study the structure and assembly of collagen has been discussed in section 4.5.3. In bones the final structure is based on an interaction between collagen and deposited apatite. Combined AFM and TEM studies on mineralised and demineralised tendon have been used to refine composite models for turkey leg tendon. TEM and AFM studies complement each other, with the TEM revealing structure within the specimen, and the AFM showing surface structure (Lees *et al* 1994). AFM tapping mode studies of *in vitro* and physiologically calcified tendon collagen have suggested that the surface structure of the fibril induces nucleation of apatite crystals, and that their subsequent growth does not markedly alter the fibril structure (Bigi *et al* 1997). Tapping mode AFM on dried collagen fibrils from rat tail tendon have allowed visualisation of proteoglycan bound to the collagen surfaces (Raspanti *et al* 1997). The distribution of the proteoglycans was determined by comparing images obtained for the native structure, samples treated with chondroitinase, and samples incubated with Cupromeronic blue, a dye specifically designed to stabilise the anionic glycosaminoglycan chains. Such studies help build a picture of the ultrastructure of these complex materials. Articular cartilage acts as a low friction bearing in synovial joints. AFM has provided a method for characterising surface and subsurface structures of freshly excised articular tissue in physiologically relevant medium (Jurvelin *et al* 1996). Cartilage discs together with a thin layer of bone were glued at the bony surface to glass coverslips and then imaged under phosphate buffered saline. AFM studies showed the articular surface to be amorphous: blurring of the images was attributed to a high surface viscosity possibly related to its lubrication properties. Surface irregularities, seen previously by SEM, were absent in the AFM images, suggesting that they are preparation artifacts. Small pits were observed but it is not clear whether these are intrinsic features of the native surface or are induced during isolation of the cartilage. Enzymatic digestion of proteoglycan revealed the fibrous substructure. The AFM allowed imaging in physiologically conditions, mechanical measurements to be made on normal and enzymatic digested material, plus measurements of the dimensions and periodicity of the collagen fibrous network. Low resolution AFM topographic and elastic images of bone are similar to those obtained by light and electron microscopy but at higher resolution the AFM revealed dramatic fluctuations of elastic properties over small (*circa* 50 nm) distances (Tao *et al* 1992).

A significant body of work has been carried out on the interaction between the biomineral scaffold of bone and the non-collagenous structural proteins which glue the composite together (via mineralised collagen fibrils) by the Hansma group in Santa Barbara (Hansma *et al* 2005; Kindt *et* al 2005; Fantner *et al* 2005; 2006). This work would require a chapter in its own right to really do it justice, but to summarise the main findings in a couple of sentences the group have shown that the 'glue' components in the composite act via a mechanism which was hitherto unknown. The secret of the strength of the 'glue' polymers involves the formation of sacrificial bonds in the polymer chains which greatly increase the energy required to disrupt them. Furthermore, it was demonstrated that these bonds can be reformed following rupture if the strain on them is removed, thus demonstrating the self-healing potential of the 'glue' (Fantner *et al* 2006). The group have also developed an analytical instrument which uses indentation testing to try to correlate fracture risk with a simple measurable mechanical response, although at the time of writing this had not been demonstrated in a clinical sense (Hansma *et al* 2008).

7.12.2. Teeth

A number of studies have demonstrated the value of using AFM to study dentin (Kinney *et al* 1993; 1996; Cassinelli and Morra, 1994; Marshall Jnr *et al* 1993; 1995) and tooth enamel (Sollbohmer *et al* 1995). The dentin body of teeth is protected from abrasive corrosion during eating, and from chemical attack from acidic foods or bacterial metabolites, by an outer coating of enamel composed of about 86% hydroxyl apatite crystals. In order to image the enamel, crowns were extracted, embedded in Epon, and polished to reveal the enamel surface. Using a specially designed cell the enamel could be rapidly exposed to acidic drinks and the AFM used to follow and quantify the etching, or demineralisation as a function of time (Sollbohmer *et al* 1995). Dentine is situated between the pulp and the enamel of the human tooth. It is mainly composed of hydroxyapatite crystals and collagen and has a microtubular structure. Current methods of dentin bonding depend on a demineralisation of the dentin to create a microporous structure which can be penetrated by bonding agents. AFM has been used to observe demineralisation, drying and bonding processes (Kinney *et al* 1993; Cassinelli and Morra, 1994; Marshall Jnr *et al* 1993; 1995).

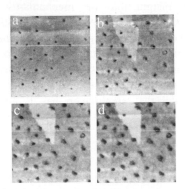

Figure 7.26. AFM studies showing the effect of demineralisation of dentine in dilute phosphoric acid (3 mM). Scan size 40 x 40 μm. Imaged after polishing in deionised water at time intervals (a) 0 s, (b) 30 s, (c) 60 s and (d) 80 s. The upper portion of each image shows a gold reference island. Data reproduced from Marshall Jnr *et al* 1995 AFM of conditioning agents on dentine, J. Biomed. Mat. Res. copyright John Wiley 1995 with permission of the authors and John Wiley & Sons, Inc.

These types of studies are illustrated by investigations of the changes occurring on exposure to a range of conditioning agents (Marshall Jnr *et al* 1995). Discs were cut from sterilised dentin perpendicular to the long axis of the tooth, thus orienting tubules perpendicular to the exposed surface. The surfaces were polished and a gold grid evaporated onto the surface as a reference for monitoring the demineralisation process. Changes in the peritubular and intertubular regions were followed as a function of the exposure time to the conditioning agent (Fig. 7.26). By replacing the normal silicon nitride tip-cantilever assembly with a diamond tip and stainless steel cantilever it has been possible to produce topographic and elasticity maps across the peritubular and intertubular regions of cut faces of dentin specimens (Kinney *et al* 1996). Transport of ions through the dentine tubules is believed to be an important mechanism for nerve stimulation in dentine hypersensitivity. Nughes and Denuault (Nughes and Denuault, 1996) have used a scanning electrochemical microscope to image ion fluxes through the tubules and to identify and monitor the effects of blocked pores. It is suggested that such investigations could help in the assessment of treatments for dentin hypersensitivity (Nughes and Denuault, 1996). AFM mapping of the surface charge on natural hydroxyapatite from skeletal tissue has been carried out by means of a novel technique termed 'chemical force titration' (Smith *et al* 2003). This method makes use of the fact that the pull-off force between a chemically modified AFM tip and a surface will vary with pH, according to the ionisation potentials pK_a of the two surfaces. Strictly speaking the measurement probes the interaction of the chemically modified tips and substrates as a function of the pH of the bulk solution, not the pH at the surface of the molecular monolayer, although this is approximately equivalent to twice the surface pK_a. Some degree of success was achieved in correlating the charge patterns that were observed on these mineral surfaces with the binding of a range of matrix proteins, providing new insight into the mechanisms of the protein–mineral interaction (Smith *et al* 2003). The same group have also studied the interactions between the enamel matrix protein amelogenin and its putative target cells, human periodontal ligament fibroblasts and a human osteosarcoma cells (Kirkham *et al* 2006). The range and magnitude of adhesion forces recorded between amelogenin functionalised AFM tips and the cells were indicative of specific receptor-ligand interactions. This suggested that amelogenin has a direct effect at target cell membranes (Kirkham *et al* 2006).

7.12.3. Shells

Shells are one of the few examples of biological materials on which it is possible to resolve detail at the atomic level (Friedbacker *et al* 1991; Manne *et al* 1994). Atomic scale resolution was achieved on pismo (*Tivela stultorum*) and sea urchin (*Stronglyocentrotus purpuratus*) shells, after powdering them and compressing them into KBr discs before imaging with AFM (Friedbacker *et al* 1991). Diatoms are unicellular algae which are frequently used to check the resolution of objectives for light microscopy. The shells are silicon based and coated with layers of polysaccharides, proteins and lipids. An investigation of a range of species of diatom surfaces by AFM showed characteristic pores and ripples, and the raphe and their central nodes were identified (Linder *et al* 1992). AFM has been used to image the matrix proteins of oyster (*Crassostrea virginica*) shells (Donachy *et al* 1992) and to study the distribution of different sized cuticular pigment granules in relation to coloured patches on the surface of Quail egg shells (Makita *et al* 1993). Nacre, the pearly layer of mollusc shells, consists of layers of crystalline argonite tablets. AFM has been used to examine the different distributions of aragonite tablets in a bivalve (*Atrina* sp.) and a gastropod (*Haliotis rufescens*). A feature of the studies was that by imaging in liquids it was possible to follow the dissolution or demineralisation of the nacre: by stripping the structure away layer by layer the structure of individual tablets and their inter-relationship with vertically adjacent tablets was revealed (Manne *et al* 1994). The majority of recent AFM work on shells has been made on diatoms, since these are relatively easily to obtain and a better understanding of the biology of planktonic species is likely to be of increasing importance in terms of their possible impact on global climate change: they are believed to play an important role in the export of carbon from oceanic surface waters (Smetacek 1985; Dugdale *et al* 1998).

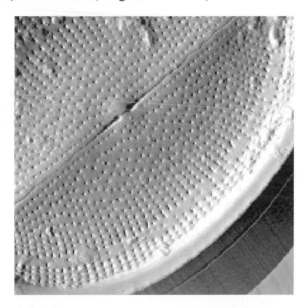

Figure 7.27. AFM image of the surface of a diatom. Scan size: 20 x 20 μm.

Furthermore, diatom communities provide a popular tool for monitoring environmental conditions, past and present, and are commonly used in studies of water quality. In terms of AFM studies the work has progressed from simple imaging of the structure of dead diatoms (fig. 7.27), to a more complete analysis of living diatoms through a combination of high resolution imaging alongside the force spectroscopy capabilities of the technique. For example, Higgins and co-workers characterised both the adhesive and elastic properties of the mucilage layers secreted by three living benthic diatoms (Higgins *et al* 2003a). By mapping the elasticity across the sample surface the study revealed large spatial variations in the location of the mucilage layers between different species. In a related study the same group investigated the adhesive mucilage and the mechanism of cell-substratum adhesion of two benthic raphid diatoms, revealing the presence of multistranded tethers that appear to arise along the raphe openings. In the force spectra these tethers were seen to extend for a considerable distance from the cell before forming a "holdfast-like" attachment with the substratum (Higgins *et al* 2003b).

References: selected reviews

Butt, H-J, Wolff, E.K, Gould, S.A.C, Northern, B.D, Peterson, C.M. and Hansma, P.K. (1990). Imaging cells with the atomic force microscope. *J. Struct. Biol.* **105**, 54-61.

Firtel, M. and Beveridge, T.J. (1995). Scanning probe microscopy in microbiology. *Micron* **26**, 347-362.

Hansma, H.G, Kim, K.J, Laney, D.E, Garcia, R.A, Argaman, M, Allen, M.J. and Parsons, S.M. (1997). Properties of biomolecules measured from atomic force microscopic images: a review. *J. Struct. Biol.* **119**, 99-108.

Henderson, E. (1994). Imaging of living cells by atomic force microscopy. *Progr. Surface Sci.* **46**, 39-60.

Ohnesorge, F.M, Horber, J.K.H, Haberle, W, Czerny, C-P, Smith, D.P.E. and Binnig, G. (1997). AFM review study on pox viruses and living cells. *Biophys. J.* **73**, 2183-2194.

Shao, Z, Mou, J, Czajkowsky, D.M, Yang, J. and Yuan, J-Y. (1996). Biological atomic force microscopy: what is achieved and what is needed. *Advances Phys.* **45**, 1-86.

Selected references

A-Hassan, E, Heinz, W.F, Antonio, M.D, D'Costa, N.P, Nageswaran, S, Schoenenberger, C-A. and Hoh, J.H. (1998). Relative microelastic mapping of living cells by atomic force microscopy. *Biophys. J.* **74**, 1564-1578.

Abu-Lail, N.I. and Camesano, T.A. (2006). Specific and Nonspecific Interaction Forces Between *Escherichia coli* and Silicon Nitride, Determined by Poisson Statistical Analysis. *Langmuir* **22**, 7296-7301.

Ahimou, F, Touhami, A. and Dufrêne, Y.F. (2003). Real time imaging of the surface topography of living yeast cells by atomic force microscopy. *Yeast* **20**, 25–30.

Ahimou, F, Semmens, M.J., Novak, P.J. and Haugstad, G. (2007). Biofilm cohesiveness measurement using a novel Atomic Force Microscopy methodology. *Appl. & Environ. Microbiol.* **73**, 2897–2904.

Allen, M.J, Lee, C, Lee, J.D, Pogany, G.C, Balloch, M, Siekhaus, W.J. and Balhorn, R. (1993). Atomic force microscopy of mammalian sperm chromatin. *Chromossoma* **102**, 623-630.

Allen, M.J, Bradbury, E.M. and Balhorn, R. (1995). The natural subcellular surface structure of the bovine sperm cell. *J. Struct. Biol.* **114**, 197-208.

Allen, M.J, Lee IV, J.D, Lee, C. and Balhorn, R. (1996). Extent of sperm chromatin hydration determined by atomic force microscopy. *Mol. Reproduction & Development* **45**, 87-92.

Almqvist, N, Backman, L. and Fredriksson, S. (1994). Imaging human erythrocyte spectrin with atomic force microscopy. *Micron* **25**, 227-232.

Almqvist, N, Bhatia, R, Primbs, G, Desai, N, Banerjee, S. and Lal, R. (2004). Elasticity and adhesion force mapping reveals real-time clustering of growth factor receptors and associated changes in local cellular rheological properties. *Biophys. J.* **86**, 1753–1762.

Amako, K, Takade, A, Umeda, A. and Yoshida, M. (1993). Imaging of the surface structures of Epon thin sections created with a glass knife and a diamond knife by the atomic force microscope. *J. Electron Microscopy* **42**, 121-123.

Antonik, M.D, D'Costa, N.P. and Hoh, J.H. (1997). A biosensor based on micromechanical interrogation of living cells. *IEEE Eng. Medicine Biol.* March/April, 66-72.

Auger-Messier, M, Turgeon, E.S, Leduc, R, Escher, E. and Guillemette, G. (2005). The constitutively active N111G-AT1 receptor for angiotensin II modifies the morphology and cytoskeletal organization of HEK-293 cells. *Exp. Cell Res.* **308**, 188–195.

van der Aa, B.C, Michel, R.M, Asther, M, Zamora, M.T, Rouxhet, P.G. and Dufrêne, Y.F. (2001). Stretching cell surface macromolecules by atomic force microscopy. *Langmuir* **17**, 3116–3119.

Barbee, K.A, Davies, P.F. and Lal, R. (1994). Shear stress-induced reorganisation of the surface topography of living endothelial cells imaged by atomic force microscopy. *Circulation Res.* **74**, 163-171.

Barbee, K.A. (1995). Changes in surface topography in endothelial monolayers with time at confluence: influence on subcellular shear stress distribution due to flow. *Biochem. Cell Biol.* **73**, 501-505.

Beckman, M, Kolb, H-A. and Lang, F. (1994). Atomic force microscopy of peritoneal macrophages after particle phagocytosis. *J. Membrane Biol.* **140**, 197-204.

Beech, I.B. (1996). The potential use of atomic force microscopy for studying corrosion of metals in the presence of bacterial biofilms-an overview. *International Biodeterioration & Biodegradation* **37**, 141-149.

Beech, I.B, Cheung, C.W.S, Johnson, D.B. and Smith, J.R. (1996). Comparative studies of bacterial biofilms on steel surfaces using atomic force microscopy and environmental scanning electron microscopy. *Biofouling* **10**, 65-77.

Beech, I. B, Smith, J. R, Steele, A. A, Penegar, I.. and Campbell, S.A. (2002).The use of atomic force microscopy for studying interactions of bacterial biofilms with surfaces. *Colloids Surf. B* **23**, 231–247.

Bernfield, M. (1999) Functions of cell surface heparin sulfate proteoglycans. *Ann. Rev. Biochem.* **68**, 729-777.

Betzig, E, Patterson, G.H, Sougrat, R, Lindwasser, O.W, Olenych, S, Bonifacino J.S, Davidson, M.W, Lippincott-Schwartz, J. and Hess, H.F. (2006). Imaging intracellular fluorescent proteins at nanometer resolution. *Science* **313**, 1642-1645.

Bigi, A, Gandolfi, M, Roveri, N. and Valdre, G. (1997). *In vitro* calcified tendon collagen: an atomic force and scanning electron microscopy investigation. *Biomaterials* **18**, 657-665.

Bolshakova, A.V, Kiselyova, O.I. and Yaminsky, I.V. (2004). Microbial surfaces investigated using atomic force microscopy. *Biotechnol. Prog.* **20**, 1615–1622.

Braet, F, Kalle, W.H.J, de Zanger, R.B, de Grooth, B.G, Raap, A.K, Tanke, H.J. and Wisse, E. (1996a). Comparative atomic force and scanning electron microscopy: an investigation on fenestrated endothelial cells in vitro. *J. Microscopy* **181**, 10-17.

Braet, F, de Zanger, R, Kalle, W, Raap, A, Tanke, H. and Wisse, E. (1996b). Comparative scanning, transmission and atomic force microscopy of the microtubular cytoskeleton in fenestrated liver endothelial cells. *Scanning Microscopy Supplement* **10**, 225-236.

Braet, F, de Zanger, R. and Wisse, E. (1997a). Drying cells for SEM, AFM and TEM by hexamethyldisilazane: a study on hepatic endothelial cells. *J. Microscopy-Oxford* **186**, 84-87.

Braet, F, de Zanger, R, Kammer, S. and Wisse, E. (1997b). Noncontact versus contact imaging: an atomic force microscopic study on hepatic endothelial cells in vitro. *Intern. J. Imaging Syst. Technol.* **8**, 162-167.

Braet, F, Seynaeve, C, de Zanger, R. and Wisse, E. (1998a). Imaging surface and submembranous structures with the atomic force microscope: a study on living cancer cells, fibroblasts and macrophages. *J. Microscopy-Oxford* **190**, 328-338.

Braet, F, Rotsch, C, Wisse, E. and Radmacher, M. (1998b). Comparison of fixed and living liver endothelial cells by atomic force microscopy. *Appl. Phys. A* **66**, S575-S578.

Braga, P.C. and Ricci, D. (1998). Atomic force microscopy: application to investigation of *Escherichia coli* morphology before and after exposure to cefodizime. *Antimicrobial Agents & Chemotherapy* **42**, 18-22.

Bremer, P.J, Geesey, G.G. and Drake, B. (1992). Atomic force microscopy examination of the topography of a hydrated bacterial biofilm on a copper surface. *Current Microbiology* **24**, 223-230.

Braunstein, D. and Spudich, A. (1994). Structure and activation dynamics of RBL-2H3 cells observed with scanning force microscopy. *Biophys. J.* **66**, 1717-1725.

Bowen, W.R, Lovitt, R.W. and Wright, C.J. (2000a). Direct Quantification of Aspergillus niger Spore Adhesion in Liquid Using an Atomic Force Microscope. *J. Colloid & Interface Sci.* **228**, 428-433.

Bowen, W.R, Lovitt, R.W. and Wright, C.J. (2000b). Direct quantification of *Aspergillus niger* spore adhesion to mica in air using an atomic force microscope. *Colloids Surf. A Physicochem. Eng. Asp.* **173**: 205-210.

Bowen, W.R, Hilal, N, Lovitt, R.W, Wright, C.J. (2000c). Application of atomic force microscopy to the study of micromechanical properties of biological materials *Biotechnol. Letts.* **22**, 893-903.

Brown, H.G. and Hoh, J.H. (1997). Entropic exclusion by neurofilament sidearms: a mechanism for maintaining neurofilament spacing. *Biochemistry* **36**, 15035-15040.

Butt, H-J, Wolff, E.K, Gould, S.A.C, Northern, B.D, Peterson, C.M. and Hansma, P.K. (1990). Imaging cells with the atomic force microscope. *J. Struct. Biol.* **105**, 54-61.

Butt, H-J. (1992). Measuring local surface charge densities in electrolyte solutions with a scanning force microscope. *Biophys. J.* **63**, 578-582.

Cail, T.L. and Hochella, M.F. (2005). The effects of solution chemistry on the sticking efficiencies of viable *Enterococcus faecalis*: An atomic force microscopy and modeling study. *Geochim. Cosmochim. Acta*, **69**, 2959-2969.

Camesano, T.A. and Logan, B. E. (2000). Probing electrostatic interactions using atomic force microscopy. *Environ. Sci. Technol.* **34**, 3354-3362.

Canet, D, Rohr, R, Chamel, A. and Guillian, F. (1996). Atomic force microscopy study of isolated ivy leaf cuticles observed directly and after embedding in Epon. *New Phytol.* **134**, 571-577.

Cassinelli, C. and Morra, M. (1994). Atomic force microscopy studies of the interaction of a dentin adhesive with tooth hard tissue. *J. Biomed. Mater. Res.* **28**, 1427-1431.

Chang, L, Kious, T, Yorgancioglu, M, Keller, D. and Pfeiffer, J. (1993). Cytoskeleton of living, unstained cells imaged by scanning force microscopy. *Biophys. J.* **64**, 1282-1286.

Charras, G.T, Lehenkari, P.P. and Horton, M.A. (2001). Atomic force microscopy can be used to mechanically stimulate osteoblasts and evaluate cellular strain distributions. *Ultramicroscopy* **86**, 85-95.

Charras, G.T. and Horton, M.A. (2002), Single cell mechanotransduction and its modulation analyzed by atomic force microscope indentation. *Biophys J.* **82**, 2970-2981.

Chen, X, Kis, A, Zettl, A. and Bertozzi, C.R. (2007). A cell nanoinjector based on carbon nanotubes. *PNAS USA* **104**, 8218-8222.

Chtcheglova, L.A, Shubeita, G.T, Sekatskii, S.K. and Dietler, G. (2004). Force Spectroscopy with a Small Dithering of AFM Tip: A Method of Direct and Continuous Measurement of the Spring Constant of Single Molecules and Molecular Complexes. *Biophys. J.* **86**, 1177-1184.

Cuerrier, C.M, Lebel, R. and Grandbois, M (2007). Single cell transfection using plasmid decorated AFM probes. *Biochem. Biophys. Res. Commun.* **355**, 632-636.

Da Silva, L.B, Trebes, J.E, Balhorn, R, Mrowka, S, Anderson, E, Attwood, D.T, Barbee Jnr, T.W, Brase, J, Corzett, M, Gray, J, Koch, J.A, Lee, C, Kern, D, London, R.A, MacGowan, B.J, Matthews, D.L. and Stone, G. X. (1992). X-ray laser microscopy of rat sperm nuclei. *Science* **258**, 269-271.

Demanet, C.M. and Sankar, K.V. (1996). Atomic force microscopy images of a pollen grain: a preliminary study. *S. African. J. Bot.* **62**, 221-223.

Doktycz, M.J, Sullivan, C.J, Hoyt, P.R, Pelletiera, D.A, S. Wu, S. and Allison, D.P. (2003). AFM imaging of bacteria in liquid media immobilized on gelatin coated mica surfaces. *Ultramicroscopy* **97**, 209-216

Domke, J, Parak, W.J, George, M, Gaub, H.E. and Radmacher, M. (1999). Mapping the mechanical pulse of single cardiomyocytes with the atomic force microscope. *European Biophys. J. Biophys. Letts.* **28**, 179-186.

Donachy, J.E, Drake, B. and Sikes, C.S. (1992). Sequence and atomic-force microscopy analysis of a matrix protein from the shell of the oyster *Crassostrea virginica*. *Marine Biol.* **114**, 423-428.

Dufrêne, Y. F., C. J. P. Boonaert, P. A. Gerin, M. Asther, and P. G. Rouxhet. (1999). Direct probing of the surface ultrastructure and molecular interactions of dormant and germinating spores of *Phanerochaete chrysosporium. J. Bacteriol.*181, 5350-5354.

Dufrêne, Y.F. (2004). Using nanotechniques to explore microbial surfaces. *Nature Rev. Microbiol.* **2**, 451-460.

Dugdale, R.C. and Wilkerson, F.P. (1998). Silicate regulation of new production in the equatorial Pacific upwelling. *Nature* **391**, 270-273.

Dupres, V, Menozzi, F.D, Locht, C, Clare, B.H, Abbott, N.L, Cuenot, S, Bompard, C, Raze, D. and Dufrêne, Y.F. (2005). Nanoscale mapping and functional analysis of individual adhesins on living bacteria. *Nature Methods* **2**, 515-520.

Dupres, V, Verbelen, C. and Dufrêne, Y.F. (2007). Probing molecular recognition sites on biosurfaces using AFM. *Biomaterials* **28**, 2393-2402.

Ebner, A, Kienberger, F, Kada, G, Stroh, C.M, Geretschläger, M, Kamruzzahan, A.S.M, Wildling, L, Johnson, W.T, Ashcroft, B, Nelson, J, Lindsay, S.M, Gruber, H.J. and Hinterdorfer, P. (2005). Localization of Single Avidin–Biotin Interactions Using Simultaneous Topography and Molecular Recognition Imaging. *CHEMPHYSCHEM.* **6**, 897–900.

Emerson, R.J. and Camesano, T.A. (2004). Nanoscale Investigation of Pathogenic Microbial Adhesion to a Biomaterial. *Appl. & Environ. Microbiol.* **70**, 6012–6022.

Eppell, S.J, Simmons, S.R, Albrecht, R.M. and Marchant, R.E. (1995). Cell-surface receptors and proteins on platelet membranes imaged by scanning force microscopy using immunogold contrast enhancement. *Biophys. J.* **68**, 671-680.

Fantner, G.E, Hassenkam, T, Kindt, J.H, Weaver, J.C, Birkedal, H., Pechenik, L, Cutroni, J.A, Cidade, G.A.G, Stucky, G.D, Morse, D.E. and Hansma, P.K. (2005). Sacrificial bonds and hidden length dissipate energy as mineralized fibrils separate during bone fracture. *Nature Materials* **4**, 612-616.

Fantner, G.E, Oroudjev, E, Schitter, G, Golde, L.S, Thurner, P, Finch, M.M, Turner, P, Gutsmann, T, Morse, D.E, Hansma, H. and Hansma, P.K. (2006). Sacrificial Bonds and Hidden Length: Unraveling Molecular Mesostructures in Tough Materials. *Biophys. J.* **90**, 1411-1418.

Friedbacher, G, Hansma, P.K, Ramli, E. and Stucky, G.D. (1991), Imaging powders with the atomic force microscope: from biominerals to commercial materials. *Science* **253**, 1261-1263.

Fritz, M, Radmacher, M. and Gaub, H.E. (1993). *In vitro* activation of human platelets triggered and probed by atomic force microscopy. *Exp. Cell Res.* **205**, 187-190.

Fritz, M, Radmacher, M. and Gaub, H.E. (1994). Granule motion and membrane spreading during activation of human platelets imaged by atomic force microscopy. *Biophys. J.* **66**, 1328-1334.

Fullwood, N.J, Hammiche, A, Pollock, H.M, Hourston, D.J. and Song, M. (1994). Atomic force microscopy of the cornea and sclera. *Current Eye Res.* **14**, 529-535.

Gabai, R, Segev, L. and Joselevich, E. (2005). Single Polymer Chains as Specific Transducers of Molecular Recognition in Scanning Probe Microscopy. *J. Amer. Chem. Soc.* **127**, 11390-11398.

Gad, M. and Ikai, A. (1996). Method for immobilising microbial cells on gel surface for dynamic AFM studies. *Biophys. J.* **69**, 2226-2233.

Gad, M, Itoh, A. and Ikai, A. (1997). Mapping cell wall polysaccharides of living microbial cells using atomic force microscopy. *Cell Biol. Internat.* **21**, 697-706.

Garcia, C.R.S, Takeuschi, M, Yoshioka, K. and Miyamoto, H. (1997). Imaging *Plasmodium falciparum* -infected ghost and parasite by atomic force microscopy. *J. Struct. Biol.* **119**, 92-98.

Geesey, G.G, Mittelman, M.W, Iwaoka, T. and Griffiths, P.R. (1986). Role of bacterial exopolymers in the deterioration of metallic copper surfaces. *Materials Perform.* **25**, 37-40.

Giebel, K-F, Bechinger, C, Herminghaus, S, Riedel, M, Leiderer, P, Weiland, U. and Bastmeyer, M. (1999). Imaging of cell/substrate contacts of living cells with surface plasmon resonance microscopy. *Biophys. J.* **76**, 509-516.

Gould, S.A.C, Drake, B, Prater, C.B, Weisenhorn, A.L, Manne, S, Hansma, H.G, Hansma, P.K, Massie, J, Longmire, M, Elings, V, Northern, B.D, Mukergee, B, Peterson, C.M, Stoeckenius, W, Albrecht, T.R. and Quate, C.F. (1990) From atoms to integrated circuit chips, blood cells, and bacteria with the atomic force microscope. *J. Vac. Sci. & Technol. A* **8**, 369-373.

Grandbois, M, Dettmann, W, Benoit, M. and Gaub, H.E. (2000). Affinity imaging of red blood cells using an atomic force microscope. J. Histochem. Cytochem. **48**, 719–724.

Gunning, P.A, Kirby, A.R, Parker, M.L, Gunning, A.P. and Morris, V.J. (1996). Comparative imaging of *Pseudomonas putida* bacterial biofilms by Scanning Electron Microscopy and both dc contact and ac non-contact Atomic Force Microscopy. *J. Appl. Bact.* **81**, 276-282.

Haberle, W, Horber, J.K.H. and Binnig, G. (1991). Force microscopy on living cells. *J. Vac. Sci. & Technol. B* **9**, 1210-1213.

Haberle, W, Horber, J.K.H, Ohnesorge, F, Smith, D.P.E. and Binnig, G. (1992). In situ investigations of single living cells infected by viruses. *Ultramicroscopy* **42-44**, 1161-1167.

Han, W, Mou, J, Sheng, J, Yang, J. and Shao, Z. (1995). Cryo atomic force microscopy: a new approach for biological imaging at high resolution. *Biochemistry* **34**, 8215-8220.

Hanley, S.J. and Gray, D.G. (1994). Atomic force microscope images of black spruce wood sections and pulp fibres. *Horzforschung* **48**, 29-34.

Hansma, P.K, Fantner, G.E, Kindt, J.H, Thurner, P.J, Schitter, G, Turner, P.J, Udwin, S.F. and Finch, M.M. (2005). Sacrificial bonds in the interfibrillar matrix of bone. *J. Musculoskelet. Neuronal Interact.* **5**, 313-315.

Hansma, P, Turner, P, Drake, B, Yurtsev, E, Proctor, A, Mathews, P, Lelujian, J, Randall, C, Adams, J, Jungmann, R, Garza-de-Leon, F, Georg Fantner, G, Mkrtchyan, H, Pontin, M, Weaver, A, Brown, M.B, Nadder Sahar, N, Rossello, R. and Kohn, D. (2008). The bone diagnostic instrument II: Indentation distance increase. *Rev. Sci. Instr.* **79**, 064303.

Hansma, H.G, Kim, K.J, Laney, D.E, Garcia, R.A, Argaman, M, Allen, M.J. and Parsons, S.M. (1997). Properties of biomolecules measured from atomic force microscopic images: a review. *J. Struct. Biol.* **119**, 99-108.

Haydon, P.G, Henderson, E. and Stanley, E.F. (1994). Localization of individual calcium channels at the release face of a presynaptic nerve terminal. *Neuron* **13**, 1275-1280.

Haydon, P.G, Lartius, R, Parpura, V. and Marchese-Ragona, S.P. (1996). Membrane deformation of living glial cells using atomic force microscopy. *J. Microscopy-Oxford* **182**, 114-120.

Heinz, W.F. and Hoh, J.H. (1999a). Spatially resolved force spectroscopy of biological surfaces using the atomic force microscope. *Trends Biotechnol.* **17**, 143-150.

Heinz, W.F. and Hoh, J.H. (1999b). Relative surface charge mapping with the atomic force microscope. *Biophys. J.* **76**, 528-538.

Hell, S. (2007) Breaking the barrier: fluorescence microscopy with diffraction-unlimited resolution. *Chromosome Res.* **15**, 55-55.

Henderson, E, Haydon, P.G. and Sakaguchi, D.S. (1992). Actin dynamics in living glial cells imaged by atomic force microscopy. *Science* **257**, 1944-1946.

Henderson, E. (1994). Imaging of living cells by atomic force microscopy. *Progr. Surface Sci.* **46**, 39-60.

Henderson, R.M, Schneider, S, Li, Q, Hornby, D, White, S.J. and Oberleithner, H. (1996a). Atomic force microscopy used to image the inwardly-rectifying ATP-sensitive potassium channel protein, ROMK 1. *Kidney Internat.* **50**, 1780.

Henderson, R.M, Schneider, S, Li, Q, Hornby, D, White, S.J. and Oberleithner, H. (1996b). Imaging ROMK 1 inwardly rectifying ATP-sensitive K+ channel protein using atomic force microscopy. *PNAS USA* **93**, 8756-8760.

Hertz, H. (1882). Uber die Beruhrung fester elastischer Korper. *J. Reine Angew Math.* **92**, 156-171.

Higgins, M.J, Sader, J.E, Mulvaney, P. and Wetherbee, R. (2003). Probing the Surface of Living Diatom Cells with Atomic Force Microscopy: The Nanostructure and Nanomechanical Properties of the Mucilage Layer. *J. Phycol.* **39**, 722-734.

Higgins, M.J, Molino, P, Mulvaney, P. and Wetherbee, R. (2003b). The structure and nanomechanical properties of the adhesive mucilage that mediates diatom-substratum adhesion and motility. *J. Phycol.* **39**, 1181-1193.

Hinterdorfer, P, Baumgartner, W, Gruber, H.J, Schilcher, K. and Schindler, H-J. (1996). Detection and localization of individual antibody-antigen recognition events by atomic force microscopy. *PNAS USA* **93**, 3477-3481.

Hinterdorfer, P. and Dufrêne, Y.F. (2006) Detection and localization of single molecular recognition events using atomic force microscopy. *Nature Methods* **3**, 347-355.

Hofmann, U.G, Rotsch, C, Parak, W.J. and Radmacher, M. (1997). Investigating the cytoskeleton of chicken cardiocytes with the atomic force microscope. *J. Struct. Biol.* **119**, 84-91.

Hoh, J.H. and Schoenenberger, C-A. (1994). Surface morphology and mechanical properties of MDCK monolayers by atomic force microscopy. *J. Cell Sci.* **107**, 1105-1114.

Hoh, J.H, Heinz, W.F. and A-Hassan, E. (1997). Force volume. *Digital Instruments Support Note No. 240*, Digital Instruments, 112 Robin Hill Road, Santa Barbara, CA 93117, California, USA.

Holstein, T.W, Benoit, M, von Herder, G, Wanner, G, David, C.N. and Gaub, H.E. (1994). Fibrous mini-collagens in hydra nematocysts. *Science* **265**, 402-404.

Horber, J.K.H, Haberle, W, Ohnesorge, F, Binnig, G, Liebich, H.G, Czerny, C.P, Mahnel, H. and Mayr, A. (1992). Investigations of living cells in the nanometer regime with the atomic force microscope. *Scanning Microscopy* **6**, 919-930.

Horton, M, Charras, G. and Lehenkari, P (2002). Analysis of ligand–receptor interactions in cells by atomic force microscopy. *J. Recept. Signal Transduct. Res.* **22**, 169–190.

Jaschke, M, Butt, H-J. and Wolff, E.K. (1994). Imaging flagella of Halobacteria by atomic force microscopy. *Analyst* **119**, 1943-1946.

Jolley, J.G, Geesey, G.G, Hankins, M.R, Wright, R.B. and Wichlacz, P.L. (1988). Auger electron spectroscopy and X-ray photoelectron spectroscopy of the biocorrosion of copper by gum arabic, BCS and *Pseudomonas atlantica* exopolymer. *J. Surface & Interface Anal.* **11**, 371-376.

Jurvelin, J.S, Muller, D.J, Wong, M, Studer, D, Engel, A. and Hunziker, E.B. (1996). Surface and subsurface morphology of bovine humeral articular cartilage as assessed by atomic force and transmission microscopy. *J. Struct. Biol.* **117**, 45-54.

Kasas, S, Gotzos, V,. and Celio, M.R. (1993). Observation of living cells using atomic force microscopy. *Biophys. J.* **64**, 539-544.

Kasas, S, Fellay, B. and Cargnello, R. (1994). Observation of the action of penicillin on Bacillus-subtilis using atomic force microscopy-techniques for the preparation of bacteria. *Surface & Interface Anal.* **21**, 400-401.

Kasas, S. and Ikai, A. (1996). A method for anchoring round shaped cells for atomic force microscope imaging. *Biophys. J.* **68**, 1678-1680.

Kasas, S, Wang, X, Hirling, H, Marsault, R, Huni, B, Yersin, A, Regazzi, R, Grenningloh, G, Riederer, B, Forro, L, Ditter, G. and Catsicas, S. (2005). Superficial and deep changes of cellular mechanical properties following cytoskeleton disassembly. *Cell Motil. Cytoskeleton* **62**, 124–132.

Kinney, J.H, Balooch, M, Marshall Jnr., G.W. and Marshall S.J. (1993). Atomic force microscope study of dimensional changes in dentine during drying. *Arch. Oral Biol.* **38**, 1003-1007.

Kinney, J.H, Balooch, M, Marshall, S.J, Marshall Jnr, G.W. and Weihs, T.P. (1996). Atomic force microscope measurements of the hardness and elasticity of peritubular and intertubular human dentin. *J. Biomechanical Eng.* 118, *133-135*.

Kirby, A.R, Fyfe, D.J, Parker, M.L, Gunning, A.P, Gunning, P.A. and Morris, V.J. (1998). Structural studies on human coleretal adenocarcinoma HT29 cells by atomic force microscopy, transmission electron microscopy and scanning electron microscopy. *Probe Microscopy* **1**, 153-162.

Kirkham, J, Andreev, I, Robinson C, Brookes S.J, Shore R.C. and Smith D.A. (2006). Evidence for direct amelogenin–target cell interactions using dynamic force spectroscopy. *Eur. J. Oral Sci.* **114** (Suppl. 1) 219–224.

Kolari, M, Schmidt, U, Kuismanen, E. and Salkinoja-Salonen, M. S. (2002). Firm but slippery attachment of Deinococcus geothermalis. *J. Bacteriol.* **184**, 2473–2480.

Kondo, H. (1984). Polyethylene glycol (PEG) embedding and subsequent de-embedding as a method for the structural and immunocytochemical examination of biological specimens by electron microscopy. *J. Electron Microsc. Tech.* **1**, 227-241.

Kordylewski, L, Goings, G. E. and Page, E. Rat atrial myocyte plasmalemmal caveolae in situ: reversible experimental increases in caveolar size and in surface density of caveolar necks. *Circ. Res.* **73**, 135-146.

Kordylewski, L, Saner, D. and Lal, R. (1994). Atomic force microscopy of freeze-fracture replicas of rat atrial tissue. *J. Microscopy-Oxford* **173**, 173-181.

Lamontagne, C-A, Cuerrier, C.M. and Grandbois, M. (2008). AFM as a tool to probe and manipulate cellular processes. *Pflugers Arch.–Eur. J. Physiol.* **456,** 61–70.

Laney, D.E, Garcia, R.A, Parsons, S.M. and Hansma, H.G. (1997). Changes in the elastic properties of cholinergic synaptic vesicles as measured by atomic force microscopy. *Biophys. J.* **72**, 806-813.

Le Grimellec, C, Lesniewska, E, Cachia, C, Schreiber, J.P, de Fornel, F. and Goudonnet, J.P. (1994). Imaging of the membrane surface of MDCK cells by atomic force microscopy. *Biophys. J.* **67**, 36-41.

Le Grimellac, C, Lesniewska, E, Giocondi, M-C, Cachia, C, Schreiber, J.P. and Goudonnet, J.P. (1995). Imaging the cytoplasmic leaflet of the plasma membrane by atomic force microscopy. *Scanning Microscopy* **9**, 401-411.

Le Grimellac, C, Lesniewska, E, Giocondi, M-C, Finot, E. and Goudonnet, J.P. (1997). Simultaneous imaging of the surface and submembraneous cytoskeleton in living cells by tapping mode atomic force microscopy. *C.R. Acad. Sci. Paris, Life Sci.* **320**, 637-643.

Lees, S, Prostak, K.S, Ingle, V.K. and Kjoller, K. (1994). The loci of mineral in turkey leg tendon as seen by atomic force microscope and electron microscopy. *Calcif. Tissue Internat.* **55**, 180-189.

Lehenkari, P.P, Charras, G.T, Nykänen, A. and Horton, M.A. (2000). Adapting atomic force microscopy for cell biology. *Ultramicroscopy* **82**, 289–295.

Leonenko, Z, Finotb, E. and Amrein, M. (2007). Adhesive interaction measured between AFM probe and lung epithelial type II cells. *Ultramicroscopy* **107**, 948–953.

Lesniewska, E, Giocondi, M-C, Vie, V, Finot, E, Goudonnet, J-P. and Le Grimellec, C. (1998). Atomic force microscopy of renal cells: limits and prospects. *Kidney Internal.* **53**, (suppl. 65), S42-S48.

Li, X. and Logan, B.E. (2004). Analysis of Bacterial Adhesion Using a Gradient Force Analysis Method and Colloid Probe Atomic Force Microscopy. *Langmuir* **20**, 8817-8822.

Li, S, Huang, N.F. and Hsu, S. (2005). Mechanotransduction in endothelial cell migration. *J. Cell Biochem.* **96**, 1110–1126.

Linder, A, Colchero, J, Apell, H-J, Marti, O. and Mlynek, J. (1992). Scanning force microscopy of diatom shells. *Ultramicroscopy* **42-44**, 329-332.

Loll, P.J. and Axelsen, P.H. (2000). The structural biology of molecular recognition by vancomycin. *Ann. ReV. Biophys. Biomol. Struct.* **29**, 265-289.

Ludwig, M, Dettmann, W. and Gaub, H.E. (1997). Atomic force microscope imaging contrast based on molecular recognition. *Biophys. J.* **72**, 445-448.

Makita, T, Ohoue, M, Yamoto, T. and Hakoi, K. (1993). Atomic force microscopy (AFM) of the cuticular pigment globules of the Quail egg shell. *J. Electron Microscopy* **42**, 189-192.

Manne, S, Zaremba, C.M, Giles, R, Huggins, L, Walters, D.A, Becher, A, Morse, D.E, Stucky, G.D, Didymus, J.M, Mann, S. and Hansma, P.K. (1994). Atomic force microscopy of the nacreous layer in mollusc shells. *Proc. Royal Soc. London B* **256**, 17-23.

Margulis, L, Ashen, J.B, Sole, M. and Guerrrero, R. (1993). "Composite, large spirochetes from microbial mats-spirochete structure review". *PNAS USA* **90**, 6966-6970.

Marshall Jnr., G.W, Balooch, M, Tench, R, Kinney, J.H. and Marshall, S.J. (1993). Atomic force microscopy of acid effects on dentin. *Dent. Mater.* **9**, 265-268.

Marshall Jnr., G.W, Balooch, M, Kinney, J.H. and Marshall, S.J. (1995). Atomic force microscopy of conditioning agents on dentin. *J. Biomedical Materials Res.* **29**, 1381-1387.

Maurice, P, Forsythe, J, Hersman, L. and Sposito, G. (1996). Application of atomic-force microscopy to studies of microbial interactions with hydrous Fe(III)-oxides. *Chemical Geology* **132**, 33-43.

McMaster, T.J, Miles, M.J. and Walsby, A.E. (1996a). Direct observation of protein secondary structure in gas vesicles by atomic force microscopy. *Biophys. J.* **70**, 2432-2436.

McMaster, T.J, Winfield, M.O, Baker, A.A, Karp, A. and Miles, M.J. (1996b). Chromosome classification by atomic force microscopy volume measurement. *J. Vac. Sci. & Technol. B* **14**, 1438-1443.

McMaster, T.J, Winfield, M.O, Karp, A. and Miles, M.J. (1996c). Analysis of cereal chromosomes by atomic force microscopy. *Genome* **39**, 439-444.

Mulhern, P.J, Blackford, B.L, Jericho, M.H, Southam. G. and Beveridge, T.J. (1992). AFM and STM studies of the interaction of antibodies with S-layer sheath of the archaeobacterium *Methanospiririllum hungatei*. *Ultramicroscopy* **42**, 1214-1224.

Meller, D, Peters, K. and Meller, K. (1997). Human cornea and sclera studied by atomic force microscopy. *Cell Tissue Res.* **288**, 111-118.

Müller, D.J, Schabert, F.A, Buldt, G. and Engel, A. (1995). Imaging purple membranes in aqueous solutions at subnanometer resolution by atomic force microscopy. *Biophys. J.* **68**, 1681-1686.

Neagu, C, van der Werf, K.O, Putman, C.A. J, Kraan, Y.M, de Grooth, B.G, van Hulst, N.F. and Greve, J. (1994). Analysis of immunolabeled cells by atomic force microscopy, optical microscopy, and flow cytometry. *J. Struct. Biol.* **112**, 32-40.

Nughes, S. and Denuault, G. (1996). Scanning electrochemical microscopy: amperometric probing of diffusional ion fluxes through porous membranes and human dentine. *J. Electroanaytical Chem.* **408**, 125-140.

Oberleithner, H, Giebisch, G. and Geibel, J. (1993). Imaging the lamellipodium of migrating epithelial cells in vivo by atomic force microscopy. *Pflugers Arch.* **425**, 506-510.

Oberleithner, H, Schwab, A, Wang, W, Giebisch, G, Hume, F. and Geibel, J. (1994). Living renal epithelial cells imaged by atomic force microscopy. *Nephron* **66**, 8-13.

Oberleithner, H, Brinkmann, E, Giebisch, G. and Geibel, J. (1995). Visualizing life on biomembranes by atomic force microscopy. *Kidney Internat.* **48**, 923-929.

Oberleithner, H, Schneider, S, Larmer, J. and Henderson, R.M. (1996). Viewing the renal epithelium with the atomic force microscope. *Kidney & Blood Pressure Res.* **19**, 142-147.

Oberleithner, H, Geibel, J, Guggino, W, Henderson, R.M, Hunter, M, Schneider, S.W, Schwab, A. and Wang, W.H. (1997a). Life on biomembranes viewed with the atomic force microscope. *Wiener Klinische Wochenschrift* **109**, 419-423.

Oberleithner, H, Schneider, S.W. and Henderson, R.M. (1997b). Structural activity of a cloned potassium channel (ROMK 1) monitored with the atomic force microscope: the "molecular sandwich" technique. *PNAS USA* **94**, 14144-14149.

Oberleithner, H, Riethmuller, C, Ludwig, T, Shahin, V, Stock, C, Schwab, A, Hausberg, M, Kusche, K. and Schillers, H. (2006). Differential action of steroid hormones on human endothelium. *J. Cell Sci.* **119**, 1926–1932.

Ohnesorge, F, Heckl, W.M, Harberle, W, Pum, D, Sara, M, SchindlerH, Schilcher, K, Kiener, A, Smith, D.P.E, Sleytr, U.B. and Binnig, G. (1992). Scanning force microscopy studies of the S-layers from *Bacillus coagulans* E38-66, *Bacillus sphaericus* CCM2177 and of an antibody binding process. *Ultramicroscopy* **42-44**, 1236-1242.

Ohnesorge, F.M, Horber, J.K.H, Haberle, W, Czerny, C-P, Smith, D.P.E. and Binnig, G. (1997). AFM review study on pox viruses and living cells. *Biophys. J.* **73**, 2183-2194.

Osada, T, Uehara, H, Kim, H. and Ikai, A. (2003). mRNA analysis of single living cells. *J. Nanobiotechnol.* **1**, 2 doi:10.1186/1477-3155-1-2.

Parker, M.L, Brocklehurst, T.F, Gunning, P.A, Coleman, H.P. and Robins, M.M. (1995). Growth of food-borne pathogenic bacteria in oil-water emulsions: I Methods for investigating the form of growth. *J. Appl. Bacteriol.* **78**, 601-608.

Parker, M.L, Kirby A.R. and Morris V.J. (2008). *In-situ* imaging of pea starch in seeds. *Food Biophysics* **3**, 66-76.

Parpura, V, Haydon, P.G. and Henderson, E. (1993a). Three-dimensional imaging of living neurons and glial with the atomic force microscope. *J. Cell Sci.* **104**, 427-432.

Parpura, V, Haydon, P.G, Sakaguchi, D.S. and Henderson, E. (1993b). Atomic force microscopy and manipulation of living glial cells. *J. Vac. Sci. & Technol. A* **11**, 773-775.

Papura, V, Doyle, R.T, Basarsky, T.A, Henderson, E. and Haydon, P.G. (1995). Dynamic imaging of purified individual synaptic vesicles. *Neuroimag.* **2**, 3-7.

Patterson, G.H, Betzig, E, Lippincott-Schwartz, J. and Hess, H.F. (2007). Developing Photoactivated Localization Microscopy (PALM). *4th IEEE International Symposium On Biomedical Imaging: Macro to Nano,* Vols 1-3 pp 940-943.

Pelling, A.E, Sehati, S, Gralla E.B, Valentine, J.S. and Gimzewski, J.K. (2004). Local nanomechanical motion of the cell wall of *Saccharomyces cerevisiae*. *Science* **305**, 1147–1150.

Pelling, A.E, Veraitch, F.S, Chu, C, Nicholls, B.M, Hemsley, A.L, Mason, C. and Horton, M.A. (2007). Mapping correlated membrane pulsations and fluctuations in human cells. *J. Molecular Recognition* **20**, 467-475.

Puech, P-H, Poole, K, Knebel, D. and Muller, D.J. (2006). A new technical approach to quantify cell–cell adhesion forces by AFM. *Ultramicroscopy* **106**, 637–644.

Pietrasanta, L.I, Schaer, A. and Jovin, T.M. (1994). Imaging subcellular structures of rat mammary carcinoma cells by scanning force microscopy. *J. Cell Sci.* **107**, 2427-2437.

Picco, L.M, Bozec, L, Ulcinas, A, Engledew, D.J, Antognozzi, M, Horton, M.A. and Miles, M.J. (2007). Breaking the speed limit with atomic force microscopy. *Nanotechnol.* **18**, 044030.

Prass, M, Jacobson, K, Mogilner, A. and Radmacher, M. (2006). Direct measurement of the lamellipodial protrusive force in a migrating cell. *J. Cell Biol.* **174**, 767–772.

Prater, C.B, Weisenhorn, A.L, Northern, B.D, Peterson, C.M, Gould, S.A.C. and Hansma, P.K. (1990). Imaging molecules and cells with the atomic force microscope. *Proc. XIIth International Congress for Electron Microscopy*, San Francisco, USA, San Francisco Press, Inc., pp 254-255.

Putman, C.A.J, van der Werf, K.O, de Grooth, B.G, van Hulst, N.F, Segerink, F.B. and Greve, J. (1992). Atomic force microscope with integrated optical microscope for biological applications. *Rev. Sci. Instrum.* **63**, 1914-1917.

Putman, C.A. J, de Grooth, B.G, Hansma, P.K. and van de Hulst, N.F. (1993). Immunogold labels: cell surface markers in atomic force microscopy, *Ultramicroscopy* **48**, 177-182.

Putman, C.A.J, van der Werf, K.O, de Grooth, B.G, van Hulst, N.F. and Greve, J. (1994). Viscoelasticity of living cells allows high resolution imaging by tapping mode atomic force microscopy. *Biophys. J.* **67**, 1749-1753.

Raab, A, Han, W, Badt, D, Smith-Gill, S.J, Lindsay, S.M,, Schindler, H. and Hinterdorfer, P. (1999). Antibody recognition imaging by force microscopy. *Nature Biotechnol.* **17**, 902-905.

Radmacher, M, Fritz, M. and Hansma, P.K. (1995). Imaging soft samples with the atomic force microscope: gelatin in water and propanol. *Biophys. J.* **69**, 264-270.

Radmacher, M, Fritz, M, Kacher, C.M, Cleveland, J.P. and Hansma, P.K. (1996). Measuring the viscoelastic properties of human platelets with the atomic force microscope. *Biophys. J.* **70**, 556-567.

Radmacher, M. (2007). Studying the mechanics of cellular processes by atomic force microscopy. *Methods Cell Biol.* **83**, 347–372.

Raspanti, M, Alessandrini, A, Ottani, V. and Ruggeri, A. (1997). Direct visualisation of collagen-bound proteoglycans by tapping-mode atomic force microscopy. *J. Struct. Biol.* **119**, 118-122.

Rotsch, C. and Radmacher, M. (1997). Mapping local electrostatic forces with the atomic force microscope. *Langmuir* **13**, 2825-2832.

Rotsch, C, Jacobson, K. and Radmacher, M. (1999). Dimensional and mechanical dynamics of active and stable edges in motile fibroblasts investigated by using atomic force microscopy. *PNAS USA* **96**, 921–926.

Rowley, J.R, Flynn, J.J. and Takahashi, M. (1995). Atomic force microscopic information on pollen exine substructure. *Nuphar. Bot. Acta* **108**, 300-308.

Rajyaguru, J.M, Kado, M, Richardson, M.C. and Musznski, M.J. (1997). X-ray micrography and imaging of *Escherichia coli* cell shape using laser plasma pulsed point X-ray sources. *Biophys. J.* **72**, 1521-1526.

Schaus, S.S. and Henderson, E.R. (1997). Cell viability and probe-cell interactions of XR1 glial cells imaged by atomic force microscopy. *Biophys. J.* **73**, 1205-1214.

Schoenenberger, C-A. and Hoh, J.H. (1994). Slow cellular dynamics in MDCK and R5 cells monitored by time-lapse atomic force microscopy. *Biophys. J.* **67**, 929-936.

Siedlecki, C.A. and Marchant, R.E. (1998). Atomic force microscopy for characterization of the biomaterial interface. *Biomaterials* **19**, 441-454.

Schabert, F.A, Hefti, A, Goldie, K, Stemmer, A, Engel, A, Meyer, E, Overney, R. and Guntherdodt, H.J. (1992). Ambient pressure scanning probe microscopy of 2D regular protein arrays. *Ultramicroscopy* **42-44**, 1118-1124.

de Souza Pereira, R, Parizotto, N.A. and Baranauskas, V. (1994). Observation of Baker's Yeast strains used in biotransformation by atomic force microscopy. *Appl. Biochem. Biotechnol.* **59**, 135-143.

Schabert, F.A. and Engel, A. (1994). Reproducible acquisition of *Escherichia coli* porin surface topographs by atomic force microscopy. *Biophys. J.* **67**, 2394-2403.

Schabert, F.A, Henn, C. and Engel, A. (1995). Native *Escherichia coli* ompF porin surfaces probed by atomic force microscopy. *Science* **268**, 92-94.

Schilcher, K, Hinterdorfer, P, Gruber, H.J. and Schindler, H. (1997). A non-invasive method for the tight anchoring of cells for scanning force microscopy. *Cell Biol. Internat.* **21**, 769–778.

Shroff, S.G, Saner, D.R. and Lal, R. (1995). Dynamic micromechanical properties of cultured rat atrial myocytes measured by atomic force microscopy. *Amer. J. Physiol.* **269** (*Cell Physiol.* **38**) C286-C292.

Smetacek, V.S. (1985). Role of sinking in diatom life-history cycles: Ecological, evolutionary and geological significance. *Marine Biol.* **84**, 239-251.

Smith, D.A, Connell, S.D, Robinson C. and Kirkham J. (2003). Chemical force microscopy: applications in surface characterisation of natural hydroxyapatite. *Anal. Chim. Acta* **479**, 39–57.

Sollbohmer, O, May, K-P. and Anders, M. (1995). Force microscopical investigation of human teeth in liquids. *Thin Solid Films* **264**, 176-183.

Soon, L.L.L, Bottema, C. and Breed, W.G. (1997). Atomic force microscopy and cytochemistry of chromatin from marsupial spermatoza with special reference to *Sminthopsis crassicaudata*. *Mol. Reproduction & Development* **48**, 367-374.

Southam, F, Firtel, M, Blackford, B.L, Jericho, M.H, Xu, W, Mulhern, P.J. and Beveridge, T.J. (1993). Transmission electron microscopy, scanning tunneling microscopy, and atomic force microscopy of the cell-envelope layers of the archaeobacterium. *Methanospirillium hungatei* GP1. *J. Bacteriol.* **175**, 1946-1955.

Steele, A, Goddard, D.T. and Beech, I.B. (1994). An atomic force microscopy study of the biodeterioration of stainless steel in the presence of bacterial biofilms. *International Biodeterioration & Biodegradation* **341**, 34-46.

Steinberger, R.E, Allen, A.R, Hansma, H.G. and Holden, P.A. (2002). Elongation correlates with nutrient deprivation in *Pseudomonas aeruginosa*—unsaturated biofilms. *Microb. Ecol.* **43**, 416–423.

Stroh, C.M, Ebner, A, Geretschläger, M, Freudenthaler, G, Kienberger, F, Kamruzzahan, A.S.M, Smith-Gill, S.J, Gruber, H.J. and Hinterdorfer, P. (2004a). Simultaneous topography and recognition imaging using force microscopy. *Biophys. J.* **87**, 1981–1990.

Stroh, C, Wang, H, Bash, R, Ashcroft, B, Nelson, J, Gruber, H.J, Lohr, D, Lindsay, S.M. and Hinterdorfer, P. (2004b). Single-molecule recognition imaging microscopy. *PNAS USA* **101**, 12503–12507.

Suo, Z, Yang, X, Avci, R, Kellerman, L, Pascual, D.W, Fries, M. and Steele, A. (2007). HEPES-Stabilized Encapsulation of *Salmonella typhimurium*. *Langmuir* **23**, 1365-1374.

Suo, Z, Avci, R, Yang, X. and Pascual, D.W. (2008). Efficient Immobilization and Patterning of Live Bacterial Cells. *Langmuir* **24**, 4161-4167.

Suo, Z, Avci, R, Deliorman, M, Yang, X. and Pascual, D.W. (2009). Bacteria survive multiple puncturings of their cell walls. *Langmuir* doi: 10.1021/la8033319.

Surman, S.B, Walker, J.T, Goddard, D.T, Morton, L.H.G, Keevil, C.W, Weaver, W, Skinner, A, Hanson, K, Caldwell, D. and Kurtz, J. (1996). Comparison of microscope techniques for the examination of biofilms. *J. Microbiol. Methods* **25**, 57-70.

Szabo, B, Selmeczi, D, Kornyei, Z, Madarasz, E. and Rozlosnik, N. (2002). Atomic force microscopy of height fluctuations of fibroblast cells. *Phys. Rev. E Stat. Nonlin. Soft Matter Phys.* **65**, 041910.

Takeout, M, Miyamoto, H, Sako, Y, Komizu, H. and Kusumi, A. (1998). Structure of the erythrocyte membrane skeleton as observed by atomic force microscopy. *Biophys. J.* **74**, 2171-2183.

Tkachenko, E. and Simons, M. (2002). Clustering induces redistribution of syndecan-4 core protein into raft membrane domains. *J. Biol. Chem.* **277**, 19946-19951.

Tao, N.J, Lindsay, S.M. and Lees, S. (1992). Measuring the microelastic properties of biological material. *Biophys. J.* **63**, 1165-1169.

Titcombe, L.A, Huson, M.G. and Turner, P.S. (1997). Imaging the internal cellular structure of Merino wool fibres using atomic force microscopy. *Micron* **28**, 69-71.

Uehara, H, Osada, T. and Ikai, A. (2004). Quantitative measurement of mRNA at different loci within an individual living cell. *Ultramicroscopy* **100**,197–201.

Ushiki, T, Shigeno, M. and Abe, K. (1994). Atomic force microscopy of embedment-free sections of cells and tissue. *Arch. Histol. Cytol.* **57**, 427-432.

Ushiki, T, Hitomi, J, Ogura, S, Umemoto, T. and Shigeno, M. (1996). Atomic force microscopy in histology and cytology. *Arch. Histol. Cytol.* **59**, 421-431.

Velegol, S.B. and Logan, B.E. (2002). Contributions of Bacterial Surface Polymers, Electrostatics, and Cell Elasticity to the Shape of AFM Force Curves. *Langmuir* **18**, 5256-5262.

Vinckier, A, Heyvaert, I, D'Hoore, A, van Haesendonck, C, Engelborghs, Y. and Hellemans, L. (1995). Immobilising and imaging microtubules by atomic force microscopy. *Ultramicroscopy* **57**, 337-343.

Van der Wel, N.H, Putman, C.A.J, Van Noort, S.J.T, de Grooth, B.G. and Emons, A.M.C. (1996). Atomic force microscopy of pollen grains, cellulose microfibrils, and protoplasts. *Protoplasma* **194**, 29-39.

Walsh, C.T, Fisher, S.L, Park, I.S, Prahalad, M. and Wu, Z. (1996). Bacterial resistance to vancomycin: five genes and one missing hydrogen bond tell the story. *Chem. Biol.* **3**, 21-28.

Willemson, O.H, Snel, M.M.E, van der Werf, K.O, de Grooth, B.G, Greve, J, Hinterdorfer, P, Gruber, H.J, Schindler, H, van Kooyk, Y. and Figdor, C.G. (1998). Simultaneous height and adhesion imaging of antibody-antigen interactions by atomic force microscopy. *Biophys. J.* **75**, 2220-2228.

Wolosewick, J.J. (1980). The application of polyethylene glycol (PEG) to electron microscopy. *J. Cell Biol.* **86**, 675-681.

Wright, C.J. and Armstrong, I. (2006). The application of atomic force microscopy force measurements to the characterisation of microbial surfaces. *Surface & Interface Anal.* **38**, 1419–1428.

Wu, H.W, Kuhn, T. and Moy, V.T. (1998). Mechanical properties of L929 cells measured by atomic force microscopy: effects of anticytoskeletal drugs and membrane crosslinking. *Scanning* **20**, 389–397.

Xu, W, Mulhern, P.J, Blackford, B.L, Jericho, M.H, Firtel, M. and Beveridge, T.J. (1996). Modeling and measuring the elastic properties of an archaeal surface, the sheath of *Methanospirillum hungatei*, and the implication for methane production. *J. Bacteriol.* **178**, 3106-3112.

Yamashina, S. and Shigeno, M. (1995). Application of atomic force microscopy to ultrastructural and histochemical studies of fixed and embedded cells. *J. Electron Microscopy* **44**, 462-466.

Yamamoto, A. and Tashiro, Y. (1994). Visualisation by an atomic force microscope of the surface of ultrathin sections of rat kidney and liver cells embedded in LR white. *J. Histochem. Cytochem.* **42**, 1463-1470.

Yokokawa, M, Wada, C, Ando, T, Sakai, N, Yagi, A, Yoshimura, S.H. and Takeyasu, K. (2006). Fast-scanning atomic force microscopy reveals the ATP/ADP-dependent conformational changes of GroEL. *EMBO J.* **25**, 4567–4576.

Yuan, Y. and Lenhoff, A.M. (1999). Characterisation of phase separation in mixed surfactant films by liquid tapping mode atomic force microscopy. *Langmuir* **15**, 3021-3025.

Zachee, P, Boogaerts, M, Hellemans, L. and Snauwaert, J. (1992). Adverse role of the spleen in hereditary spherocytosis: evidence by the use of the atomic force microscope. *Brit. J. Haematology* **80**, 264-265.

Zachee, P, Boogaerts, M, Snauwaert, J. and Hellemans, L. (1994). Imaging uremic red blood cells with the atomic force microscope. *Amer. J. Nephrology* **14**, 197-200.

Zachee, P, Snauwaert, J, Vandenberghe, P, Hellemans, L. and Boogaerts. (1996). *Brit. J. Haematology* **95**, 472-481.

Zhang, P-C, Bai, C, Huang, Y-M, Zhao, H, Fang, Y, Wang, N-X. and Li, Q. (1995). Atomic force microscopy study of fine structures of the entire surface of red blood cells. *Scanning Microscopy* **9**, 981-988.

Zhang P-C, Bai, C, Cheng, Y, Fang, Y, Liming, F. and Pan, H. (1996). Direct observation of uncoated spectrin with the atomic force microscope. *Science in China Ser. B* **39**, 378-385.

Zhang, Y, Sheng, S.J. and Shao, Z. (1996). Imaging biological structures with the cryo atomic force microscope. *Biophys. J.* **71**, 2168-2176.

Zhang, X, Chen, A, De Leon, D, Li, H, Noiri, E, Moy, V.T. and Goligorsky, M.S. (2004). Atomic force microscopy measurement of leukocyte–endothelial interaction. *Amer. J. Physiol. Heart Circ. Physiol.* **286**, H359–367.

CHAPTER 8

OTHER PROBE MICROSCOPES

8.1. Overview

There are a wide variety of other types of probe microscope. They differ principally in the type of surface interaction detected, and therefore the type of sensor used. In most other respects the instruments are similar - for example, they all use piezoelectric devices to initiate tip/sample movement and consequently all utilise high voltage amplifiers. Even the instrument software contains many common elements. The following list gives a flavour of the instruments available, although there are almost certainly others, particularly as prototypes.

Scanning Tunnelling Microscope (STM)
Photon Scanning Tunnelling Microscope (PSTM)
Scanning Capacitance Microscope (SCAM)
Atomic Force Microscope (AFM)
Magnetic Force Microscope (MFM)
Photonic Force Microscope (PFM)
Scanning Near-Field Optical Microscope (SNOM)
Scanning Near-Field Infrared Microscope (SNIM)
Scanning Near-Field Microwave Microscope (SNMM)
Scanning Near-Field Acoustic Microscope (SNAM)
Scanning Near-Field Thermal Microscope (SNTM)
Scanning Plasmon Near-Field Microscope (SPNM)
Scanning Ion Conductance Microscope (SICM)
Scanning Electrochemical Microscope (SECM)
Scanning Thermal Microscope (SThM)

Only some of the instrument types are relevant to biological experiments, with the more promising ones discussed below.

8.2. Scanning tunnelling microscope (STM)

Although this book deals almost exclusively with AFM, it was not the first type of probe microscope. That accolade goes to the STM, which was designed to examine conducting surfaces, particularly semiconductor materials such as silicon under ultra high vacuum (UHV), Fig 8.1. It was such a significant achievement (Binnig *et al*, 1982) that it won its creators, Gerd Binnig and Heinrich Rohrer, the Nobel Prize in 1986. The STM uses a sharp conducting tip which is either mechanically cut or electrochemically etched from metals such as gold, tungsten, or platinum-iridium alloy.

Figure 8.1. A UHV STM head manufactured by WA Technology Ltd, Cambridge, UK. The circular flange enables the head to be bolted onto other UHV apparatus. The stack of horizontal plates (A) is used to provide vibration isolation where the damping is provided by viton rubber beads positioned between each plate. The sample holder (B) secures the sample vertically, with the piezoelectric tube sited horizontally. Alignment of the tip and sample can be observed by the optical stage (C) - it is not, as some visitors to our lab believe, "......for observing the atoms"! Tip/sample separation is effected by the coarse and fine micrometers (D).

A small 'bias' voltage (V_b) of about 1 V is applied between the tip and sample, with the sample usually, but not exclusively, being at a positive potential, Fig 8.2. When the separation between the tip and sample is reduced to about 1 nm, a quantum mechanical effect occurs and a 'tunnelling' current (I_t) is generated and flows between them. This tunnelling current is extremely small - only a few nanoamps, yet despite its size, it is easily detected using a low noise amplifier. The interesting feature of the tunnelling current, is that it decays exponentially with increasing distance between the tip and the sample i.e. its magnitude changes hugely with variation in tip-sample separation. In fact, if the tip-sample separation is reduced by a mere 1 Å, the tunnel current can increase by about an order of magnitude. This makes it ideal for detecting small features on a surface, and it is so sensitive that it can easily resolve atomic lattices. This highlights an important difference when compared with AFM, namely the detection mechanism in STM has an exponential dependence on tip-sample separation, whereas in AFM this dependence is linear. This is why STM is intrinsically more sensitive than AFM. Interestingly, the extreme vertical sensitivity of the STM dictates that only the atom at the very apex of the tip participates significantly in the tunnelling process. Therefore, an STM tip behaves as though it were atomically sharp - quite some improvement over an AFM tip. Although all these factors add up to what seems like a considerable advantage, the STM has actually been rendered virtually obsolete for biological samples because it requires a conducting substrate. Despite this, it is useful to be familiar with the technique because it set the foundations on which all modern probe microscopes are built.

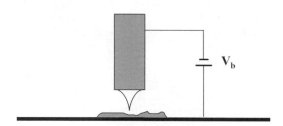

Figure 8.2. Schematic diagram showing the small separation between tip and sample. The bias voltage (V_b) is necessary to initiate the tunnelling current (I_t).

Graphite was a popular choice of substrate in bio-STM, but it is unfortunately hydrophobic and therefore repels any biological samples containing water, leading to poor binding of samples to the substrate. This drastically reduces the probability of locating (biological) samples because they are easily swept aside by the tip, particularly if adhesive forces are present, and as the tip-sample separation in STM is small, there is a significant Van der Waals attractive force between the two. Unfortunately as the control loop monitors the tunnelling current and not the force, it can become surprisingly large, thus leading to movement or damage of the sample. Some of the earliest bio-STM papers contain what are now recognised as image artifacts that actually show details of flaws in the substrate. Another limitation with bio-STM is that as the tunnelling current decays extremely rapidly away from the sample, it quickly becomes difficult to detect - the signal gets lost in the electronic noise. This sets a ridiculously low maximum height on any potential sample, and realistically restricts bio-STM to flat, film-like materials or individual molecules.

Perhaps the most common mode of operating the STM is called 'constant current' as the control loop moves the conducting tip up and down, relative to the sample, in order to maintain a constant tunnelling current. This is analogous to dc mode in AFM where the sample is imaged under a constant force.

Q: *"If the STM maps the variation in tunnel current over the sample, just what does the final image represent? Is it topographical data?"*

A: *The final image is only an approximation of the surface topography. More precisely, the STM records the number of full or empty electron states, commonly known as the 'local density of states' (LDOS). The size of the bias voltage determines exactly which states participate in the tunnelling process, and therefore influences the image contrast. Whilst for metallic conductors the final image is closely related to the topography, semiconductors, and particularly biological materials, may generate an image that is substantially unrelated to the topography. Additionally, the presence of delocalised electrons, such as those found in benzene rings, can promote enhanced tunnelling. This results in the area of the image appearing brighter, which falsely suggests that the area is taller than it really is.*

The tunnelling phenomenon that occurs between an STM tip and a conducting or semiconducting sample is reasonably well understood. However, biological samples are generally very poor conductors, so exactly why these materials should allow the passage of a tunnelling current is unclear. Conducting and semiconducting samples can be imaged reasonably routinely with STM. However, biological samples give a much poorer success rate which could be due to either sample displacement, as outlined earlier, or that the samples will not always sustain a measurable tunnel current. Whatever the reason, users of bio-STM will no doubt be familiar with the "Death or Glory" scenario, where the sample images extremely well or not at all!

8.3. Scanning near-field optical microscope (SNOM)

The performance of *conventional* optical microscopes is limited by lens aberrations and by diffraction effects, although recent developments in fluorescence microscopy have overcome this (see section 7.1.2). The 'Abbé diffraction limit', states that the maximum achievable resolution is approximately equal to half the wavelength of the light source. Green light, in the middle of the visible spectrum, has a wavelength of 550 nm. Within this constraint therefore, the theoretical spatial resolution is 275 nm (0.275 μm). However this assumes the optics are faultless and that 100 % of the light is perfectly in focus, which is clearly unrealistic. The practical resolution limit for a high quality instrument is in fact closer to 400 nm (0.4 μm). With SNOM (also known as NSOM), a laser acts as a single wavelength light source, which is transmitted through a fine optical fibre onto the sample. To circumvent diffraction problems, the SNOM uses a piezoelectric tube to position the optical fibre in the 'near-field', i.e. so close to the sample that it is illuminated before diffraction causes the light to spread out. As a rule of thumb, the separation between the sample and the optical fibre should be less than about one third of the diameter of the aperture. This ensures that you really are operating in the near-field (Fig. 8.3). The optical fibres are effectively sharpened by heating then drawing-out to a point of around 50 nm in diameter. Unfortunately this is some 10 times less than the wavelength of the light source, which causes the laser light to exit not just from the very tip of the optical fibre, but also from the sides. In order to avoid this, the exterior of the fibre is coated with a thin layer of metal, usually aluminium, and consequently light only escapes through the aperture at the very end of the fibre.

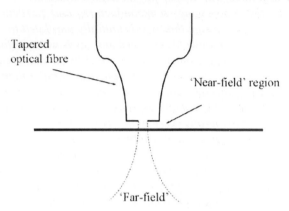

Figure 8.3. By positioning the optical fibre in the 'near-field' region the light interacts with the sample before it diffracts. The resolution of the SNOM is therefore no longer diffraction limited but depends on just how small the aperture at the end of the optical fibre can be manufactured

When operating in the near field regime the resolution of the instrument is approximately equal to the aperture diameter, and is largely independent of wavelength of the illumination. This equates to an improvement in resolution by an order of magnitude over a conventional high quality optical instrument. In the future, and with new developments, the aperture size of the optical fibre could be reduced in an effort to improve the resolution even further.

Figure 8.4. A combined SNOM/AFM produced by Thermomicroscopes, California, USA. The SPM stage is mounted onto an inverted optical microscope which permits both sample observation and the use of fluorescent markers.

The laser light illuminates the sample via the optical fibre, scattering evanescent light which is intimately associated with the surface of the sample, and allowing it to be collected by an objective lens, in what is known as the 'far-field'. The objective lens can either be in the reflection or the transmission position, with the latter generally being preferred for semi-transparent biological samples (Fig. 8.4). The light intensity collected by the objective lens provides the information used to construct the optical image. Obviously it is necessary, as with other probe microscopes, to utilise a control loop to keep the optical fibre at the correct height above the sample. In order to provide the control loop with a suitable input signal, a technique known as 'shear-force feedback' is often used. This is similar to non-contact ac mode in AFM, except with SNOM the vertical optical fibre is oscillated laterally - as it approaches the sample the amplitude of oscillation is reduced due to the influence of the attractive Van der Waals force. This information can be used to record the topography of the surface, so that optical and topographical images of the same region can be compared. Since the optical fibre is not in contact with the sample it is free to resonate to any disturbance, therefore a quiet operating environment, with some degree of acoustic isolation is essential. It is also possible to label areas of interest on the sample with fluorescent markers, which are then excited by the laser light and recorded as bright areas. Of course, the laser must emit light of the correct wavelength in order for the marker to fluoresce. As an alternative to using fibre illumination it is possible to fabricate AFM style probes but with a small aperture milled at the apex to serve as a waveguide. Additionally, 'apertureless SNOM' is possible by using an AFM tip to scatter incident laser light. In this case the tip is usually metallic, or silicon coated with metal, as it promotes enhanced light scattering. The resolution of the apertureless method is potentially higher because the probe apex is not inadvertently blunted by manufacturing the waveguide.

8.4. Scanning ion conductance microscope (SICM)

This non-contact instrument, with a sensor rather like a micro-pH probe, was developed by Hansma and co-workers (Hansma *et al* 1989) in order to image insulating materials that were immersed under various electrolytes. A micropipette containing an electrode and filled with electrolyte, is positioned close to the sample, Figs. 8.5 & 8.6. Its proximity to the sample restricts the flow of ions through the aperture at the end of the micropipette. Consequently, there is a decrease in the electrical conductance between the electrode in the micropipette and a reference electrode situated in the bulk of the electrolyte. To obtain topographical information, the micropipette is scanned over the sample using a piezoelectric actuator, whilst the control loop adjusts its vertical (Z) position in order to maintain a constant electrical conductance. Alternatively, it is possible to record local variations in ion-flow (i.e. ionic current I_{DC}) by scanning the micropipette over the sample at a constant height. The ionic current can be monitored using a patch-clamp amplifier. This can be used to provide electrophysiological data on biological systems, for example the ionic currents through pores or channels in the surfaces of membranes.

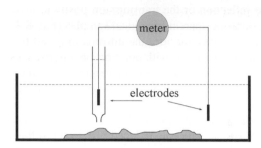

Figure 8.5. The electrical conductance between the two electrodes decreases as the micropipette closely approaches the sample. This is because the flow of ions through the tiny aperture is restricted.

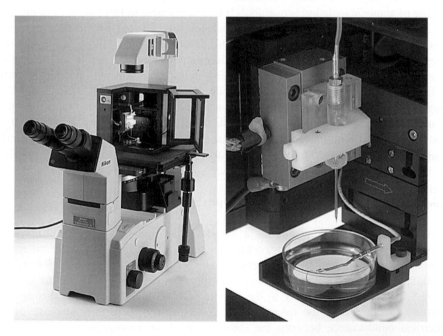

Figure 8.6. Overall view of the instrument (left) mounted on an inverted optical microscope. The Faraday cage enclosure is to eliminate any electrical interference. The 'business end' of the apparatus (right) shows the micropipette and the reference electrode. Image courtesy of Noah Freedman, Ionscope Ltd, Cambridge, UK.

Later designs (Shevchuk *et al*, 2001) have used a modulation of the pipette's vertical position by adding an AC signal to the Z piezo. This introduces an AC component to the ionic current called I_{MOD} which can now be used as the control signal instead of I_{DC}. This provides a much more robust control signal for the feedback loop that is not affected by partial pipette blockage or drift in the ion concentration caused by evaporation of the buffer. Live cells can now be studied over extended periods (hours) allowing the possibility of capturing dynamic events.

The achievable resolution is governed by just how small the aperture at the end of the glass micropipette can be manufactured - typical values are similar to those for SNOM optical fibres made using a laser heated pipette puller, at about 50 nm in diameter. Apertures as small as 10 nm can be made using quartz capillaries. The SICM has some advantages for imaging live cells over conventional AFM. With an AFM being operated in contact or tapping mode there is inevitably some deformation of the delicate cell membrane by the tip. This interaction may also cause the cell to exhibit some type of stress response. With the non-contact operation of the SICM, these deformation generated artifacts can be avoided.

8.5. Scanning thermal microscope (SThM)

This type of instrument allows the user to visualise a sample in terms of its thermal properties in addition to its topography. The thermal sensor, first described by Dinwiddie and co-workers (Dinwiddie *et al*, 1994), appears similar to an AFM cantilever but is in fact manufactured from 'Wollaston wire' - fine platinum wire encased in a sheath of silver. At the very end of the sensor the silver is etched away with acid to reveal the platinum core, and this portion is angled downwards to form a tip. The tip of the thermal sensor (Fig. 8.7.) is in contact with the sample during a scan and the topography can be recorded with the aid of a reflective surface attached to legs of the sensor. The sensor acts as both a local resistive heater and as a thermometer. The exposed platinum core at the tip is only around 5 μm in diameter, contributing the bulk of the probe's electrical resistance.

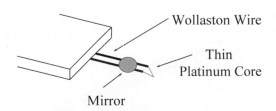

Figure 8.7. The Wollaston wire is etched away to reveal its core of platinum, which then forms the tip of the thermal sensor. The platinum core is extremely fine so the tip is the only part of the sensor with a significant electrical resistance.

This gives the advantages of providing a small localised heat source, and also minimises any thermal drift caused by temperature changes in the flexible legs of the probe. Despite their relatively crude appearance, sensors can be manufactured with low spring constants. With the application of a small dc electrical signal, the tip of the sensor heats up and some of this heat is transferred to the sample. The imaging technique is based on keeping the sensor tip at a constant temperature by altering the applied electrical signal as necessary during a scan. It is then possible to determine the electrical power required to maintain a certain temperature across different regions of the sample, and therefore obtain an image illustrating the relative variation in thermal conductivity. In addition, it is possible to add an ac component to the electrical signal so that a short heat pulse is applied to the surface, usually at a frequency of a few kHz. This provides information on the 'thermal diffusivity' i.e. how the heat penetrates into the sample at various depths. The low mass of the probe ensures a rapid response to changes in thermal properties.

The beauty of this technique is that all three types of data, topography, thermal conductivity and thermal diffusivity, can be acquired simultaneously over the same area. If the tip is held in a fixed x-y position, it is possible to explore the effects of localised sample heating by recording the sensor's z-displacement. This can provide thermomechanical information, such as melting and expansion. Early probes were somewhat crude, but more recent designs are much more like a typical AFM tip with very much sharper apexes Fig 8.8.

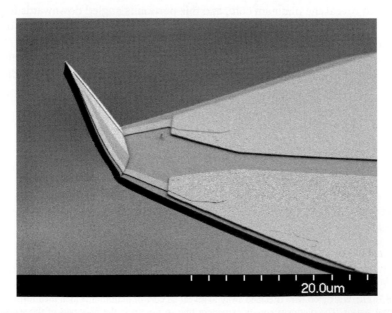

Figure 8.8. Newer thermal probe designs offer sharper tips and therefore improved resolution. Image courtesy of Anasys Instruments.

8.6. Optical tweezers and the photonic force microscope (PFM)

A remarkable device (Ashkin *et al*, 1985; 1986; 1987a; 1987b.) lies at the heart of this instrument which is known by a variety of names; either as an 'optical trap', 'optical tweezers', or even 'laser tweezers'. Whilst is not a conventional probe microscope, it is so important and powerful in its own right that it is worthwhile understanding how it operates.

Biological materials are almost invariably low density and therefore mainly semi-transparent. In the presence of light, small particulate biomaterials such as cells and bacteria can act as miniature lenses, causing the light to 'refract' i.e. to bend or deflect. If the light is particularly intense, say from a laser, it can lead to the generation of a sideways force on the illuminated object. This seems rather bizarre as there is no obvious physical interaction to cause this force, much less a force that is actually strong enough to move the object around. To understand how this force arises first refer to Fig. 8.9.

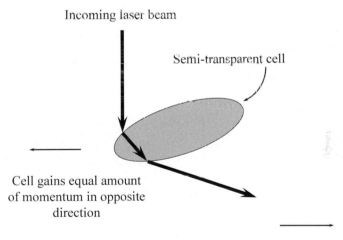

Incoming laser beam

Semi-transparent cell

Cell gains equal amount of momentum in opposite direction

Deflected laser beam gains x-component of momentum

Figure 8.9. The law of conservation of momentum dictates that any momentum gained from the deflection of the laser beam must be exactly balanced by the cell gaining momentum in the opposite direction.

As the laser beam is refracted by the lens-like behaviour of the cell, it gains a little momentum in the positive x-direction. As a fundamental physical quantity, momentum is said to be 'conserved' i.e. the total momentum of the whole system always remains constant. The only way that this can remain true is if the cell gains an equal amount of momentum in the opposite (negative x) direction - hence the cell moves. For simplicity Fig. 8.9 only shows one possible light path. For a broad and unfocussed laser beam, wider than the target cell, it seems reasonable to expect that the individual forces arising from all the various light paths would cancel out so the cell would remain stationary. However, in reality even a laser beam is not homogeneous since the intensity at the centre of the beam is somewhat higher than that at its edge. Fig. 8.10 shows two light paths A and B through a cell. The variation in laser intensity with distance from the optical axis is illustrated by the graduated shading, and the beam is shown travelling from left to right. Path A involves lower intensity light than path B, therefore the momentum gain and subsequent force on the cell is lower for A than for B. This means that the cell will always move to where the light intensity is highest, in this case towards the optical axis. Once at the optical axis the opposing forces arising from the two light paths exactly cancel each other out, and the cell remains trapped at the centre of the beam. The result of a seemingly continuous steam of photons striking the left hand side of the cell is the generation of a force gently pushing the cell to the right. This force, known as 'radiation pressure', is only significant for intense sources of illumination. However when a lens is put into a parallel beam, an extremely concentrated region of light is generated at the focal point. Any small object within a few microns of the focal point will be drawn towards it since that is where the light intensity is highest. With well focussed light the radiation pressure is overcome by the force pushing the object towards the focal point. This means that, depending on its starting position, the object can even travel towards the light source!   Once the object arrives at the focal point it remains trapped there and can be positioned by moving the laser beam.

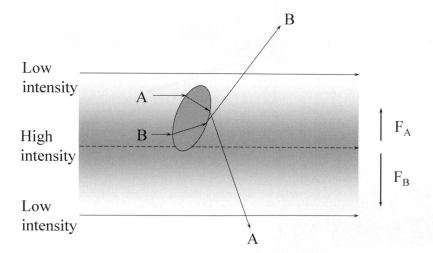

Figure 8.10. The very centre of the laser beam is more intense than at its edge. Therefore, the refracted beam (A) gains less momentum than refracted beam (B). Consequently the balancing force (F_A) on the cell as a result of beam (A) is smaller than the force (F_B) due to beam (B), and the net force pushes the cell towards the optical axis.

The first probe microscope based on optical tweezers was described by Ghislain and Webb (Ghislain and Webb, 1993). A glass shard was scanned over the sample, with scattered laser light being used to determine the displacement of the shard due to sample topography. Later Florin and co-workers (Florin *et al*, 1997) developed another variant of the instrument. In this case it is a fluorescently labelled latex bead that is captured at the focus with optical tweezers and employed as the imaging tip. Additionally, the laser doubles as an excitation source for the fluorescent label. Consequently, when the latex bead is displaced by features on the surface of the sample, it moves relative to the optical focus and the fluorescence intensity decreases - providing an elegant means of recording the sample topography.

A major disadvantage of optical tweezers has traditionally been the complexity of the instrumental set-up (Fig 8.11). The device had to be assembled 'from the ground up' by the user, which involved an optical bench featuring a laser and a large collection of lenses, mirrors, polarizers and so on. The software to control the instrument had to be written 'in house'. These factors presented a rather steep learning curve for the biologist or biophysicist wanting to enter into the field.

Figure 8.11. A traditional open architecture optical tweezers set-up. Image courtesy of John Yukich, Department of Physics, Davidson College, NC, USA.

In 2008 an all in one package was developed and marketed by a German company (JPK Instruments AG) that promises a much more user friendly alternative (Fig

8.12). It can be employed as a force sensor with sensitivities in the range of 0.1-100 pN, somewhat more sensitive (but limited) than what can be achieved with a cantilever based system (10-10,000 pN). The 'grip' that optical tweezers are capable of exerting on the sample is known as the 'stiffness of the optical trap'. This increases as the laser power rises, although clearly there is a limit if sample heating and photobleaching are to be avoided.

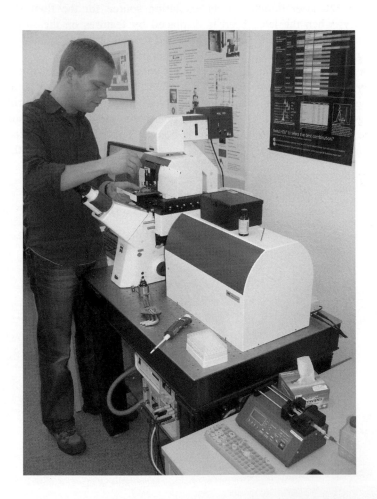

Fig 8.12. An optical tweezers set-up in a relatively compact instrument (JPK Instruments AG)

References

Ashkin, A. and Dziedzic, J.M. (1985). Observation of radiation-pressure of particles by alternating light beams. *Phys. Rev. Letts.* **54**, 1245-1248.

Ashkin, A. Dziedzic, J.M. Bjorkholm, J.E. and Chu, S. (1986). Observation of a single-beam gradient force optical trap for dielectric particles. *Optics Letts.* **11**, 288-290.

Ashkin, A. Dziedzic, J.M. and Yamane, T. (1987a). Optical trapping and manipulation of single cells using infrared laser beams. *Nature* **330**, 769-771.

Ashkin, A. and Dziedzic, J.M. (1987b). Optical trapping and manipulation of viruses and bacteria. *Science* **235**, 1517-1520.

Binnig, G. Rohrer, H. Gerber, C. and Weibel, E. (1982). Surface studies by scanning tunnelling microscopy. *Phys. Rev. Letts.* **49**, 57-61.

Dinwiddie, R.B. Pylkki, R.J. and West, P.E. (1994). *Thermal conductivity contrast imaging with a scanning thermal microscope.* In Thermal Conductivity (ed. T.W. Tong,) **22**, 668-677. Technomics: Lancaster PA.

Florin, E-L. Pralle, A. Hörber, J.K.H. and Stelzer, E.H.K. (1997). Photonic force microscope based on optical tweezers and two-photon excitation for biological applications. *J. Struct. Biol.* **119**, 202-211.

Ghislain, L.P. and Webb, W.W. (1993). Scanning-force microscope based on an optical trap. *Optics Letts.* **18**, 1678-1680.

Hansma, P.K. Drake, B. Marti, O. Gould. A.C. and Prater, C.B. (1989). The scanning ion-conductance microscope. *Science.* **243**, 641-643.

Shevchuk, A.I. Gorelik, J. Harding, S.E. Lab, M.J. Klenerman, D. and Korchev, Y.E. (2001). Simultaneous measurement of Ca2+ and cellular dynamics: combined scanning ion conductance and optical microscopy to study contracting cardiac myocytes. *Biophys. J.* **81**, 1759-1764.

CHAPTER 9

FORCE SPECTROSCOPY

9.1. Force measurement with the AFM

Probably the most unique aspect of the atomic force microscope is its ability to physically interact with the sample. Returning to the analogy of a blind person visualising samples by touching them, it takes no great leap of imagination to realise that with the sense of touch comes the ability to push, pull, deform and manipulate objects. In short the AFM can do much more than most people might imagine from their traditional notion of microscopy, of simply producing images of a sample. Factor in the extraordinary sensitivity of the AFM and you start to appreciate the power that the dimension of touch gives to the AFM. The range of forces that the AFM can sense in this mode of operation spans the range from tens of piconewtons to tens of nanonewtons. This range encompasses some of the stronger interactions which can occur between single molecules, such as covalent and electrostatic bonds, and even receptor-ligand and antibody-antigen recognition, which arise due to multiple hydrogen bonding, but is not quite sensitive enough to detect more subtle interactions, such as single hydrogen bonds. These very weak interactions remain in the realm of the optical tweezers (section 8.6) for the time being. However the AFM is not constrained by the relatively small upper limit of optical tweezers (about 100 pN) and can quantify much larger forces, revealing a huge range of new applications in biology. For example, the AFM can easily cope with the multiple event interactions likely to occur in biological processes such as inter-cellular binding or adhesion. Furthermore, the force spectroscopy measurements can be performed in a liquid environment and, for biological systems, under physiological conditions at unparalleled spatial resolution. The literature in this area is truly vast and yet the technique of force spectroscopy using AFM could still be considered as an emergent technology, due to the relative lack of traceable methods and approaches. For the beginner there appears to be a rather bewildering array of methodologies. Force spectroscopy involves stretching or compressing molecules between the tip and the substrate. Some force spectroscopy experiments employ functionalised tips and surfaces in order to control the place of attachment to the molecules, whilst others simply involve non-specific attachment of the molecule of interest to the AFM tip and stretching of the molecule, followed by data mining of the subsequent force spectra to separate out the 'good' data. This latter approach has led to criticism from practitioners of the more traditional force measurement techniques, such as the surface force apparatus (SFA). Despite these concerns, the ongoing process of careful validation in AFM force spectroscopy is helping it to become an established method, and undoubtedly AFM will continue to play an important role in solving previously intractable scientific problems. The purpose of this chapter is to explain the steps involved in performing force spectroscopy experiments with the AFM, and to summarise the various methods that have been applied to tackle different problems, so that you the reader can decide which technique is appropriate for your situation. This approach will be illustrated with examples from the scientific literature. The intention is to clarify certain aspects rather than to review the entire literature. Specific examples of force spectroscopy that have led to new insights in particular areas are described and cited in the relevant subject chapters.

The use of an AFM in the force mode can be divided roughly into two categories. The first category we will term "pulling" experiments; where the data stream comes mainly from the retraction part of a force-distance cycle. The second category will be termed "pushing" experiments, where the approach part of the force-distance cycle generates the majority of the information. Before discussing both these areas further, a brief background on the necessary first steps required in force spectroscopy is detailed below.

9.2. First steps in force spectroscopy: from raw data to force-distance curves

9.2.1 Quantifying cantilever displacement

Assuming that the AFM uses optical lever detection (which virtually all AFMs will do) then, in addition to measuring k, the sensitivity of the optical response to cantilever bending, seen at the photodetector, must be determined (Meyer and Amer, 1988). This so-called *InvOLS* (inverse optical lever sensitivity) factor can be determined in a variety of ways, most of which involve deflecting the lever by a known distance and recording the subsequent electrical signal generated by the photodetector. The most straightforward way to achieve this is to simply engage the AFM tip onto a hard surface (i.e. one which will not deform in response to the force exerted by the AFM tip) and perform a 'force versus distance' experiment. In this mode of operation the x,y channels of the AFM scanners are frozen, the feedback loop suspended, and the tip and sample pushed together (and normally also pulled apart again), by ramping the z piezo channel of the scanner, and plotting the resultant photodiode signal generated by the bending of the cantilever. The resulting raw output data generates a graph of photodiode output (potential difference or current in some cases) versus scanner displacement. The slope of this plot yields the *InvOLS* factor that should be used to calculate the *actual* distance that the cantilever deflects, s, since it relates the potential difference from the photodiode output, V, to the distance, s, moved by the cantilever thus:

$$InvOLS = \frac{V}{s} \qquad (9.1)$$

It is important to remember to carry out the *InvOLS* measurement in the liquid in which the subsequent experiments are performed, because the refractive index of the liquid will affect the properties of the optical lever detection system in non-trivial fashion. This method of pressing the tip against a hard surface is fine, provided a plain AFM tip is being used in a 'tack and pull' mode of operation (see section 9.3), but it is not a good idea if you've just spent a considerable amount of time carefully functionalising your tip with a particular molecular species in order to map specific sites on a sample. In this case performing an *InvOLS* run against a hard surface will probably damage or destroy the functionality of your tip. In the best case scenario, after the *InvOLS* run the mapping experiment simply doesn't work and no specific interactions are obtained – frustrating, but not as bad as the worse alternative which involves collecting a wealth of non-specific interaction data and several weeks of pointless data analysis. The need to make *InvOLS* measurements without ramming the tip into a hard surface and risking destroying it's functionality has therefore become a major driver for the development of alternative methods. To this end a method utilising the hydrodynamic response of cantilever - colloidal sphere assemblies in fluids has been proposed (Craig and Neto, 2001). This method was adapted from a procedure suggested for measuring *InvOLS* for standard cantilevers (Proksch and Cleveland, 2001). More recently, a more practical, alternative method for determining *InvOLS* of rectangular cantilevers, without the need for contacting a surface, has been developed (Higgins *et al* 2006). As the cantilever is sitting in the liquid it is buffeted by the thermal motion of molecules which excite into oscillation all of the inherent flexural modes. By measuring the resultant thermal noise spectrum (see section 3.2.2) of the cantilever, as revealed by the output from the photodiode, and then fitting the fundamental resonant mode to the power response function $S(f)$ of a simple harmonic oscillator, unknown parameters in the following equation can be obtained:

$$S(f) \cong P_{white} + \frac{P_{dc} f_R^4}{(f^2 - f_R^2)^2 + \dfrac{f^2 f_R^2}{Q^2}} \qquad (9.2)$$

Higgins and co-workers have derived equation 9.3 below which can then be used to determine *InvOLS*.

$$InvOLS = \sqrt{\frac{2k_B T}{\pi k f_r P_{dc} Q}} \qquad (9.3)$$

where, k_B, is Boltzmann's constant, T, absolute temperature, f_r, resonant frequency of the cantilever, and P_{dc}, is the dc power response ($V^2.Hz^{-1}$) of the cantilever measured from the photodetector.

9.2.2. Determining cantilever spring constants

Having determined *InvOLS* the next step in the procedure is to measure the spring constant of the AFM cantilever. This must be determined experimentally, since the quoted value is only an approximation based on the as-manufactured length and thickness and can be out by a factor of two or three! This seemingly obvious step can turn out to be one of the most problematic parts of the whole process, and the problems associated with this task are discussed later in this section. For now let us concern ourselves with the basic physics of cantilever mechanics. The resonant frequency of a simple harmonic oscillator, for example a mass on the end of a spring, can be determined using the following equation, which can be found in any basic physics textbook.

$$f = \frac{1}{2\pi}\sqrt{\frac{k}{m}} \qquad (9.4)$$

where k is the spring constant of the spring and m is the mass of the attached weight. But of course an AFM cantilever beam is not a simple point mass, but has its weight distributed along its length, so that the above equation needs to be modified slightly by using an *effective* mass, the unit m_0 seen in Eq. 9.5 below, which is governed by the lever geometry. Also, rather than just measuring the resonant frequency of an AFM cantilever alone, the spring constant can be determined more accurately by measuring the changes in resonant frequency as small masses (in the form of tungsten spheres) are added, using the following equation (Cleveland *et al* 1993).

$$\omega^2 = \frac{k}{m^* + m_0} \qquad \text{since} \qquad f = \frac{\omega}{2\pi} \qquad (9.5)$$

where k is the spring constant, m^* the end loaded mass, m_0 the effective cantilever mass, and ω the angular frequency of the lever. A plot of added mass, m^*, versus ω^2 has a slope equal to k, and an intercept equal to the effective cantilever mass m_0. By carefully performing the measurements described above Cleveland and co-workers (Cleveland *et al* 1993) derived the following equation which allows calculation of k with reasonable accuracy, by just measuring the unloaded resonance of the cantilever, assuming that one has accurate information on the length and width but is unsure of the thickness.

$$k = 2w\left(\pi l f_r\right)^3 \sqrt{\frac{\rho^3}{E}} \qquad (9.6)$$

where l is the length of the cantilever, w its width (not to be confused with ω in Eq. 9.5 above), ρ the density of the cantilever material, E the elastic modulus or (Young's modulus) of the cantilever material, and f_r is the measured resonant frequency. This is a useful equation because the length and width of micro-fabricated AFM cantilevers are well controlled by the high accuracy of the photolithography process (sub-micron), but thickness can vary significantly.

A more accurate method known as the 'Sader' method, requiring only that the unloaded resonance of the cantilever is measured, has been developed for rectangular cantilevers (Sader *et al* 1995; Sader *et* al 1999*)*. These authors present a detailed comparison of previous methods and, in addition, correct for the effects that air-damping and gold coating have on the measurement of the resonant frequency of AFM cantilevers. Perhaps most importantly they also demonstrate the importance of load position in relation to spring constant. The 'Sader' method has been extended to enable simultaneous calibration of the torsional spring constant (k_t) of rectangular AFM cantilevers (Green *et al* 2004). This extension, referred to as the 'Torsional Sader' method, enables calibration of the torsional spring constant using an identical experimental methodology. With the advances in the application of force spectroscopy to frictional measurements which have occurred in recent times, new definitions need to be considered as to what is actually meant by the spring constant. The normal spring constant of an AFM cantilever is generally referred to with a z subscript thus (k_z) and the torsional spring constant with a t subscript (k_t). The measured spring constant of an AFM cantilever depends quite strongly on where it is being loaded since, amongst other things, this determines the effective mass and effective length. For this reason other methods advocate a rather more direct measurement of the spring constant of a cantilever by pushing on the actual tip itself (i.e. where the force acts upon the sample) with a reference spring (Fig. 9.1), and monitoring the relative deflection (Li *et al* 1993).

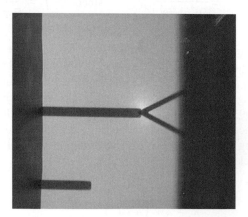

Figure 9.1 Determining the spring constant of an unknown V-shaped cantilever by pressing it against a calibrated rectangular reference lever.

This technique has benefited from advances in microelectromechanical systems (MEMS) to produce an array of springs whose geometry can be tailored to enable the measurement of normal, torsional and longitudinal spring constants with an uncertainty of less than 7% (Cumpson and Hedley, 2003; Cumpson *et al* 2004; 2005).

A fairly comprehensive comparison of many of the new methods, both mechanical (static deflection of cantilevers by pushing against a known spring) and dynamic (based upon analysis of thermal or driven motion spectra to model resonance characteristics), has been carried out recently (Clifford and Seah, 2005). For rectangular cantilevers most of the recent dynamic methods provide reasonably good estimates of spring constant, with an uncertainty level around the 15-20% mark. This arises principally from uncertainty in the value of the Young's modulus of the cantilever material, since the use of bulk values for a beam geometry which has a significant surface to volume ratio may lead to errors, since it ignores the potential effects of surface stress or cracks, and assumes perfect material homogeneity (Petersen and Guarnieri, 1979). However, when Clifford and Seah used finite element analysis to model the flexure of the more complex V-shaped cantilevers to predict spring constants the news was not so good. Comparison of the FEA predictions with measured values of k_z, obtained from application of the most commonly applied methods, revealed significant problems, with none of the dynamic methods (which fit resonance data) to determine k_z being deemed effective. With a modification of one of the equations (to correct for the effect of bending of a clamped triangular end-plate where the force is not applied at the apex) the method described by Neumeister and Ducker produced the best match with the FEA data, within a remarkable 1% (Neumeister and Ducker, 1994; Clifford and Seah, 2005), but this requires the highly problematic measurement of cantilever thickness, and a value for Young's modulus.

One of the main problems from a users' point of view is in the actual implementation of the various published methods, which are by their very nature highly mathematical and somewhat daunting for the non-physicist. To remedy this situation Cook and co-workers have written a comprehensive paper which provides practical advice on implementing the two most common dynamic methods (the 'Sader' and Thermal noise method). This excellent paper reviews the theory behind each method and discusses practical experimental means for obtaining the required parameters (Cook *et al* 2007). A positive recent development in the quest for cantilever calibration is the push to develop methods which are traceable to the international system of units (SI units). To this end force standards based upon electrostatic devices have been constructed (Chung *et al* 2008). Chung and co-workers speculate that this approach may yield reference forces in the hundreds of piconewtons range with an uncertainty level of around 1%, and so transform spring constant calibration in AFM. For the time being though the take-home lesson would appear to be that rectangular levers are the way to go for force spectroscopy applications if a dynamic non-contact method is used, where accurate knowledge of spring constant is needed but without the risk of tip damage which comes with a contact technique. Having said all of the above, in many of the latest generation of AFMs the determination of *InvOLS* and *k* tends to be included in software modules as a semi or fully automated procedure, which one engages at the start of each force spectroscopy experiment: in this case even the 'raw' output will appear as force versus scanner distance.

Irrespective of how you arrive at the values for *InvOLS* and the spring constant of the cantilever, the force, *F*, exerted by the lever on a sample is calculated using the reassuringly simple formulae given by Hooke's law, which multiplies the distance that the cantilever deflects, *s*, by it's spring constant, *k*, thus;

$$F = k.s \qquad (9.7)$$

9.2.3 Anatomy of a force-distance curve

When "force-distance" curves are presented in papers they are modified and optimised in all sorts of ways to make them easier to understand for the lay reader. This can give rise to confusion among beginners who, when doing experiments for the first time, can be presented with 'raw' data which appears less than intuitive. The purpose of this section is to explain what the different parts of a force-distance curve are and how they are obtained from the raw data. In order to familiarise you, the reader, with a raw force curve, a typical example is displayed in figure 9.2, with the pertinent features labelled.

Probably the most confusing aspect of raw force-distance plots is the fact that the data stream starts at the right hand side of the *x* axis and works back towards the origin before returning towards the starting point. Most experiments display their data the other way around, with the origin defining the starting point. Once this break with convention is understood most of the rest falls into place. There are many possible variations, but a typical force-distance cycle begins by pulling the AFM tip away from the sample surface to a predetermined distance, to ensure that the tip and sample are not in contact at the start of the cycle: this defines the zero force. Next the *z* piezo scanner ramps the tip and sample back together again until the tip and the sample surface come into contact, causing the lever to begin to deflect. The tip and sample are then pushed together, often to a predetermined maximum extent, in order to control the magnitude of the contact force and to avoid indentation of the sample. After a user-specifiable time delay the scanner direction is then reversed, pulling the tip and the sample apart and back to the starting point. If there is no plastic deformation of the sample, and no adhesion between the tip and the sample, then the return path of the cantilever should be (virtually) identical to that traversed on approach, and the data for the two stages lie on top of one another. However if, as in the example shown in figure 9.2, there is adhesion between the tip and sample, then a negative peak appears in the retract portion of the data. The depth of the negative peak is a measure of the rupture force required to pull the tip free of the surface, and the (shaded) area enclosed by the peak defines the work done (or energy consumed) in breaking the adhesive interaction. The gradient of the curve in the contact region provides information on the mechanical properties of the sample (see section 9.5.3).

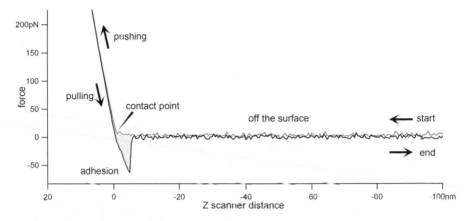

Figure 9.2 A raw force - scanner displacement graph (approach-grey line, retract-black line) obtained at a constant (approach, retract) velocity of 1 μm.s^{-1}. The cycle starts at the right hand side of the plot and the sample is moved toward to the cantilever. Initially the cantilever is above the surface, unperturbed and straight: no net force is applied. Once contact is made the 'pushing' regime begins, and the cantilever is deflected upwards. A positive force is generated which increases in a linear fashion as the lever and sample are driven together. At just over 200 pN the scanner reverses direction to begin the 'pulling' regime: The force reduces at the same rate as it increased as the lever relaxes back toward its straight position. Initially the data obtained upon pullback overlays perfectly the approach data. However, in this example the force then becomes negative due to adhesion between the lever and the sample surface. After further pulling tip-sample contact is broken and the downward bent cantilever snaps back to its straight, zero net force position, where it remains until the end of the cycle. The net force required to break contact is the adhesive force and the shaded area defines the work done in breaking the adhesion

In the raw display format, such as shown in figure 9.2, the *x* axes of force-distance graphs may sometimes have negative values to the right hand side of the zero point; this simply reflects the direction in which the z piezo scanner moves during the experimental cycle, with negative values being defined as motion of the tip away from the surface. For presentation purposes the data is then usually plotted as a force – extension curve in order to remove this anomaly. Extension is obtained by defining the contact point between tip and sample as zero separation and then displacement from the zero position becomes by definition a positive value. For raw data obtained at different points on the sample surface the contact point can appear anywhere along the *x* axis, as it is dependant upon the gap between the tip and sample at the start of the cycle, and this can vary across a sample surface, particularly if its rough. Finally, force-distance plots are frequently inverted so that the rupture force is positive, in order to make the data more intuitively pleasing (see examples in the next section).

9.3 Pulling Methods

9.3.1 Intrinsic elastic properties of molecules

Pulling methods represent the largest application of force spectroscopy with the AFM. One aspect of this is the study of the intrinsic mechanical properties of individual polymers, including both synthetic and biological macromolecules. In order to measure the mechanical properties of individual polymers it is necessary to stretch them: the polymers need to be attached to a surface and then to the AFM tip. The simplest method of achieving this, at least experimentally, is a method which might be termed 'tack and pull'. Many single molecule force spectroscopy studies on biopolymers (proteins and polysaccharides) use this method: these biopolymers tend to be pretty good at sticking to surfaces and the AFM tip, without the need for functionalisation. However, other macromolecules, such as synthetic polymers, and more precise studies on biopolymers, often require the functionalisation of tips and substrate surfaces. Details of how this is achieved are discussed in the next section (9.3.2).

Applications include fundamental elastic properties of synthetic polymers (Gianotti & Vancso, 2007), the effects of primary and secondary structure on the mechanical properties of proteins (Ng *et al* 2007), double and single stranded nucleic acids (Strunz *et al* 1999), and polysaccharides (Marszalek *et al.* 1998, 2001).

Force spectroscopy of single molecules is best illustrated by considering the simpler "tack and pull" approach and its use to study forced protein unfolding. This area provides an excellent demonstration of the power of AFM to study single molecule mechanics, and will help to explain the various steps, both practical and theoretical, in the collection and analysis of data. In this method the protein molecules of interest are allowed to adsorb from dilute solution onto a substrate for which they have some affinity. For most proteins the substrate of choice is gold-coated mica, freshly prepared to ensure surface cleanliness (Rounsevell *et al* 2004). Linkage onto the gold surface occurs via sulphur groups present in the protein to form the relatively strong S-Au bond. Sulphur is present in the amino acid cysteine, but if this is absent then it can be engineered into the sequence by genetic modification. This method brings the advantage that it can be placed in particular positions in the protein structure in a highly controlled fashion, usually at or close to either terminus, and thus give added control over the stretching measurements (Rounsevell *et al* 2004). The sample is then placed into the liquid cell of the AFM and the tip pressed down onto the surface in random positions and a force curve recorded as the tip is retracted. Occasionally a molecule will be picked up by the tip, and stretched as the sample and tip are pulled apart. Some practitioners advocate the use of relatively high contact forces (10-40 nN) and long contact duration times (1-3 s) (Marszalek *et* al 1998), whilst others find that just going through a regular force-distance cycle, in which the tip is in transient contact with the surface, is sufficient to pick up molecules (Rounsevell *et al* 2004). An everyday analogy is a bird pecking the ground in search of worms; most attempts are unsuccessful but, every so often, it strikes lucky, and a worm is pulled out of the soil. For this method to yield useful data in an AFM experiment the sample preparation conditions need to be optimised (think of it like the bird stamping its feet to simulate rain and trick the worms into coming close to the surface). Unlike the bird, which would be happy to pull up as many worms as it could at each attempt, in the AFM experiment one needs to be more selective. The coverage of the molecules must be kept low enough to avoid more than one molecule binding to the tip at the same time. The ideal preparative conditions will vary for different samples. For IgG type protein domains, incubation on a gold surface with a solution concentration of up to 5 μM, followed by rinsing in clean buffer has been recommended (Rounsevell *et al* 2004). For polysaccharides, the slightly different regime of drop deposition and drying onto clean glass from 0.001-0.1 % solutions, followed by extensive rinsing was preferred (Marszalek *et al* 1998; 2001). A good compromise between sample coverage and successful attachment to the AFM tip should produce a 'hit rate' no higher than 10 % in the retraction curves. A higher value than this greatly increases the likelihood of polluting the data with events which are due to multiple molecules being pulled in tandem (Rounsevell *et al* 2004).

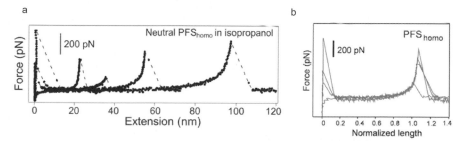

Figure 9.3 (a) Force-extension curves of individual, neutral poly(ferrocenylsilane) homopolymer stretched in isopropanol. (b) superposition of force curves normalised to a force value of 250 pN. The dotted line indicates the force chosen for normalisation. Reproduced with permission from Zhou *et al* 2006 and Elsevier.

Unfortunately, as can be seen by the large range of sample concentrations used in the studies cited above, the actual molecular concentration required to meet these criteria needs to be determined empirically. Different systems will have different affinities for both the substrate and the tip, potentially stretching your patience long before you adequately stretch the molecules! But it could be worse, spare a thought for the poor old bird: here success or failure is a matter of life and death. An obvious downside of this "tack and pull" technique is that the tip can attach anywhere along the molecule leading to a range of 'apparent' contour lengths. However, problems associated with the analysis of such data can be resolved by normalising the data at a selected force value onto a single curve as illustrated in figure 9.3 (Zou *et al.* 2006). This process works if the data comes from stretching single molecules, since it relies upon the fact that molecular elasticity should scale linearly with contour length. Incidentally this provides a means of selecting force spectra from individual molecules, because data from multiple molecules pulled in tandem will not conform to this behaviour (Zou *et al.* 2006). Nowadays, significant progress has been made in the modelling and interpretation of force spectra obtained from molecules pulled in tandem (Fantner *et al* 2006; Gu *et al* 2008). It should be appreciated that in the early days of AFM force spectroscopy this complex mixed data hampered attempts at interpretation of mechanical properties. In the studies of protein folding the complications were such that it wasn't until "polyproteins", tandem repeats of single domains constructed by protein engineering, were used that meaningful progress could be made (Carrion-Vazquez *et al* 1999). These samples allowed the unequivocal assignment of unfolding events seen in the AFM force spectra (figure 9.4). The general idea is that with such samples each domain should unfold with a near identical characteristic signature in the force curve and, because the number of domains in each "polyprotein" is known, occasions where more than one molecule is stretched during a measurement also become obvious, as the curve exhibits too many unfolding events.

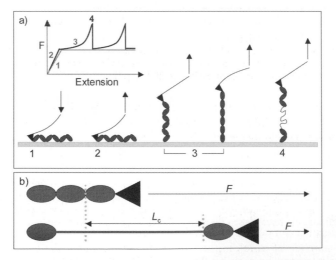

Figure 9.4 (a) Response of the cantilever during "tack and pull" force spectroscopy experiment. (1) Tip makes contact with surface and is deflected. (2) Protein is adsorbed to tip. (3) Tip is retracted from surface at fixed pulling speed and the force regime begins; the cantilever deflects in response to the entropic restoring force of the protein being extended. (4) A domain unfolding event occurs and the system relaxes. Inset is a schematic force trace illustrating the approach trace (grey) and the retraction trace (black) exhibiting force peaks expected from an ideal pull. (b) A protein domain unfolding as shown in part 4 results in an increased contour length (ΔL_c). Reproduced with permission from Rounsevell *et al.* 2004 and Elsevier.

This provided a rational basis for post-acquisition scrutiny of data, such that 'good' data could be sifted from 'bad' data. An example of real data, both "good" and "bad" is shown in figure 9.5 (Rounsevell *et al.* 2004)

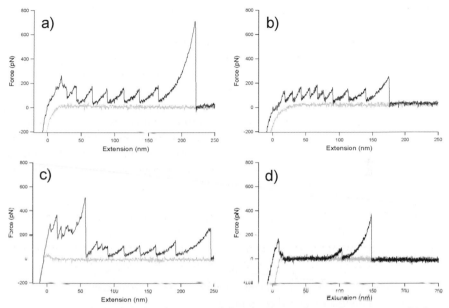

Figure 9.5 Data obtained during the course of force spectroscopy experiment on a multi-domain polyprotein. (Approach-grey, retract-black). (a) An ideal trace. (b) Multiple proteins pulled in tandem. (c) High initial non-specific forces. (d) Protein is unfolded before pulling cycle commences. Reproduced with permission from Rounsevell *et al.* 2004 and Elsevier.

As a result a generic set of selection criterion, reproduced in table 9.1, was suggested (Best *et al* 2003). For protein unfolding or polymer stretching studies, multiple repeat measurements must be carried to provide statistically adequate data. The data should be obtained at several different pulling speeds, because subsequent modelling of the data (section 9.5) incorporates load-rate as one of the parameters. Once collection of the data is optimised, refinements can be incorporated into the experiments. For example, after determining the maximum distance that a particular molecule can be stretched before it detaches from the tip (or substrate), subsequent tacked molecules can be kept within this limit and stretched repeatedly. By counting the initial number of unfolding events, and then allowing the molecule to relax, before stretching it once more, measurement of domain refolding becomes possible (Carrion-Vazquez *et al* 1999). If this is done as a function of time then it is possible to determine refolding rates directly. Furthermore, such clamped molecules have paved the way for a new force technique known as 'force-clamping' (Braun *et al* 2004; Oberhauser *et al* 2001; Fernandez & Li, 2004; Humphris *et al* 2000; 2002), In this method the molecules can be subjected to a series of step-wise force ramps under force-based feedback control. By employing force feedback control the force-clamp method has the advantage that it makes it possible to maintain either a constant force on the molecule being stretched throughout the measurement, or to subject it to a truly constant rate of increase in applied force. This simplifies subsequent interpretation of the data, which appears as a series of steps, each of which represents a single unfolding event, rather than the sawtooth patterns seen in the more traditional approach of simply pulling on the molecules. In the traditional mode of simple pulling at a fixed rate determined by the *z* piezo, domain unfolding causes the *actual* loading rate on the protein to vary throughout the course of the stretch. Another alternative procedure is that clamped molecules can be held at a given (usually low) force and the thermal noise spectra of the cantilever monitored with time (Kawakami *et al* 2004), in a similar fashion to that employed in analogous optical tweezers measurements. In this way thermally induced unfolding events can be detected, a process which some have argued is closer to the natural course of events likely to occur in solution, rather than the forced stretching data obtained from the more active pulling method (Kawakami *et al* 2004).

Table 9.1. Selection criteria for force spectra in protein unfolding studies (reproduced with permission from Best *et al* 2003 and Elsevier)

1	The non-specific/non-assignable forces at the beginning of a trace, if present, must be low (i.e. in the same force range or below that of the force peaks)
2	Force peaks must be equidistant, and the distance between them must be approximately the expected contour length of the protein
3	The trace must include three or more peaks to be confident in distinguishing force peaks from detachment events of proteins from the AFM tip
4	The base of each successive peak is higher than the base of the previous peak
5	An obvious tip-separation event (i.e. a peak of greater force than the previous peaks) must be observed
6	The pulling cycle must terminate with the complete relaxation of the cantilever, and the approach and retraction baselines must overlay, indicating that there was no drift during the course of the measurement
7	Once a trace is collected then all the peaks in the trace should be measured excluding the first peak and the final, detachment peak

Whichever method is chosen the question in many non-AFM practitioners minds remained about whether the data obtained by forcible stretching of proteins bears any relation to the real-world dynamics which govern the behaviour of proteins? Carrion-Vazquez and colleagues carried out careful comparisons of experimental data gathered by mechanical unfolding with the AFM versus chemical denaturation of a polyprotein. Monte-Carlo modelling of the AFM data yielded virtually identical values for unfolding rate constants and also virtually identical transition states to those obtained with chemical denaturation (Carrion-Vazquez *et al* 1999). These results demonstrated, at least for the protein studied, that mechanical unfolding by AFM reflects the same events that are observed in traditional chemical unfolding experiments. However, those authors also revealed that measured refolding kinetics of the polyprotein were markedly different for the traditional chemical and mechanical methods. This difference was suggested to arise from limitations that tethering placed on a molecule's degrees of freedom. Furthermore, they showed that even a very modest tethering force of 5pN provides sufficient entropic restoring force to dramatically diminish the refolding kinetics of the molecule (Carrion-Vazquez 1999), a factor likely to affect even the gentlest 'force-clamp' methods of force spectroscopy with the AFM.

There is also theoretical work which supports the validity of forced stretching of molecules with the AFM. In 1997 the physicist Christopher Jarzynski derived an identity, relating the irreversible work of multiple measurements to equilibrium-free energy differences (Jarzynski 1997). This work demonstrated the possibility of obtaining equilibrium thermodynamic parameters from processes carried out arbitrarily far from equilibrium, so long as multiple measurements at a sufficiently high signal/noise ratio are available. The work was later adapted to the analysis of single molecule pulling experiments (Hummer & Szabo, 2001) and validated experimentally with measurements of the mechanical unfolding of RNA (Liphardt 2002).

9.3.2. Molecular recognition force spectroscopy

The second main application of pulling force spectroscopy differs from that described previously, in that it is not concerned with the intrinsic mechanical behaviour of the molecule under study, but rather with using mechanical studies to probe possible interactions between molecules. Biological examples include receptor-ligand binding (Lee *et al* 1994) and antibody-antigen recognition (Dammer *et al* 1996) to name but two. Information can be obtained on factors such as kinetic rates, affinities and even the dynamic structure of the binding pocket involved in molecular recognition interactions (Hinterdorfer *et al* 2002). For this type of study it is necessary to functionalise the AFM tip with one of the molecules of interest. This is normally achieved through covalent attachment, rather than relying upon the vagaries of physisorption. Force - distance cycles are then carried out by using the functionalised AFM tip to fish for the complementary ligand or receptor on a sample surface and measuring the magnitude and frequency of adhesive interactions seen upon retraction of the tip. Once tip-sample binding has been established in a repeatable manner, the specificity of the interaction can be unequivocally established by adding the complementary partner for the active species, or a suitable inhibitor, as a free molecule into the solution. This should eliminate adhesion (figure 9.6), or at the very least a dramatically reduce the frequency of adhesive events; receptor- ligand binding is a stochastic process and some 'hits' will still be possible, but much less likely. Although the selective tethering of molecules offers significantly better control over what's being stretched than the 'tack and pull' method, the larger number of preparatory steps involved means there are many subtle aspects of the experiment which need to be optimised in order to obtain good results.

The fact that this mode of force spectroscopy is concerned with measuring specific interaction between different molecules means that their mode of presentation is crucial. As a general first step strategies are needed to protect the binding sites of the molecules of interest during attachment to the tip or in some cases the substrate or sample surface. This can be achieved by using a complexed form of the molecule (e.g. molecule – inhibitor complex) during the incubation step in the covalent attachment process, followed by break-up of the complex after attachment, often by washing or dialysis. This will yield a tethered molecule which retains its functional affinity for its complement ligand. For example, if one wishes to probe carbohydrate-lectin binding using a lectin-functionalised tip, then the carbohydrate recognition domain (CRD) of the lectin can be occupied with its target sugar whilst covalent attachment is carried out, using tethers designed to couple to free groups (Gunning *et al* 2009). If this strategy is not practical there are alternative methods to ensure that the correct sites on the molecular species are accessible for interaction following tethering (Hinterdorfer *et al* 2002). Strategies for tip functionalisation have been extensively reviewed recently (Barratin and Voyer, 2008).

Probably the most significant step forward in molecular recognition force spectroscopy was pioneered by Peter Hinterdorfer and colleagues, and is based upon attaching the probe molecule to one end of a flexible spacer consisting of a polyethylene glycol (PEG) chain (Hinterdorfer *et al* 1996, 2002, Kienberger *et al* 2006). PEG is a synthetic linear polymer which is relatively easy to synthesise in a controlled manner, such that very monodisperse chain lengths can be produced. Hinterdorfer's method involves chemical modification of PEG to produce a hetero-bifunctional linker with different reactive chemistries at either end (Haselgrübler *et al* 1995; Hinterdorfer *et al* 2002). One end of the PEG spacer is functionalised to link to the AFM tip and the other to the probe molecule under investigation. The result is that the AFM cantilever becomes akin to a fishing rod, the PEG spacer representing the line and the probe-molecule the baited hook. The list of available end-group chemistries which can be incorporated onto PEG chains is almost limitless, but the most commonly used are the amine reactive ester N-hydroxysuccinimide (NHS) to link to an amine-modified AFM tip, and the thiol reactive group 2-(pyridyldithio)propionyl (PDP) to couple to free cysteines in the probe molecule (Hinterdorfer *et al* 1996).

Amine modification of tips involves treatment of the (cleaned) silicon or silicon nitride tips with ethanolamine which, through an esterification reaction, produces a uniform film of amine groups on the silicon surface at a density such that only one or two amine molecules are present on the apex of the tip (a detailed description of the method is given by Ebner *et al* 2007). This prevents multiple PEG chain attachment to the tip apex. Furthermore, amine-coated tips also exhibit no adhesion toward mica surfaces in the absence of further functionalisation.

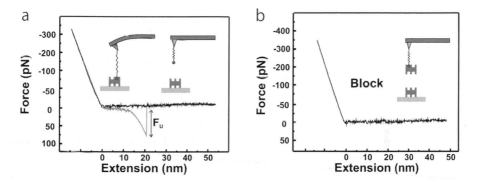

Figure 9.6 Validation of specific recognition events in force spectra (approach -black, retract–grey). (a) Adhesion (of magnitude F_u) occurs upon retraction of the tip because the ligand is bound to the receptor on the sample surface and the tip must be torn free. (b) If an excess of free receptor is introduced into the surrounding liquid the ligand on the tip is capped and thus cannot interact with the surface-bound receptor molecules: No adhesion occurs upon retraction of the tip, and the approach and retract curves overlay perfectly. Adapted by permission from Macmillan Publishers Ltd (Hinterdorfer, P. and Dufrêne, Y.F. 2006. *Nature Methods* **3**, 347-355) copyright 2006.

The use of PEG spacers provides several critical advantages in molecular recognition force spectroscopy. The probe molecule has a greatly enhanced degree of freedom over a tightly-tethered molecule, and is thus free to sample many more interaction modes and a larger interaction area on each attempt to dock with its intended target. The soft and non-linear elastic properties of PEG make it easy to discriminate between the non-specific and receptor-specific interaction events in the subsequent force-distance curves (see figure 9.7). Furthermore, spacing of the active probe molecule away from the tip reduces the likelihood that it will be damaged (crushed) beneath the tip during contact with the often hard sample surface during the force-distance cycle. This improves the repeatability of the data, a factor which is crucial for the validation step following addition of free ligand to the experiments. *A-priori* knowledge of the PEG spacer length also allows the filtering of adhesion data in terms of separation between tip and sample: this is quantified by measuring the distance between the tip-surface pull-off point in a force - distance curve and the position of the adhesive event minima along the *x* axis (see figure 9.7). This is very useful because there can be many potential sources of non-specific adhesion when a sample and tip are pulled apart, such as electrostatic and hydrophobic interactions, and these can be difficult to exclude if no spacer is used (error bars for adhesion are frequently of the same magnitude as the measured modal value for receptor-ligand type systems without spacers).

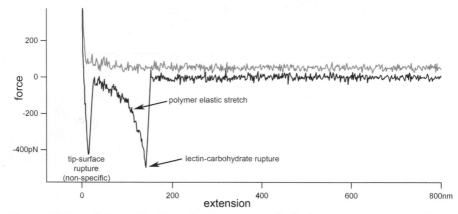

Figure 9.7 Force-distance data illustrating two sources of adhesive interaction (non-specific and polymer stretching). Data was obtained by stretching a surface-bound glucomannan polysaccharide with a lectin functionalised tip. Note that the spacer effect of the long-chain polysaccharide moves the lectin-carbohydrate rupture event away from the tip-surface separation event (approach-grey, retract-black).

Figure 9.7 illustrates how the use of a polymer spacer enables such discrimination. Upon initial separation of the tip and surface there is, in this case, a non-specific adhesive interaction. Although non-ideal (it is obviously better not to have any non-specific adhesion) the data illustrates how to discriminate between specific and non-specific adhesive events when a polymer linker is used. The non-specific adhesion is obvious from the gradient of the first negative seen peak upon retraction - it's linear and identical to that of the tip-surface contact region indicating that the tip and sample surface are still in contact with each other and moving in perfect unison. After a few tens of nanometres the energy stored in the negatively bent lever is sufficient to break the tip free of the surface and the lever returns to its zero force (straight) position. However, with further retraction the lever begins again to deflect downwards but, this time, the gradient looks very different: it deviates significantly from the linear form exhibited by the initial peak because elastic deformation is occurring. This is the characteristic signature which confirms that this peak comes from the stretching of the spacer molecule which is spanning the gap between the tip and surface. In this case the 'spacer' molecule is a relatively large, semi-flexible polysaccharide, and so this stretch region is long and pronounced. A typical force recognition experiment will consist of tens, hundreds or even thousands of pulling cycles. By binning the resulting adhesion data in terms of appropriate separation distances, one can discriminate between receptor-specific and non-specific events in an objective manner, an advantage whose significance is difficult to overstate. Interestingly there do appear to be optimal spacer lengths; it was noted that a PEG spacer of 35 nm produced fewer 'hits' for a biotinilated-PEG tip probing an avidin test system than the same system utilising an 8 nm PEG spacer (Gabai *et al* 2005). Problems associated with the non-linear loading by polymer tethers will be exacerbated as the tether length increases (Ray *et al* 2007 and also see section 9.5.2). In addition to providing information on the magnitude of specific interactions, specific molecular recognition force spectroscopy can be combined with ac modes of AFM imaging (section 3.2.2) to produce what is termed 'molecular recognition images'. In this technique the extra damping of the tip-sample contact region of the cycle due to receptor-ligand interaction is used to map the location of receptors or ligands on a surface (Raab *et al* 1999; Ebner *et al* 2005). The idea is that the free amplitude of cantilever oscillation is kept within the extended length of the spacer molecule so that the if the probe molecule (ligand) encounters a target molecule (receptor) the pair will remain bound as the tip passes over, causing a measurable damping of the oscillating cantilever. If the free amplitude of cantilever oscillation exceeds the extended length of the spacer molecule then any ligand-receptor interactions will be ruptured *during* the oscillation cycle and any damping of the tip as it passes over the target molecule becomes undetectable (Ebner *et al* 2005). Thus pure topographical imaging can be compared to 'recognition' imaging by selecting an appropriate value for the oscillation amplitude. As a means of mapping adhesive interactions across a sample surface this is a much faster and more convenient alternative to conventional 'force-volume' imaging (section 7.1.2) in which a force-distance curve is performed at each imaging point. An important point to note though is that this technique will only work properly to produce independent topography and adhesion maps of the sample surface with cantilevers with a low quality factor ($Q < 1$) in liquid, operated below resonance (Hinterdorfer & Dufrêne, 2006). In fact dynamic methods of mapping are becoming more widespread, even for quantitative molecular recognition force spectroscopy, due to their speed advantage over the quasi-static approach of performing force-distance curves at discrete points on a sample (Janovjak *et al* 2005).

9.3.3. Chemical force microscopy (CFM)

Chemical force microscopy is yet another facet of AFM force spectroscopy. It could be considered a subset of molecular recognition force spectroscopy in that functionalised AFM tips are used to probe surfaces for which some affinity is expected. It differs in that in general CFM tends to involve interactions between much smaller molecules than the macromolecular samples of molecular recognition force spectroscopy. CFM often utilises the friction generated between a functionalised tip and a surface for which it has some chemical affinity, in order to generate image contrast, as an alternative to actually quantifying rupture forces directly. Research using CFM is mainly focussed in two areas. The technique was first developed with the intention of adding chemical specificity to AFM imaging and for high-resolution chemical mapping of surfaces (Frisbie *et al* 1994), and this focus still acts as the main driver for some aspects of the technique. The second and growing application of CFM is in the study of fundamental aspects of chemical bonding and molecular tribology at surfaces on the nanoscale (Noy 2008). In this respect CFM has made significant advances into the understanding of these processes, which have in turn fed directly into improved understanding of AFM force spectroscopy. In fact to label CFM as a different technique is purely a matter of semantics, since the boundaries between this second mode of CFM and molecular recognition force spectroscopy are fuzzy: a biological molecular recognition force spectroscopy experiment may be designed to map and quantify the interaction between an antibiotic and its target peptide (Gilbert *et al* 2007) and could therefore be described equally well as an example of CFM.

9.4 Pushing Methods

9.4.1. Colloidal probe microscopy (CPM)

A huge area of interest accessible to AFM force spectroscopy is the study of interactions which govern the behaviour of colloidal particles in solution. Understanding the interactions between colloidal particles is important to a range of industries from the food and pharmaceutical industries through to oil recovery and mineral flotation. The term colloid is generally accepted to describe discrete objects (the dispersed phase), which are in the size range of nanometres to micrometres, dispersed in a liquid (the continuous phase). This diverse subject now known as 'Soft Matter' includes emulsions, foams, suspensions, through to biological systems such as the cells circulating in your bloodstream. The dispersed phases can vary in texture from soft and deformable particles though semi-solid, semi-rigid all the way to solid rigid particles. The principal issue in colloidal systems is to understand how to ensure that the particles remain dispersed within what is a non-solvent and don't simply aggregate, coalesce, sediment or cream. The interfaces created in the production of foams and emulsions are stabilised by the adsorption of surface-active molecules, usually amphiphilic molecules. These molecules adsorbed at the surface of droplets, or onto the surface of solid particles, will determine the interactions - with electrostatic and steric effects being the two most common processes - between the particles in the continuous phase, and hence the stability of the system.

Historically many techniques have been developed to measure colloidal particle interactions (Craig 1997). In recent years the advent of the surface force apparatus (SFA) (Israelachvili 1978), the atomic force microscope (AFM) (Binnig 1986) and total internal reflectance microscopy (TIRM) (Prieve 1990) have produced dramatic experimental progress in this field. Each technique has its own advantages and disadvantages (Hodges 2002). TIRM enables observation of the interaction between dynamic particles which are free to rotate and translate. This method has the disadvantage that large quantities of salt are often needed to screen repulsive interactions, in order to bring the particle into close contact with the surface of interest. The SFA is limited by its cross cylinder geometry (Hodges

2002). In order to determine absolute separation between the surfaces by interferometry the substrate is limited to silvered mica sheets. Although SFA has been adapted to increase the range of substrates that may be studied, it is not possible to directly determine the absolute separation between surfaces in such systems.

The versatility of AFM, and the ability to attach colloid probes of interest to the cantilever tip (which is how the term colloid probe microscopy was coined), has allowed a wider range of substrates to be investigated and ultimately, unlike SFA, its versatility allows direct measurement between two particles of colloidal dimensions. Furthermore, AFM allows two surfaces of interest to be mechanically driven together so there is no need to shield the very interactions that one wishes to measure, as is sometimes necessary when using TIRM. Similarly to SFA, AFM studies were initially limited to investigating rigid model surfaces i.e. particle-substrate or particle-particle interactions where the absolute separation is defined as the point at which 'hard' contact is made between these two surfaces. However, this has not hampered efforts into measuring interactions where one or both of the surfaces are deformable, which is vital to understand systems such as foams or emulsions. Reported work has evolved in recent years from trying to fit the interaction of deformable oil droplets (Hartley 1999) and air bubbles (Butt 1994) to DLVO theory, to reporting the measurement (Gunning *et al* 2004; Dagastine *et al* 2004) of deformable systems and the development and/or application of models to explain the deformability of the surfaces under investigation in these systems (Chan *et al* 2001; 2002; Gillies *et al* 2002; Aston & Berg, 2002). Other studies are focussing on external ways of accurately determining the separation between two surfaces or, more precisely, how to define the point of contact in the case of deformable samples. For example Clark and co-workers have developed an evanescent wave – AFM (EW-AFM) that is able to resolve the inter-particle distance between surfaces using a methodology similar to that currently used by the TIRM (Clark *et al* 2004; Mckee *et al* 2005). The main limitations of CPM are the AFM's rather poor ability to access very low speed regimes due to piezo drift and creep in the scanners, and the practical issues of maintaining the scrupulous standards of "surface cleanliness" (see chapter 5.2) required in the confines of an often fiddlingly small AFM liquid cell. Nevertheless, applications of CPM continue to grow and have already led to new insights into the behaviour of colloidal systems. A comprehensive review of this emerging area has been written by Hans-Jurgen Butt (Butt *et al* 2005) and Liang and co-workers have produced a more recent review of the comparison of CPM with traditional colloidal techniques (Liang *et al* 2007).

A difference between a CPM experiment and the stretching type of experiments described in the preceding sections is that in the former both parts of the force - distance curve are utilised. Pushing is just as interesting as pulling to a colloid scientist; hence it has been included under the section on "pushing" methods. The basic physical processes which occur when colloidal particles are brought into close proximity with each other, or with another surface, are similar to those described in chapter 3 (section 3.1) which occur between atoms. Interactions between charged colloidal particles are described by the DLVO theory (Derjaguin and Landau, 1941; Verwey and Overbeek, 1948), which is discussed in more detail in section 9.5.7. However, there are other effects that influence the interactions which are unique to the fluid environment in which the colloidal particles are situated. Principal of these is the hydrodynamic force which acts upon the particles, originating from the drag exerted upon them as they move through the liquid medium. When widely separated this limits the translational diffusion of the particles in a manner dictated simply by their size and the viscosity of the continuous phase. However as in human relations, when the particles start to get cosy things become more interesting! Imagine two particles suspended in a liquid being brought together. The ease with which the particles can be forced together changes as the intervening liquid film gets thinner. The surrounding liquid has a finite viscosity and this limits the speed with which it can drain out of the closing gap between the approaching particles. This opposes their coming together and they experience a repulsive force which grows in size as they get closer together. Now imagine that the particles are together and we want to pull them apart. The very same limitations on liquid drainage occur as the liquid medium tries to fill the expanding gap between the particles, and this time they experience an adhesive force. The magnitude of this so-called hydrodynamic force is clearly dependent upon the size of the particles, the speed at which they approach or try to move away from each other, and the viscosity of the liquid medium. For this reason the measured force between interacting colloidal particles is often normalised with respect to particle size, since this is usually a fixed quantity. Hence most force plots in CPM papers display force divided by particle radius versus distance. The magnitude of the hydrodynamic force can also be affected by the nature of the particles, whether they are soft or rigid, and the nature of the interfacial or surface layer surrounding each particle. Other forces which act upon colloidal particles can also arise from the nature of the particle or the surrounding medium and include steric interactions between the molecules adsorbed to surfaces or interfaces on neighbouring particles and osmotic effects such as hydration and depletion forces. In addition to studying the fundamental aspects of colloidal interactions, colloid probe cantilevers also find widespread application in quantitative mechanical force spectroscopy where sample deformation is being studied.

The first step in carrying out a CPM experiment is to obtain or make a colloid probe: an AFM cantilever that has a colloidal particle glued next to, or in place of, the pyramidal tip. These can be purchased (expensive) or "home-made" (cheaper). A common but complicated method to produce colloidal probes uses fine wires and a micromanipulator to transfer tiny drops of glue onto the end of an AFM cantilever followed by the addition of a colloidal particle. However, provided you have a means of viewing the tip then most commercial AFMs have all the requirements needed to go down the simpler 'do-it-yourself' route, without the need for an expensive micromanipulator; namely some form of crude approach control, via a motor or manual screw, plus fine x,y positional adjusters for the sample stage or AFM head.

9.4.2. *How to make a colloid probe cantilever assembly*

Most CPM experiments use colloidal silica particles, mainly because they are very robust, readily available in well controlled monodisperse sizes, and amenable to further functionalisation using well-established silane chemistry. Silica particles, or "beads" as they are commonly referred to, generally come dispersed in an aqueous solution. Sometimes the larger size ranges (>10 μm which tend to be glass rather than pure silica) come as powders. The 'solution' forms nearly always contain some preservative, and often a little surfactant to prevent aggregation, and must be washed before use.

     To wash the beads remove a few tens of microlitres of the stock silica bead suspension and place it in an Ependorf tube and then top up the sample to around the 1 mL mark with pure water. Next pellet out the beads using a bench-top centrifuge (30s at 6000rpm is fine) and wash through a series of water (3 x 1 ml), followed by ethanol or propan-2-ol (3 x 1 mL), resuspending the pellet by vortexing (or sonication if required) between each wash. Now that the beads are suspended in alcohol you are ready to deposit them onto a surface prior to transferring them onto the tip. For plain silica, which is hydrophilic, the lid of a plastic Petri dish is ideal since it is hydrophobic and the particles are easy to pick off the surface in the next step (to avoid frustration it is worth noting that they are virtually impossible to remove from glass following drop deposition). Simply pipette a few microlitres of the bead suspension onto the surface of the Petri dish and wait a few minutes for the alcohol to evaporate. If you're using dry beads an efficient way to transfer them onto a surface is to cut a wedge out of circular filter paper, dip the thin end lightly into your jar of beads and then touch it gently onto the substrate (in this case a glass slide works as well as plastic). This simple method, developed by Gleb Yakubov (Unilever Research, Colworth, UK), has the advantage of conveniently reducing greatly the number of beads being handled, and thus minimising the hazard of inhaling them, which can occur if you try to dust the dry beads 'salt-wise' onto the surface. Now that the colloidal beads are ready you need to prepare the AFM cantilever that you intend to stick the bead onto: insert the cantilever into the AFM. Begin the attachment process by mixing a small amount of two-part epoxy adhesive (the slow-setting varieties are best for the beginner). Next streak a tiny amount of the freshly-mixed adhesive onto a clean glass slide, using a micro-litre pipette tip or pin. Place the slide onto the AFM sample stage and lower the AFM head with its pre-mounted cantilever so that its apex just overlaps the end of the glue streak. Next gently lower the tip towards the glue until a snap-in event is seen. This is quite easy to detect as a rapid twitching of the lever (seen as an abrupt change in the reflected illumination coming from the back of the tip if looking from above). As soon as this occurs reverse the lever away from the glue whereupon a clear negative tug-down of the lever will be seen. Even if no glue can be seen on the tip following the procedure this tug-down upon retraction of the lever confirms that some glue will have been transferred onto the tip (in fact if you can see the glue on the tip after transfer it is almost certainly too much, as shown in figure 9.8b). Next replace the slide containing the glue with your pre-prepared colloid particle sample and repeat the process, taking care to align the particle as close to the apex of the lever as possible before picking it up. The snap-in twitch of the lever is usually sufficient to just catch the colloidal particle (figure 9.8a) without engulfing it completely in the epoxy resin (figure 9.8b) and, as before, the lever should be reversed as soon as this event is seen.

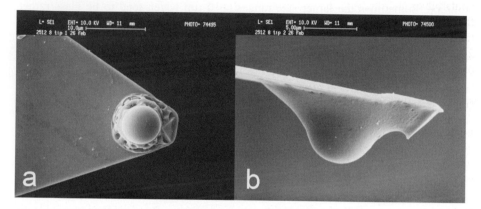

Figure 9.8 SEM images of an ideal (a) and non-ideal (b) colloid probe, where too much glue was used and the silica bead is engulfed. Images taken by Kathryn Cross.

Chemically-functionalised colloid particles often require a little more encouragement to persuade them to dislodge from the surface. In this case when you get to the snap-in point in the procedure you need to 'tweak' the AFM head with the *x* or *y* adjuster to provide a little shear. At the same time as this is being done pull the tip away from the surface. If you can't manage this, ask someone younger - any self-respecting gamer should be able to cope with this dexterous challenge pretty easily! It may seem obvious, but the best way to confirm that you have actually stuck a bead onto the tip is to check for its absence on the Petri dish/slide following transfer. It is surprisingly difficult to visualise them optically once they're stuck onto the underside of the cantilever, especially with the V-shaped varieties: even with an inverted microscope it's hard to get the lighting right. The final step is to rest on your laurels while the epoxy cures; typically overnight for slow-setting types but over a pub-lunch once you're practised enough to use quicker setting types of resin. For the reasons outlined above verification of the quality of the probe is best done by SEM. If you have access to a modern variable SEM this can be done before the use of the probes because these SEMs can image the probe without the need to coat it with metal. If not, this check can be done after the use. Not only does the SEM provide unequivocal proof of the presence of the bead and its position on the cantilever, it also easily shows-up any large surface asperities which can invalidate modelling assumptions in subsequent data analysis.

For deformable colloidal systems oil droplets can be attached to the cantilever (Gunning *et al* 2004). The first step is to thoroughly clean a glass microscope slide. Begin by flaming the glass in a Bunsen burner (blue flame) until it just begins to produce a yellow flame, indicating the release of sodium ions from the surface. Allow the slide to cool then clean thoroughly with a solution of laboratory grade detergent (e.g. 10% Micro-90, International Products Corporation, NJ, USA) using a toothbrush to scrub the slide (wear gloves to prevent contamination from your hands). Next rinse both the slide and your gloved hands thoroughly with tap water (ions are needed to rinse away the detergent properly) then demineralised water followed by ultrapure water (18.2 MΩ) and then propan-2-ol. Finally the slide should be rinsed again with ultrapure water and placed in an oven (120°C) to dry. Once dry allow the slide to cool to room temperature; now it's ready for coating with a spray of oil droplets. This can be conveniently achieved by spraying a fine mist of oil (n-tetradecane) onto the slide using a pipette tip as a venturi nozzle and an air-line. A couple of puffs are usually sufficient to deposit a good number of droplets onto the glass (you need a pretty good covering of droplets but try to avoid generating a continuous film of oil). The next step is to place water onto the slide and this has to be done in a particular manner in order to leave some of the oil droplets stuck to the slide. With the oiled slide placed on the stage of an inverted microscope, the water (typically 100 μL) should be squeezed drop-wise out of a pipette held a few cm above the slide so that it falls onto it vertically. This seems to minimise shear forces (which occur if the water is dispensed from the pipette placed directly onto the glass) allowing some of the droplets to remain attached to the glass. Having said that many of the droplets will become detached and float freely on the water surface.

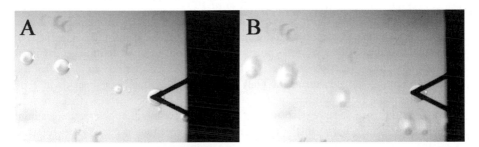

Figure 9.9 Oil droplet capture onto a 200 μm long SiN₃ cantilever. (A) lever is driven into sessile droplet which is attached to the glass surface (note: the out of focus droplets are free floating on the water surface) (B) The droplet is transferred onto the cantilever as it is retracted away from the glass - the focus tracks the lever-bound droplet, hence the remaining droplets now appear blurred.

Using a low power objective (10x) look through the microscope to find the attached droplets (the focus should be trained on the surface of the glass, not the liquid). Apart from their different focal position, attached droplets appear slightly flattened and can be distinguished from floating droplets by tapping the microscope – floating droplets will move when the microscope is tapped, attached ones won't. The next step is to place the AFM head into position and transfer the droplets onto the end of the AFM cantilever. This is done by simply positioning the apex of the lever over an attached droplet and driving the tip into it (figure 9.9a). The lever jumps into contact with a distinct twitch as the drop makes contact with the cantilever, whereupon the tip should be reversed pulling the droplet off the glass surface (figure 9.9b). A movie of the process is available at (www.ifr.ac.uk/SPM/).

This method works best for V-shaped silicon nitride cantilevers but will also work with rectangular silicon levers, although the droplet has less lateral stability on these. The lever-droplet assembly can now be positioned over one of the remaining sessile (attached) droplets for inter-droplet force measurements. Adding surfactant to the liquid cell allows the production of an interfacial film by self-assembly at the oil-water interface of the droplets. This should be done after droplet capture though, as once coated in this way they are impossible to transfer onto the lever. By making further additions to the liquid cell a whole host of environmental factors can be studied (i.e. the effects of changing ionic strength, adsorbing or non-adsorbing polymers, competing surfactants).

9.4.3. Deformation and indentation methods

Another main application for "pushing" type force spectroscopy is the measurement of the compressional properties of materials at the nanoscale. The principle involves pushing the material with an AFM tip until it deforms under the applied load and quantifying this deformation relative to an incompressible standard. This methodology can take two forms which in reality are only distinct in terms of how hard the sample is pushed. For many biological applications, such as measuring the turgor pressure of bacterial cells (Arnoldi *et al* 1998; Gaboriaud *et al* 2008) or the mechanical characterisation of mammalian cells (Li *et al* 2008) the pushing force is normally limited to ensure that the compression remains within the elastic (i.e. reversible) regime, although some plastic deformation can be difficult to avoid in practice. For more fundamental studies of material properties familiar to the material scientist, the measurement goes deliberately beyond this limit to include plastic deformation and ultimately indentation of the sample surface (Calabri *et al* 2008). In either case sample deformation is revealed by the changes in the gradient of the force-distance curve in the contact region (sometimes misleadingly referred to as the 'constant compliance' region). As mentioned at the beginning of this chapter the contact region of the force-distance curve contains information on the viscoelasticity of the sample surface because the loading force, f is proportional to the deflection of the cantilever, s:

where k is the spring constant of the cantilever. For hard samples the magnitude of the cantilever deflection, s is directly proportional to the scanner displacement, z of the sample (i.e. $s = z$). For soft samples the same scanner displacement, z will cause a smaller deflection, s of the cantilever as a result of elastic deformation of the sample, δ (i.e. $s = z\text{-}\delta$). Thus the force-distance curve for the soft sample will be of the form;

$$f = k.(z - \delta) \qquad (9.9)$$

By comparing the data for the soft sample with that obtained for the hard reference material (substrate) it is possible to generate an indentation-force curve which can then be modelled to determine the elastic modulus of the sample. Raw data can be fitted to models such as Hertzian contact theory (see section 9.5.3). The most challenging aspect of the process is obtaining an accurate measure of the size and shape of the indenter (i.e. the tip). For indentation measurements this is commonly done by SEM imaging of the resultant pits and relies upon the fact that the unknown part of the tip (the apex) makes up only a small proportion of the depth profile. The biggest problem in using AFM for indentation measurements is however that the load is not applied purely normal to the surface, but rather at an angle in most AFM tip-sample geometries. This causes the tip to plough sideways through the material to some extent during indentation, introducing potential errors for subsequent analysis.

For the 'non-indenting' class of measurements the converse is true, since penetration of the sample is deliberately limited and the main region of the tip which interacts with the sample is the difficult-to-define apex. For this reason most studies in this class tend to use a colloid probe tip to define better the interaction (Li *et al* 2008): the contact region presented by a standard AFM pyramidal tip is extremely difficult to characterise with sufficient accuracy. Furthermore, there is a very real possibility that its shape might alter during the course of the experiment by blunting or contamination of the tip, and this renders the use of standard AFM tips in such studies problematic to say the least. Add to this the uncertainties about the exact surface chemistry and physical properties of the apex of a standard tip and one can see why replacing it with a colloidal sphere of traceable characteristics becomes desirable for quantitative measurement.

9.5 Analysis of force-distance curves

Although some theoretical considerations have been discussed already, it is useful to discuss the general underlying principles and an exposition of models commonly applied to analyse the data generated by force spectroscopy: a brief introduction to the principal mathematical treatments is given in the following sections.

9.5.1. Worm-like chain and freely jointed chain models

The simplest linear polymer can be described as a random coil: a string of monomer units in which there is free rotation about the inter-monomer linkages. For real polymers the rotation about the inter-monomer linkage is restricted and the volume of the polymer increases. This can arise from the chemical structure of the monomer unit, the nature of the linkage, or the adoption of an ordered secondary structure: these polymers are called semi-flexible coils The stretching behaviour of semi-flexible polymers is best described by two statistical models, the Kratky-Porod worm-like chain model (WLC) (Kratky and Porod, 1949; Flory 1998) and the freely-jointed chain (FJC) (Beuche 1962; Smith *et al* 1996) models. These models are equivalent but use different concepts to describe stiff chains. The WLC model introduces the concept of a persistence length l_P. If one starts at a certain position on the polymer chain and walks along the chain then l_P is a measure of the length of the journey required for the start and end points to be randomly distributed in space. The model provides the best intuitive description of biopolymers such as DNA and certain helical polysaccharide chains. The WLC model has been extended to deal with particular limitations of the original model and several versions appear in the literature. The most commonly applied formula describes the extension, z, of a worm-like chain with contour length L_c and persistence length, l_P, in response to a stretching force, F, as;

$$F(z) = \frac{k_B T}{l_P}\left[\frac{1}{4}\left(1 - \frac{z}{L_c}\right)^{-2} + \frac{z}{L_c} - \frac{1}{4}\right]$$ (9.10)

where, K_B, is Boltzmann's constant and T, absolute temperature (Bustamante *et al* 1994; Marko and Siggia, 1995). In the particular case of DNA an extra term representing a stretching modulus, K_0, has been added (Wang *et al* 1997) to give the equation:

$$F(z) = \frac{k_B T}{l_P}\left[\frac{1}{4}\left(1 - \frac{z}{L_c}\right)^{-2} + \frac{z}{L_c} - \frac{F}{K_0} - \frac{1}{4}\right]$$ (9.11)

In the FJC model the stiffness is described by effectively replacing the monomer lengths by Kuhn statistical segment lengths, l_K connected by flexible linkages, in order to account for the increased volume. For semi-flexible polymers the Kuhn length is twice the persistence length. The FJC model is particularly suited for describing polymers lacking secondary structure but showing restricted rotation of the monomeric units about the inter-monomer linkages. The general form is given in equation 9.12 below:

$$Z(F) = L_C \left[\coth\left(\frac{Fl_K}{K_BT} \right) - \frac{K_BT}{Fl_K} \right] \tag{9.12}$$

To extend this model to the higher force ranges more commonly employed in force spectroscopy experiments of ordered stiffer polymers such as DNA, a modification was developed which imbues the segments with a finite elasticity so that the chain is treated as a series of springs (Smith *et al.* 1996):

$$Z(F) = L_C \left[\coth\left(\frac{Fl_K}{K_BT} \right) - \frac{K_BT}{Fl_K} \right] \left(1 + \frac{F}{K_s l_K} \right) \tag{9.13}$$

Where, l_k *is* Kuhn length and, K_S an elasticity parameter.

Provided the contour length of the polymer is large compared to l_K and the degree of extension is moderate then the extension is largely entropic in nature. By fitting the experimental force-extension data to either of the models the two fundamental polymer properties of contour length and persistence or Kuhn length can be determined (figure 9.10).

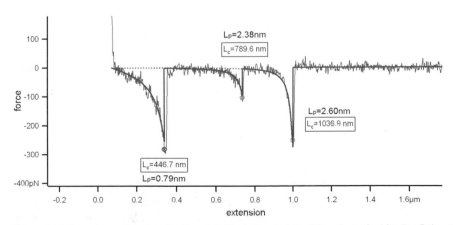

Figure 9.10 Force-extension data for the stretching of a semi-flexible polysaccharide. By fitting the experimental data (noisy grey line) to the WLC model (smooth black line) the chain characteristics of contour length, L_c and persistence length, L_p are obtained.

In the example shown in figure 9.10 three stretching events are seen. These could correspond to three polysaccharide chains of different length being stretched in one measurement cycle. However, a more likely explanation is that it is due to a couple of loops of the same molecule being pulled from the surface, since it is not necessarily just the end of the chain that is bound to the surface as illustrated in figure 9.11.

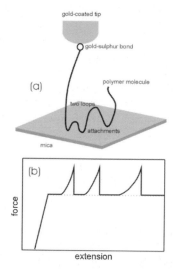

Figure 9.11 (a) Schematic diagram illustrating how multiple events can be seen in a typical polymer pulling experiment. (b) Three stretching events are seen in the force spectrum which correspond to two loops (which are pinned by intervening train sections of the polymer) being pulled off the surface. Reproduced with permission from Haschke *et al* (2004) and the American Chemical Society.

At larger extensions enthalpic terms need to be considered. The modifications introduced in equations 9.11 and 9.13 provide a first approximation to describing such effects. At even larger extensions more detailed models are required to describe events such as the unzipping of helices, configurational changes in the shape of the sugar rings, or the localised and sequential intrinsic unfolding of protein domains.

9.5.2 Molecular interactions

Receptor-ligand binding is a stochastic process in which the bound and unbound states exist in equilibrium. The equations describing the process basically quantify the ratio of on-rates and off-rates for the complex. Even at zero applied force any given bond has finite lifetime, τ_0, and will dissociate at longer timescales. The average lifetime of a bond (τ_0) is given by the inverse of the off-rate constant (k_{off}): $\tau_0 = 1/k_{off}$. However, if the timescale over which an attempt is made to break the bond is less than the average lifetime of the bond then it will resist rupture and a finite force will be required. This is basically what happens in a force-spectroscopy experiment, and following the logic of this argument one can see that the magnitude of the rupture force for any given bond is not a single value but varies with the rate at which the complex is ruptured. Hence the pulling speed is a parameter which is varied in specific recognition experiments to produce plots that relate the magnitude of the unbinding force to the loading rate. It is important to note though that the actual loading rate, r, when a polymer spacer is used in the experiments is a combination of the pulling speed set in the instrument software and the non-linear rate of extension shown in the force distance profile of the stretching region of the adhesion curve (i.e. $r=df/dt$). Indeed it has been shown recently that the non-linear loading rate which polymer tethers exert upon the bonds being investigated produces systematic errors in the subsequent calculation of dissociation rates which require correction (Ray *et al* 2007). On the timescale of a typical AFM experiment (milliseconds to seconds) thermal activation governs the unbinding processes. In this case the lifetime of the complex can be approximated by a Boltzmann expression (Bell 1978; Hinterdorfer *et al* 2002; Keinberger *et al* 2006):

$$\tau(0) = \tau_{osc} \exp\left(\frac{E_{barr}}{k_B T}\right) \qquad (9.14)$$

where τ_{osc} is the inverse of the natural oscillation frequency of the complex and E_{barr} is the energy barrier for dissociation. This gives a simple Arrhenius dependency of dissociation rate on barrier height. Imposed force acting on the complex deforms the interaction energy landscape, lowering the activation energy barrier. The lifetime of a bond being stretched at constant force, f, is given by (Evans and Ritchie, 1997):

$$\tau(f) = \tau_{osc} \exp\left(E_{barr} - \frac{fx}{k_B T}\right) \qquad (9.15)$$

where x is the distance of the energy barrier from the minimum along the direction of applied force. The lifetime of the complex under constant force compares to lifetime under zero force thus;

$$\tau(f) = \tau(0) \exp\left(-\frac{fx}{k_B T}\right) \qquad (9.16)$$

It is possible to predict how the distribution of unbinding force will vary with loading rate (Strunz *et al* 2000). The maximum of each distribution, $f^*(r)$, gives the most probable unbinding force for its respective loading rate, r, and is given by;

$$f^*(r) = \frac{k_B T}{x} \ln \frac{rx}{k_B T k_{off}} \qquad (9.17)$$

The unbinding force scales linearly with the logarithm of the loading rate, so for a single energy barrier this will result in a single gradient in a plot of f^* versus r. For complexes which have more than one energy barrier, the plot will exhibit a series of linear regimes, one for each barrier. Thus, by varying the loading rate in a force spectroscopy experiment one can obtain both structural and kinetic information about the interaction of the complexed species. The slope of the measured rupture force versus loading-rate plots yield the length of the energy barrier (the inverse of the slope represents the distance from the energy minimum to the maximum of the activation barrier along the projection of the applied force vector). Extrapolation of the data to zero force yields the off-rate constant for dissociation of the complex.

A problem which commonly plagues force spectroscopy data for receptor-ligand interactions is the presence of an unexpected high force tail in the histogram of measured rupture forces. This causes problems in the analysis of the data because it skews the fit obtained from the distributions, and this can lead to errors in the calculation of important parameters. The origins of this high rupture force 'tail' have been investigated by Gu and co-workers and have been attributed to the (nearly) simultaneous rupture of more than one bond (Gu *et al* 2008). An uneven distribution of rupture forces can arise due to several factors; polydispersity in the spacer length, variations in the spacer attachment position on the AFM tip and sample roughness. As a consequence of any of these factors, when multiple bond breakage is measured in a pulling cycle, one bond will always experience higher loading than the other(s) (i.e. one spacer will be the most stretched) and so will be the first to break. When rupture occurs the load will immediately transfer onto the bond connected by the new most-stretched spacer and this too may rupture. However this second bond will rupture at a significantly lower 'apparent' force than the first and occur almost instantly. The problem is that the second rupture will be irresolvable as a separate event and so its magnitude will simply add to that of the initial rupture, but as explained not in a quantised fashion (i.e. it won't be a simple multiple of the typical single bond strength). This leads to apparent 'single' event peaks in the data which have larger magnitude than that typical of the rupture of a single bond (Gu *et al* 2008). Gu and co-workers developed a model top take account of these ideas, and the group went on in a second paper to show that AFM data on biotin - streptavidin complexes could be analysed to yield kinetic parameters in agreement with the energy landscape predicted by molecular dynamics simulations (Guo *et al* 2008). They also showed that, in situations when more than one molecular bond might rupture during the pulling measurements, there is a noise-limited range of pulling velocities where the kinetic parameters measured by force spectroscopy correspond to the true energy landscape. Outside this range of velocities, the kinetic parameters extracted by using the standard most probable force approach might be interpreted as artificial energy barriers that are not present in the actual energy landscape (Guo *et al* 2008).

There is a real need to automate or semi-automate the analysis of force spectroscopy data. A typical molecular recognition experiment can generate hundreds or thousands of force-distance curves and detailed procedures for analysis have been described by Baumgarter and co-workers (Baumgartner *et al* 2000). In molecular recognition force spectroscopy the objective is quantification of the magnitude of the unbinding or rupture force from the force-extension data. This can be done by simple histographic analysis of the adhesive events gathered from the multiple pulling-cycles. A smarter method has been developed which constructs empirical probability density functions from the unbinding force data (Baumgartner *et al* 2000). Single Gaussian functions of unitary area are calculated from the mean and variances of every value for the rupture force. The rupture forces are obtained by measuring the jump heights of the adhesive events in the retraction part of the curve. Individual Gaussian functions are added up and normalised to yield a plot where the data are weighted by their accuracy (Baumgartner *et al* 2000; Hinterdorfer *et al* 2002).

A more recent automated procedure has been developed which is available to download as freeware (Kuhn *et al* 2005). As well as automating the analysis of force spectra it extends the capabilities of the technique by employing pattern recognition to pick out unusual events in the data, which can be indicative of sub-populations within a sample set. This can reveal rare cases with different unbinding/unfolding pathways which could otherwise be lost by averaging (Kuhn *et al* 2005).

9.5.3. Deformation analysis

Modelling of the deformation which occurs when an AFM tip (or colloidal tip) is pressed into a sample surface is based upon Hertz's theories of contact mechanics (Hertz 1882). Hertz observed that a glass sphere placed on a lens generated elliptical Newton's rings. He realised that this meant the pressure exerted by the sphere on the lens must also assume an elliptical distribution. His theory assumes that the surface being deformed is homogeneous and ignores adhesion between the probe and the surface. The two most commonly used models of contact mechanics, the JKR ((Johnson, Kendall and Roberts, 1971) and DMT (Derjaguin, Muller and Toporov, 1975) are based on the Hertz model. The pertinent aspects of the JKR and DMT models in terms of AFM contact mechanics were summarised in very useful paper (Piétrement and Troyon, 2000) and are reproduced overleaf:

9.5.4 Adhesive force at pull-off

The two models include the effect of adhesion between the probe and the surface and the relevant equations for the adhesive (rupture) force are:

$$F_{ad} = \frac{3}{2}\pi\gamma R \quad (JKR) \tag{9.18}$$

$$F_{ad} = 2\pi\gamma R \quad (DMT) \tag{9.19}$$

where R is radius of curvature of the indenting probe and γ is the interfacial energy.

9.5.5 Elastic indentation depth, δ, and contact radius, a, during deformation

Both models generate expressions for the indentation depth (δ) and contact radius (a) during deformation in terms of the radius of the probe and the adhesive force.

$$a(JKR) = \left(\frac{R}{K}\left(\sqrt{F_{ad(JKR)}} + \sqrt{F_n + F_{ad(JKR)}}\right)^2\right)^{\frac{1}{3}} \tag{9.20}$$

$$\delta(JKR) = \frac{a^2(JKR)}{R} - \frac{4}{3}\sqrt{\frac{a(JKR)F_{ad}(JKR)}{RK}} \tag{9.21}$$

$$a(DMT) = \left(\frac{R}{K}\left(F_n + F_{ad(DMT)}\right)\right)^{\frac{1}{3}} \tag{9.22}$$

$$\delta(DMT) = \frac{a^2(DMT)}{R} \tag{9.23}$$

where K is a reduced elastic modulus of the tip and sample given by the term;

$$K = \frac{4}{3}\left(\left(1 - v_1^2\right)/E_1 + \left(1 - v_2^2\right)/E_2\right)^{-1} \tag{9.24}$$

where E_1 and E_2 are the Young's moduli for the tip and sample, and v_1 and v_2 the tip and sample Poisson ratios.

9.5.6 Contact radius at zero load

The contact radius at zero load (a_0) is given by the expressions:

$$a_0(JKR) = \left(\frac{6\pi R^2}{K}\right)^{\frac{1}{3}}$$ (9.25)

$$a_0(DMT) = \left(\frac{2\pi R^2}{K}\right)^{\frac{1}{3}}$$ (9.26)

The JKR model is best suited to compliant samples, strong adhesive forces and large tip radii, whereas DMT is better suited for stiff materials, weak adhesion forces and small tip radii. The transition region between the JKR and DMT models is covered by the 'M-D' model which includes a parameter that can be used to decide whether the JKR or DMT model is most applicable in a given case (Tabor 1977; Maugis 1992; Dugdale 1960). However, this can be difficult to apply to experimental data because numerical solutions to the equations have to be derived by iteration (Piétrement and Troyon, 2000).

Recently, a set of generalised solutions to the M-D model have been developed, which enable comparison with experimental data and also accurately accommodate both the JKR and MDT models, thus enabling the modelling of data from intermediate samples (Piétrement and Troyon, 2000). Another popular model for nano-indentation, which treats the unloading data as purely elastic, is the Oliver-Pharr model (Oliver and Pharr, 1992).

9.5.7 Colloidal forces

The stability of colloidal particles can be understood in terms of the DLVO theory which describes the interaction between spherical colloidal particles in an aqueous medium in terms of the attractive van der Waals (vdW) forces and repulsive Coulomb interactions (see for example Israelachvili 2009). Figure 9.12 shows a plot of the effective pairwise interaction energy (W) versus particle surface separation (D) derived using DLVO theory.

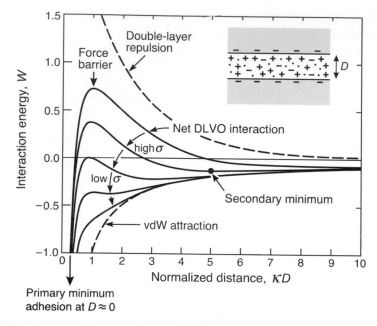

Figure 9.12 Energy versus distance profiles for DLVO interactions. As the surface charge (σ) decreases so does the height of the repulsive energy barrier. The interaction becomes purely attractive and colloidal stability is lost. Note that one unit on the horizontal scale is the effective distance over which the dissolved electrolyte screens a unit charge since κ is the surface dissociation constant (i.e. reciprocal of the Debye length). Reproduced with kind permission from Israelachvili (2009).

The net potential curves (solid lines) can be used to predict stability against coalescence of the particles for different scenarios. All of the curves have a primary minimum at very short range where van der Waals forces dominate. Colloidal particles in sufficiently close proximity will fall into this energy well and suffer irreversible aggregation. In a stable system the presence of surface charge on the particles causes a repulsive energy barrier to appear. At low electrolyte concentration and high surface charge the Brownian motion of the colloidal particles at ambient temperature will not provide sufficient energy to breach the relatively large energy barrier. However, as the electrolyte concentration of the continuous phase increases the effective range of the surface charge repulsion will be reduced, due to screening by mobile counterions (Figure 9.12 inset), and the energy barrier is lowered. At sufficiently high salt (and/or low surface charge) the barrier is removed completely and the dispersion becomes unstable. Particles are attracted to each other at long range and will aggregate/coalesce very rapidly as they fall irreversibly into the deep primary minimum. Between these two extremes (i.e. moderate ionic strength or surface charge), there exists a secondary minimum in the net energy curve which can cause reversible aggregation of the particles. They stick to one another but are kept out of the primary minimum by the remaining energy barrier (think of it like standing on a trampoline covering a well, you'll sag in a little way to reach an apparent equilibrium position but still be able to climb out with a moderate amount of work). Experimental data obtained from colloid probe microscopy can be compared to DVLO theory (Ralston *et al* 2005; Liang *et al* 2007) and an inexpensive DLVO fitting routine is available from Clayton McKee (Department Chemical Engineering, Virginia Tech. VA). Deviations from DLVO behaviour can indicate the presence of non-DLVO interactions in the system which may occur due to hydrodynamic, steric, osmotic or hydrophobic effects. However even in the absence of 'non-DLVO' effects the model can sometimes fail for an electrostatically dominated system. Like any analytical model it is subject to mathematical approximations. The Coulomb contributions in particular are considered at a mean-field level that ignores charge-charge correlations of the dissolved ions. Each ion only feels the collective average field of its neighbours rather than their individual interactions. Moreover, the equilibrium distribution of these ions is treated in a linear response approximation. These neglected effects become important at high electrolyte concentrations or in the presence of polyvalent ions and give rise to significant deviation from DLVO predictions. Further, the Derjaguin approximation is used to simplify the geometry of curved surfaces and is limited to an interaction range which is much less than the radius of the colloidal particles. Consequently, DLVO theory can also fail for small particles where surface curvature is comparable to characteristic separation distances.

References

Arnoldi, M, Kacher, C.M, Bäuerlein, E, Radmacher, M. and Fritz, M. (1998). Elastic properties of the cell wall of Magnetospirillum gryphiswaldense investigated by atomic force microscopy. *Appl. Phys. A*, **66**, S613–S617.

Aston, D.E. and Berg, J.C. (2002). Thin-film hydrodynamics in fluid interface-atomic force microscopy. *Ind.Eng. Chem. Res.* **41**, 389-396.

Barratin, R. and Voyer, N. (2008). Chemical modifications of AFM tips for the study of molecular recognition events. *Chem. Commun.* 1513-1532.

Baumgartner, W, Hinterdorfer, P. and Schindler, H. (2000). Data analysis of interaction forces measured with the atomic force microscope. *Ultramicroscopy* **82**, 85-95.

Bell, G.I. (1978) Models for the specific adhesion of cells to cells. *Science* **200**, 618-627.

Best, R.B, Brockwell D.J, Toca-Herrera J.L, Blake A.W, Smith D.A, Radford S.E,.and Clarke J. (2003). Force mode atomic force microscopy as a tool for protein folding studies. *Analytica Chimica Acta* **479**, 87-105.

Beuche, F. (1962) *Physical Properties of Polymers*. Interscience, New York.

Binnig, G, Quate, C.F. and Gerber, C. (1986). Atomic Force Microscope, *Phys. Rev.Letts.* **56**, 930.

Braun, O, Hanke, A. and Seifert, U. (2004). Probing molecular free energy landscapes by periodic loading. *Phys. Rev. Letts.* **93**, Article number 158105.

Bustamante, C, Marko, J.F, Siggia, E.D. and Smith, S. (1994). Entropic elasticity of lambda-phage DNA. *Science*, **265**, 1599-1600.

Butt, H.J. (1994). A technique for measuring the force between a colloidal particle in water and a bubble. *J.Coll.Interface Sci.* **166**, 109-117.

Butt, H.J, Cappella, B. and Kappl, M. (2005). Force measurements with the atomic force microscope: Technique, interpretation and applications. *Surface Science Reports* **59**, 1-152.

Calabri, L, Pugno, N, Menozzi, C. and Valeri, S. (2008). AFM nanoindentation: tip shape and tip radius of curvature effect on the hardness measurement. *J. Phys.: Condensed Matter* **20**, Article number 474208.

Carrion-Vazquez, M, Oberhauser, A.F, Fowler, S.B, Marszalek, P.E, Broedel, S.E, Clark, J. and Fernandez, J.M. (1999). Mechanical and chemical unfolding of a single protein: A comparison. *PNAS USA* **96**, 3694-3699.

Chan, D.Y.C, Dagastine, R.R. and White, L.R. (2001). Forces between a rigid probe particle and a liquid interface: I. The repulsive case, *J. Colloid & Interface Sci.* **236**, 141-154.

Chan, D.Y.C, Dagastine, R.R. and White, L.R. (2002). Forces between a rigid probe and a liquid interface: II. The general case. *J. Colloid & Interface Sci.* **247**, 310-320.

Chung, K-H, Scholz, S, Shaw, G.A, Kramar, J.A and Pratt, J.R. (2008). SI traceable calibration of an instrumented indentation sensor spring constant using electrostatic force. *Rev. Sci. Instrum.* **79**, Article number 095105.

Clark, S.C, Walz, J.Y. and Ducker, W.A. (2004). Atomic force microscopy colloid-probe measurements with explicit measurement of particle-solid separation, *Langmuir*, **20**, 7616-7622.

Cleveland, J.P, Manne, S, Bocek, D. and Hansma, P.K. (1993). A non-destructive method for determining the spring constant of cantilevers for scanning force microscopy. *Rev. Sci. Instrum.* **64**, 403-405.

Clifford, C.A. and Seah, M.P. (2005). The determination of atomic force microscope cantilever spring constants via dimensional methods for nanomechanical analysis. *Nanotechnology* **16**, 1666-1680.

Cook, S.M, Schäffer, T.E, Chynoweth, K.M, Wigton, M., Simmonds, R.W and Lang, K.M. (2006). Practical implementation of dynamic methods for measuring atomic force microscope cantilever spring constants. *Nanotechnology* **17**, 2135-2145.

Craig, V.S.J. (1997). An historical view of surface force measurement techniques, *Colloids & Surfaces A: Physicochem.Eng. Aspects* **129-130**, 75-93.

Craig, V.S.J. and Neto, C. (2001). *In-situ* calibration of colloid probe cantilevers in force microscopy: Hydrodynamic drag on a sphere approaching a wall. *Langmuir* **17**, 6018-6022.

Cumpson P.J. and Hedley J. (2003). Accurate analytical measurements in the atomic force microscope: a microfabricated spring constant standard potentially traceable to the SI. *Nanotechnology* **14**, 1279-1288.

Cumpson PJ, Zhdan P. and Hedley J. (2004). Calibration of AFM cantilever stiffness: a microfabricated array of reflective springs. *Ultramicroscopy* **100**, 241-251.

Cumpson PJ, Hedley J, Clifford CA. (2005). Microelectromechanical device for lateral force calibration in the atomic force microscope: Lateral electrical nanobalance. *J. Vac. Sci. Technol. B* **23**, 1992-1997.

Dagastine, R.R, Stevens, G.W, Chan, D.Y.C. and Grieser, F. (2004). Forces between two oil droplets in aqueous solution measured by AFM. *J. Colloid Interface Sci.* **273**, 339-342.

Dammer, U, Hegner, M, Anselmetti, D, Wagner, P, Dreier, M, Huber, W. and Güntherodt, H-J. (1996). Specific antigen/antibody interactions measured by force microscopy. *Biophys. J.* **70**, 2437-2441.

Derjaguin, B.V. and Landau, L. (1941). Theory of the stability of strongly charged lyophobic sols and of the adhesion of strongly charged particles in solutions of electrolytes, *Acta Physico Chemica URSS* **14**, 633-662.

Derjaguin, B.V, Muller V.M. and Toporov, Y.P. (1975). Effect of contact deformations on the adhesion of particles, *J. Colloid & Interface Sci.* **53**, 314-325.

Dugdale, D.S. (1960) Yielding of steel sheets containing slits. *J. Mech. Phys. Solids* **8**, 100-104.

Ebner, A, Kienberger, F, Kada, G, Stroh, C.M, Geretschläger, M, Kamruzzahan, A.S.M, Wildling, L, Johnson, W.T, Ashcroft, B, Nelson, J, Lindsay, S.M, Gruber, H.J. and Hinterdorfer, P. (2005). Localization of Single Avidin–Biotin Interactions Using Simultaneous Topography and Molecular Recognition Imaging. *CHEMPHYSCHEM.* **6**, 897 900.

Ebner, A, Hinterdorfer, P. and Gruber, H.J. (2007). Comparison of different aminofunctionalization strategies for attachment of single antibodies to AFM cantilevers. *Ultramicroscopy* **107**, 922-927.

Evans, E. and Ritchie, K. (2007). Dynamic strength of molecular adhesion bonds. *Biophys. J.* **72**, 1541-1555.

Fantner, G.E, Oroudjev, E, Schitter, G, Golde, L.S, Thurner, P, Finch, M.M, Turner, P, Gutsmann, T, Morse, D.E, Hansma, H. and Hansma, P.K. (2006). Sacrificial Bonds and Hidden Length: Unraveling Molecular Mesostructures in Tough Materials. *Biophys. J.* **90**, 1411–1418.

Flory, P. (1998) Statistical Mechanics of Chain Molecules. *Hanser, Munchen.*

Frisbie, C.D, Royzsnyai, L.W, Noy, A, Wrighton, M.S. and Lieber, C.M. (1994). Functional group imaging by chemical force microscopy. *Science* **265**, 2071-2074.

Fernandez, J.M. and Li, H. (2004). Force-Clamp Spectroscopy Monitors the Folding Trajectory of a Single Protein. *Science* **303**, 1674-1678.

Gabai, R, Segev, L. and Joselevich, E. (2005). Single Polymer Chains as Specific Transducers of Molecular Recognition in Scanning Probe Microscopy. *J. Amer. Chem. Soc.* **127**, 11390–11398.

Gaboriaud, F, Gee, M.L, Strugnell, R. and Duval, J.F.L (2008). Coupled Electrostatic, Hydrodynamic, and Mechanical Properties of Bacterial Interfaces in Aqueous Media. *Langmuir* **24**, 10988-10995.

Giannotti, M.I. and Vancso, G.J. (2007). Interrogation of Single Synthetic Polymer Chains and Polysaccharides by AFM-Based Force Spectroscopy. *CHEMPHYSCHEM.* **8**, 2290-2307.

Gilbert, Y, Deghorain, M, Wang, L, Xu, B, Pollheimer, P.D, Gruber, H.J, Errington, J, Hallet, B, Haulot, X, Verbelen, C, Hols, P. and Dufrene, Y.F. (2007). Single-molecule force spectroscopy and imaging of the vancomycin/D-Ala-D-Ala interaction. *Nano Letters* **7**, 796-801.

Gillies, G, Prestidge, C.A. and Attard, P. (2002). An AFM study of the deformation and nanorheology of cross-linked PDMS droplets. *Langmuir* **18**, 1674-1679.

Green, C.P, Lioe, H., Cleveland, J.P, Proksch, R, Mulvaney, P, and Sader, J.E. (2004) Normal and torsional spring constants of atomic force microscope cantilevers. *Rev. Sci. Instrum.* **75**, 1988-1996.

Gu, C, Kirkpatrick, A, Ray, C, Guo, S. and Akhremitchev, B.B. (2008). Effects of multiple-bond ruptures in force spectroscopy measurements of interactions between fullerene C60 molecules in water. *J. Phys. Chem. C* **112**, 5085-5092.

Gunning, A.P, Mackie, A.R, Wilde, P.J. and Morris, V.J. (2004). Atomic force microscopy of emulsion droplets: Probing droplet-droplet interactions. *Langmuir* **20**, 116-122.

Gunning, A.P, Chambers, S, Pin, C, Man, A.L, Morris, V.J. and Nicoletti, C. (2008). Mapping specific adhesive interactions on living human intestinal epithelial cells with atomic force microscopy. *FASEB J.* **22**, 2331–2339.

Gunning, A.P, Bongaerts, R.J.M. and Morris, V.J. (2009). Recognition of galactan components of pectin by galectin-3. *FASEB J.* **23**, 415–424.

Guo, S, Ray, C, Kirkpatrick, A, Lad, N. and Akhremitchev B.B. (2008). Effects of Multiple-Bond Ruptures on Kinetic Parameters Extracted from Force Spectroscopy Measurements: Revisiting Biotin-Streptavidin Interactions. *Biophys. J.* **95**, 3964–3976.

Hartley, P.G, Grieser, F, Mulvaney P, and Stevens G.W. (1999). Surface forces and deformation at the oil-water interface probed using AFM force microscopy. *Langmuir* **15**, 72827289.

Haschke, H, Miles, M.J. and Koutsos, V. (2004). Conformation of a single polyacrylamide molecule adsorbed onto a mica surface studied with atomic force microscopy. *Macromolecules* **37**, 3799-3803.

Haselgrübler, T, Amerstorfer, A, Schindler, H. and Gruber, H.J. (1995). Synthesis and Applications of a New Poly(ethylene glycol) Derivative for the Crosslinking of Amines with Thiols. *Bioconjugate Chem.* **6**, 242-248.

Hertz, H. (1882). Uber die Beruhrung fester elastischer Korper. *J. Reine Angew Math.* **92**, 156-171.

Higgins, M.J, Proksch, R, Sader, J.E, Polcik, M., McEndoo, S, Cleveland J.P. and Jarvis S.P. (2006). Noninvasive determination of optical lever sensitivity in atomic force microscopy. *Rev. Sci. Instrum.* **77**, Article number 013701.

Hinterdorfer, P, Baumgartner, W, Gruber, H.J, Schilcher, K. and Schindler, H. (1996). Detection and localization of individual antibody-antigen recognition events by atomic force microscopy. *PNAS USA.* **93**, 3477–3481.

Hinterdorfer, P, Gruber, H.J, Kienberger, F, Kada, G, Riener, C, Broken, C. and Schindler, H. (2002). Surface attachment of ligands and receptors for molecular recognition force microscopy. *Colloids Surf. B* **23**, 115– 123.

Hinterdorfer, P. and Dufrêne, Y.F. (2006). Detection and localization of single molecular recognition events using atomic force microscopy. *Nature Methods* **3**, 347-355.

Hodges, C.S. (2002). Measuring forces with the AFM: polymeric surfaces in liquids. *Adv. Colloid & Interface Sci.* **99**, 13-75.

Hummer, G, and Szabo, A. (2001). Free energy reconstruction from nonequilibrium single-molecule pulling experiments. *PNAS USA.* **98**, 3658–3661.

Humphris, A.D.L, Tamayo, J, Miles, M.J. (2000). Active Quality Factor Control in Liquids for Force Spectroscopy. *Langmuir* **16**, 7891-7894.

Humphris, A.D.L, Antognozzi, M, McMaster, T.J. and Miles, M.J. (2002). Transverse Dynamic Force Spectroscopy: A Novel Approach to Determining the Complex Stiffness of a Single Molecule. *Langmuir*, 18, 1729-1733.

Israelachvili, J.N. and Adams, G. (1978). Measurement of forces between two mica surfaces in aqueous electrolyte solutions in the range 0-100 nm, *J. Chem. Soc. Faraday Trans. 1* **74**, 975-977.

Israelachvili, J.N. (2009). Intermolecular and Surface Forces, 3rd Edition (to be published 2009)

Janovjak, H, Müller, D.J. and Humphris, A. (2005). Molecular Force Modulation Spectroscopy Revealing the Dynamic Response of Single Bacteriorhodopsins. *Biophys. J.l* **88**, 1423–1431.

Jarzynski, C. (1997). Nonequillibrium equality for free energy differences. *Phys. Rev. Letts.* **78**, 2690–2693.

Johnson, K.L, Kendall, K. and Roberts, A.D. (1971). Surface Energy And Contact Of Elastic Solids. *Proc. Royal Soc. London A* **324**, 301-313.

Kienberger, F, Kada, G, Gruber, H.J, Pastuschenko, V.P, Riener, C, Trieb, M, Knaus, H-G, Schindler, H. and Hinterdorfer, P. (2000). Recognition Force Spectroscopy Studies of the NTA-His6 Bond. *Single Moecules* **1**, 59-65.

Kratky, O. and Porod, G. (1949). Röntgenuntersuchung gelöster Fadenmoleküle. *Recl. Trav. Chim.* **68**, 1106-1122.

Kawakami, M, Byrne, K, Khatri, B, Mcleish, T.C.B, Radford, S.E. and Smith, D.A. (2004). Viscoelastic Properties of Single Polysaccharide Molecules Determined by Analysis of Thermally Driven Oscillations of an Atomic Force Microscope Cantilever. *Langmiur* **20**, 9299-9303.

Kienberger, F, Ebner, A, Gruber, H.J. and Hinterdorfer, P. (2006). Molecular Recognition Imaging and Force Spectroscopy of Single Biomolecules. *Accounts Chem. Res.* **39**, 29-36.

Kuhn, M, Janovjak, H, Hubain, M. and Müller, D.J. (2005). Automated alignment and pattern recognition of single molecule force spectroscopy data. *J. Microscopy-Oxford* **218**, 125-132.

Lee, G.U, Kidwell, D.A. and Colton, R.J. (1994). Sensing Discrete Streptavidin-Biotin Interactions with Atomic Force Microscopy. *Langmuir* **10**, 354-357.

Li, Y.Q, Tao, N.J, Pan, J, Garcia, A.A and Lindsey, S.M. (1993). Direct measurement of interaction forces between colloidal particles using the scanning force microscope. *Langmuir* **9**, 637-641.

Li, Q.S, Lee, G.Y.H, Ong, C.N. and Lim, C.T. (2008). AFM indentation study of breast cancer cells. *Biochem. & Biophys. Res. Commun.* **374**, 609-613.

Liang, Y, Hilal, N, Langston, P. and Starov, V. (2007). Interaction forces between colloidal particles in liquid: Theory and experiment. *Adv. Colloid & Interface Sci.* **134–135**, 151–166.

Liphardt, J, Dumont, S, Smith, S.B, Tinoco Jr, I, and Bustamante, C. (2002). Equilibrium information from nonequilibrium measurements in an experimental test of Jarzynski's equality. *Science* **296**, 1832–1835.

Marko, J.F. and Siggia, E.D. (1995). Stretching DNA. *Macromolecules* **28**, 8759-8770.

Marszalek, P.E, Oberhauser, A.F, Pang, Y.P. and Fernandes, J.M. (1998). Polysaccharide elasticity governed by chair-boat transitions of the glucopyranose ring. *Nature* **396**, 661-664.

Marszalek, P.E, Li, H. and Fernandes, J.M. (2001). Fingerprinting polysaccharides with single molecule atomic force microscopy. *Nature Biotechnol.* **19**, 258-262.

Maugis, D. (1992) Adhesion of spheres. The JKR-DMT transition using a Dugdale model, *J. Colloid & Interface Sci.* **150**, 243-269.

McKee, C.T, Clark, S.C, Walz, J.Y. and Ducker, W.A. (2005). Relationship between Scattered Intensity and Separation for Particles in an Evanescent Field. *Langmuir* **21**, 5783-5789.

Meyer, G. and Amer, N.M. (1988). Novel approach to atomic force microscopy. *Appl. Phys. Letts.* **53** 1045-1047.

Neumeister, J.M. and Ducker, W.A. (1994). Lateral, normal, and longitudinal spring constants of atomic force microscopy cantilevers. *Rev. Sci. Instrum.* **65** 2527-2531.

Ng, S.P, Randles, L.G. and Clarke, J. (2007). Single studies of protein folding using atomic force microscopy. *Methods Mol. Biol.* **350**, 139-167.

Noy, A. (2008). Strength in Numbers: Probing and Understanding Intermolecular Bonding with Chemical Force Microscopy. *Scanning* **30**, 96–105.

Oberhauser, A.F, Hansma, P.K, Carrion-Vazquez, M. and Fernandez, J.M. (2001). Stepwise unfolding of titin under force-clamp atomic force microscopy. *PNAS. USA* **98**, 468-472.

Oliver, W.C. and Pharr, G.M. (1992). An improved technique for determining hardness and elastic modulus using load and displacement sensing indentation experiments. *J. Mat. Res.* **7**, 1564-1583.

Petersen, K.E and Guarnieri, C.R (1979). Young's modulus measurements of thin films using micromechanics. *J. Appl. Phys.* **50**, 6761-6766.

Piétrement, O. and Troyon, M. (2000). General Equations Describing Elastic Indentation Depth and Normal Contact Stiffness versus Load. *J. Colloid & Interface Sci.* **226**, 166–171.

Prieve, D.C. and Frej, N.A. (1990). Total Internal Reflectance Microscopy - A quantitative tool for the measurement of colloidal forces. *Langmuir* **6**, 396-403.

Proksch, R. and Cleveland, J.P. (2001). STM'01 Conference Abstract, British Columbia University, Vancouver, Canada.

Raab, A, Han, W, Badt, D, Smith-Gill, S.J, Lindsay, S.M, Schindler, H. and Hinterdorfer, P. (1999) Antibody recognition imaging by force microscopy. *Nature Biotechnology* **17**, 902-905.

Ralston, J, Larson, I, Rutland, M.V.V, Feiller, A.A, and Kleijn, M. (2005). Atomic force microscopy and direct surface force measurements–(IUPAC technical report). *Pure & Appl. Chem.* **77**, 2149-2170.

Ray, C, Brown, J.R. and Akhremitchev B.B. (2007). Rupture Force Analysis and the Associated Systematic Errors in Force Spectroscopy by AFM. *Langmuir* **23**, 6076-6083.

Rounsevell, R, Forman, J.R. and Clarke, J. (2004). Atomic Force Microscopy: mechanical unfolding of proteins. *Methods* **34**, 100-111.

Sader, J.E, Larson, I, Mulvaney, P. and White L.R. (1995). Method for calibration of atomic force cantilevers. *Rev. Sci. Instrum.* **66**, 3789-3798.

Sader, J.E, Chon, J.W.M, and Mulvaney, P. (1999) Calibration of rectangular atomic force microscope cantilevers. *Rev. Sci. Instrum.* **70**, 3967-3969.

Smith, S.B, Cui, Y. and Bustamante, C. (1996). Overstretching B-DNA: The Elastic Response of Individual Double-Stranded and Single-Stranded DNA Molecules. *Science* **271**, 795-799.

Strunz, T, Oroszlan, K, Schäfer, R. and Güntherodt, H-J. (1999). Dynamic force spectroscopy of single DNA molecules. *PNAS USA* **96**, 11277–11282.

Strunz, T, Oroszlan, K, Schumakovitch, I., Güntherodt, H-J. and Hegner, M. (2000). Model energy landscapes and the force-induced dissociation of ligand-receptor bonds. *Biophys. J.* **79**, 1206-1212.

Tabor, D. (1977). The hardness of solids. *J. Colloid & Interface Sci.* **58**, 145-179

Verwey, E.J.W. and Overbeek, J.Th.G. (1948). Theory of the stability of lyophobic colloids, Elsevier, Amsterdam.

Wang, M.D, Yin, H, Landick, R, Gelles, J. and Block, S.M. (1997). Stretching DNA with optical tweezers. *Biophys. J.* **72**, 1335-1346.

Zhou, S, Korczagin, I, Hempenius, M.A, Schönherr, H. and Vancso, G.J. (2006). Single molecule force spectroscopy of smart poly(ferrocenylsilane) macromolecules: Towards highly controlled redox-driven single chain motors. *Polymer* **47**, 2483–2492.

SPM BOOKS

Bai. C. (1995). *Scanning Tunnelling Microscopy and its application*. Springer-Verlag.

Batteas, J.D. (2005). *Applications of Scanned Probe Microscopy to Polymers* (ACS Symposium Series) American Chemical Society.

Behm, R.J. *et al.* (1990). *Scanning tunnelling microscopy and related methods; proceedings*. Kluwer Academic Publishers.

Bhushan, B., Fuchs H. (2005). *Applied Scanning Probe Methods II : Scanning Probe Microscopy Techniques*. Springer.

Bhushan, B., Fuchs, H., Tomitori, M. (2008). *Applied Scanning Probe Methods VIII: Scanning Probe Microscopy Techniques*. Springer.

Birdi, K. S. (2003). *Scanning Probe Microscopes: Applications in Science and Technology*. CRC Press.

Bonnell. D.A. (1993) *Scanning tunnelling microscopy and spectroscopy: Theory, techniques and applications*. VCH.

Chen. C.J. (1993). *Introduction to scanning tunnelling microscopy*. Oxford University Press (NY).

Cohen, S.H. *et al.* (1995). *Atomic Force Microscopy/Scanning Tunnelling Microscopy; Proceedings*. Plenum.

Colton, R.J. *et al.* (1998). *Procedures in scanning probe microscopies*. Wiley.

Drelich, J., Mittal K. (2005). *Atomic Force Microscopy in Adhesion Studies*. Brill Academic Publishers.

Guntherodt, H-J., (1995). Forces in scanning probe methods: Proceedings of the NATO advanced study. Kluwer Academic Publishers.

Guntherodt, H-J., and Weisendanger, R. (1995). Scanning tunnelling microscopy; Further applications and related scanning techniques. Springer-Verlag.

Guntherodt, H-J., and Weisendanger, R. (1994). Scanning tunnelling microscopy; General principles and applications to adsorbate covered surfaces. Springer-Verlag.

Guntherodt, H-J., and Weisendanger, R. (1993). Scanning tunnelling microscopy; Theory of STM and related scanning probe methods. Springer-Verlag.

Jena, B.P., Horber, H.J.K. (2002). *Atomic Force Microscopy in Cell Biology* Academic Press.

Jena, B.P., Heinrich, J. K. (2006). *Force Microscopy: Applications in Biology and Medicine*. Wiley-Liss.

Kaupp, G. (2005). *Atomic Force Microscopy, Scanning Nearfield Optical Microscopy and Nanoscratching : Application to Rough and Natural Surfaces*. Springer.

Marti, O., and Amrein, M. (1993). *STM and SFM in biology*. Academic Press.

Magnov, S.N., Whangbo, M-H. (1995). *Surface Analysis with STM and AFM: Experimental and Theoretical Aspects of image analysis*. VCH.

Meyer, E., Hug, H.J., Bennewitz R. (2003) *Scanning Probe Microscopy: The Lab on a Tip*. Springer-Verlag.

Minne, S.C. *et al*. (1999). *Bringing scanning probe microscopy up to speed*. Kluwer Academic Publishers.

Morita, S., Wiesendanger, R., Meyer, E. (2002). *Noncontact Atomic Force Microscopy*. Springer Verlag.

Neddermeyer, H. (1993). *Scanning tunnelling microscopy*. Kluwer Academic Publishers.

Samori, P. (2006). *Scanning Probe Microscopies Beyond Imaging: Manipulation of Molecules and Nanostructures*. Wiley-VCH.

Sarid. D. (1994). *Scanning Force Microscopy with applications to electric, magnetic, and atomic forces*. Oxford University Press (NY).

Stroscio. J.A., Kaiser. W.J. (1994). *Scanning Tunnelling Microscopy*. Academic Press.

Weisendanger, R. (1994). *Scanning probe microscopy and spectroscopy: methods and applications*. Cambridge University Press.

INDEX

scleroglucan 110-111
sectioning 63, 321-322
shear force feedback 347
shells 276, 328
signal to noise ratio 16, 18, 28, 61,
  66, 73
silanes 83-84, 128, 377
silicon 188, 190
silicon nitride 6
snap in 42, 377-378
soya 109, 132-133
spectrin 141, 302-303
sperm cells 316
spores 290-291, 300
spray deposition 77-78, 110
spring constant 6, 11-12, 17, 20, 35,
  38, 359-361
SSB 97
staining 35, 103, 135, 313-314, 317,
  320, 324
starch 63, 116-117, 125-128
  granules 68, 127-128,
    320-321
static charge 14
stereographs 73, 324
strain gauge 31
streptavidin-biotin 386
stylus 5, 24
substrates 58
  binding to 81
  derivatised 81-84
surface
  activity 181
  charge 280, 283, 328, 390
  free energy 181
  pressure 185
  tension 181-182
surface plasmon resonance 36, 307
surfactant 182-184, 220
  pulmonary 216

tapping mode
  in air 45, 48, 52, 56, 61
  in liquid 48, 52

teeth 327
tendon 325
thaumatin 233-234
thermal
  conductivity 350
  diffusivity 350
  drift 18, 31, 59, 67, 264
  noise 17
  sensor 349
  spectrum 50-52
tip 2, 5-6
  aspect ratio 13
  contamination 59-60
  convolution 14, 31-32
  deconvolution 32
  e-beam (supertips) 263-264
  functionalisation 14
  quality 246, 265
  radius of curvature 14
  shape 31-33
tissue 321
titin 131, 139-141
tobacco mosaic virus 33, 238
topography imaging 55
transcription factors 98
transmission electron microscopy
  241-242, 302, 307, 318, 321, 323,
  325-326
tubules (lipid) 194-195
tungsten
  spheres 359
tuning (cantilever) 50-52
tunnelling current 19, 343
twinning 242

ultra high vacuum 20, 342

van der Waals force 41, 49, 181,
  266-268, 279, 295
vesicles 193-194, 205, 211-213, 260
  collapsed 212
  gas 261
  synaptic 307